D1599133

Vavilov-Cherenkov and Synchrotron Radiation

Fundamental Theories of Physics

An International Book Series on The Fundamental Theories of Physics:
Their Clarification, Development and Application

Editor:
ALWYN VAN DER MERWE, *University of Denver, U.S.A.*

Editorial Advisory Board:
GIANCARLO GHIRARDI, *University of Trieste, Italy*
LAWRENCE P. HORWITZ, *Tel-Aviv University, Israel*
BRIAN D. JOSEPHSON, *University of Cambridge, U.K.*
CLIVE KILMISTER, *University of London, U.K.*
PEKKA J. LAHTI, *University of Turku, Finland*
ASHER PERES, *Israel Institute of Technology, Israel*
EDUARD PRUGOVECKI, *University of Toronto, Canada*
FRANCO SELLERI, *Università di Bara, Italy*
TONY SUDBURY, *University of York, U.K.*
HANS-JÜRGEN TREDER, *Zentralinstitut für Astrophysik der Akademie der Wissenschaften, Germany*

Volume 142

Vavilov-Cherenkov and Synchrotron Radiation

Foundations and Applications

by

G.N. Afanasiev

Bogoliubov Laboratory of Theoretical Physics,
Joint Institute for Nuclear Research, Dubna,
Moscow Region, Russia

KLUWER ACADEMIC PUBLISHERS
DORDRECHT / BOSTON / LONDON

A C.I.P. Catalogue record for this book is available from the Library of Congress.

ISBN 1-4020-2410-X (HB)
ISBN 1-4020-2411-8 (e-book)

Published by Kluwer Academic Publishers,
P.O. Box 17, 3300 AA Dordrecht, The Netherlands.

Sold and distributed in North, Central and South America
by Kluwer Academic Publishers,
101 Philip Drive, Norwell, MA 02061, U.S.A.

In all other countries, sold and distributed
by Kluwer Academic Publishers,
P.O. Box 322, 3300 AH Dordrecht, The Netherlands.

Printed on acid-free paper

All Rights Reserved
© 2004 Kluwer Academic Publishers
No part of this work may be reproduced, stored in a retrieval system, or transmitted
in any form or by any means, electronic, mechanical, photocopying, microfilming, recording
or otherwise, without written permission from the Publisher, with the exception
of any material supplied specifically for the purpose of being entered
and executed on a computer system, for exclusive use by the purchaser of the work.

Printed in the Netherlands.

CONTENTS

PREFACE

The importance of the Vavilov-Cherenkov radiation stems from the property that a charge moving uniformly in a medium emits γ quanta at the angle uniquely related to its energy. This has numerous applications. We mention only the neutrino experiments in which the neutrino energy is estimated by the angle at which the electron originating from the decay of neutrino is observed.

This book is intended for students of the third year and higher, for postgraduates, and professional scientists, both experimentalists and theoreticians. The Landau and Lifschitz treatises *Quantum Mechanics*, *Classical Field Theory* and *Electrodynamics of Continuous Media* are more than enough for the understanding of the text.

There are three monographs devoted to the Vavilov-Cherenkov radiation. Jelly's book *Cherenkov Radiation and its Applications* published in 1958 contains a short theoretical review of the Vavilov-Cherenkov radiation and a rather extensive description of experimental technique. Ten years later, the two-volume Zrelov monograph *Vavilov-Cherenkov Radiation and Its Application in High-Energy Physics* appeared. Its first volume is a quite extensive review of experimental and theoretical results known up to 1968. The second volume is devoted to the construction of the Cherenkov counters. In 1988, the Frank monograph *Vavilov-Cherenkov Radiation. Theoretical Aspects* was published. It presents mainly a collection of Frank's papers with valuable short commentaries describing their present status. It is highly desirable to translate this book into English.

The main goal of this book is to present new developments in the theory of the Vavilov-Cherenkov effect for the 15 years following the appearance of Frank's monograph. We briefly mention the main questions treated:

1) The Vavilov-Cherenkov radiation for the unbounded charge motion in a medium (the so-called Tamm-Frank problem);

2) Exact solutions for semi-infinite and finite charge motions in a non-dispersive medium. Their study allows one to identify how the Cherenkov shock waves and the bremsstrahlung shock waves are distributed in space;

3) Accelerated and decelerated charge motions in a medium. Their study allows one to observe the formation and time evolution of the singular shock

xi

waves (including the finite Cherenkov shock wave) arising when the charge velocity coincides with the velocity of light in a medium;

4) The consideration of the Vavilov-Cherenkov radiation in dispersive media with and without damping supports Fermi's claim that a charge moving uniformly in a dispersive medium radiates at each velocity. It turns out that the position and magnitude of the maximum of the frequency distribution depend crucially on the damping parameter value;

5) The measurement of the radiation intensities at finite observational distances leads to the appearance of plateau in some angular interval. The linear (not angular) dimensions of this plateau on the observational sphere do not depend on the sphere radius. Inside this plateau the radiation intensity is not described by the Tamm formula at any observational distance;

6) The taking into account of the finite dimensions of a moving charge or the medium dispersion leads to the finite energy radiated by a moving charge for the entire time of its motion. This in turn allows one to determine how a charge should move if all its energy losses were owed to the Cherenkov radiation;

7) The Vavilov-Cherenkov radiation for a charge moving in a finite medium interval. This includes the consideration of the original Tamm problem (having instantaneous velocity jumps at the beginning and the end of the charge motion), the smooth Tamm problem (in which there are no discontinuities of the charge velocity) and the absolutely continuous charge motion (for which the charge velocity and all its time derivatives are continuous functions of time) in a finite spatial interval. This permits one to relate the asymptotic behaviour of the radiation intensities to the discontinuities of the charge trajectory;

8) It is studied how the radiation intensity changes when a charge moves in one medium while the observations are made in another, with different dielectric properties (in fact, this is a typical experimental situation);

9) The Vavilov-Cherenkov and transition radiations for the spherical interface between two media (previously, only the plane interface was considered in the physical literature);

10) The radiation of electric, magnetic, and toroidal dipoles moving in a medium. This allows one to study the radiation arising from the moving neutral particles (e.g., neutrons, neutrinos, etc.);

11) The fine structure of the Cherenkov rings is studied. We mean under this term the plateau in the radiation intensity (which is due to the Cherenkov shock wave), sharp maxima at the ends of this plateau (we associate them with bremsstrahlung shock waves arising at the accelerated and decelerated parts of the charge trajectory) and small oscillations inside this plateau (they are due to the interference of the Cherenkov and bremsstrahlung shock waves);

12) The kinematics of the two-photon simultaneous emission for a charge moving uniformly in medium. It turns out that under certain circumstances the photon emission angles are fixed. The radiation intensity should have sharp maxima at these angles (similarly to the single-photon Cherenkov emission). This creates favourable conditions for the observation of the two-photon Cherenkov effect.

The importance of the synchrotron radiation is because it is extensively used for the study of nuclear and particle reactions, astrophysical problems, and has a variety of biological and medical applications. There are a few books of the Moscow State University School, and the recently (2002) published book *Radiation Theory of Relativistic Particles* (Ed. V.A. Bordovitsyn) which, in fact, presents a collection of papers of various authors devoted to the questions related to the synchrotron radiation. In the present monograph we study the synchrotron radiation in a medium, and the synchrotron radiation in vacuum, in the near zone. These questions were not considered in the references just mentioned.

The questions considered in this monograph were reported in a number of seminars of the Joint Institute for Nuclear Research, and in various international scientific conferences and symposia.

My deep gratitude is owed to the administration of the Laboratory of Theoretical Physics of the Joint Institute for Nuclear Research which has created nice conditions for the scientific activity, and to my co-authors without whom this monograph could not have appeared. Particular gratitude is owed to Dr. V.M. Shilov for the technical assistance in the preparation of this manuscript.

CHAPTER 1

INTRODUCTION

The Vavilov-Cherenkov (VC) effect and synchrotron radiation (SR) are two of the most prominent phenomena discovered in the 20th century. The VC effect arises when a charged particle moves in a medium with a velocity v greater than the velocity of light c_n in a medium. Here $c_n = c/n$, c is the velocity of light in vacuum and n is the medium refractive index. It should be noted that the acoustic analogue of the VC effect has been known from the middle of 19th century. A bullet or shell, moving in the air with the velocity greater than the velocity of sound in air creates a shock wave of conical form with its apex approximately at the position of the moving body. This conical shock wave is usually referred to as the Mach shock wave after the name of the Austrian scientist Ernst Mach who, while experimentally studying supersonic air streaming past the body at rest, obtained remarkable photographs showing the distribution of the velocity of the air around the body. Similar photographs can be found in [1].

To best of our knowledge, the electromagnetic field (EMF) of a charge moving uniformly in a dispersion-free medium was first obtained by Oliver Heaviside in 1889. We quote him ([2] p.335):

The question now suggests itself, What is the state of things when $u > v$? It is clear, in the first place, that there can be no disturbance at all in front of the moving charge (at a point, for simplicity). Next, considering that the spherical waves emitted by a charge in its motion along the z axis travel at speed v, the locus of their fronts is a conical surface whose apex is at the charge itself, whose axis is that of z, and whose semiangle θ is given by $\sin\theta = v/u$.

(Here u and v are the charge velocity and the velocity of light in medium, resp.). The Heaviside findings concerning this problem were summarized in Volume 3 of his *Electromagnetic Theory* published in 1905 ([3]).

Further, Lord Kelvin on p.4 of his paper *Nineteenth Century Clouds over the Dynamical Theory of Heat and Light* ([4]) wrote:

If this uniform final velocity of the atom exceeds the velocity of light, by ever so little, a non-periodic conical wave of equi-voluminal motion is produced, according to the same principle as that illustrated for sound by Mach's beautiful photographs of illumination by electric sparks, showing, by changed refractivity, the condensational-rarefactional disturbance produced in air by the motion through it of a rifle bullet. The semi-vertical angle of the cone, whether in air or ether, is equal to the angle whose sine is the ratio of the wave velocity to the velocity of the moving body.

In the footnote to this remark Lord Kelvin states:

On the same principle we see that a body moving steadily (and, with little error, we may say also that a fish or water fowl propelling itself by fins or web-feet) through calm water, either floating on the surface or wholly submerged at some moderate distance below the surface, produces no wave disturbance if its velocity is less than the minimum wave velocity due to gravity and surface tension (being about 23 cms. per second, or 0.44 of a nautical mile per hour, whether for sea or fresh water); and if its velocity exceeds the minimum wave velocity, it produces a wave disturbance bounded by two lines inclined on each side of its wake at angles each equal to the angle whose sine is the minimum wave velocity divided by the velocity of the moving body.

Unfortunately, these investigations were forgotten for many years. For example, the information about the Heaviside searches appeared only in 1974 as a result of historical findings by Kaiser ([5]) and Tyapkin ([6]).

The modern history of the VC effect begins with the Cherenkov experiments (1934-1937) (see their nice exposition in his Doctor of Science dissertation [7]) performed at the suggestion of his teacher S.I. Vavilov. In them the γ quanta from an RaE source trapping into a vessel filled with water, induced the blue light detected by the observer outside the vessel. Later it was associated with the radiation of the Compton electrons knocked out by the incoming γ quanta from the water molecules. Since electrons in the Cherenkov experiments were completely absorbed in the water, S.I. Vavilov attributed the above blue light to the deceleration of electrons ([8]):

We think that the most probable reason for the γ luminescence is the radiation arising from the deceleration of Compton electrons. The hardness and intensity of γ rays in the experiments of P.A. Cherenkov were very large. Therefore the number of Compton scattering events and the number of scattered electrons should be very considerable in fluids. The free electrons in a dense fluid should be decelerated within negligible distances. This should be followed by the radiation of a continuous spectrum. Thus weak visible radiation may arise, although the boundary of bremsstrahlung and its maximum should be located somewhere in the Roentgen region. It follows from this that the energy distribution in the visible region should rise towards the violet part of spectrum, and the blue-violet part of spectrum should be especially intensive.

At first, P.A. Cherenkov was a follower of Vavilov's explanation of the nature of radiation observed in his experiments. We quote him [9]:

All the above-stated facts unambiguously testify that the nature of the γ luminescence is owed to the electromagnetic deceleration of electrons moving in a fluid. The facts that γ luminescence is partially polarized and that its brightness has a highly pronounced asymmetry strongly resemble the similar picture for the bremsstrahlung of fast electrons in the Roentgen region. However, in the case of γ luminescence the complete theoretical interpretation encounters with a number of difficulties.

(our translation from the Russian).

In 1937 the famous paper by Frank and Tamm [10] appeared in which the electromagnetic field strengths of a charge moving uniformly in medium were evaluated in the spectral representation. It was shown there that radiation intensities of an electron moving uniformly in medium are added in the direction defined by the so-called Cherenkov angle θ_c ($\cos \theta_c = 1/\beta n$, $\beta = v/c$, n is the medium refractive index). Tamm and Frank also found the energy radiated by an electron, per unit length of its path through a cylinder surface coaxial with the motion axis. These quantities were in agreement with Cherenkov's experiments. Owing to the dependence of the refractive index on the frequency, the velocity of light $c_n = c/n$ in the medium is also frequency-dependent. This leads to the disappearance of the singular Cherenkov cone in the time representation.

In 1938 the experiment by Collins and Reiling [11] was performed in which a 2 Mev electron beam was used to study the VC radiation in various substances. The pronounced Cherenkov rings were observed at the angles given by the Tamm-Frank theory.

In 1939, in the Tamm paper [12], the motion of an electron in a finite spatial interval was considered. Under certain approximations he obtained the formula for the angular radiation intensity which is frequently used by experimentalists for the identification of the charge velocity. This formula is now known as the Tamm formula.

After that Cherenkov changed his opinion in a favour of the Tamm-Frank theory. The reasons for this are analysed in Chapter 5.

The next important step was made by Fermi [13] who considered a charge moving uniformly in a medium with dielectric constant chosen in a standard form extensively used in optics. From his calculations it follows that for this choice of dielectric permittivity a charge moving uniformly in medium should radiate at each velocity. This, in its turn, means that for any velocity there exists a frequency interval for which the Tamm-Frank radiation condition is satisfied.

The first quantum consideration of the VC effect was given by V.L. Ginzburg [14]. The formula obtained by him up to terms of the order $\hbar\omega/m_0 c^2$ (m_0 is the mass of a moving charge in its rest frame and ω is the frequency of an emitted quantum) coincides with the classical expression given by Tamm and Frank in [10].

After the appearance of these classical papers the studies of the VC effect developed very quickly. There are three monographs devoted to this subject. The first one was published in 1958 and was written by Jelley [15]. This book presents a review of experimental and theoretical investigations of the VC effect. The second one is Zrelov's two-volume treatise [16]. The second volume is devoted to Cherenkov counters, and the first volume is the review of experimental and theoretical studies of the VC radiation. The Frank book [17] stays slightly aside of two just mentioned monographs. Its author, one of the founders of the theory of the VC effect and a Nobel prize winner, does not fear to declare that he does not understand something in a particular problem, or that something is not very clear to him in a question discussed. This fair position of Frank has stimulated a lot of investigations and, in particular, ours.

We briefly review the contents of this book.

Chapter 2 is devoted to the so-called Tamm problem considered by Tamm in 1939. In this problem, the charge motion in a finite spatial interval is studied. For the radiation intensity Tamm obtained a remarkably simple formula. Usually it is believed that for the charge velocity smaller than the velocity of light in the medium the Tamm formula describes the bremsstrahlung, whilst for the charge velocity exceeding the velocity of light in the medium it describes both the bremsstrahlung and the radiation arising from the charge uniform motion. In 1989 and 1992 two papers by Ruzicka and Zrelov appeared ([18,19]) in which it was claimed that the radiation observed in the Tamm problem is owed to the instantaneous velocity jumps at the start and end of the motion. We quote them:

> Summing up, one can say that the radiation of a charge moving with a constant velocity along a limited section of its path (the Tamm problem) is the result of two bremsstrahlungs produced at the beginning and the end of motion.

And, further,

> Since the Tamm-Frank theory is a limiting case of the Tamm theory one can consider that the above conclusion is valid for it as well.

On the other hand, it was shown in [20] that in the time representation for the dispersion-free medium the Cherenkov shock wave (this term means the shock wave produced by a charge uniformly moving in medium with a velocity greater than the velocity of light in a medium) exists side by side with the bremsstrahlung shock waves and cannot be reduced to them. Then the question arises, how to reconcile results of [18,19] and [20]. The answer is that the authors of [18,19] analysed the Tamm problem in terms of the Tamm approximate formula. However, it was shown in [21,22] that the Tamm formula, owing to approximations involved in its derivation, does not describe the Cherenkov shock wave properly. In this chapter, to clarify this conflicting situation, we analyse this problem in four different ways.

In chapter 3, based on the references [23,26], it is investigated in the time representation, how a charge moving non-uniformly in a dispersion-free medium radiates. It is shown that for the semi-infinite accelerated motion, beginning from the state of rest, an indivisible complex consisting of the Cherenkov shock wave and the shock wave closing the Cherenkov cone arises at the instant, when the charge velocity v coincides with the velocity of light c_n in medium. The apex of the Cherenkov shock wave attached to a moving charge, moves with the charge velocity, while the mentioned-above shock wave closing the Cherenkov cone propagates with the velocity of light in medium. This results in an increase of the above complex dimensions. For the semi-infinite decelerated motion, terminating with the state of rest, it is shown how the Cherenkov shock wave is transformed into the blunt shock wave which detaches the charge at the instant when the charge velocity coincides with the velocity of light in medium. In the same chapter, there is investigated, in the time representation, the so-called smoothed Tamm problem. In it the charge velocity changes linearly from zero at the initial instant up to the value v_0 with which it moves in a finite spatial interval. After that a charge is linearly decelerated, reaching the state of rest at some other instant of time. The bremsstrahlung shock waves arise at the start and end of motion. If $v_0 > c_n$, a complex, consisting of the Cherenkov shock wave and the shock wave enclosing the Cherenkov cone, arises at the accelerated part of a charge trajectory, when the charge velocity v coincides with c_n. This complex detaches from a charge at the decelerated part of its trajectory when the charge velocity v again coincides with c_n. The above complex does not arise if $v_0 < c_n$.

Chapter 4 deals with an unbounded charge motion in a dispersive medium. The radiation intensities are evaluated [26-28] in the time and the spectral representations for the dielectric constant chosen in a standard one-pole form. In the time representation, in the absence of damping there is a critical charge velocity v_c, independent of frequency, below and above which the behaviour of radiation intensities is essentially different. Above v_c the radiation intensity consists of a number of maxima, the largest of them is at the same position at which the singular Cherenkov cone lies in the absence of dispersion. Below v_c there is a bunch of radiation intensity maxima separated from a moving charge and lying at a quite large distance from it. The quasi-classical estimations for the position of this bunch and of particular maxima composing it agree with exact calculations. These predictions were recently confirmed experimentally [29]. It is shown in the same chapter that for $v > v_c$ the switching on the medium damping leads to a decrease of the maxima of the radiation intensity except for those lying in the neighbourhood of $\cos\theta_c = c/vn$. On the other hand, for $v < v_c$ the radiation intensities are much more affected by the switching on the damping: they disappear almost completely, even for quite small values of a damping parameter. In the same chapter, the radiation intensities are also evaluated in the spectral representation, which is more frequently used by experimentalists than the time representation. It is shown that both the value (which is not surprising since the medium is absorptive) and position of the maximum of the radiation intensity depend crucially on the observational distance and the damping parameter. This raises uneasy questions about the interpretation of the VC radiation spectra presented by experimentalists.

The chapter 5 is devoted to the evaluation of the radiation intensities at finite observational distances and to taking into account the effects of accelerated motion [22,30-32]. This chapter may be viewed as the translation of chapters 2 and 3 into the frequency language. In fact, experimentalists measure the number of photons with a given frequency and the energy radiated by a moving charge at the given frequency. Usually the VC radiation is observed in the frequency interval corresponding to the visible light. There are only a few experiments (such as [29]) dealing with the VC radiation in the time representation. Certainly, frequencies lying outside the frequency interval of a visible light also contribute to the radiation intensity in the time representation. Turning to the observation of the VC radiation at finite distances we observe that for the Tamm problem the radiation intensities evaluated on an observational sphere of finite radius r (the Tamm approximate formula corresponds to an infinite observational distance) have a plateau in the angular range surrounding the Cherenkov angle θ_c. Physically this may be explained as follows. A charge moving in a finite medium interval emits photons under the Cherenkov angle θ_c towards the motion axis. A particular photon, emitted at a given instant, intersects the observational sphere at a particular angle which depends on the charge position in the interval of motion. Since the transition to the frequency representation involves integration over the whole time of a charge motion, one obtains the above angular plateau. The appearance of the angular plateau is also supported by the analytic consideration of Chapter 2. The need for formulae working at finite distances is because the Tamm approximate formula for the angular radiation intensity does not work at realistic observational distances.

In the same Chapter closed analytical expressions are obtained for the radiation intensities of a charge moving with deceleration in a finite spatial interval and valid at a finite distance from a moving charge. The taking into account of the deceleration effects is needed for describing the recent experiments with heavy ions, where pronounced Cherenkov rings were observed [33]. The large velocity losses for heavy ions are owed to their large atomic number (energy losses are proportional to the square of the charge). The above analytical formulae are valid for relatively small accelerations for which the change of a velocity is much smaller than the velocity itself.

Closed analytical expressions for radiation intensities are obtained also for arbitrary charge deceleration for which the so-called Tamm condition, allowing us to disregard the acceleration effects, is strongly violated. Unfortunately, these analytic formulae are valid only at infinite observational distances. An important case for applications corresponds to the complete charge stopping in a medium (this was realized in the original Cherenkov experiments). When the final velocity is zero and the initial velocity is greater than the velocity of light c_n in medium, the pronounced maximum in the angular distribution appears at the Cherenkov angle corresponding to the initial charge velocity.

Using the spectral representation we consider the smooth Tamm problem in which the charge velocity changes smoothly from zero up to some value $v > c_n$, with which it moves for some time. After that a charge is smoothly decelerated down to reaching the state of rest. When non-uniform parts of the charge interval of motion tend to zero, their contribution to the radiation intensity also tends to zero, and only the uniform part of the charge motion interval contributes to the total radiation intensity. However, according to Chapter 2 the bremsstrahlung shock waves exist even for the instantaneous velocity jumps. The possible outcome of this controversy is that not only the velocity jumps but the acceleration jumps as well contribute to the radiation intensity. In fact, for the smooth Tamm problem treated there are no velocity jumps but there are acceleration jumps at the start and end of the motion, and at the instants when the uniform and non-uniform motions meet each other. To see this explicitly we have considered two kinds of absolutely continuous charge motion in a finite spatial interval. Although the velocity behaviour is visually indistinguishable from the velocity behaviour in the original Tamm problem (with velocity and acceleration discontinuities) and in the smoothed Tamm problem (without the velocity discontinuities, but with the acceleration ones), the corresponding intensities differ appreciably: for the absolutely continuous charge motion the radiation intensities are exponentially small outside some angular interval. This points out that not only the velocity discontinuities are essential, but the discontinuities of higher derivatives of the charge trajectory as well.

Chapter 6 treats the radiation arising from electric, magnetic, and toroidal dipoles moving in medium. As far as we know, Frank was the first to evaluate the electromagnetic field (EMF) strengths and the energy flux per unit frequency and per unit length of cylinder surface coaxial with the motion axis [34]. These quantities depend on the dipole spatial orientation. Frank postulated that the moments of electric and magnetic dipoles moving in a medium are related to those in their rest frame by the same transformations as in vacuum. For an electric dipole and for a magnetic dipole parallel to the velocity, he obtained expressions which satisfied him. For a magnetic dipole perpendicular to the velocity the radiated energy did not disappear for $v = c_n$. Its vanishing is intuitively expected and is satisfied, e.g., for an electric charge and dipole and for a magnetic dipole parallel to the velocity. On these grounds Frank declared [35] the formula for the radiation intensity of the magnetic dipole perpendicular to the velocity to be incorrect. He also admitted that the correct expression for the above intensity is obtained if the above transformation law is changed slightly. This claim was supported by Ginzburg in [36], who pointed out that the internal structure of a moving magnetic dipole and the polarization induced inside it are essential. This idea was further elaborated in [37]. In [38], the radiation of toroidal dipoles (i.e., elementary (infinitesimally small) toroidal solenoids (TS)) moving uniformly in a medium was considered. It was shown that the EMF of the TS moving in a medium extends beyond its boundaries. This seemed surprising since the EMF of a TS resting either in the medium (or vacuum) or moving in the vacuum is confined to its interior. After many years Frank returned in [39,40] to the original transformation laws. In particular, in [40] he considered the rectilinear current frame moving uniformly in a medium. The evaluated electric moment of the current distribution moving in the medium was in agreement with that obtained by the law postulated in [34].

The goal of this Chapter consideration is to obtain EMF potentials and strengths for point-like electric and magnetic dipoles and an elementary toroidal dipole moving in the medium with arbitrary velocity v greater or smaller than the velocity of light c_n in medium. In the reference frame attached to a moving source there are finite static distributions of charge and current densities. We postulate that charge and current densities in the laboratory frame, relative to which the source moves with a constant velocity, can be obtained from the rest frame densities via Lorentz transformations, the same as in vacuum. The further procedure is to tend the dimensions of the charge and current sources in the laboratory frame to zero, in a straightforward solution of the Maxwell equations for the EMF potentials in the laboratory frame, with the point-like charge and current densities in the r.h.s. of these equations, and in a subsequent evaluation of the EMF strengths. In the time and spectral representations, this was done in [41,42]. The reason for using the spectral representation, which is extensively used by experimentalists, is to compare our results with those of [34-40] written in the frequency representation.

In Chapter 7, there is discussed how the VC radiation affects the charge motion. Usually the VC radiation is associated with the radiation of a charge moving uniformly in medium with the velocity greater than the velocity of light in medium. Owing to the radiation a moving charge inevitably loses its energy. The self-energy of a point-like charge is infinite. A moving point-like charge emits all frequencies. In a dispersion-free medium all frequencies propagate without damping if the charge velocity is greater than the velocity of light in medium. The total energy radiated per unit length, obtained by integration of the spectral energy over all frequencies, is infinite. There are several ways of overcoming this difficulty. The first is to consider a charge of finite dimension. Its self-energy $\mathcal{E}_f$ is finite. Therefore there is maximal frequency $\mathcal{E}_f/\hbar$ which can be emitted. The energy radiated by a moving charge per unit length is also finite. Equating it to the loss of kinetic energy one finds how the velocity of a finite charge moving in a non-dispersive medium changes as a result of the VC radiation. Another way of achieving the finite energy losses is to consider the charge motion in a dispersive medium. For this medium with a dielectric constant approximated by the one-pole formula, the Tamm-Frank radiation condition $\beta^e n^2(\omega) > 1$ is satisfied in a finite frequency interval. Integrating the radiated energy over this interval one obtains a finite value for the energy radiated per unit length. Equating it to the kinetic energy loss one finds how the VC radiation affects the velocity of a point-like charge moving in a dispersive medium. In reality these processes compete with each other and with ionization energy losses. All these questions are also discussed.

The following problem is also discussed in this Chapter. It deals with an electric charge moving inside a spatial region S filled with a medium of refractive index n_1, while measurements are made outside this region, in a medium of refractive index n_2. In fact, this is a typical situation in experiments with VC radiation. For example, in the original Cherenkov experiments [7] the γ quanta emitted by electrons moving in a vessel filled with a water were observed outside this vessel by a human eye. The case in which S was a dielectric cylinder C was considered by Frank and Ginzburg in 1947 [43] who showed that there will be no radiation flux outside C for $1/n_1 < \beta < 1/n_2$. Under the radiation flux they realized the radial one (that is, in the direction perpendicular to the axis of motion). We have evaluated the energy flux along the axis of motion and have shown that outside the dielectric cylinder S it is zero everywhere except for the discrete set of observational frequencies at which it is infinite.

We have considered two other problems corresponding to the radiation of a charge moving uniformly in a spherically symmetric dielectric sample. The first of them deals with a charge moving inside a dielectric sphere S of refractive index n_1 (medium 1), while observations of the energy flux are made outside this sphere in a medium of refractive index n_2 (medium 2). It is shown that the angular spectrum broadens in comparison with the Tamm angular spectrum corresponding to the charge motion in a finite interval lying inside the unbounded medium 1. There is also observed a rise in the angular intensities at large angles. We associate them with the reflection of the VC radiation from the internal side of the sphere S. The second problem treats a charge whose motion begins and ends in a medium 2 of refractive index n_2 and which during its motion penetrates the dielectric sphere S of refractive index n_1 (medium 1). In addition to the VC radiations in medium 1 (if the condition $\beta n_1 > 1$ is satisfied) and in medium 2 (if the condition $\beta n_2 > 1$ is satisfied), and to the bremsstrahlung arising at the beginning and end of a charge motion in medium 2, there is the so-called transition radiation arising when a moving charge crosses the surface of the sphere S separating the media 1 and 2. For the plane boundary between the media 1 and 2, transition radiation was first considered by Frank and Ginzburg in 1946 [44]. In the problems treated (spherical boundary between two media) the frequency radiation spectrum exhibits the characteristic oscillations. Probably, they are of the same nature as those for the dielectric cylinder.

Chapter 8 is devoted to the synchrotron radiation, which is such well-known phenomenon that it seems to be almost impossible to add something essential in this field. Schott was probably the first person who extensively studied SR. His findings were summarized in the encyclopedic treatise *Electromagnetic Radiation* [45]. He developed the electromagnetic field (EMF) into Fourier series and found solutions of Maxwell's equations describing the field of a charge moving in a vacuum along the circular orbit. The infinite series of EMF strengths had a very poor convergence in the most interesting case $v \sim c$. Fortunately Schott succeeded in an analytical summation of these series and obtained closed expressions for the radiation intensity averaged over the azimuthal angle ([45], p.125). Further development is owed to Moscow State University school (see, e.g., books [46]-[49] and review [50]) and to Schwinger et al. [51] who considered the polarization properties of SR and its quantum aspects. The instantaneous (i.e., taken at the same instant of a proper time) distribution of SR on the surface of observational sphere was obtained by Bagrov et al. ([52,53]) and Smolyakov [54]. They showed that the instantaneous distribution of SR in a vacuum possesses the so-called projector effect (that is, the SR has the form of a beam which is very thin for $v \sim c$).

Much less is known about SR in a medium. The papers by Schwinger, Erber et al, [55,56] should be mentioned in this connection. Yet they limited themselves to an EMF in a spectral representation and did not succeed in obtaining the EMF strengths and radiation flux in the time representation. It should be noted that Schott's summation procedure does not work if the charge velocity exceeds the velocity of light in medium. The formulae obtained by Schott and Schwinger are valid at observational distances r much larger than the radius a of the charge orbit. In modern electron and proton accelerators this radius reaches a few hundred meters and even a few kilometers, respectively. This means that large observational distances are unachievable in experiments performed on modern accelerators and that formulae describing the radiation intensity at moderate distances and near the charge orbit are needed. In the past, time-averaged radiation intensities in the near zone were studied in ([57-59]). However, their consideration was based on the expansion of field strengths in powers of a/r. The convergence of this expansion is rather poor in the neighbourhood of the charge orbit. SR has numerous applications in nuclear physics (nuclear reactions with γ quanta), solid state physics (see, e.g., [60]), astronomy ([61,62]), etc.. There are monographs and special issues of journals devoted to application of SR ([62-64]). The book [65] the major part of which is devoted to the SR should be also mentioned.

The goal of this Chapter is to study SR in a vacuum and in a medium. In the latter case, the charge velocity v can be less or greater than the velocity of light c_n in medium. We limit ourselves to consideration in the time representation. We analyse radiation arising from the charge circular motion both in the far and near zones, in a vacuum and in a medium. For synchrotron motion in a medium with the charge velocity greater than the velocity of light in the medium the singular contours are found on which the electromagnetic field strengths are infinite. For the charge motion in a vacuum the contours are found on which electromagnetic field strengths and radiation intensities acquire maximal values.

Chapter 9 deals with experiments in which the fine structure of the Cherenkov rings was observed. Under it we mean the existence of the Cherenkov shock wave of finite extension manifesting as a plateau in the observed radiation intensity and of the shock wave associated with the exceeding the light velocity barrier and manifesting as the intensity bursts at the end of the plateau. Small oscillations of the radiation intensity inside the plateau are owed to the interference of the VC radiation and bremsstrahlungs.

There should be also mentioned the intriguing experiments [66] in which the Cherenkov rings with anomalous large radii (corresponding to the charge velocity greater than the velocity of light in the vacuum) were observed.

The possibility of the two-photon Cherenkov effect was predicted by Frank and Tamm in [67] who showed that the conservation of the energy and momentum does not prohibit the process in which a moving charge emits simultaneously two photons. There is no experimental confirmation of this effect up to now. The calculations of the two-photon radiation intensity are known, but they were performed without paying enough consideration to the exact kinematical relations. The goal of this Chapter treatment is to point out that the two-photon Cherenkov effect will be strongly enhanced for special orientations of photons and the recoil charge. This makes easier the experimental search of the 2-photon Cherenkov effect.

References

1. Hayes W.D. and Probstein R.F. (1959) *Hypersonic Flow Theory*, Academic Press, New York, and London.
2. Heaviside O. (1889) On the Electromagnetic Effects due to the Motion of Electrification through a Dielectric, *Phil.Mag.*, **27**, 5-th series, pp.324-339.
3. Heaviside O. (1922) *Electromagnetic Theory*, Benn Brothers Ltd, London (Reprinted Edition).
4. Lord Kelvin (1901) Nineteenth Century Clouds over the Dynamical Theory of Heat Light, *Phil.Mag.*, **2**, 6-th series, pp. 1-40.
5. Kaiser T.R. (1974), Heaviside Radiation, *Nature*, **247**, pp. 400-401 .
6. Tyapkin A.A. (1974) First Theoretical Prediction of Radiation Discovered by Vavilov and Cherenkov, *Usp. Fiz. Nauk*, **112**, p. 735.
7. Cherenkov P.A. (1944) Radiation of Electrons Moving in Medium with Superluminal Velocity, *Trudy FIAN*, **2, No 4**, pp. 3-62.
8. Vavilov S.I. (1934) On Possible Reasons for the Blue γ Radiation in Fluids, *Dokl. Akad, Nauk*, **2, 8**, pp. 457-459.
9. Cherenkov P.A. (1936) Influence of Magnetic Field on the Observed Luminescence of Fluids Induced by Gamma Rays, *Dokl. Akad, Nauk*, **3, 9**, pp. 413-416.
10. Frank I.M. and Tamm I.E. (1937) Coherent Radiation of Fast Electron in Medium, *Dokl. Akad. Nauk*, **14**, pp. 107-113.
11. Collins G.B. and Reiling V.G. (1938) Cherenkov Radiation *Phys. Rev*, **54**, pp. 499-503.
12. Tamm I.E. (1939) Radiation Induced by Uniformly Moving Electrons, *J. Phys. USSR*, **1, No 5-6**, pp. 439-461.
13. Fermi E. (1940) The Ionization Loss of Energy in Gases and in Condensed Materials, *Phys. Rev*, **57**, pp. 485-493.
14. Ginzburg V.L. (1940) Quantum Theory of Radiation of Electron Uniformly Moving in Medium, *Zh. Eksp. Teor. Fiz.*, **10** pp. 589-.
15. Jelley J.V., (1958) *Cherenkov Radiation and its Applications*, Pergamon, London.
16. Zrelov V.P. (1970) *Vavilov-Cherenkov Radiation in High-Energy Physics*, vols. 1 and 2, Israel Program for Scientific Translations.
17. Frank I.M. (1988) *Vavilov-Cherenkov Radiation*, Nauka, Moscow.
18. Zrelov V.P. and Ruzicka J. (1989) Analysis of Tamm's Problem on Charge Radiation at its Uniform Motion over a Finite Trajectory *Czech. J. Phys.*, **B 39**, pp. 368-383.
19. Zrelov V.P. and Ruzicka J. (1992) Optical Bremsstrahlung of Relativistic Particles in a Transparent Medium and its Relation to the Vavilov-Cherenkov Radiation *Czech. J. Phys.*, **42**, pp. 45-57.
20. Afanasiev G.N., Beshtoev Kh. and Stepanovsky Yu.P. (1996) Vavilov-Cherenkov Radiation in a Finite Region of Space *Helv. Phys. Acta*, **69**, pp. 111-129.
21. Afanasiev G.N., Kartavenko V.G. and Stepanovsky Yu.P. (1999) On Tamm's Problem in the Vavilov-Cherenkov Radiation Theory *J.Phys. D: Applied Physics*, **32**, pp. 2029-2043.
22. Afanasiev G.N. and Shilov V.M. (2002) Cherenkov Radiation versus Bremsstrahlung in the Tamm Problem *J.Phys. D: Applied Physics*, **35**, pp. 854-866.
23. Afanasiev G.N., Eliseev S.M. and Stepanovsky Yu.P. (1998) Transition of the Light Velocity in the Vavilov-Cherenkov Effect *Proc. Roy. Soc. London*, **A 454**, pp. 1049-1072.

24. Afanasiev G.N. and Kartavenko V.G. (1999) *Cherenkov-like shock waves associated with surpassing the light velocity barrier Canadian J. Phys.*, **77**, pp. 561-569.
25. Afanasiev G.N. and Shilov V.M. (2000) On the Smoothed Tamm Problem *Physica Scripta*, **62**, pp. 326-330.
26. Afanasiev G.N. and Kartavenko V.G. (1998) Radiation of a Point Charge Uniformly Moving in a Dielectric Medium *J. Phys. D: Applied Physics*, **31**, pp.2760-2776.
27. Afanasiev G.N., Kartavenko V.G. and Magar E.N. (1999) Vavilov-Cherenkov Radiation in Dispersive Medium *Physica*, **B 269**, pp. 95-113.
28. Afanasiev G.N., Eliseev S.M and Stepanovsky Yu.P. (1999) Semi-Analytic Treatment of the Vavilov-Cherenkov Radiation *Physica Scripta*, **60**, pp. 535-546.
29. Stevens T.E., Wahlstrand J.K., Kuhl J. and Merlin R. (2001) Cherenkov Radiation at Speeds below the Light Threshold: Photon-Assisted Phase Matching *Science*, **291**, pp. 627-630.
30. Afanasiev G.N., Kartavenko V.G. and Ruzicka J, (2000) Tamm's Problem in the Schwinger and Exact Approaches *J. Phys. A: Mathematical and General*, **33**, pp. 7585-7606.
31. Afanasiev G.N. and Shilov V.M. (2000) New Formulae for the Radiation Intensity in the Tamm Problem *J. Phys.D: Applied Physics*, **33**, pp. 2931-2940.
32. Afanasiev G.N., Shilov V.M., Stepanovsky Yu.P. (2002) New Analytic Results in the Vavilov-Cherenkov Radiation Theory *Nuovo Cimento*, **B 117**, pp. 815-838; Afanasiev G.N., Shilov V.M., Stepanovsky Yu.P. (2003) Numerical and Analytical Treatment of the Smoothed Tamm Problem *Ann.Phys. (Leipzig)*, **12**, pp. 51-79
33. Ruzicka J. et al. (1999) The Vavilov-Cherenkov Radiation Arising at Deceleration of Heavy Ions in a Transparent Medium *Nuclear Instruments and Methods*, **A 431**, pp. 148-153.
34. Frank I.M. (1942) Doppler Effect in Refractive Medium *Izv. Acad. Nauk SSSR, ser.fiz.*, **6**, pp.3-31.
35. Frank I.M. (1952) Cherenkov Radiation for Multipoles, In the book: *To the memory of S.I.Vavilov*, pp.172-192, Izdat. AN SSSR, Moscow (in Russian).
36. Ginzburg V.L., (1952) On the Cherenkov Radiation for Magnetic Dipole, In the book: *To the Memory of S.I.Vavilov*, pp.193-199, Izdat. AN SSSR, (in Russian).
37. Ginzburg V.L. (1984) On Fields and Radiation of 'true' and current magnetic Dipoles in a Medium, *Izv. Vuz., ser. Radiofizika*, **27**, pp. 852-872.
38. Ginzburg V.L. and Tsytovich V.N. (1985) Fields and Radiation of Toroidal Dipole Moments Moving Uniformly in a Medium *Zh. Eksp. Theor. Phys.*, bf 88, pp. 84-95.
39. Frank I.M. (1984) Vavilov-Cherenkov Radiation for Electric and Magnetic Multipoles *Usp.Fiz.Nauk*, **144**, pp. 251-275.
40. Frank I.M. (1989) On Moments of magnetic Dipole Moving in Medium *Usp.Fiz.Nauk* **158**, pp. 135-138.
41. Afanasiev G.N. and Stepanovsky Yu.P. (2000) Electromagnetic Fields of Electric, Magnetic and Toroidal Dipoles Moving in Medium *Physica Scripta*, **61**, pp.704-716.
42. Afanasiev G.N. and Stepanovsky Yu.P. (2002) On the Radiation of Electric, Magnetic and Toroidal Dipoles *JINR Preprint*,**E2-2002-142**, pp. 1-30.
43. Ginzburg V.L. and Frank I.M. (1947) Radiation of Electron and Atom Moving on the Channel Axis in a dense Medium *Dokl. Akad. Nauk SSSR*, **56**, pp. 699-702.
44. Ginzburg V.L. and Frank I.M. (1946) Radiation of a Uniformly Moving Electron due to its Transition from one Medium to Another *Zh. Eksp. Theor. Phys.*, **16**, pp. 15-28.
45. Schott G.A. (1912) *Electromagnetic Radiation*, Cambridge, Cambridge Univ.Press.
46. Sokolov A.A. and Ternov I.M. (1974) *The Relativistic Electron* Moscow, Nauka (in Russian).
47. Ternov I.M., Mikhailin V.V. and Khalilov V.R. (1985) *Synchrotron radiation and its applications*, Moscow, Moscow Univ. Publ., (in Russian).
48. Ternov I.M. and Mikhailin V.V. (1986) *Synchrotron radiation. Theory and experiment.*, Moscow, Energoatomizdat (in Russian).

49. Bordovitsyn V.A. (Ed.) (1999) *Synchrotron radiation theory and its developments. In memory of I.M. Ternov*, Singapore, World Scientific.
50. Ternov I.M. (1995) Synchrotron Radiation *Usp. Fiz. Nauk*, **165**, pp. 429-456.
51. Schwinger J. (1949) On the Classical Radiation of Accelerated Electrons *Phys.Rev.*,**A 75**, pp. 1912-1925.
52. Bagrov V.G., (1965) Indicatrix of the Charge Radiation External Field According to Classical Theory it Optika i Spectroscopija, **28, No 4**, pp. 541-544, (In Russian).
53. Sokolov A.A., Ternov I.M. and Bagrov V.G. (1966) Classical theory of synchrotron radiation, in: *Synchrotron Radiation* (Eds.:Sokolov A.A. and Ternov I.M.),pp. 18-71, Moscow, Nauka, (in Russian).
54. Smolyakov N.V. (1998) Wave-Optical Properties of Synchrotron Radiation *Nucl. Instr. and Methods*,**A 405**, pp. 235-238.
55. Schwinger J., Tsai W.Y. and Erber T. (1976) Classical and Quantum Theory of Synergic Synchrotron-Cherenkov Radiation, *Ann. of Phys.*, **96**, pp.303-352.
56. Erber T., White D., Tsai W.Y. and Latal H.G. (1976) Experimental Aspects of Synchrotron-Cherenkov Radiation, *Ann. of Phys.*, **102**, pp. 405-447.
57. Villaroel D. and Fuenzalida V. (1987) A Study of Synchrotron Radiation near the Orbit, *J.Phys. A: Mathematical and General*, **20**, pp. 1387-1400.
58. Villaroel D. (1987) Focusing Effect in Synchrotron Radiation, *Phys.Rev.*, *A 36*, pp. 2980-2983.
59. Villaroel D. and Milan C. (1987) Synchrotron Radiation along the Radial Direction *Phys.Rev.*, **D38**, pp. 383-390.
60. Ovchinnikov S.G. (1999) Application of Synchrotron Radiation to the Study of Magnetic Materials *Usp. Fiz. Nauk*, **169**, pp. 869-887.
61. Jackson J.D., (1975) *Classical Electrodynamics*, New York, Wiley.
62. Ryabov B.P. (1994) Jovian S emission: Model of Radiation Source *J. Geophys. Res.*, **99, No E4**, pp. 8441-8449.
63. Synchrotron Radiation (Kunz C.,edit.), (Springer, Berlin, 1979).
64. Nuclear Instr. & Methods, A 359, No 1-2 (1995); Nuclear Instr. & Methods, A 405, No 2-3 (1998).
65. (2002) *Radiation Theory of Relativistic Particles* (Ed. Bordovitsyn V.A.), Moscow, Fizmatlit.
66. Vodopianov A.S., Zrelov V.P. and Tyapkin A.A. (2000) Analysis of the Anomalous Cherenkov Radiation Obtained in the Relativistic Lead Ion Beam at CERN SPS *Particles and Nuclear Letters*, **No 2[99]-2000**, pp. 35-41.
67. Tamm I.E. and Frank I.M. (1944) Radiation of Electron Moving Uniformly in Refractive Medium *Trudy FIAN*, **2, No 4**. pp. 63-68.

CHAPTER 2

THE TAMM PROBLEM IN THE VAVILOV-CHERENKOV RADIATION THEORY

2.1. Vavilov-Cherenkov radiation in a finite region of space

The Vavilov-Cherenkov (VC) effect is a well established phenomenon widely used in physics and technology. A nice exposition of it may be found in Frank's book [1]. In most textbooks and scientific papers the VC effect is considered in the spectral representation. To obtain an answer in the time representation an inverse Fourier transform should be performed. The divergent integrals occurring obscure the physical picture. As far as we know, there are only a few attempts in which the VC effect is treated without using the spectral representation. First, we should mention Sommerfeld's paper [2] in which the hypothetical motion of an extended charged particle in a vacuum with a velocity $v > c$ was considered. Although the relativity principle prohibits such a motion in the vacuum, all the equations of [2] are valid in a medium if we identify c with the velocity of light in the medium. Unfortunately, owing to the finite dimensions of the charge, the equations describing the field strengths are so complicated that they are not suitable for physical analysis. The other reference treating the VC effect without recourse to the Fourier transform is Heaviside's book [3] in which the superluminal motions of a point charge both in a vacuum and an infinitely extended medium were considered. Heaviside was not aware of Sommerfeld's paper [2], just as Tamm and Frank [4,5] did not know about Heaviside's investigations. It should be noted that Frank and Tamm formulated their results in the spectral representation. The results of Heaviside (without referring to them) were translated into modern physical language in [6].

It is our goal to investigate electromagnetic effects arising from the motion of a point-like charged particle in a medium, in a finite spatial interval.

2.1.1. MATHEMATICAL PRELIMINARIES

Let a point-like particle with a charge e move in a dispersion-free medium with polarizabilities ϵ and μ along the given trajectory $\vec{\xi}(t)$. Its electromagnetic field (EMF) at the observational point $(\vec{r}, t)$ is then given by the

Liénard-Wiechert retarded potentials

$$\Phi(\vec{r},t) = \frac{e}{\epsilon}\sum \frac{1}{Z_i}, \quad \vec{A}(\vec{r},t) = \frac{e\mu}{c}\sum \vec{v}_i/Z_i, \quad (div\vec{A} + \frac{\epsilon\mu}{c}\dot{\Phi} = 0). \quad (2.1)$$

Here $\vec{v}_i = (d\vec{\xi}/dt)|_{t=t_i}$, $Z_i = ||\vec{r} - \vec{\xi}(t_i)| - \vec{v}_i(\vec{r} - \vec{\xi}(t_i))/c_n|$, and c_n is the velocity of light inside the medium $(c_n = c/\sqrt{\epsilon\mu})$. The sum in (2.1) is performed over all physical roots of the equation

$$c_n(t - t') = |\vec{r} - \vec{\xi}(t')| \qquad (2.2)$$

which tells us that the radiation from a moving charge propagates with the light velocity c_n in medium. To preserve causality the time of radiation t' should be smaller than the time of observation t. Obviously t' depends on the coordinates $\vec{r}, t$ of the point P at which the EMF is observed. Let a particle move with a constant velocity v along the z axis $(\xi = vt)$. Equation (2.2) then has two roots

$$c_n t' = \frac{c_n t - \beta_n z}{1 - \beta_n^2} \mp \frac{r_m}{|1 - \beta_n^2|}. \qquad (2.3)$$

Here $r_m = \sqrt{(z - vt)^2 + \rho^2(1 - \beta_n^2)}$, $\rho^2 = x^2 + y^2$, $\beta_n = v/c_n$. In what follows we also need $c_n(t - t')$ which is given by

$$c_n(t - t') = \beta_n \frac{vt - z}{\beta_n^2 - 1} \pm \frac{r_m}{|\beta_n^2 - 1|}. \qquad (2.4)$$

We shall denote the t' corresponding to the upper and lower signs in (2.3) and (2.4) as t_1' and t_2', resp.. It is easy to check that

$$c_n^2(t - t_1')(t - t_2') = \frac{r^2}{\beta_n^2 - 1}, \quad r^2 = (z - vt)^2 + \rho^2. \qquad (2.5)$$

Consider a few particular cases.

2.1.2. PARTICULAR CASES.

The uniformly moving charge with a velocity $v < c_n$.
It follows from (2.5) that $t - t_1'$ and $t - t_2'$ have different signs for $\beta_n < 1$. As only a positive $t - t'$ corresponds to the physical situation, one should choose the plus sign in (2.4) which corresponds to the upper signs both in (2.3) and (2.4). For the electromagnetic potentials one obtains the well-known expressions

$$\epsilon\Phi = \frac{e}{r_m}, \quad A_z = \frac{e\beta\mu}{r_m} \quad (\beta = v/c). \qquad (2.6)$$

It follows from this that a uniformly moving charge carries the EMF with itself. In fact, EMF strengths decrease as $1/r^2$ as $r \to \infty$, and therefore no energy is radiated into the surrounding space.

A uniformly moving charge with a velocity $v > c_n$.
This section briefly reproduces the contents of [6]. It follows from (2.5) that for the case treated $(t - t'_1)$ and $(t - t'_2)$ are of the same sign which coincides with the sign of the first term in (2.4). It is positive if

$$t > z/v. \tag{2.7}$$

The two physical roots are

$$c_n t'_{1,2} = -\frac{c_n t - \beta_n z \pm r_m}{\beta_n^2 - 1}.$$

The positivity of the expression staying under the square root in r_m requires

$$\mathcal{M} = vt - z - \rho/\gamma_n > 0 \quad \text{or} \quad t > \frac{z}{v} + \frac{\rho}{v\gamma_n}. \tag{2.8}$$

Here $\gamma_n = 1/\sqrt{|\beta_n^2 - 1|}$. Since this inequality is stronger than (2.7) one may use only (2.8), which shows that the EMF is enclosed inside the Cherenkov cone given by (2.8). Its analogy in acoustics is the Mach cone. For the electromagnetic potentials one finds

$$\epsilon\Phi = \frac{2e}{r_m}\Theta(\mathcal{M}), \quad A_z = \frac{2e\mu\beta}{r_m}\Theta(\mathcal{M}) \tag{2.9}$$

(the factor 2 appears because there are two physical roots satisfying (2.2)). Here $\Theta(x)$ is a step function. It equals 1 for $x > 0$ and 0 for $x < 0$. It is seen that $r_m = 0$ on the surface of the Cherenkov cone where $\mathcal{M} = 0$. Therefore electromagnetic potentials are zero outside the Cherenkov cone ($\mathcal{M} < 0$), differ from zero inside it ($\mathcal{M} > 0$), and are infinite on its surface ($\mathcal{M} = 0$). The electromagnetic strengths ($\vec{D} = \epsilon\vec{E}$, $\vec{E} = -\text{grad}\Phi - \dot{\vec{A}}/c$, $\vec{B} = \mu\vec{H} = \text{curl}\vec{A}$) are given by

$$H_\phi = -\frac{2e\rho\beta}{\gamma_n^2 r_m^3}\Theta(\mathcal{M}) + \frac{2e\beta}{\gamma_n r_m}\delta(\mathcal{M}),$$

$$\epsilon\vec{E} = -\frac{2er}{\gamma_n^2 r_m^3}\vec{n}_r \cdot \Theta(\mathcal{M}) + \frac{2e\beta}{\gamma_n r_m} \cdot \delta(\mathcal{M})\vec{n}_m \tag{2.10}$$

Here $n_r = (\rho\vec{n}_\rho + (z - vt)\vec{n}_z)/r$ is the unit radial vector directed inside the Cherenkov cone from the charge current position and $\vec{n}_m = \vec{n}_\rho/\beta_n - \vec{n}_z/\beta_n\gamma_n$ is the unit vector lying on the surface of the Cherenkov cone (Fig. 2.1). The δ function terms in these equations corresponding to the Cherenkov shock wave (CSW, for short) differ from zero only on the surface of the Cherenkov cone.

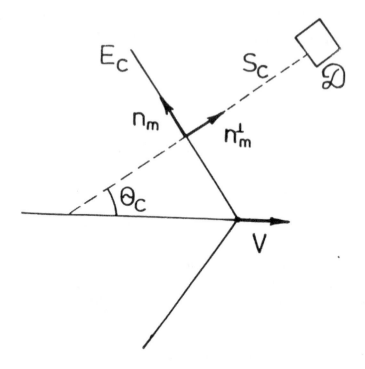

Figure 2.1. CSW propagating in an infinite medium. There is no EMF in front of the Cherenkov cone. Behind it there is the EMF of the moving charge. At the Cherenkov cone itself there are singular electric, E, and magnetic, H, fields. The latter having only the ϕ component is perpendicular to the plane of figure.

We observe that both terms in $\vec{E}$ and $\vec{H}$ are singular on the Cherenkov cone (since r_m vanishes there). On the other hand, according to the Gauss theorem the integral from ϵE taken over the sphere surrounding the charge should be equal to $4\pi e$. The integrals from each of the terms entering into E are divergent. Only their sum is finite (take into account their different signs). This was explicitly shown in [6].

The observer at the (ρ, z) point will see the following picture. There is no EMF for $c_n t < R_m$ ($R_m = (z + \rho/\gamma_n)/\beta_n$). At the time $c_n t = R_m$ the Cherenkov shock wave (CSW) reaches the observer. At this instant the actual and two coinciding retarded charge positions are $z_a = z + \rho/\gamma_n$ and $z_r = z - \rho\gamma_n$, resp.. For $c_n t > R_m$ the observer sees the EMF of the charged particle originating from the retarded positions of the particle lying to the left and right from z_r.

At large distances the terms with the Θ functions die out everywhere except on the Cherenkov cone, and for the electromagnetic field strengths one has

$$\epsilon\vec{E} = \frac{2e\beta_n}{\gamma_n r_m}\delta(\mathcal{M})\cdot\vec{n}_m, \quad \vec{H} = \frac{2e\beta}{\gamma_n r_m}\delta(\mathcal{M})\cdot\vec{n}_\phi. \tag{2.11}$$

The Poynting vector is equal to

$$\vec{S}_c = \frac{c}{4\pi}\vec{E}\times\vec{H} = \frac{c}{4\pi}\sqrt{\frac{\mu}{\epsilon}}\cdot[\frac{2e\beta}{r_m\gamma}\delta(\mathcal{M})]^2\cdot\vec{n}_m^\perp \tag{2.12}$$

Here $\vec{n}_m^\perp = n_\rho/\beta_n\gamma_n + n_z/\beta_n$ is the unit vector normal to the surface of the Cherenkov cone (Fig.2.1). An observer being placed at the ρ, z point will detect the CSW at the instant $t = (z + \rho/\gamma_n)/v$. The beam of charged particles propagating along the z axis with a velocity $v > c_n$ produces an energy flux in the $\vec{n}_m^\perp$ direction with the electric vector in the $\vec{n}_m$ direction.

Uniform motion with $v < c_n$ in the finite spatial interval.
Let a charge be at rest at the point $z = -z_0$ for $t < -t_0$ ($t_0 = z_0/v$). During the time interval $-t_0 < t < t_0$ the charge moves with the constant velocity $v < c_n$. For $t > t_0$ the charge is again at rest at the point $z = z_0$. The electromagnetic potentials are equal to

$$\epsilon\Phi = \frac{e}{r_1}\Theta[r_1 - c_n(t + t_0)] + \frac{e}{r_2}\Theta[c_n(t - t_0) - r_2]$$

$$+\frac{e}{r_m}\Theta[c_n(t + t_0) - r_1]\Theta[r_2 - c_n(t - t_0)],$$

$$A_z = \frac{e\beta\mu}{r_m}\Theta[c_n(t + t_0) - r_1]\Theta[r_2 - c_n(t - t_0)] \tag{2.13}$$

where we put $r_1 = [\rho^2 + (z + z_0)^2]^{1/2}$, $r_2 = [\rho^2 + (z - z_0)^2]^{1/2}$. The particular terms in (2.13) have a simple interpretation. The information about the beginning of the charge motion has not reached the points for which $t < -t_0 + r_1/c_n$. At these spatial points there is a field of the charge resting at $z = -z_0$ (the first term in Φ). The information on the ending of the motion has passed through the points for which $t > t_0 + r_2/c_n$. At those space-time points there is a field of the charge which is at rest at $z = z_0$ (second term in Φ). Finally, at the space-time points for which $-t_0 + r_1/c_n < t < t_0 + r_2/c_n$ there is the field of the uniformly moving charge (last term in Φ). The magnetic field strength is equal to

$$H_\phi = \frac{e\beta(1 - \beta_n^2)\rho}{r_m^3}\Theta[c_n(t + t_0) - r_1]\Theta[r_2 - c_n(t - t_0)]$$

$$+\frac{e\beta\rho}{r_1 r_m}\delta[c_n(t + t_0) - r_1] - \frac{e\beta\rho}{r_2 r_m}\delta[c_n(t - t_0) - r_2].$$

Before writing out the electric field strength in a general form we give its ρ component

$$E_\rho = -\frac{\partial \Phi}{\partial \rho} = \frac{e\rho}{\epsilon r_1^3} \Theta[r_1 - c_n(t + t_0)] + \frac{e\rho}{\epsilon r_2^3} \Theta[c_n(t - t_0) - r_2]$$

$$+\frac{e\rho(1 - \beta_n^2)}{\epsilon r_m^3} \Theta[c_n(t + t_0) - r_1]\Theta[r_2 - c_n(t - t_0)]$$

$$-\delta[c_n(t + t_0) - r_1]\frac{e\rho}{\epsilon r_1}\left(\frac{1}{r_1} - \frac{1}{r_m}\right)$$

$$+\delta[c_n(t - t_0) - r_2]\frac{e\rho}{\epsilon r_2}\left(\frac{1}{r_2} - \frac{1}{r_m}\right).$$

We now clarify the physical meaning of particular terms entering into this equation. The first term in the first line describes the electrostatic field of a charge resting at the point $z = -z_0$ up to an instant $t = -t_0$. It differs from zero outside the sphere S_1 of radius $c_n(t + t_0)$ with its center at $z = -z_0$. The second term in the same line describes the electrostatic field of a charge at rest at the point $z = z_0$ after the instant $t = t_0$. It differs from zero inside the sphere S_2 of radius $c_n(t - t_0)$ with its center at $z = z_0$. It is easy to check that for $\beta_n < 1$ the sphere S_2 is always inside S_1. The term in the second line corresponds to the electrostatic component of the EMF produced by a charge moving in the interval $(-z_0, z_0)$. The presence of the denominator r_m^3 supports this claim. This term differs from zero between the spheres S_2 and S_1. Since the terms just mentioned decrease as $1/r^2$ as $r \to \infty$, they do not contribute to the radiation field. The two terms in the third line (with $1/r_1$ and $1/r_m$ in their denominators) describe the BS shock wave arising at the beginning of motion. Finally, the two terms in the fourth line describe the BS shock wave arising at the end of motion.

In a vector form, the electric field strength is given by

$$\vec{E} = \frac{e}{\epsilon r_1^2}\vec{n}_r^{(1)}\Theta[r_1 - c_n(t + t_0)] + \frac{e}{\epsilon r_2^2}\vec{n}_r^{(2)}\Theta[c_n(t - t_0) - r_2]$$

$$+\frac{er(1 - \beta_n^2)}{\epsilon r_m^3}\vec{n}_r\Theta[c_n(t + t_0) - r_1]\Theta[r_2 - c_n(t - t_0)]$$

$$+\frac{e\rho\delta[c_n(t + t_0) - r_1]}{\epsilon r_1 r_m}\beta_n\vec{n}_\theta^{(1)} - \frac{e\rho\delta[c_n(t - t_0) - r_2]}{\epsilon r_2 r_m}\beta_n\vec{n}_\theta^{(2)}, \qquad (2.14)$$

Here $\vec{n}_r^{(1)}$, $\vec{n}_\theta^{(1)}$, $\vec{n}_r^{(2)}$ and $\vec{n}_\theta^{(2)}$ are the radial and polar unit vectors lying on the spheres S_1 and S_2 (defined by $c_n(t + t_0) = r_1$ and $c_n(t - t_0) =$

r_2, respectively) with their centers at the points $z = -z_0$ and $z = z_0$, respectively:

$$\vec{n}_r^{(1)} = \frac{1}{r_1}[\rho\vec{n}_\rho + (z + z_0)\vec{n}_z] = \frac{1}{r_1}[\vec{n}_r(r + z_0\cos\theta) - \vec{n}_\theta z_0\sin\theta],$$

$$\vec{n}_\theta^{(1)} = \frac{1}{r_1}[\vec{n}_\rho(z + z_0) - \vec{n}_z\rho] = \frac{1}{r_1}[\vec{n}_\theta(r + z_0\cos\theta) + \vec{n}_r z_0\sin\theta],$$

$$\vec{n}_r^{(2)} = \frac{1}{r_2}[\rho\vec{n}_\rho + (z - z_0)\vec{n}_z] = \frac{1}{r_2}[\vec{n}_r(r - z_0\cos\theta) + \vec{n}_\theta z_0\sin\theta],$$

$$\vec{n}_\theta^{(2)} = \frac{1}{r_2}[\vec{n}_\rho(z - z_0) - \vec{n}_z\rho] = \frac{1}{r_2}[\vec{n}_\theta(r - z_0\cos\theta) - \vec{n}_r z_0\sin\theta].$$

When obtaining (2.14) it was taken into account that $r_m = |r_1 - \beta_n(z + z_0)|$ for $c_n(t + t_0) = r_1$ and $r_m = |r_2 - \beta_n(z - z_0)|$ for $c_n(t - t_0) = r_2$. For $\beta_n < 1$, these expressions are reduced to $r_m = r_1 - \beta_n(z + z_0)$ and $r_m = r_2 - \beta_n(z - z_0)$, respectively. At the observational distances large compared with the interval of motion $(r \gg 2z_0)$

$$\vec{n}_r^{(1)} \approx \vec{n}_r^{(2)} \approx \vec{n}_r, \quad \vec{n}_\theta^{(1)} \approx \vec{n}_\theta^{(2)} \approx \vec{n}_\theta.$$

For a distant observer the radiation field is given by

$$\vec{E} = \frac{e\beta_n\rho}{\epsilon r_1 r_m}\delta[c_n(t + t_0) - r_1] \cdot \vec{n}_\theta^{(1)} - \frac{e\beta_n\rho}{\epsilon r_2 r_m}\delta[c_n(t - t_0) - r_2] \cdot \vec{n}_\theta^{(2)}$$

$$\vec{H} = \vec{n}_\phi e\beta\rho\left\{\frac{\delta[c_n(t + t_0) - r_1]}{r_1 r_m} - \frac{\delta[c_n(t - t_0) - r_2 n]}{r_2 r_m}\right\}. \qquad (2.15)$$

An observer at the (ρ, z) point will detect the radiation arising from the particle instantaneous acceleration and deceleration at the instants $t = -t_0 + r_1/c_n$ and $t = t_0 + r_2/c_n$, respectively.

The total Poynting vector is equal to the sum of energy fluxes emitted at the points $z = \pm z_0$:

$$\vec{S} = \vec{S}_1 + \vec{S}_2, \qquad (2.16)$$

$$\vec{S}_1 = \frac{c}{4\pi}\sqrt{\frac{\mu}{\epsilon}} \cdot \left\{\frac{e\beta\rho\delta[c_n(t + t_0) - r_1]}{r_1 r_m}\right\}^2 \cdot \vec{n}_r^{(1)},$$

$$\vec{S}_2 = \frac{c}{4\pi}\sqrt{\frac{\mu}{\epsilon}} \cdot \left\{\frac{e\beta\rho\delta[c_n(t - t_0) - r_2]}{r_2 r_m}\right\}^2 \cdot \vec{n}_r^{(2)}.$$

Here $\vec{n}_r^{(1)}$ and $\vec{n}_r^{(2)}$ are the unit vectors normal to S_1 and S_2, respectively. EMF strengths (2.15) are obtained from (2.14) by dropping the terms which decrease as $1/r^2$ at infinity. This is possible since r_m is nowhere zero for $\beta_n < 1$. It turns out that the vector $\vec{S}$ differs from zero only on the surfaces of the spheres S_1 and S_2. This means that it describes (for $r \to \infty$) only divergent spherical waves emitted at the $z = z_0$ and $z = -z_0$ points.

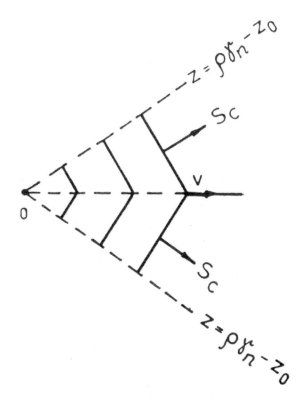

Figure 2.2. The superluminal motion of a charge in a medium begins from the state of rest at $z = -z_0$. In the $z < \rho\gamma_n - z_0$ region an observer sees (consecutively in time) the EMF of the charge at rest, the BS shock wave and the EMF of the moving charge. There is no VCR in this spatial region. In the $z > \rho\gamma_n - z_0$ region the observer sees consecutively the EMF of the charge at rest, the CSW, the EMF from two retarded positions of the charge, the BS and the EMF from the retarded position of the charge moving away. The BS shock wave (not shown here) is tangential to S_c at the point where S_c intersects the surface $z = \rho\gamma_n - z_0$.

Uniform motion with $v > c_n$ in a semi-finite spatial interval.

a) The charge motion begins from the state of rest (Fig. 2.2). Let a charge be at rest at the point $z = -z_0$ up to an instant $t' = -t_0$. For $t' > -t_0$ it moves with a velocity $v > c_n$. For an observer at the point (ρ, z) the condition for the particle to be at rest is $c_n(t + t_0) < r_1$. The condition $t' > -t_0$ for the beginning of the charge motion is different for upper and lower signs in (2.3) (see [7]). The solution corresponding to the upper sign exists only if $z > \rho\gamma_n - z_0$ and $R_m/c_n < t < -t_0 + r_1/c_n$ ($R_m = (z + \rho/\gamma_n)/\beta_n$). The solution corresponding to the lower sign exists both for $z < \rho\gamma_n - z_0$ and $z > \rho\gamma_n - z_0$. It is easy to check that $t > -t_0 + r_1/c_n$ for $z < \rho\gamma_n - z_0$ and $t > R_m/c_n$ for $z > \rho\gamma_n - z_0$. The electric scalar and magnetic vector potentials are given by

$$\epsilon\Phi = \frac{e}{r_1}\Theta[r_1 - c_n(t+t_0)] + \frac{e}{r_m}\Theta(z+z_0 - \rho\gamma_n)\Theta[r_1 - c_n(t+t_0)]\Theta(c_nt - R_m)$$

$$+\frac{e}{r_m}\Theta(\rho\gamma_n - z_0 - z)\Theta[c_n(t+t_0) - r_1] + \frac{e}{r_m}\Theta(z+z_0 - \rho\gamma_n)\Theta(c_nt - R_m),$$

$$A_z = \frac{e\beta\mu}{r_m}\{\Theta(z+z_0 - \rho\gamma_n)\Theta[r_1 - c_n(t+t_0)]\Theta(c_nt - R_m)$$

$$+\Theta(\rho\gamma_n - z_0 - z)\Theta[c_n(t+t_0) - r_1] + \Theta(z+z_0 - \rho\gamma_n)\Theta(c_nt - R_m)\}. \quad (2.17)$$

As a result the observer at the point (ρ, z) will see the following picture:

Let $z < \rho\gamma_n - z_0$. Then, for $t < -t_0 + r_1/c_n$ the observer sees the electrostatic field of a charge at rest at $z = -z_0$ (the first term in Φ which differs from zero outside the sphere S_1 defined by $c_n(t+t_0) = r_1$). The third term in Φ and the second term in A_z describe the charge radiation from particular points of its trajectory. They are confined to the interior of the sphere S_1. There is no CSW in this spatial region.

Let the observer be in the spatial region where $z > \rho\gamma_n - z_0$. In this case, for $t < R_m/c_n$, he sees the electrostatic field of the charge at rest at $z = -z_0$ (the first term in Φ). At the time $t = R_m/c_n$ the CSW reaches him. At this instant the retarded positions of a charge coincide and are given by $z' = z - \rho\gamma_n$. In the time interval $R_m/c_n < t < -t_0 + r_1/c_n$ the solution corresponding to the upper sign (the second term in Φ and the first in A_z) gives the EMF from the retarded positions of the particle in the interval $-z_0 < z' < z - \rho\gamma_n$. On the other hand, the solution corresponding to the lower sign (last terms in Φ and A_z), describes for $t > R_m/c_n$ the EMF from the retarded position of the charged particle lying to the right of the $z' = z - \rho\gamma_n$. Thus in the time interval $R_m/c_n < t < -t_0 + r_1/c_n$ the observer sees simultaneously the electrostatic field of a charge at rest at $z = -z_0$, and the EMF from two retarded positions of a charge lying to the left and right of $z' = z - \rho\gamma_n$. At the instant $t = -t_0 + r_1/c_n$ the BS shock wave from the $z = -z_0$ point reaches the observer. After this instant he sees the EMF from the charge retarded position lying to the right of $z' = z - \rho\gamma_n$. As the time advances the distance between the observational point and the particle retarded position increases. Correspondingly the EMF diminishes at the observational point.

For a distant observer only the singular parts of the field strengths survive

$$\vec{E} = -\frac{e\beta_n\rho\delta(c_n(t+t_0) - r_1)}{\epsilon(\beta_n(z+z_0) - r_1)r_1} \cdot \vec{n}_\theta^{(1)}$$

$$+\frac{2e}{\epsilon\gamma_n r_m}\Theta(z+z_0 - \rho\gamma_n)\delta(c_nt - R_m) \cdot \vec{n}_m,$$

$$\vec{H} = \vec{n}_\phi H, \quad H = -\frac{e\beta\rho\delta(c_n(t+t_0) - r_1)}{(\beta_n(z+z_0) - r_1)r_1}$$

$$+\frac{2e}{\gamma_n r_m n}\Theta(z + z_0 - \rho\gamma_n)\delta(c_n t - R_m). \tag{2.18}$$

When obtaining these expressions, we omitted the terms which do not contain delta functions and which decrease as $1/r^2$ as $r \to \infty$ (they do not contribute to the radiation). For the terms containing r_m^3 in their denominators, this is not valid on the Cherenkov cone (since $r_m = 0$ on it). For the spatial region $z < \rho\gamma_n - z_0$ the singular EMF is confined to the surface of a sphere S_1 of radius $r_1 = c_n(t + t_0)$. A distant observer detects the BS shock wave at the instant $t = -t_0 + r_1/c_n$. There is no CSW in this region of space. For $z > \rho\gamma_n - z_0$ a distant observer detects the CSW at $t = R_m/c_n$ and the BS shock wave at $t = -t_0 + r_1/c_n$.

The Poynting vector is equal to $\vec{S} = \vec{S}_1 + \vec{S}_c$, where

$$\vec{S}_1 = \frac{c}{4\pi}\sqrt{\frac{\mu}{\epsilon}} \cdot \left[\frac{e\beta\rho\delta(c_n(t+t_0) - r_1)}{(\beta_n(z+z_0) - r_1)r_1}\right]^2 \cdot \vec{n}_r^{(1)}$$

is the BS shock wave different from zero at the surface of the shock wave emitted at the beginning of motion and

$$\vec{S}_c = \frac{c}{4\pi}\sqrt{\frac{\mu}{\epsilon}} \cdot [\frac{2e\beta}{\gamma_n r_m}\Theta(z + z_0 - \rho\gamma_n)\delta(\mathcal{M})]^2 \cdot \vec{n}_m^\perp \tag{2.19}$$

is the CSW different from zero at the surface of the Cherenkov cone.

b) The charge motion ends in a state of rest (Fig. 2.3). Let a charge move with a velocity $v > c_n$ from $z = -\infty$ up to a point $z = z_0$. After that it remains at rest there. The condition for the charge to be at rest is $c_n(t - t_0) > r_2$. The solution corresponding to the lower sign in (2.3) exists only for $z < z_0 + \rho\gamma_n$ and $R_m/c_n < t < t_0 + r_2/c_n$ (see [7]). The solution corresponding to the upper sign in (2.3) exists both for $z > z_0 + \rho\gamma_n$ if $t > t_0 + r_2/c_n$ and for $z < z_0 + \rho\gamma_n$ if $t > R_m/c_n$. The electromagnetic potentials are equal to :

$$\Phi = \frac{e}{\epsilon r_2}\Theta[c_n(t - t_0) - r_2] + \frac{e}{\epsilon r_m}\Theta(z - z_0 - \rho\gamma_n)\Theta[c_n(t - t_0) - r_2]$$

$$+\frac{e}{\epsilon r_m}\Theta(c_n t - R_m)\Theta(z_0 + \rho\gamma_n - z)\{1 + \Theta[r_2 - c_n(t - t_0)]\},$$

$$A_z = \mu\beta\frac{e}{r_m}\Theta(z - z_0 - \rho\gamma_n)\Theta[c_n(t - t_0) - r_2]$$

$$+\mu\beta\frac{e}{r_m}\Theta(c_n t - R_m)\Theta(z_0 + \rho\gamma_n - z)\{1 + \Theta[r_2 - c_n(t - t_0)]\}. \tag{2.20}$$

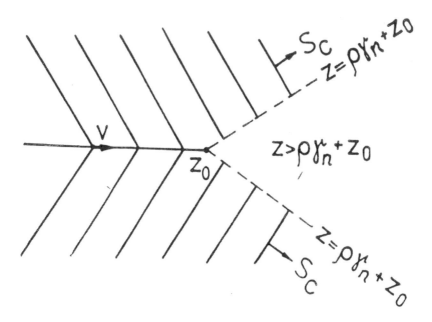

Figure 2.3. The superluminal motion ends in the state of rest at $z = z_0$. In the region $z > \rho\gamma_n + z_0$ the observer sees no field up to some instant, when the shock BS wave reaches him. Later he sees the EMF of the charge at rest and the EMF from one retarded position of the charge. In the region $z < \rho\gamma_n + z_0$ the EMF is equal to zero up to some instant when the CSW reaches the observer. After that he sees the EMF from two retarded positions of the charge up to the instant when the BS shock wave reaches him. Later the observer sees simultaneously the field of the charge at rest and that of the retarded positions of the charge. The BS shock wave (not shown here) is tangential to S_c at the point where S_c intersects the surface $z = \rho\gamma_n + z_0$.

For an observer in the $z > \rho\gamma_n + z_0$ region there is no EMF for $c_n(t-t_0) < r_2$. At $t = t_0 + r_2/c_n$ he detects the BS shock wave. For $t > t_0 + r_2/c_n$ the observer sees the EMF of the charge at rest at the $z = z_0$ point and the EMF of the retarded positions of the charge trajectory lying to the left of the $z = z_0$ point. There is no CSW in this spatial region despite the presence of the radiation associated with the charge superluminal motion. For an observer in the $z < \rho\gamma_n + z_0$ region, the EMF is equal to zero for $c_n t < R_m$. At $t = R_m/c_n$ the CSW reaches the observational point. At this instant two retarded charge positions coincide and are equal to $z' = z - \rho\gamma_n$. For $R_m/c_n < t < t_0 + r_2/c_n$ the solution corresponding to the lower sign gives the EMF emitted from the points of the charge trajectory that lie in the interval $(z - \rho\gamma_n < z' < z_0)$. At $t = t_0 + r_2/c_n$, the BS shock wave emitted from the $z = z_0$ point reaches the observer. After that, the solution corresponding to the lower sign gives the EMF of the charge at rest at the $z = z_0$ point. On the other hand, the solution corresponding to the upper sign for $c_n t > R_m$ gives EMF from the charge retarded positions lying to the left of $z - \rho\gamma_n$ point. The EMF at the observational point diminishes as the radiation arrives from more remote points.

The field strengths and Poynting vector in the wave zone are:

$$\vec{E} = e\frac{\delta(c_n(t - t_0) - r_2 n)}{\beta_n(z - z_0) - r_2}\frac{\rho\beta_n}{\epsilon r_2}\cdot\vec{n}_\theta^{(2)} + e\delta(c_n t - R_m)\frac{2}{\epsilon r_m\gamma_n}\Theta(\rho\gamma_n + z_0 - z)\cdot\vec{n}_m,$$

$$\vec{H} = e\left[\frac{\delta(c_n(t - t_0) - r_2)}{\beta_n(z - z_0) - r_2}\frac{\beta}{r_2} + \frac{2}{r_m\gamma_n\sqrt{\epsilon\mu}}\delta(c_n t - R_m)\right]\cdot\vec{n}_\phi, \qquad (2.21)$$

$$S = S_2 + S_c, \quad S_2 = \frac{c}{4\pi}\sqrt{\frac{\mu}{\epsilon}}\cdot\left[\frac{\delta(c_n(t - t_0) - r_2)}{\beta_n(z - z_0) - r_2}\frac{\rho\beta}{r_2}\right]^2\cdot\vec{n}_r^{(2)},$$

$$S_c = \frac{c}{4\pi}\sqrt{\frac{\mu}{\epsilon}}\cdot\left[\frac{2\beta}{r_m\gamma_n}\delta(\mathcal{M})\Theta(z_0 + \rho\gamma_n - z)\right]^2\cdot\vec{n}_m^\perp.$$

In the spatial region $z > \rho\gamma_n + z_0$ a distant observer detects the BS shock wave corresponding to the termination of motion at $t = t_0 + r_2/c_n$. There is no CSW there. For $z < \rho\gamma_n + z_0$ the observer sees the CSW at $t = R_m/c_n$ and the BS shock wave at $t = t_0 + r_2/c_n$.

Uniform motion with $v > c_n$ in a finite spatial interval.
Let a charge be at rest at the point $z = -z_0$ up to an instant $t = -t_0$ ($t_0 = z_0/v$). In the time interval $-t_0 < t < t_0$ the particle moves with a constant velocity $v > c_n$. For $t > t_0$ the particle is again at rest at the point $z = z_0$ (Fig. 2.4). According to [1,8] the physical realization of this model is, e.g., β decay followed by nuclear capture. An observer in various space-time regions will detect the following physical situations:

    i) $z < \rho\gamma_n - z_0$.
For $t < -t_0 + r_1/c_n$ the observer sees the EMF of the charge at rest at $z = -z_0$. At $t = -t_0 + r_1/c_n$ the BS shock wave originating from the $z = -z_0$ point (BS$_1$ shock wave for short) reaches him. For $-t_0 + r_1/c_n < t < t_0 + r_2/c_n$ the observer sees the EMF of the charge moving with the superluminal velocity (the lower sign in (2.3)). At $t = t_0 + r_2/c_n$ the BS shock wave originating from the $z = z_0$ point (BS$_2$ shock wave for short) reaches him. Finally, for $t > t_0 + r_2/c_n$ the observer sees the EMF of the charge at rest at $z = z_0$. There is no CSW in this spatial region despite the observation of superluminal motion.

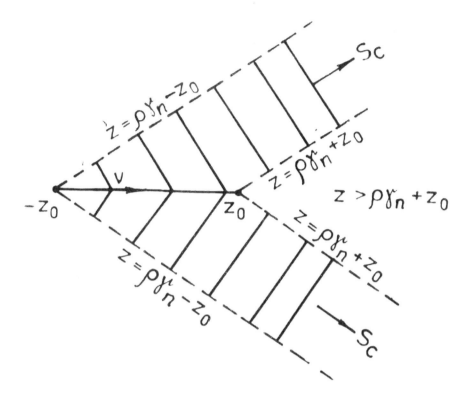

Figure 2.4. The superluminal motion begins from the state of rest at the point $z = -z_0$ and ends by the state of rest at the point $z = z_0$. For the finite distances the space-time distribution of EMF is rather complicated (see the text). The distant observer will see the following space-time picture. In the region $z < \rho\gamma_n - z_0$ he detects the BS_1 shock wave (from the $z = -z_0$ point) first and BS_2 shock wave (from the $z = z_0$ point) later. In the $z > \rho\gamma_n + z_0$ region these waves arrive in the reverse order. In the $\rho\gamma - z_0 < z < (\rho^2\gamma_n^2 + z_0^2/\beta_n^2)^{1/2}$ region the observer consecutively detects the CSW, BS_1 shock wave and the BS_2 shock wave. In the region $(\rho^2\gamma_n^2 + z_0^2/\beta_n^2)^{1/2} < z < \rho\gamma_n + z_0$ the latter two waves arrive in the reverse order. The CSW S_c is tangential to the BS_1 shock wave at the point where S_c intersects the surface $z = \rho\gamma - z_0$ and to the BS_2 shock wave at the point where S_c intersects the surface $z = \rho\gamma + z_0$ (see Fig. 2.7).

ii) $\rho\gamma_n - z_0 < z < (\rho^2\gamma_n^2 + z_0^2/\beta_n^2)^{1/2}$.
For $t < R_m/c_n$ the observer sees the EMF of the charge at rest at $z = -z_0$. At $t = R_m/c_n$ the CSW reaches him. For $R_m/c_n < t < -t_0 + r_1/c_n$ the observer simultaneously sees the EMF of the charge at rest at $z = -z_0$ and the EMF of the moving charge (both signs give contribution). At $t = -t_0 + r_1/c_n$ the BS_1 shock wave reaches him. For $-t_0 + r_1/c_n < t < t_0 + r_2/c_n$ the observer will see the EMF of the moving charge (the lower sign in (2.3)). At $tt_0 + r_2/c_n$ the BS_2 shock wave reaches him. Lastly, for $t > t_0 + r_2/c_n$ the observer sees the EMF of the charge resting at $z = z_0$

iii) $[\rho^2\gamma_n^2 + z_0^2/\beta_n^2]^{1/2} < z < z_0 + \rho\gamma_n$.

For $t < R_m/c_n$ the observer sees the EMF of the charge at rest at the $z = -z_0$ point. At $t = R_m/c_n$ the CSW reaches him. For $R_m/c_n < t < t_0 + r_2/c_n$ the observer sees the EMF of the charge at rest at $z = -z_0$ and the EMF of the moving charge (both signs of Eq.(2.3) give a contribution). At $t = t_0 + r_2/c_n$ the BS$_2$ shock wave reaches the observational point. For $t_0 + r_2/c_n < t < -t_0 + r_1/c_n$ the observer simultaneously sees the EMF of the charge at rest at $z = -z_0$, the EMF of the charge at rest at $z = z_0$, and the EMF of the moving charge (upper sign in (2.3)). At $t = -t_0 + r_1/c_n$ the BS$_1$ shock wave reaches him. Finally, for $t > -t_0 + r_1/c_n$ the observer sees the EMF of the charge at rest at $z = z_0$.

iv) $z > z_0 + \rho\gamma_n$.

For $t < t_0 + r_2/c_n$ the observer will see the EMF of the charge at rest at the $z = -z_0$ point. At $t = t_0 + r_2/c_n$ the BS$_2$ shock wave reaches him. For $t_0 + r_2/c_n < t < -t_0 + r_1/c_n$ he sees the EMF of the charge at rest at the $z = \pm z_0$ points and the EMF of the moving charge (the upper sign in (2.3)). At $t = -t_0 + r_1/c_n$ the BS$_1$ shock wave reaches him. Lastly, for $t > -t_0 + r_1/c_n$ the observer sees the EMF of the charge at rest at $z = z_0$. There is no CSW in this spatial region.

The electromagnetic potentials are equal to

$$\Phi = \Phi_1 + \Phi_2 + \Phi_m, \quad A_z = \beta\mu\epsilon\Phi_m. \tag{2.22}$$

Here

$$\Phi_1 = \frac{e}{\epsilon r_1}\Theta(r_1 - c_n(t + t_0)), \quad \Phi_2 = \frac{e}{\epsilon r_2}\Theta(c_n(t - t_0) - r_2),$$

$$\Phi_m = \frac{e}{\epsilon r_m}\{\Theta(z_0 - z + \rho\gamma_n)\Theta(z_0 + z - \rho\gamma_n)\Theta(c_n t - R_m)$$

$$\times[\Theta(r_1 - c_n(t + t_0)) + \Theta(r_2 - c_n(t - t_0))]$$

$$+\Theta(z - z_0 - \rho\gamma_n)\Theta(r_1 - c_n(t + t_0))\Theta(c_n(t - t_0) - r_2)$$

$$+\Theta(\rho\gamma_n - z - z_0)\Theta(c_n(t + t_0) - r_1)\Theta(r_2 - c_n(t - t_0))\}.$$

At large distances the field strengths are

$$\epsilon\vec{E} = -\frac{\delta(c_n(t + t_0) - r_1))}{\beta_n(z + z_0) - r_1}\frac{e\rho\beta_n}{r_1}\cdot\vec{n}_\theta^{(1)} + \frac{\delta(c_n(t - t_0) - r_2)}{\beta_n(z - z_0) - r_2}\frac{e\rho\beta_n}{r_2}\cdot\vec{n}_\theta^{(2)}$$

$$+\delta(c_n t - R_m)\frac{2e}{r_m\gamma_n}\Theta(\rho\gamma_n + z_0 - z)\Theta(z + z_0 - \rho\gamma_n)\cdot\vec{n}_m,$$

$$\vec{H} = H_\phi\vec{n}_\phi, \quad H_\phi = -\frac{\delta(c_n(t + t_0) - r_1)}{\beta_n(z + z_0) - r_1}\frac{e\rho\beta}{r_1} + \frac{\delta(c_n(t - t_0) - r_2)}{\beta_n(z - z_0) - r_2}\frac{e\rho\beta}{r_2}$$

$$+\frac{2e}{r_m\gamma_n\sqrt{\epsilon\mu}}\delta(c_nt-R_m)\Theta(\rho\gamma_n+z_0-z)\Theta(z+z_0-\rho\gamma_n). \qquad (2.23)$$

When obtaining (2.23), the terms decreasing as $1/r^2$ at infinity were omitted. Amongst them, there are terms proportional to $1/r_m^3$. For large observational distances they are small everywhere except for the Cherenkov cone, where $r_m = 0$. If one tries to obtain the Fourier field components from (2.23) one gets the divergent expressions. On the other hand, Fourier components will be finite if one includes into (2.23) the terms proportional to $1/r_m^3$ mentioned above. The total Poynting vector reduces to the sum of energy fluxes radiated at the $z = \pm z_0$ points, and to the Cherenkov flux:

$$\vec{S} = \vec{S}_1 + \vec{S}_c + \vec{S}_2, \qquad (2.24)$$

$$\vec{S}_1 = \frac{c}{4\pi}\sqrt{\frac{\mu}{\epsilon}}\cdot\left[\frac{\delta(c_n(t+t_0)-r_1)}{\beta_n(z+z_0)-r_1}\frac{e\rho\beta}{r_1}\right]^2\cdot\vec{n}_r^{(1)},$$

$$\vec{S}_2 = \frac{c}{4\pi}\sqrt{\frac{\mu}{\epsilon}}\cdot\left[\frac{\delta(c_n(t-t_0)-r_2)}{\beta_n(z-z_0)-r_2}\frac{e\rho\beta}{r_2}\right]^2\cdot\vec{n}_r^{(2)},$$

$$\vec{S}_c = \frac{c}{4\pi}\sqrt{\frac{\mu}{\epsilon}}\cdot\left[\frac{2e}{r_m\gamma_n}\delta(\mathcal{M})\Theta(z+z_0-\rho\gamma_n)\Theta(z_0+\rho\gamma_n-z)\right]^2\cdot\vec{n}_m^\perp.$$

It is seen that $\vec{S}_1$ is infinite on the spherical surface $c_n(t+t_0) = r_1$. The factor $\beta_n(z+z_0) - r_1$ in the denominator vanishes at the point where BS$_1$ intersects the CSW. Correspondingly, $\vec{S}_2$ is infinite on the spherical surface $c_n(t-t_0) = r_2$. The factor $\beta_n(z-z_0) - r_2$ in the denominator vanishes at the point where BS$_2$ intersects the CSW. Finally, $\vec{S}_c$ is infinite on the CSW. The factor r_m in the denominator vanishes on the CSW.

For a distant observer the radiation field looks different in various spatial regions (Fig. 2.5).

i) $z < \rho\gamma_n - z_0$

At the instant $-t_0+r_1/c_n$ the observer detects the BS$_1$ shock wave. At the later time $t = t_0 + r_2/c_n$ he detects the BS$_2$ shock wave. There is no CSW in this spatial region.

ii) $\rho\gamma_n - z_0 < z < (\rho^2\gamma_n^2 + z_0^2/\beta_n^2)^{1/2}$

The observer detects (consecutively in time) the CSW at $t = R_m/c_n$, the BS$_1$ shock wave at the instant $-t_0 + r_1/c_n$ and the BS$_2$ shock wave at the instant $t = t_0 + r_2/c_n$.

iii) $(\rho^2\gamma_n^2 + z_0^2/\beta_n^2)^{1/2} < z < \rho\gamma_n + z_0$

The observer sees the CSW at the instant $-t_0+R_m/c_n$, the BS$_2$ shock wave at the instant t_0+r_2/c_n, and the BS$_1$ shock wave at the instant $-t_0+r_1/c_n$.

iv) $z > \rho\gamma_n + z_0$.

At the instant t_0+r_2/c_n the observer fixes the BS$_2$ shock wave. At the later

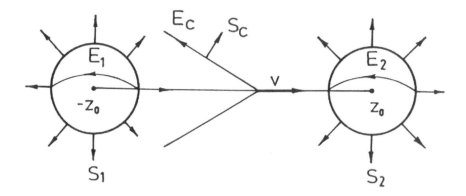

Figure 2.5. The schematic presentation of the EMF for a superluminal motion in a finite spatial interval. The magnetic field of the BSs and of the moving charge has only a ϕ component. The electric field of the BSs has only the θ_1 and θ_2 components. The electric field of the moving charge has singular and non-singular parts. The singular part $\vec{E}_c$ lies on the Cherenkov cone. The non-singular part lies on the radius directed from the particle actual position inwards the Cherenkov cone.

instant $-t_0 + r_1/c_n$ he detects the BS_1 shock wave. As in case i), there is no CSW in this spatial region.

However, some reservation is needed. In the next chapter the instantaneous jumps in velocity in the original Tamm problem will be changed by the velocity linearly rising (or decreasing) with time. It will be shown there that, in addition to the BS shock waves arising at the beginning (BS_1) and at the end (BS_2) of motion, two new shock waves arise at the instant when the charge velocity coincides with the velocity of light in medium. One of them is the Cherenkov shock wave of finite extensions (C_M), whilst the other shock wave closes the Cherenkov cone (C_L) (see Fig. 3.8). Owing to the instantaneous jumps in velocity in the original Tamm problem, the above three shock waves are created simultaneously. When discussing the BS shock waves throughout this chapter, we keep in mind the mixture of these three shock waves (BS_1, BS_2 and C_L). In particular, they are mixed in electromagnetic field strengths (2.23). The traces of these shock waves are contained in electromagnetic potentials (2.22). We observe that Φ_1 and Φ_2 contain terms with r_1 and r_2 in their denominators. The electric field strengths corresponding to them contain δ functions $\delta[(c_n t + t_0) - r_1]$ and

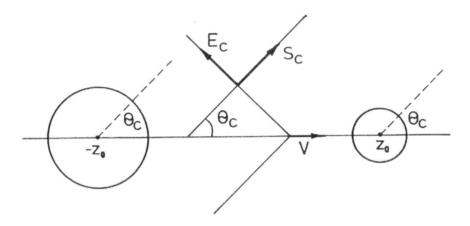

Figure 2.6.   An observer not very far from the z axis sees the BS maximum at an angle different from the Cherenkov angle θ_c. Thus angular resolution is possible for him. For a distant observer the time resolution between the VCR and bremsstrahlungs is still possible.

$\delta[(c_n t - t_0) - r_2]$ with r_1^2 and r_2^2 in their denominators. It is essential that these electric field strengths are uniformly distributed over the spheres S_1 and S_2 of the radii $r_1 = c_n(t+t_0)$ and $r_2 = c_n(t-t_0)$ and do not have a maximum at the Cherenkov angle θ_c ($\cos\theta_c = 1/\beta n$). On the other hand, Φ_m and A_z contain terms with r_m in their denominators. The electric and magnetic fields corresponding to them contain the same δ functions as above but with denominators r_m vanishing at the Cherenkov cone. Thus the BS shock waves treated in this section describe not only the transition of a charge from the state of rest to the state of motion, but also its exceeding the velocity of light in medium.

The BS shock waves from the $z = \pm z_0$ points have maxima at the angles θ_1 and θ_2 slightly different from the Cherenkov angle θ_c. They are defined by

$$\cos\theta_{1,2} = \mp\frac{\epsilon_0}{\beta_n^2\gamma_n^2} + \frac{1}{\beta_n}\left[1 - (\epsilon_0/\beta_n\gamma_n)^2\right]^{1/2}.$$

Let the distance from the observational point be comparable with the motion distance $2z_0$. This observer then will detect the maximum of the BS at the angles θ_1 and θ_2 different from θ_c, and for him the CSW will be clearly separated from the BS shock wave. On the other hand, if the observer is at a distance much larger than $2z_0$, the BS from the $z = \pm z_0$ points and the CSW will have a maximum at almost the same angle θ_c. In this case angular separation of the VCR and BS is hardly possible.

On the observational sphere S of radius r the VCR fills a band of the finite width $r(\theta_1 - \theta_2)$ enclosed between these angles whilst the BS differs from zero on the whole observational sphere. The observation of the VCR on the sphere of large radius is masked by the smallness of the angular region to which the VCR is confined. On the other hand, in the observational $z =$const plane the VCR fills the ring $R_1 < \rho < R_2$ where $R_1 = (z - z_0)/\gamma_n$ and $R_2 = (z + z_0)/\gamma_n$ whilst the intensity of BS has pronounced maxima at $\rho = R_1$ and $\rho = R_2$ (see Chapter 9).

If the intensity of the charged particles is so low that inside the interval $(-z_0, z_0)$ there is only one charged particle at each instant of time, the time resolution between the Cherenkov photons and the BS photons is still possible. We conclude that the description of the VCR in the time representation by direct solving of the Maxwell equations greatly simplifies the consideration. In particular, the prescriptions are easily obtained when and where the CSW should be observed in order to discriminate it from the BS shock wave. This is contrasted with the consideration in terms of the spectral representation where (owing to the lack of the exact analytical solution) the discrimination of the VCR from the BS presents a problem (see, e.g., [1,8-10]). On the other hand, if the frequency dependence of ϵ and μ is essential, an analysis via the Fourier method seems to be more appropriate. In this sense these two methods complement each other.

2.1.3. ORIGINAL TAMM PROBLEM

Tamm considered the following problem. A point charge is at rest at the point $z = -z_0$ of the z axis up to an instant $t = -t_0$. In the time interval $-t_0 < t < t_0$ it moves uniformly along the z axis with a velocity v greater than the velocity of light c_n in medium. For $t > t_0$ the charge is again at rest at the point $z = z_0$. In the spectral representation the non-vanishing z of the vector potential (VP) is given by

$$A_\omega = \frac{\mu}{c} \int \frac{1}{R} j_\omega(x', y', z') \exp\left(-in\omega R/c\right) dx' dy' dz',$$

where $R = [(x-x')^2 + (y-y')^2 + (z-z')^2]^{1/2}$ and j_ω is the Fourier component of the current density defined as

$$j_\omega = \frac{1}{2\pi} \int j(t) \exp(-i\omega t) dt.$$

For a charge moving uniformly in the interval $(-z_0, z_0)$ one finds

$$j(t) = ev\delta(x)\delta(y)\delta(z - vt)\Theta(z + z_0)\Theta(z_0 - z)$$

and

$$j_\omega = \frac{e}{2\pi}\delta(x)\delta(y)\exp(-i\omega z/v)\Theta(z + z_0)\Theta(z_0 - z).$$

Inserting all this into A_ω and integrating over x' and y' one finds

$$A_\omega(x, y, z) = \frac{e\mu}{2\pi c}\int_{-z_0}^{z_0}\frac{dz'}{R}\exp\left[-ik_n\left(\frac{z'}{\beta_n} + R\right)\right],$$

$$R = [\rho^2 + (z - z')^2]^{1/2}, \quad \rho^2 = x^2 + y^2, \quad k_n = kn, \quad k = \frac{\omega}{c}. \tag{2.25}$$

At large distances from the charge ($r \gg z_0$) one has $R = r - z'\cos\theta$, $\cos\theta = z/r$. Inserting this into (2.25) and integrating over z' one obtains

$$A_\omega^{\mathrm{T}}(\rho, z) = \frac{e\beta\mu}{\pi r\omega}\exp\left(-ik_n r\right)q(\omega), \quad H_\phi^{\mathrm{T}} = \frac{ien\beta\sin\theta}{\pi c r}\exp\left(-ik_n r\right)q,$$

$$E_\theta^{\mathrm{T}} = \frac{ie\mu\beta\sin\theta}{\pi c r}\exp\left(-ik_n r\right)q, \quad q(\omega) = \frac{\sin\left[\omega t_0(1 - \beta_n\cos\theta)\right]}{1 - \beta_n\cos\theta}. \tag{2.26}$$

Superscript T means that these expressions were obtained by Tamm. In the limit $kz_0 \to \infty$

$$q \to \frac{\pi}{\beta n}\delta(\cos\theta - 1/\beta_n), \quad A_\omega^{\mathrm{T}}(\rho, z) = \frac{e}{nr\omega}\delta(\cos\theta - 1/\beta_n)\exp\left(-ik_n r\right),$$

$$H_\phi^{\mathrm{T}} = \frac{ie\sin\theta}{cr}\exp\left(-ik_n r\right)\delta(\cos\theta - 1/\beta_n),$$

$$E_\theta^{\mathrm{T}} = \frac{ie\mu\sin\theta}{ncr}\exp\left(-ik_n r\right)\delta(\cos\theta - 1/\beta_n). \tag{2.26'}$$

Now we evaluate the field strengths in the time representation. They are given by

$$H_\phi^{\mathrm{T}} = -\frac{2e\beta}{\pi c r}\sin\theta\int_0^\infty nq(\omega)\sin[\omega(t - r/c_n)]d\omega,$$

$$E_\rho^{\mathrm{T}} = -\frac{2e\mu\beta}{\pi c r}\sin\theta\cos\theta\int_0^\infty q(\omega)\sin[\omega(t - r/c_n)]d\omega,$$

$$E_z^{\mathrm{T}} = \frac{2e\mu\beta}{\pi c r}\sin^2\theta\int_0^\infty q(\omega)\sin[\omega(t - r/c_n)]d\omega. \tag{2.27}$$

It should be noted that only the spherical θ component of $\vec{E}$ differs from zero

$$E_r^{\mathrm{T}} = 0, \quad E_\theta^{\mathrm{T}} = -\frac{2e\mu\beta}{\pi cr} \sin\theta \int_0^\infty q(\omega) \sin[\omega(t - r/c_n)]d\omega.$$

Consider now the function $q(\omega)$. For $\omega t_0 \gg 1$ it becomes $\pi\delta(1 - \beta_n \cos\theta)$. This means that under these conditions $\vec{E}_\omega$ and $\vec{H}_\omega$ have a sharp maximum at $1 - \beta_n \cos\theta = 0$. Or, in other words, photons with the energy $\hbar\omega$ should be observed at an angle $\cos\theta = 1/\beta_n$.

The energy flux through the sphere of the radius r for the entire motion of the charge is

$$\mathcal{E} = r^2 \int S_r(t)dtd\Omega, \quad S_r = \frac{c}{4\pi}E_\theta(t)H_\phi(t).$$

Expressing $E_\theta(t)$ and $H_\phi(t)$ through their Fourier transforms

$$E_\theta(t) = \int E_\theta(\omega)\exp(i\omega t)d\omega, \quad H_\phi(t) = \int H_\phi(\omega)\exp(i\omega t)d\omega$$

and integrating over t, one presents $\mathcal{E}$ in the form

$$\mathcal{E} = \int \frac{d^2\mathcal{E}}{d\omega d\Omega}d\omega d\Omega,$$

where

$$\frac{d^2\mathcal{E}}{d\omega d\Omega} = \frac{cr^2}{2}[E_\theta(\omega)H_\phi^*(\omega) + \mathrm{c.c.}] \tag{2.28}$$

is the energy radiated into unit solid angle and per frequency unit. Substituting here $E_\theta(\omega)$ and $H_\phi(\omega)$, from (2.26) one finds

$$\frac{d^2\mathcal{E}}{d\omega d\Omega} = \frac{e^2\mu n\beta^2 \sin^2\theta}{\pi^2 c}q^2. \tag{2.29}$$

This is the famous Tamm formula frequently used by experimentalists for the identification of the charge velocity. Using the relation

$$\left(\frac{\sin\alpha x}{x}\right)^2 \to \pi\alpha\delta(x) \quad \text{for} \quad \alpha \to \infty,$$

one obtains in the limit $\omega t_0 \to \infty$

$$q^2 \to \frac{\pi k z_0}{\beta^2 n}\delta\left(\cos\theta - \frac{1}{\beta n}\right)$$

and

$$\frac{d^2\mathcal{E}}{d\omega d\Omega} \rightarrow \frac{e^2 \mu k z_0}{\pi c}\left(1 - \frac{1}{\beta_n^2}\right)\delta\left(\cos\theta - \frac{1}{\beta n}\right). \tag{2.30}$$

The energy flux per frequency unit through a sphere S of the radius $r \gg z_0$ is

$$\frac{d\mathcal{E}}{d\omega} = \int d\Omega \frac{d^2\mathcal{E}}{d\omega d\Omega}.$$

Integrating (2.29) over the solid angle $d\Omega$, one obtains for large $k z_0$

$$\frac{d\mathcal{E}}{d\omega} = W_{\mathrm{BS}}$$

for $v < c_n$ and

$$\frac{d\mathcal{E}}{d\omega} = W_{\mathrm{BS}} + W_{\mathrm{Ch}} \tag{2.31}$$

for $v > c_n$. Here

$$W_{\mathrm{BS}} = \frac{2e^2\mu}{\pi c\beta n^2}\left(\ln\frac{1+\beta_n}{|1-\beta_n|} - 2\beta_n\right) \quad \text{and} \quad W_{\mathrm{Ch}} = \frac{e^2 \mu k L}{c}\left(1 - \frac{1}{\beta_n^2}\right).$$

Here $L = 2z_0$ is the charge interval of motion. Tamm identified W_{BS} with the spectral distribution of the BS, arising from the instantaneous acceleration and deceleration of the charge at the instants $\pm t_0$, respectively. On the other hand, W_{Ch} was identified with the spectral distribution of the VCR. This is supported by the fact that W_{Ch} being related to the charge interval of motion

$$\frac{d^2\mathcal{E}}{d\omega dL} = \frac{e^2}{c^2}\left(1 - \frac{1}{\beta_n^2}\right) \tag{2.32}$$

coincides with the famous Frank-Tamm formula describing the energy losses per unit length and per unit frequency for a charge unbounded motion [5]. In the absence of dispersion, the Tamm field strengths (2.27) are easily integrated:

$$H_\phi^{\mathrm{T}}(t) = -\frac{e\beta\sin\theta}{r(1 - \beta_n\cos\theta)}\{\delta[c_n(t-t_0)-r+z_0\cos\theta]-\delta[c_n(t+t_0)-r-z_0\cos\theta]\},$$

$$E_\theta^{\mathrm{T}}(t) = -\frac{e\beta\sin\theta}{rn(1 - \beta_n\cos\theta)}\times$$

$$\times\{\delta[c_n(t-t_0)-r+z_0\cos\theta]-\delta[c_n(t+t_0)-r-z_0\cos\theta]\}. \tag{2.33}$$

The Tamm field strengths in the time representation are needed to compare them with the exact ones given by (2.22) and (2.23). This, in turn, may shed light on the physical meaning of the Tamm radiation intensity (2.29).

2.1.4. COMPARISON OF THE TAMM AND EXACT SOLUTIONS

Exact solution
Above (Eqs.(2.22) and (2.23)), we obtained an exact solution of the treated problem (i.e., the superluminal charge motion in a finite spatial interval) in the absence of dispersion. For convenience we shall refer to the BS shock waves emitted at the beginning of the charge motion ($t = -t_0$) and at its termination ($t = t_0$) as to the BS_1 and BS_2 shock waves, respectively.

In the wave zone we rewrite the field strengths in the form

$$\vec{E} = \vec{E}_{BS} + \vec{E}_{Ch}, \quad \vec{E}_{BS} = \vec{E}_{BS}^{(1)} + \vec{E}_{BS}^{(2)} \quad \vec{H} = \vec{H}_{BS} + \vec{H}_{Ch},$$

$$\vec{H} = H_\phi \vec{n}_\phi, \quad H_\phi = H_{BS} + H_{Ch}, \quad H_{BS} = H_{BS}^{(1)} + H_{BS}^{(2)}. \qquad (2.34)$$

Here

$$\vec{E}_{BS}^{(1)} = -\frac{e\beta}{n} \frac{\delta[c_n(t + t_0) - r_1]}{\beta_n(z + z_0) - r_1} \frac{r \sin\theta}{r_1} \vec{n}_\theta^{(1)},$$

$$\vec{E}_{BS}^{(2)} = \frac{e\beta}{n} \frac{\delta[c_n(t - t_0) - r_2]}{\beta_n(z - z_0) - r_2} \frac{r \sin\theta}{r_2} \vec{n}_\theta^{(2)},$$

$$\vec{E}_{Ch} = \frac{2}{\epsilon r_m \gamma_n} \delta(c_n t - R_m) \Theta(\rho\gamma_n + z_0 - z) \Theta(-\rho\gamma_n + z_0 + z) \vec{n}_m,$$

$$H_{BS}^{(1)} = -e\beta \frac{\delta[c_n(t + t_0) - r_1]}{\beta_n(z + z_0) - r_1} \frac{r \sin\theta}{r_1}, \quad H_{BS}^{(2)} = e\beta \frac{\delta[c_n(t - t_0) - r_2]}{\beta_n(z - z_0) - r_2} \frac{r \sin\theta}{r_2},$$

$$H_{Ch} = \frac{2}{r_m \gamma_n \sqrt{\epsilon\mu}} \delta(c_n t - R_m) \Theta(\rho\gamma_n + z_0 - z) \Theta(-\rho\gamma_n + z_0 + z).$$

Here γ_n, r_1, r_2, r_m, $\vec{n}_\theta^{(1)}$, $\vec{n}_\theta^{(1)}$ and $\vec{n}_m$ are the same as above. The delta functions $\delta[c_n(t + t_0) - r_1]$ and $\delta[c_n(t - t_0) - r_2]$ entering (2.34) describe spherical BS shock waves emitted at the instants $t = -t_0$ and $t = t_0$; $n_\theta^{(1)}$ and $n_\theta^{(1)}$ are the unit vectors tangential to the above spherical waves and lying in the $\phi = $ const plane; $\vec{E}_{BS}^{(1)}$, $\vec{E}_{BS}^{(2)}$, $\vec{H}_{BS}^{(1)}$ and $\vec{H}_{BS}^{(2)}$ are the electric and magnetic field strengths of the BS_1 and BS_2 shock waves, respectively As we have learned, owing to the charge instantaneous deceleration, BS_1 and BS_2 include effects originated at the beginning of motion and those associated with exceeding the velocity of light barrier. The function $\delta(c_n t - R_m)$ describes the position of the CSW. The inequalities $R_m < c_n t$ and $R_m > c_n t$ correspond to the points lying inside the VC cone and outside it, respectively; $\vec{n}_m$ is the vector lying on the surface of the VC cone; r_m is the so-called Cherenkov singularity: $r_m = 0$ on the VC cone surface; $\vec{E}_{Ch}$ and $\vec{H}_{Ch}$ are the electric and magnetic field strengths describing CSW, They originate from the charge uniform motion in the interval $(-z_0, z_0)$; $\vec{E}_{Ch}$ and $\vec{H}_{Ch}$ are infinite on the surface of the VC cone and vanish outside it. Inside the VC cone $\vec{E}_{\text{Ch}}$ and $\vec{H}_{\text{Ch}}$ decrease as r^{-2} at large distances, and therefore do not give a contribution in the wave zone. These terms are not included in (2.34), but they are easily restored from the exact electromagnetic potentials (2.22).

Comparison with the Tamm solution
At large distances one can expand r_1 and r_2 in (2.34) $r_1 = r + z_0 \cos\theta$, $r_2 = r - z_0 \cos\theta$. Here $r = [\rho^2 + z^2]^{1/2}$. Neglecting z_0 in comparison with r in the denominators of $\vec{E}_{\text{BS}}$ and $\vec{H}_{\text{BS}}$ in (2.34), one finds

$$\vec{E}_{\text{T}} = \vec{E}_{\text{BS}}, \quad \vec{H}_{\text{T}} = \vec{H}_{\text{BS}}, \quad \vec{E} = \vec{E}_{\text{T}} + \vec{E}_{\text{Ch}}, \quad \vec{H} = \vec{H}_{\text{T}} + \vec{H}_{\text{Ch}},$$

where $\vec{E}_{\text{T}}$ and $\vec{H}_{\text{T}}$ are the same as in (2.33). This means that the Tamm field strengths (2.33) describe only the BS shock waves (in the generalized sense mentioned above) and do not contain the CSW originating from the charge uniform motion in the interval $(-z_0, z_0)$. Correspondingly, the maxima of their Fourier transforms (2.26) refer to the traces of the CSW in the BS arising from the charge instantaneous deceleration.

To elucidate why the CSW is absent in (2.27) we consider the product of two Θ functions entering into the definition (2.34) of Cherenkov field strengths $\vec{E}_{\text{Ch}}$ and $\vec{H}_{\text{Ch}}$:

$$\Theta(\rho\gamma_n + z_0 - z)\Theta(-\rho\gamma_n + z_0 + z). \tag{2.35}$$

It is seen that the CSW of the length $\Delta L = L/\beta_n\gamma_n, \gamma_n = 1/\sqrt{\beta_n^2 - 1}, L = 2z_0$ is enclosed between two straight lines L_1 and L_2 originating from the ends of the interval of motion and inclined at the angle θ_c towards the motion axis. The CSW, being perpendicular to these straight lines, propagates along them with its normal inclined at the angle θ_c towards the motion axis. We rewrite (2.35) in spherical coordinates

$$\Theta(\theta - \theta_2)\Theta(\theta_1 - \theta), \tag{2.35'}$$

where θ_1 and θ_2 are defined by

$$\cos\theta_1 = -\frac{\epsilon_0}{\beta_n^2\gamma_n^2} + \frac{1}{\beta_n}\left[1 - \left(\frac{\epsilon_0}{\beta_n\gamma_n}\right)^2\right]^{1/2},$$

$$\cos\theta_2 = \frac{\epsilon_0}{\beta_n^2\gamma_n^2} + \frac{1}{\beta_n}\left[1 - \left(\frac{\epsilon_0}{\beta_n\gamma_n}\right)^2\right]^{1/2}$$

and $\epsilon_0 = z_0/r$. The CSW intersects the observational sphere S of the radius r in the angular interval $\Delta\theta = \theta_1 - \theta_2$. With the increase of the observational distance r, the angular region $\Delta\theta$, to which the CSW is confined, diminishes (since $\theta_1 \rightarrow \theta_2$), although the transverse extension ΔL of CSW remains the same. The CSW associated with the charge uniform motion in the interval $(-z_0, z_0)$ drops out if for $\Delta\theta \ll 1$, one naively neglects the term (2.35') with the product of two Θ functions.

We prove now that essentially the same approximation was implicitly made during the transition from (2.25) to (2.26). When changing R in the exponential in (2.25) to $r - z' \cos\theta$ it was implicitly assumed that the quadratic term in the expansion of R is small compared to the linear term. Consider this more carefully. We expand R up to the second order:

$$R \approx r - z' \cos\theta + \frac{z'^2}{2r} \sin^2\theta.$$

In the exponential in (2.25) the following terms then appear

$$\frac{z'}{v} + \frac{1}{c_n}\left(r - z'\cos\theta + \frac{z'^2}{2r}\sin^2\theta\right).$$

We collect terms involving z'

$$\frac{z'}{c_n}[(\frac{1}{\beta_n} - \cos\theta) + \frac{z'}{2r}\sin^2\theta].$$

Taking for z' its maximal value z_0, we present the condition for the second term in the expansion of R to be small in the form

$$\epsilon_0 \ll 2\left(\frac{1}{\beta_n} - \cos\theta\right)/\sin^2\theta$$

It is seen that the right hand side of this equation and that of Eq.(2.35) vanish for $\cos\theta = 1/\beta_n$, i.e., at the angle at which the CSW exists. This means that the absence of the CSW in Eqs. (2.27) is owed to the omission of second-order terms in the expansion of R in the exponential entering (2.25).

2.1.5. SPATIAL DISTRIBUTION OF SHOCK WAVES

Consider the spatial distribution of the electromagnetic field (EMF) at a fixed instant of time. It is convenient to deal with the spatial distribution of electromagnetic potentials rather than with that of field strengths, which are the space-time derivatives of electromagnetic potentials.

We rewrite electromagnetic potentials (2.22) in the form

$$\Phi = \Phi_1 + \Phi_2 + \Phi_m. \tag{2.36}$$

Here

$$\Phi_1 = \frac{e}{\epsilon r_1}\Theta\left(r_1 - c_n t - \frac{z_0}{\beta_n}\right), \quad \Phi_2 = \frac{e}{\epsilon r_2}\Theta\left(c_n t - r_2 - \frac{z_0}{\beta_n}\right),$$

$$\Phi_m = \Phi_m^{(1)} + \Phi_m^{(2)} + \Phi_m^{(3)}, \quad A_z = A_z^{(1)} + A_z^{(2)} + A_z^{(3)}, \quad A_z^{(i)} = \mu\epsilon\beta\Phi_m^{(i)},$$

$$\Phi_m^{(1)} = \frac{e}{\epsilon r_m}\Theta(\rho\gamma_n - z - z_0)\Theta\left(\frac{z_0}{\beta_n} + r_2 - c_n t\right)\Theta\left(c_n t + \frac{z_0}{\beta_n} - r_1\right),$$

$$\Phi_m^{(2)} = \frac{e}{\epsilon r_m}\Theta(z - z_0 - \rho\gamma_n)\Theta\left(r_1 - c_n t - \frac{z_0}{\beta_n}\right)\Theta\left(c_n t - \frac{z_0}{\beta_n} - r_2\right),$$

$$\Phi_m^{(3)} = \frac{e}{\epsilon r_m}\Theta(z_0 + \rho\gamma_n - z)\Theta(z + z_0 - \rho\gamma_n)\Theta(c_n t - R_m)$$
$$\times\left[\Theta\left(r_1 - c_n t - \frac{z_0}{\beta_n}\right) + \Theta\left(\frac{z_0}{\beta_n} + r_2 - c_n t\right)\right].$$

The theta functions

$$\Theta\left(c_n t + \frac{z_0}{\beta_n} - r_1\right) \quad \text{and} \quad \Theta\left(r_1 - c_n t - \frac{z_0}{\beta_n}\right)$$

define spatial regions which, correspondingly, have and have not been reached by the BS_1 shock wave. Similarly, the theta functions

$$\Theta\left(c_n t - \frac{z_0}{\beta_n} - r_2\right) \quad \text{and} \quad \Theta\left(r_2 - c_n t + \frac{z_0}{\beta_n}\right)$$

define spatial regions which correspondingly have and have not been reached by the BS_2 shock wave. Finally, the theta function

$$\Theta(c_n t - R_m)$$

defines spatial region that has been reached by the CSW.

The potentials Φ_1 and Φ_2 correspond to the electrostatic fields of the charge at rest $z = -z_0$ up to an instant $-t_0$ and at $z = z_0$ after the instant t_0. They differ from zero outside BS_1 and inside BS_2, respectively. On the other hand, Φ_m and A_z describe the field of a moving charge. A schematic representation of the shock waves position at the fixed instant of time is shown in Fig. 2.7.

In the spatial regions 1 and 2 corresponding to $z < \rho\gamma_n - z_0$ and $z > \rho\gamma_n + z_0$, respectively, there are observed only BS shock waves. In the spatial region 1 (where $A_z^{(1)} \neq 0$, $A_z^{(2)} = A_z^{(3)} = 0$), at the fixed observational point the BS_1 shock wave (defined by $c_n t + z_0/\beta_n = r_1$) arrives first and BS_2 shock wave (defined by $c_n t - z_0/\beta_n = r_2$) later. In the spatial region 2

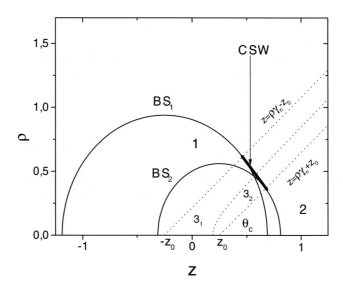

Figure 2.7.   Position of shock waves at the fixed instant of time for $\beta = 0.99$ and $\beta_c = 0.75$. BS$_1$ and BS$_2$ are BS shock waves emitted at the points $\mp z_0$ of the z axis. The solid segment between the lines $z = \rho\gamma_n - z_0$ and $z = \rho\gamma_n + z_0$ is the CSW. The inclination angle of the Cherenkov beam and its width are $\cos\theta_c = 1/\beta_c$ and $2z_0/\beta_n\gamma_n$, respectively.

(where $A_z^{(2)} \neq 0$, $A_z^{(1)} = A_z^{(3)} = 0$), these waves arrive in the reverse order. In the spatial region 3 (where $A_z^{(3)} \neq 0$, $A_z^{(1)} = A_z^{(2)} = 0$), defined by $\rho\gamma_n - z_0 < z < \rho\gamma_n + z_0$, there are BS$_1$, BS$_2$ and CSW shock waves. The latter is defined by the equation $c_n t = R_m$. Before the arrival of the CSW (i.e., for $R_m > c_n t$) there is an electrostatic field of a charge which is at rest at $z = -z_0$. After the arrival of the last of the BS shock waves there is an electrostatic field of a charge which is at rest at $z = z_0$. The spatial region where Φ_m and A_z (and, therefore, the field of a moving charge) differ from zero, lies between the BS$_1$ and BS$_2$ shock waves in the regions 1 and 2 and between CSW and one of the BS shock waves in the region 3. The spatial region 3 in its turn consists of two sub-regions 3_1 and 3_2 defined by the equations $\rho\gamma_n - z_0 < z < (\rho^2\gamma_n^2 + z_0^2/\beta_n^2)^{1/2}$ and $(\rho^2\gamma_n^2 + z_0^2/\beta_n^2)^{1/2} < z < \rho\gamma_n + z_0$, respectively. In the region 3_1 the CSW arrives first, then BS$_1$, and finally, BS$_2$. In region 3_2 BS$_1$ and BS$_2$ arrive in the reverse order.

In brief, $A_z^{(1)}$ and $A_z^{(2)}$ describe the BS in the spatial regions 1 and 2, respectively, while $A_z^{(3)}$ describe BS and VCR in the spatial region 3.

The polarization vectors of BSs are tangential to the spheres BS$_1$ and BS$_2$ and lie in the $\phi = $ const plane coinciding with the plane of Fig. 2.7. They are directed along the unit vectors $\vec{n}_\theta^{(1)}$ and $\vec{n}_\theta^{(2)}$, respectively. The polarization vector of CSW (directed along $\vec{n}_m$) lies on the CSW. It is shown by the solid line in Fig. 2.7 and also lies in the $\phi = \text{const}$ plane. The magnetic field having only the ϕ non-vanishing component is normal to the plane of figure. The Poynting vectors defining the direction of the energy transfer are normal to BS_1, BS_2 and CSW, respectively.

The VCR in the (ρ, z) plane differs from zero inside a beam of width $2z_0 \sin \theta_c$, where θ_c is the inclination of the beam towards the motion axis ($\cos \theta_c = 1/\beta_n$). When the charge velocity tends to the velocity of light in the medium the width of the above beam, as well as the inclination angle, tend to zero. That is, in this case the beam propagates in a nearly forward direction. It is essentially that the Cherenkov beam exists for any interval of motion z_0.

2.1.6. TIME EVOLUTION OF THE ELECTROMAGNETIC FIELD ON THE SURFACE OF A SPHERE

Consider the distribution of VP (in units of e/r) on a sphere S_0 of radius r at various instants of time. There is no EMF on S_0 up to an instant $T_n = 1 - \epsilon_0(1 + 1/\beta_n)$. Here $T_n = c_n t/r$. In the time interval

$$1 - \epsilon_0 \left(1 + \frac{1}{\beta_n} \right) \le T_n \le 1 - \epsilon_0 \left(1 - \frac{1}{\beta_n} \right) \tag{2.37}$$

BS radiation begins to fill the back part of S_0 corresponding to the angles

$$-1 < \cos \theta < \frac{1}{2\epsilon_0} \left[\left(T_n + \frac{\epsilon_0}{\beta_n} \right)^2 - 1 - \epsilon_0^2 \right] \tag{2.38}$$

(Fig. 2.8 (a), curve 1). In the time interval

$$1 - \epsilon_0 \left(1 - \frac{1}{\beta_n} \right) \le T_n \le \left[1 - \left(\frac{\epsilon_0}{\beta_n \gamma_n} \right)^2 \right]^{1/2} \tag{2.39}$$

BS radiation begins to fill the front part of S_0 as well:

$$\frac{1}{2\epsilon_0} \left[1 + \epsilon_0^2 - \left(T_n - \frac{\epsilon_0}{\beta_n} \right)^2 \right] \le \cos \theta \le 1.$$

The illuminated back part of S_0 is still given by (2.38) (Fig. 2.8 (a), curve 2). The finite jumps of VP shown in these figures lead to the δ function singularities in Eqs. (2.34) defining BS electromagnetic strengths. In the time intervals (2.37) and (2.39) these jumps have a finite height. The vector

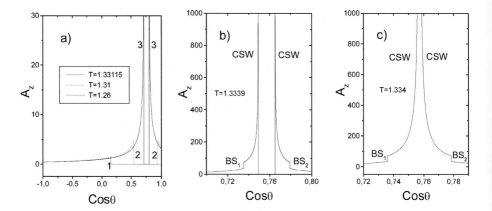

Figure 2.8. Time evolution of shock waves on the surface of the sphere S_0 for $n = 1.333$, $\beta = 0.99$, $\epsilon_0 = 0.1$. The vector potential A_z is in units of e/r, time $T = ct/r$: (a): For small times the BS shock wave occupies only the back part of S_0 (curve 1). For larger times the BS shock wave begins to fill the front part of S_0 as well (curve 2). The jumps of BS shock waves are finite. The jump becomes infinite when the BS shock wave meets the CSW (curve 3); (b): The amplitude of the CSW is infinite while BS shock waves exhibit finite jumps; (c): Position of CSW and BS shock waves at the instant when CSW touches the sphere S_0 at only one point.

potential is maximal at the angle at which the jump occurs. The value of VP is infinite at the angles defined by

$$\cos\theta_1 = -\frac{\epsilon_0}{\beta_n^2\gamma_n^2} + \frac{1}{\beta_n}\left[1 - \left(\frac{\epsilon_0}{\beta_n\gamma_n}\right)^2\right]^{1/2}$$

and

$$\cos\theta_2 = \frac{\epsilon_0}{\beta_n^2\gamma_n^2} + \frac{1}{\beta_n}\left[1 - \left(\frac{\epsilon_0}{\beta_n\gamma_n}\right)^2\right]^{1/2}. \tag{2.40}$$

which are reached at the time

$$T_{\text{Ch}} = \frac{c_n t_{\text{Ch}}}{r} = \left[1 - \left(\frac{\epsilon_0}{\beta_n\gamma_n}\right)^2\right]^{1/2}$$

(Fig. 2.8 (a), curve 3). At this instant, and at these angles, the CSW intersects S_0 first time. Or, in other words, the intersection of S_0 by the lines $z = \rho\gamma_n - z_0$ and $z = \rho\gamma_n + z_0$ (Fig. 2.7) occurs at the angles θ_1 and θ_2. At this instant the illuminated front and back parts of S_0 are given by $0 < \theta < \theta_2$ and $\theta_1 < \theta < \pi$, respectively. Beginning from this instant the

CSW intersects the sphere S_0 at the angles defined by (see Fig. 2.8 (b))

$$\cos\theta_{\text{Ch}}^{(1)}(T) = \frac{T_n}{\beta_n} - \frac{1}{\beta_n\gamma_n}(1 - T_n^2)^{1/2}$$

and

$$\cos\theta_{\text{Ch}}^{(2)}(T) = \frac{T_n}{\beta_n} + \frac{1}{\beta_n\gamma_n}(1 - T_n^2)^{1/2}.$$

The positions of the BS$_1$ and BS$_2$ shock waves are given by

$$\cos\theta_{\text{BS}}^{(1)}(T) = \frac{1}{2\epsilon_0}\left[\left(T_n + \frac{\epsilon_0}{\beta_n}\right)^2 - 1 - \epsilon_0^2\right]$$

and

$$\cos\theta_{\text{BS}}^{(2)}(T) = \frac{1}{2\epsilon_0}\left[1 + \epsilon_0^2 - \left(T_n - \frac{\epsilon_0}{\beta_n}\right)^2\right],$$

respectively (i.e., the BS shock waves follow after the CSW). Therefore, at this instant BS fills the angular regions

$$\theta_{\text{BS}}^{(1)}(T) \le \theta \le \pi \quad \text{and} \quad 0 \le \theta \le \theta_{\text{BS}}^{(2)}(T)$$

whilst the VC radiation occupies the angle interval

$$\theta_{\text{Ch}}^{(1)}(T) \le \theta \le \theta_1 \quad \text{and} \quad \theta_2 \le \theta \le \theta_{\text{Ch}}^{(2)}(T).$$

Therefore the VC radiation field and BS overlap in the regions

$$\theta_{\text{BS}}^{(1)}(T) \le \theta \le \theta_1 \quad \text{and} \quad \theta_2 \le \theta \le \theta_{\text{BS}}^{(2)}(T).$$

BS$_1$ and BS$_2$ have finite jumps in this angular interval (Fig. 2.8 (b)). The non-illuminated part of S_0 is

$$\theta_{\text{Ch}}^{(2)}(T) \le \theta \le \theta_{\text{Ch}}^{(1)}(T).$$

This lasts up to an instant $T_n = 1$ when the CSW intersects S_0 only once at the point corresponding to the angle $\cos\theta = 1/\beta_n$ (Fig. 2.8 (c)). The positions of the BS$_1$ and BS$_2$ shock waves at this instant ($T_n = 1$) are given by

$$\cos\theta = \frac{1}{\beta_n} - \frac{\epsilon_0}{2\beta_n^2\gamma_n^2} \quad \text{and} \quad \cos\theta = \frac{1}{\beta_n} + \frac{\epsilon_0}{2\beta_n^2\gamma_n^2},$$

respectively. Again, the jumps of BS waves have finite heights whilst the Cherenkov term $\Phi_m^{(3)}$ is infinite at the angle $\cos\theta = 1/\beta_n$ at which the CSW intersects S_0. After the instant $T_n = 1$, CSW leaves S_0. However, the Cherenkov post-action still remains (Fig. 2.9 (a)). In the subsequent

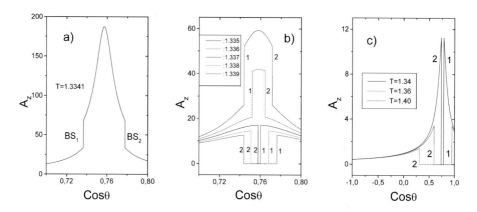

Figure 2.9. Further time evolution of shock waves on the surface of the sphere S_0: (a): The Cherenkov post-action and BS shock waves after the instant when CSW has left S_0.; (b): BS shock waves approach and pass through each other leaving after themselves a zero electromagnetic field. Numbers 1 and 2 mean BS_1 and BS_2 shock waves, respectively; (c): After some instant the BS shock wave begins to fill only the back part of S_0. Numbers 1 and 2 mean BS_1 and BS_2 shock waves, respectively.

time the BS_1 and BS_2 shock waves approach each other. They meet at the instant

$$T_n = \left[1 + \left(\frac{\epsilon_0}{\beta_n \gamma_n}\right)^2\right]^{1/2}. \tag{2.41}$$

at the angle

$$\cos\theta = \frac{1}{\beta_n}\left[1 + \left(\frac{\epsilon_0}{\beta_n \gamma_n}\right)^2\right]^{1/2}.$$

After this instant BS shock waves pass through each other and diverge (Fig. 2.9 (b)). Now BS_1 and BS_2 move along the front and back semi-spheres, respectively. There is no EMF on the part of S_0 lying between them. The illuminated parts of S_0 are now given by

$$\theta_{BS}^{(2)}(T) \leq \theta \leq \pi \quad \text{and} \quad 0 \leq \theta \leq \theta_{BS}^{(1)}(T).$$

The electromagnetic field is zero inside the angle interval

$$\theta_{BS}^{(1)}(T) \leq \theta \leq \theta_{BS}^{(2)}(T).$$

After the instant of time (2.41), BS_1 and BS_2 may occupy the same angular positions $\cos\theta_2$ and $\cos\theta_1$ like BS_2 and BS_1 shown by curve 3 in Fig. 2.8

(a). But now their jumps are finite. After the instant

$$T_n = 1 + \epsilon_0 \left(1 - \frac{1}{\beta_n}\right)$$

the front part of S_0 begins not to be illuminated (Fig. 2.9 (c)). At this instant the illuminated back part of S_0 is given by

$$-1 \leq \cos\theta \leq -1 + \frac{2(1 + \epsilon_0)}{\beta_n} - \frac{2\epsilon_0}{\beta_n^2}.$$

In the subsequent time the illuminated part of S_0 is given by

$$-1 \leq \cos\theta \leq \frac{1}{2\epsilon_0}\left[1 + \epsilon_0^2 - \left(T_n - \frac{\epsilon_0}{\beta_n}\right)^2\right].$$

As time advances, the illuminated part of S_0 diminishes. Finally, after the instant

$$T_n = 1 + \epsilon_0 \left(1 + \frac{1}{\beta_n}\right)$$

the EMF radiation leaves the surface of S_0 (and its interior).

We summarize here the main differences between VCR and BS: On the sphere S_0 the VC radiation runs over the angular region

$$\theta_2 \leq \theta \leq \theta_1,$$

where θ_1 and θ_2 are defined by Eqs. (2.40). At each particular instant of time T_n in the interval

$$\left[1 - \left(\frac{\epsilon_0}{\beta_n\gamma_n}\right)^2\right]^{1/2} \leq T_n \leq 1$$

the VC electromagnetic potentials and field strengths are infinite at the angles $\theta_{\text{Ch}}^{(1)}(T)$ and $\theta_{\text{Ch}}^{(2)}(T)$ at which the CSW intersects S_0.

After the instant $T_n = 1$ the Cherenkov singularity leaves the sphere S_0, but the Cherenkov post-action still remains. This lasts up to the instant $T_n = [1 + (\epsilon_0/\beta_n\gamma_n)^2]^{1/2}$.

On the other hand, BS runs over the whole sphere S_0 in the time interval

$$1 - \epsilon_0\left(1 + \frac{1}{\beta_n}\right) \leq T_n \leq 1 + \epsilon_0\left(1 + \frac{1}{\beta_n}\right).$$

The vector potential of BS is infinite only at the angles θ_1 and θ_2 at the particular instant of time $T_n = \sqrt{1 - \epsilon_0^2/\beta_n^2\gamma_n^2}$ when the CSW intersects S_0 for the first time. For other times the VP of BS exhibits finite jumps in the angular interval $-\pi \leq \theta \leq \pi$. The BS electromagnetic field strengths (as spatial-time derivatives of electromagnetic potentials) are infinite at those angles. Therefore Cherenkov singularities of the vector potential run over the region $\theta_2 \leq \theta \leq \theta_1$ of the sphere S_0, whilst the BS vector potential is infinite only at the angles θ_1 and θ_2 at which BS shock waves meet CSW.

The following particular cases are of special interest. For small $\epsilon_0 = z_0/r$ (the observational distance is large compared with the interval of motion) the Cherenkov singular radiation occupies the narrow angular region

$$\arccos \frac{1}{\beta_n} - \frac{\epsilon_0}{\beta_n \gamma_n} \leq \theta \leq \arccos \frac{1}{\beta_n} + \frac{\epsilon_0}{\beta_n \gamma_n},$$

whilst the BS is infinite at the boundary points of this interval. In the opposite case $\epsilon_0 \approx 1$ (this corresponds to the near zone) the singular VCR field is confined to the angular region

$$\frac{2}{\beta_n^2} - 1 \leq \cos\theta \leq 1,$$

whilst the BS is singular at $\cos\theta = 2/\beta_n^2 - 1$, and $\cos\theta = 1$ is reached at the instant $T_n = 1/\beta_n$.

When the charge velocity is close to the velocity of light in medium ($\beta_n \approx 1$), one has:

$$\cos\theta_1 \approx \frac{1}{\beta_n} - \frac{\epsilon_0}{\beta_n^2 \gamma^2} \left(1 + \frac{1}{2}\epsilon_0\right) \approx 1, \quad \cos\theta_2 \approx \frac{1}{\beta_n} - \frac{\epsilon_0}{\beta_n^2 \gamma_n^2} \left(1 - \frac{1}{2}\epsilon_0\right) \approx 1,$$

i.e., there is a narrow Cherenkov beam in a nearly forward direction.

2.1.7. COMPARISON WITH THE TAMM VECTOR POTENTIAL

Now we evaluate the Tamm vector potential

$$A_{\mathrm{T}} = \int\limits_{-\infty}^{\infty} d\omega \exp\left(i\omega t\right) A_\omega$$

Substituting here A_ω given by (2.26), we find in the absence of dispersion

$$A_{\mathrm{T}} = \frac{e\mu}{rn|\cos\theta - 1/\beta_n|} \Theta(|\cos\theta - 1/\beta_n| - |T_n - 1|/\epsilon_0). \qquad (2.42)$$

This VP can be also obtained from A_z given by (2.36) if we leave in it the terms $A_z^{(1)}$ and $A_z^{(2)}$ describing BS in the regions 1 and 2 (see Fig. 2.7) (with omitting z_0 in the factors $\Theta(\rho\gamma_n - z - z_0)$ and $\Theta(z - z_0 - \rho\gamma_n)$ entering $A_z^{(1)}$ and $A_z^{(2)}$) and drop the term $A_z^{(3)}$ which is responsible (as we have learned from the previous section) for the BS and VC radiation in region 3 and which describes the Cherenkov beam of the width $2z_0/\beta_n\gamma_n$. It is seen that A_T is infinite only at

$$T_n = 1, \quad \cos\theta = \frac{1}{\beta_n}. \tag{2.43}$$

This may be compared with the exact consideration of the previous section which shows that the BS part of A_z is infinite at the instant

$$T_{Ch} = \frac{c_n t_{Ch}}{r} = \left[1 - \left(\frac{\epsilon_0}{\beta_n\gamma_n}\right)^2\right]^{1/2} \tag{2.44}$$

at the angles θ_1 and θ_2 defined by (2.40). The $\cos\theta_1$ and $\cos\theta_2$ defined by (2.40) and T_{Ch} given by (2.44) are transformed into $\cos\theta$ and T_n given by (2.43) in the limit $\epsilon_0 \to 0$. Owing to the dropping of the $A_z^{(3)}$ term in (2.36) (describing BS and VCR in the spatial region 3) and the omission of terms containing ϵ_0 in $\cos\theta_1$ and $\cos\theta_2$, BS$_1$ and BS$_2$ waves now have the common maximum of the infinite height at the angle given by $\cos\theta = 1/\beta_n$ at which the Tamm approximation fails.

The analysis of (2.42) shows that the Tamm VP is distributed over S_0 in the following way. There is no EMF of the moving charge up to the instant $T_n = 1 - \epsilon_0(1 + 1/\beta_n)$. For

$$1 - \epsilon_0\left(1 + \frac{1}{\beta_n}\right) < T_n < 1 - \epsilon_0\left(1 - \frac{1}{\beta_n}\right)$$

the EMF fills only the back part of S_0

$$-1 < \cos\theta < \frac{1}{\beta_n} - \frac{1}{\epsilon_0}(1 - T_n)$$

(Fig. 2. 10 a, curve 1). In the time interval

$$1 - \epsilon_0\left(1 - \frac{1}{\beta_n}\right) < T_n < 1 + \epsilon_0\left(1 - \frac{1}{\beta_n}\right)$$

the illuminated parts of S_0 are given by

$$-1 < \cos\theta < \frac{1}{\beta_n} - \frac{1}{\epsilon_0}(1 - T_n) \quad \text{and} \quad \frac{1}{\beta_n} + \frac{1}{\epsilon_0}(1 - T_n) < \cos\theta < 1$$

(Fig. 2.10 a, curves 2 and 3).

The jumps of the BS$_1$ and BS$_2$ shock waves are finite. As T_n tends to 1 the BS$_1$ and BS$_2$ shock waves approach each other and fuse at $T_n = 1$.

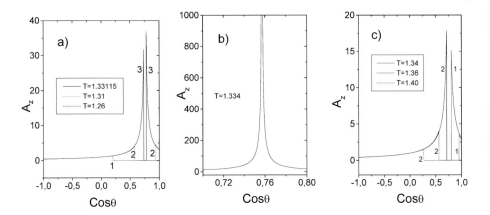

Figure 2.10. Time evolution of shock waves according to the Tamm approximate picture: a) The jumps of BS shock waves are finite. After some instant BS shock waves fill both the back and front parts of S_0 (curves 2 and 3); b) Position of the BS shock wave at the instant when its jump is infinite; c) BS shock waves pass through each other and diverge leaving after themselves a zero EMF. After some instant BS shock waves fill only the back part of S_0. Numbers 1 and 2 mean BS_1 and BS_2 shock waves, respectively.

Tamm's VP is infinite at this instant at the angle given by $\cos\theta = 1/\beta_n$ (Fig. 2.10 b). For

$$1 < T_n < 1 + \epsilon_0 \left(1 - \frac{1}{\beta_n}\right)$$

the BS shock waves pass through each other and begin to diverge, BS_1 and BS_2 filling the front and back parts of S_0, respectively (Fig. 2.10 c):

$$\frac{1}{\beta_n} + \frac{1}{\epsilon_0}(T_n - 1) < \cos\theta < 1 \quad (BS_1)$$

and

$$-1 < \cos\theta < \frac{1}{\beta_n} - \frac{1}{\epsilon_0}(T_n - 1) \quad (BS_2).$$

For larger times

$$1 + \epsilon_0 \left(1 - \frac{1}{\beta_n}\right) < T_n < 1 + \epsilon_0 \left(1 + \frac{1}{\beta_n}\right)$$

only the back part of S_0 is illuminated:

$$-1 < \cos\theta < \frac{1}{\beta_n} - \frac{1}{\epsilon_0}(T_n - 1) \quad (BS_2).$$

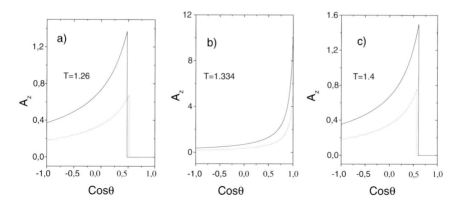

Figure 2.11.   Time evolution of BS shock waves for the charge velocity ($\beta = 0.7$) less than the velocity of light in medium ($c_n = 0.75$). Solid and dashed lines are related to the exact (2.25) and approximate (2.26) vector potentials; a) BS shock waves fill only the back part of S_0; b) The whole sphere S_0 is illuminated during some time interval; c) At later times BS again fills only the back part of S_0. When evaluating the Tamm VP the extra $1/2$ factor was occasionally included. After multiplication the dashed curve by 2 it almost coincides with exact solid curve.

Finally, for $T_n > 1 + \epsilon_0(1 + 1/\beta_n)$ there is no radiation field on and inside the S_0.

It is seen that the behaviour of the exact and approximate Tamm potentials is very alike in the spatial regions 1 and 2 where VCR is absent and differs appreciably in the spatial region 3 where it exists. Roughly speaking, the Tamm vector potential (2.42) describing evolution of BS shock waves in the absence of CSW imitates the latter in the neighborhood of $\cos \theta = 1/\beta_n$ where, as we know, the Tamm approximate VP is not correct.

This complication is absent if the charge velocity is less than the velocity of light c_n in medium. In this case one the exact VP is (see (2.13)):

$$A_z = \frac{e\beta\mu}{r_m}\Theta[c_n(t + t_0) - r_1]\Theta[r_2 - c_n(t - t_0)],$$

while the Tamm VP A_T is still given by (2.42). The results of calculations for $\beta = 0.7$, $c_n \approx 0.75$ are presented in Fig. 2.11. We see on it the exact and the Tamm VPs for three typical times: $T = 1.26$; $T = 1.334$ and $T = 1.4$. In general, the EMF distribution on the sphere surface is as follows. There is no field on S_0 up to some instant of time. Later, only the back part of S_0 is illuminated (see Fig. 2.11 a). In the subsequent times the EMF fills the whole sphere (Fig. 2.11 b). After some instant the EMF again fills only the back part of S_0 (Fig. 2.11 c). Finally, the EMF leaves S_0.

Now we analyze the behaviour of the Tamm VP for small and large motion intervals z_0. For small $\epsilon_0 = z_0/r$ it follows from (2.42) that

$$A_z = \frac{e\mu}{rn|(1/\beta_n) - \cos\theta|}\delta(1 - T_n). \qquad (2.45)$$

On the other hand, if we pass to the limit $\epsilon_0 \to 0$ in Eq.(2.26), i.e., prior to the integration, then

$$A_\omega \to \frac{e\epsilon_0\mu}{\pi c}\exp(-ik_nr), \quad A_z \to \frac{e\epsilon_0\mu}{\pi nr}\delta(T_n - 1), \qquad (2.46)$$

i.e., there is no angular dependence in (2.46). The distinction of (2.46) from (2.45) is due to the fact that integration takes place for all ω in the interval $(-\infty, +\infty)$. For large ω the condition $\omega z_0/v \ll 1$ is violated. This means that Eq. (2.45) involves the contribution of high frequencies.

For large z_0 one obtains from (2.42)

$$A_z = \frac{e\mu}{rn|(1/\beta_n) - \cos\theta|}. \qquad (2.47)$$

If we take the limit $z_0 \to \infty$ in Eq.(2.26), then

$$A_\omega \approx \frac{e\mu\beta}{r\omega}\exp(-i\omega r/c_n)\delta(1 - \beta_n\cos\theta), \quad A_z(t) \sim \delta(1 - \beta_n\cos\theta). \quad (2.48)$$

Although Eqs.(2.47) and (2.48) reproduce the position of the Cherenkov singularity at $\cos\theta = 1/\beta_n$, they do not describe the Cherenkov cone. The reason for this is that the Tamm VP (2.26) is obtained under the condition $z_0 \ll r$, and therefore it is not legitimate to take the limit $z_0 \to \infty$ in the expressions following from it (and, in particular, in Eq. (2.42)).

On the other hand, taking the limit $z_0 \to \infty$ in the exact expression (2.36) we obtain the well-known expressions for the electromagnetic potentials describing superluminal motion of charge in an infinite medium:

$$A_z = \frac{2e\beta\mu}{r_m}\Theta(vt - z - \rho/\gamma_n), \quad \Phi = \frac{2e}{\epsilon r_m}\Theta(vt - z - \rho/\gamma_n).$$

The very fact that the Tamm VPs in the spectral (2.26) and time (2.42) representations are valid both for $v < c_n$ and $v > c_n$ has given rise to the extensive discussion in the physical literature concerning the discrimination between the BS and VCR [9-10].

As follows from our consideration, the physical reason for this is the absence of the Cherenkov shock wave in (2.26) and (2.42). Exact electromagnetic potentials (2.36) and field strengths (2.34) contain CSW for any motion interval. The induced Cherenkov beam being very thin for $z_0 \to 0$ and broad for large z_0 in no case can be reduced to the BS.

2.2. Spatial distribution of Fourier components

The Fourier transform of the vector potential on the sphere S_0 of radius r is given by (2.25)

$$A_\omega = \frac{e\mu}{2\pi c} \int_{-z_0}^{z_0} \frac{dz'}{R} \exp\left[-ik\left(\frac{z'}{\beta} + nR\right)\right].$$

Here $R = [\rho^2 + (z - z')^2]^{1/2}$. Making the change of integration variable $z' = z + \rho\sinh\chi$, one obtains

$$A_\omega = \frac{e\mu}{2\pi c} \exp\left(-\frac{ikz}{\beta}\right) \int_{\chi_1}^{\chi_2} \exp\left[-\frac{ik\rho}{\beta}(\sinh\chi + \beta n\cosh\chi)\right] d\chi, \qquad (2.49)$$

where $\sinh\chi_1 = -(z_0 + z)/\rho$ and $\sinh\chi_2 = (z_0 - z)/\rho$.

2.2.1. QUASI-CLASSICAL APPROXIMATION

The stationary point of the vector potential (2.49) satisfies the equation $\cosh\chi_c + \beta n\sin\chi_c = 0$. This gives $\cosh\chi_c = \beta n\gamma_n$, $\sinh\chi_c = -\gamma_n$ for $\beta > 1/n$. It is seen that $\chi_c < \chi_1$ for $z < \rho\gamma_n - z_0$, $\chi_c > \chi_2$ for $z > \rho\gamma_n + z_0$, and $\chi_1 < \chi_c < \chi_2$ for $\rho\gamma_n - z_0 < z < \rho\gamma_n + z_0$. For $z < \rho\gamma_n - z_0$ and $z > \rho\gamma_n + z_0$ one finds

$$A_z^{out} = \frac{ie\mu\beta}{2\pi ck\rho} \exp\left(-\frac{ikz}{\beta}\right)(A_2 - A_1),$$

where

$$A_2 = \frac{1}{\cosh\chi_2 + \beta n\sinh\chi_2} \exp\left[-\frac{ik\rho}{\beta}(\sinh\chi_2 + \beta n\cosh\chi_2)\right]$$

$$= \frac{r\sin\theta}{r_2 - \beta n(z - z_0)} \exp\left[-\frac{ik}{\beta}(\beta n r_2 - z + z_0)\right],$$

and

$$A_1 = \frac{1}{\cosh\chi_1 + \beta n\sinh\chi_1} \exp\left[-\frac{ik\rho}{\beta}(\sinh\chi_1 + \beta n\cosh\chi_1)\right]$$

$$= \frac{r\sin\theta}{r_1 - \beta n(z + z_0)} \exp\left[-\frac{ik}{\beta}(\beta n r_1 - z - z_0)\right].$$

Therefore

$$A_z^{out} = \frac{ie\mu\beta\sin\theta}{2\pi ck}\left\{\frac{1}{r_2 - \beta n(z - z_0)} \exp\left[-\frac{ik}{\beta}(\beta n r_2 + z_0)\right]\right.$$

$$- \frac{1}{r_1 - \beta_n(z + z_0)} \exp\left[-\frac{ik}{\beta}(\beta n r_1 - z_0)\right]\Big\}.$$

Inside the interval $\rho\gamma_n - z_0 < z < \rho\gamma_n + z_0$ the vector potential is equal to

$$A_z^{\text{in}} = A_z^{\text{out}} + \frac{e\mu}{2\pi c} \exp\left(-\frac{ikz}{\beta}\right) \sqrt{\frac{2\pi\beta\gamma_n}{kr\sin\theta}} \exp\left(-i\frac{\pi}{4}\right) \exp\left(-\frac{ikr\sin\theta}{\beta\gamma_n}\right).$$

It is seen that A_z^{out} is infinite at $z = \rho\gamma_n \pm z_0$ (these infinities are due to the quasi-classical approximation). Therefore the exact radiation intensity should have maxima at $z = \rho\gamma_n \pm z_0$, with a kind of plateau for $\rho\gamma_n - z_0 < z < \rho\gamma_n + z_0$ and a sharp decreasing for $z < \rho\gamma_n - z_0$ and $z > \rho\gamma_n + z_0$. At the observational distances much larger than the motion length ($r \gg z_0$)

$$r_1 - \beta_n(z + z_0) \approx r(1 - \beta_n\cos\theta), \quad r_2 - \beta_n(z - z_0) \approx r(1 - \beta_n\cos\theta),$$

$$\beta_n r_1 - z_0 = \beta_n r - z_0(1 - \beta_n\cos\theta), \quad \beta_n r_2 + z_0 = \beta_n r + z_0(1 - \beta_n\cos\theta).$$

Then

$$A_z^{\text{out}} = \frac{e\mu\beta}{\pi ckr} \exp(-iknr)\frac{\sin[\omega t_0(1 - \beta n\cos\theta)]}{1 - \beta n\cos\theta},$$

which (for $r \gg z_0$) coincides with the Tamm vector potential A_z^{T} entering (2.26). Inside the interval $\rho\gamma_n - z_0 < z < \rho\gamma_n + z_0$

$$A_z^{\text{in}} = A_z^{\text{T}} + \frac{e\mu}{2\pi c} \exp\left(-\frac{ikz}{\beta}\right) \sqrt{\frac{2\pi\beta\gamma_n}{kr\sin\theta}} \exp\left(-\frac{i\pi}{4}\right) \exp\left(-\frac{ikr\sin\theta}{\beta\gamma_n}\right).$$

We observe that the infinities of A_z^{out} have disappeared owing to the approximations involved. It is seen that for $kr \gg 1$, A_z^{in} and A_z^{T} behave like $1/\sqrt{kr}$ and $1/kr$, respectively. It follows from this that the radiation intensity in the spatial regions $z > \rho\gamma_n + z_0$ and $z < \rho\gamma_n - z_0$ is described by the Tamm formula (2.29). On the other hand, inside the spatial region $\rho\gamma_n - z_0 < z < \rho\gamma_n + z_0$, the radiation intensity differs appreciably from the Tamm intensity. In fact, the second term in A_z^{in} is much larger than the first one (A_z^T) for $kr \gg 1$ (since they decrease as $1/\sqrt{kr}$ and $1/kr$ for $kr \to \infty$, respectively). It is easy to check that on the surface of the sphere of radius r the interval $\rho\gamma_n - z_0 < z < \rho\gamma_n + z_0$ corresponds to the angular interval $\theta_2 < \theta < \theta_1$, where θ_2 and θ_1 are defined by Eq.(2.40). Therefore, inside this angular interval there should be observed the maximum of the radiation intensity with its amplitude proportional to the observational distance r. In the limit $r \to \infty$ the above θ interval diminishes and for the radiation intensity one gets the δ singularity at $\cos\theta = 1/\beta n$. Probably, this singularity is owed to the quasi-classical approximation used.

2.2.2. NUMERICAL CALCULATIONS

We separate in (2.25) real and imaginary parts

$$
\mathrm{Re}A_\omega = \frac{e\mu}{2\pi c} \int\limits_{-z_0}^{z_0} \frac{dz'}{R} \cos\left[k\left(\frac{z'}{\beta} + nR\right)\right],
$$

$$
\mathrm{Im}A_\omega = -\frac{e\mu}{2\pi c} \int\limits_{-\epsilon_0}^{\epsilon_0} \frac{dz}{R} \sin\left[k\left(\frac{z'}{\beta} + nR\right)\right]. \tag{2.50}
$$

For $z_0 \ll r$ these expressions should be compared with the real and imaginary parts of the Tamm approximate VP (2.26):

$$
\mathrm{Re}A_\omega = \frac{e\beta\mu q}{\pi r\omega} \cos(k_n r), \quad \mathrm{Im}A_\omega = -\frac{e\beta\mu q}{\pi r\omega} \sin(k_n r). \tag{2.51}
$$

These quantities are evaluated (in units of $e/2\pi c$) for

$$
k_n r = 100, \quad \beta = 0.99, \quad n = 1.334, \quad \epsilon_0 = 0.1
$$

(see Figs. 2.12 a, b).

We observe that angular distributions of the VPs (2.50) and (2.51) practically coincide, having maxima on the small part of S_0 in the neighborhood of $\cos\theta = 1/\beta_n$. It is this minor difference between (2.50) and (2.51) that is responsible for the CSW which is described only by Eq. (2.50).

Now we evaluate the angular dependence of VP (2.50) on the sphere S_0 for the case in which z_0 practically coincides with r ($\epsilon_0 = 0.98$). Other parameters remain the same. We see (Fig. 2.12 c) that the angular distribution fills the whole sphere S_0. There is no pronounced maximum in the vicinity of $\cos\theta = 1/\beta_n$.

We cannot extend these results to larger z_0 as the interval of motion will partly lie outside S_0. To consider a charge motion in an arbitrary finite interval, we evaluate the distribution of VP on the cylinder surface C coaxial with the motion axis. Let the radius of this cylinder be ρ. Separating real and imaginary parts in (2.49), one obtains

$$
\mathrm{Re}A_\omega = \frac{e\mu}{2\pi c} \int\limits_{\chi_1}^{\chi_2} \cos\left[k\left(\frac{z}{\beta} + \frac{\rho}{\beta}\sinh\chi + n\rho\cosh\chi\right)\right] d\chi,
$$

$$
\mathrm{Im}A_\omega = -\frac{e\mu}{2\pi c} \int\limits_{\chi_1}^{\chi_2} \sin\left[k\left(\frac{z}{\beta} + \frac{\rho}{\beta}\sinh\chi + n\rho\cosh\chi\right)\right] d\chi. \tag{2.52}
$$

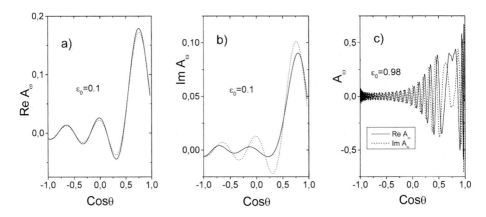

Figure 2.12.   The real (a) and imaginary (b) parts of the VP in the spectral representation (in units of $e/2\pi c$) on the surface of the sphere S_0 for $\epsilon_0 = z_0/r = 0.1$. The radiation field differs essentially from zero in the neighborhood of the Cherenkov critical angle defined by $\cos\theta_c = 1/\beta_n$. The solid and dotted curves refer to the exact and approximate formulae (2.5o) and (2.51), respectively. It turns out that a small difference between the Fourier transforms is responsible for the appearance of the VCR in the space-time representation; (c): The real and imaginary parts of A_ω for $\epsilon_0 = 0.98$. The electromagnetic radiation is distributed over the whole sphere S_0.

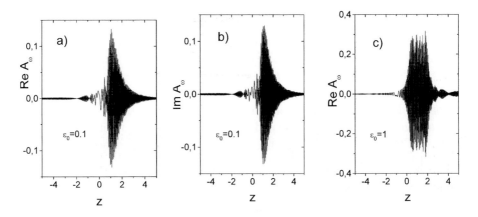

Figure 2.13.   The real (a) and imaginary (b) parts of A_ω on the surface of the cylinder C for the ratio of the interval motion to the cylinder radius $\epsilon_0 = 0.1$. The electromagnetic radiation differs from zero in the neighbourhood of $z = \gamma_n$, which corresponds to $\cos\theta_c = 1/\beta_n$ on the sphere (z is in units of ρ, A_ω in units of $e/2\pi c$); (c): The real part of A_ω for $\epsilon_0 = 1$. There is no sharp radiation maximum in the neighborhood of $z = \gamma_n$.

The distributions of $\mathrm{Re}A_\omega$ and $\mathrm{Im}A_\omega$ (in units of $e/2\pi c$) on the surface of C as a function of $\tilde z = z/\rho$ are shown in Figs. 2.13 and 2.14 for various values of $\epsilon_0 = z_0/\rho$ and ρ fixed. The calculations were made for $\beta = 0.99$ and $k\rho = 100$. We observe that for small ϵ_0 the electromagnetic field differs from zero only in the vicinity $\tilde z = \gamma_n$, which corresponds to $\cos\theta = 1/\beta_n$ (Figs. 2.13 (a),(b)). As ϵ_0 increases, the VP begin to diffuse over the cylinder surface. This is illustrated in Figs. 2.13(c) and 2.14(a) where only the real parts of A_ω for $\epsilon_0 = 1$ and $\epsilon_0 = 10$ are presented. Since the behaviour of $\mathrm{Re}A_\omega$ and $\mathrm{Im}A_\omega$ is very much alike (Figs. 2.12 and 2.13 (a),(b) clearly demonstrate this), we limit ourselves to the consideration of $\mathrm{Re}A_\omega$). We observe the disappearance of pronounced maxima at $\cos\theta = 1/\beta_n$.

For the infinite motion ($z_0 \to \infty$), Eqs. (2.52) are reduced to

$$\mathrm{Re}A_\omega = \frac{e}{2\pi c} \int_{-\infty}^{\infty} \cos\left[k\left(\frac{z}{\beta} + \frac{\rho}{\beta}\sinh\chi + n\rho\cosh\chi \right) \right] d\chi,$$

$$\mathrm{Im}A_\omega = -\frac{e}{2\pi c} \int_{-\infty}^{\infty} \sin\left[k\left(\frac{z}{\beta} + \frac{\rho}{\beta}\sinh\chi + n\rho\cosh\chi \right) \right] d\chi, \qquad (2.52').$$

These expressions can be evaluated in the analytical form (see below)

$$\frac{\mathrm{Re}A_\omega}{e\mu/2\pi c} = -\pi\left[J_0\left(\frac{\omega\rho}{v\gamma_n} \right)\sin\left(\frac{\omega z}{v} \right) + N_0\left(\frac{\omega\rho}{v\gamma_n} \right)\cos\left(\frac{\omega z}{v} \right) \right],$$

$$\frac{\mathrm{Im}A_\omega}{e\mu/2\pi c} = \pi\left[N_0\left(\frac{\omega\rho}{v\gamma_n} \right)\sin\left(\frac{\omega z}{v} \right) - J_0\left(\frac{\omega\rho}{v\gamma_n} \right)\cos\left(\frac{\omega z}{v} \right) \right] \qquad (2.53)$$

for $v > c_n$ and

$$\frac{\mathrm{Re}A_\omega}{e\mu/2\pi c} = 2\cos\left(\frac{\omega z}{v} \right) K_0\left(\frac{\rho\omega}{v\gamma_n} \right),$$

$$\frac{\mathrm{Im}A_\omega}{e/2\pi c} = -2\sin\left(\frac{\omega z}{v} \right) K_0\left(\frac{\rho\omega}{v\gamma_n} \right) \qquad (2.54)$$

for $v < c_n$ ($\gamma_n = |1 - \beta_n^2|^{-1/2}$). We see that for the infinite charge motion the A_ω is a pure periodic function of z (and therefore of the angle θ). This assertion does not depend on the values of ρ and ω. For example, for $\omega\rho/v\gamma_n \gg 1$ one has

$$\mathrm{Re}A_\omega = -\frac{e\mu}{2\pi c}\sqrt{\frac{2v\pi\gamma_n}{\rho\omega}}\sin\left[\frac{\omega}{v}\left(z + \frac{\rho}{\gamma_n} \right) - \frac{\pi}{4} \right],$$

$$\mathrm{Im}A_\omega = -\frac{e\mu}{2\pi c}\sqrt{\frac{2v\pi\gamma_n}{\rho\omega}}\cos\left[\frac{\omega}{v}\left(z + \frac{\rho}{\gamma_n} \right) - \frac{\pi}{4} \right]$$

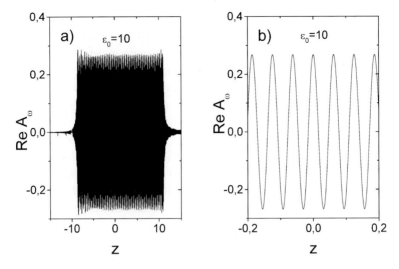

Figure 2.14.   The real part of A_ω for $\epsilon_0 = 10$; (a): There is no radiation maximum in the neighborhood of $z = \gamma_n$ and the radiation is distributed over the large z interval; (b): For a small z interval, ReA$_\omega$ evaluated according to Eq.(2.51) for $\epsilon_0 = 10$ and according to Eq.(2.53) for an infinite interval of motion are indistinguishable.

for $v > c_n$ and

$$\mathrm{ReA}_\omega = \frac{e\mu}{2\pi c}\sqrt{\frac{2v\pi\gamma_n}{\rho\omega}}\cos\left(\frac{\omega z}{v}\right)\exp\left(-\frac{\rho\omega}{v\gamma_n}\right),$$

$$\mathrm{ImA}_\omega = -\frac{e\mu}{2\pi c}\sqrt{\frac{2v\pi\gamma_n}{\rho\omega}}\sin\left(\frac{\omega z}{v}\right)\exp\left(-\frac{\rho\omega}{v\gamma_n}\right)$$

for $v < c_n$. In Fig. 2.14 (b), by comparing the real part of A_ω evaluated according to Eq.(2.51) for $\epsilon_0 = 10$ with the analytical expression (2.53) valid for $\epsilon_0 \to \infty$ we observe their perfect agreement on the small interval of surface of the cylinder C (they are indistinguishable on the interval treated). The same coincidence is valid for the imaginary part of A_ω.

To prove (2.53) and (2.54), we start from the Green function expansion in the cylindrical coordinates

$$G_\omega(\vec{r},\vec{r}') = -\frac{1}{4\pi}\frac{\exp(-ik_n|\vec{r}-\vec{r}'|)}{|\vec{r}-\vec{r}'|} = -\sum_{m=0}^{\infty}\epsilon_m\cos m(\phi-\phi')$$

$$\times\{\frac{1}{4\pi i}\int_{-k_n}^{k_n}dk_z\exp[ik_z(z-z')]G_m^{(1)}(\rho,\rho')$$

$$+\frac{1}{2\pi^2}\left(\int_{-\infty}^{-k_n}+\int_{k_n}^{\infty}\right)dk_z\exp[ik_z(z-z')]G_m^{(2)}(\rho,\rho')\},\qquad(2.55)$$

where $\epsilon_m=1/(1+\delta_{m0})$,

$$G_m^{(1)}(\rho_<,\rho_>)=J_m\left(\sqrt{k_n^2-k_z^2}\rho_<\right)H_m^{(2)}\left(\sqrt{k_n^2-k_z^2}\rho_>\right),$$

$$G_m^{(2)}(\rho_<,\rho_>)=I_m\left(\sqrt{k_z^2-k_n^2}\rho_<\right)K_m\left(\sqrt{k_z^2-k_n^2}\rho_>\right).$$

The Fourier component of VP satisfies the equation

$$(\Delta+k_n^2)A_\omega=-\frac{4\pi\mu}{c}j_\omega,\qquad(2.56)$$

where $k_n=\omega/c_n>0$ and $j_\omega=\delta(x)\delta(y)\exp(-i\omega z/v)/2\pi$ The solution of (2.56) is given by

$$A_\omega=\frac{\mu}{c}\int G_\omega(\vec{r},\vec{r}')j_\omega(\vec{r}')dV'=-i\pi\mu\exp(-i\omega z/v)H_0^{(2)}\left(\frac{\omega\rho}{v}\sqrt{\beta_n^2-1}\right)$$

for $\beta_n>1$ and

$$=2\mu\exp(-i\omega z/v)K_0\left(\frac{\omega\rho}{v}\sqrt{1-\beta_n^2}\right)\qquad(2.57)$$

for $\beta_n<1$. Separating the real and imaginary parts, we arrive at (2.53) and (2.54). Collecting in (2.52'), (2.53) and (2.54) the terms at $\sin(\omega z/v)$ and $\cos(\omega z/v)$, we get the integrals

$$\int_0^\infty\cos\left(\frac{\omega\rho}{v}\sinh\chi\right)\sin\left(\frac{\omega\rho}{c_n}\cosh\chi\right)d\chi$$

$$=\int_0^\infty\cos\left(\frac{\omega\rho}{v}x\right)\sin\left(\frac{\omega\rho}{c_n}\sqrt{x^2+1}\right)\frac{dx}{\sqrt{x^2+1}}$$

$$=\int_1^\infty\cos\left(\frac{\omega\rho}{v}\sqrt{x^2-1}\right)\sin\left(\frac{\omega\rho}{c_n}x\right)\frac{dx}{\sqrt{x^2-1}}=\frac{\pi}{2}J_0\left(\frac{\omega\rho}{v}\sqrt{\beta_n^2-1}\right)\quad(2.58)$$

for $v>c_n$ and $=0$ for $v<c_n$. In addition

$$\int_0^\infty\cos\left(\frac{\omega\rho}{v}\sinh\chi\right)\cos\left(\frac{\omega\rho}{c_n}\cosh\chi\right)d\chi$$

$$= \int_0^\infty \cos\left(\frac{\omega\rho}{v}x\right) \cos\left(\frac{\omega\rho}{c_n}\sqrt{x^2+1}\right) \frac{dx}{\sqrt{x^2+1}}$$

$$= \int_1^\infty \cos\left(\frac{\omega\rho}{v}\sqrt{x^2-1}\right) \cos\left(\frac{\omega\rho}{c_n}x\right) \frac{dx}{\sqrt{x^2-1}}$$

$$= -\frac{\pi}{2}N_0\left(\frac{\omega\rho}{v}\sqrt{\beta_n^2-1}\right) \tag{2.59}$$

for $v > c_n$ and $= K_0((\omega\rho)/v)\sqrt{1-\beta_n^2})$ for $v < c_n$. Here $\beta_n = v/c_n$.

In the limit cases these integrals pass into the tabular ones. For example, for $v \to \infty$ Eqs. (2.58) are transformed into

$$\int_0^\infty \sin\left(\frac{\omega\rho}{c_n}\cosh\chi\right) d\chi = \frac{\pi}{2}J_0\left(\frac{\omega\rho}{c_n}\right)$$

and

$$\int_0^\infty \cos\left(\frac{\omega\rho}{c_n}\cosh\chi\right) d\chi = -\frac{\pi}{2}N_0\left(\frac{\omega\rho}{c_n}\right),$$

whilst Eq. (2.59) for $c_n \to \infty$ goes into

$$\int_0^\infty \cos\left(\frac{\omega\rho}{v}\sinh\chi\right) d\chi = K_0\left(\frac{\omega\rho}{v}\right).$$

2.3. Quantum analysis of the Tamm formula

We turn now to the quantum consideration of the Tamm formula. The usual approach proceeds as follows [11]. Consider the uniform rectilinear (say, along the z axis) motion of a point charged particle with the velocity v. The conservation of energy-momentum is written as

$$\vec{p} = \vec{p}' + \hbar\vec{k}, \quad \mathcal{E} = \mathcal{E}' + \hbar\omega, \tag{2.60}$$

where $\vec{p},\mathcal{E}$ and $\vec{p}',\mathcal{E}'$ are the 3-momentum and energy of the initial and final states of the moving charge; $\hbar\vec{k}$ and $\hbar\omega$ are the 3-momentum and energy of the emitted photon. We present (2.60) in the 4-dimensional form

$$p - \hbar k = p', \quad p = (\vec{p}, \mathcal{E}/c). \tag{2.61}$$

Squaring both sides of this equation and taking into account that $p^2 = p'^2 = -m^2c^2$ (m is the rest mass of a moving charge) one obtains

$$(pk) = \hbar k^2/2, \quad k = \left(\vec{k}, \frac{\omega}{c_n}\right). \tag{2.62}$$

Or, in a more manifest form

$$\cos \theta_k = \frac{1}{\beta_n} \left(1 + \frac{n^2 - 1}{2} \frac{\hbar \omega}{\mathcal{E}} \right). \tag{2.63}$$

Here $\beta_n = v/c_n$, $c_n = c/n$ is the velocity of light in medium, n is its refractive index. When deriving (2.63) it was implicitly suggested that the absolute value of the photon 3-momentum and its energy are related by the Minkowski formula: $|\vec{k}| = \omega/c_n$.

When the energy of the emitted Cherenkov photon is much smaller than the energy of a moving charge, Eq.(2.63) reduces to

$$\cos \theta_k = 1/\beta_n, \tag{2.64}$$

which can be written in a manifestly covariant form

$$(pk) = 0. \tag{2.65}$$

Up to now we have suggested that the emitted photon has definite energy and momentum. According to [12], the wave function of a photon propagating in vacuum is described by the following expression

$$iN\vec{e} \exp\left[i(\vec{k}\vec{r} - \omega t) \right], \quad (\vec{e}\vec{k}) = 0, \quad (\vec{e})^2 = 1, \tag{2.66}$$

where N is the real normalization constant and $\vec{e}$ is the photon polarization vector lying in the plane passing through $\vec{k}$ and $\vec{p}$:

$$(\vec{e})_\rho = -\cos \theta_k, \quad (\vec{e})_z = \sin \theta_k, \quad (\vec{e})_\phi = 0, \quad (ek) = 0. \tag{2.67}$$

The photon wave function (2.66) identified with the classical vector potential is obtained in the following way. We take the positive frequency part of the second-quantized vector potential operator and apply it to the coherent state with the fixed $\vec{k}$. The eigenvalue of this VP operator is just (2.66). Now we show that the gauge invariance permits one to present a wave function in the form having the form of a classical vector potential

$$iN'p_\mu \exp\left(ikx \right), \quad (pk) = 0. \tag{2.68}$$

where N' is another real constant. The electromagnetic potentials satisfy the following equations

$$\left(\Delta - \frac{1}{c_n^2} \frac{\partial^2}{\partial t^2} \right) \vec{A} = -\frac{4\pi\mu}{c} \vec{j}, \quad \left(\Delta - \frac{1}{c_n^2} \frac{\partial^2}{\partial t^2} \right) \Phi = -\frac{4\pi}{\epsilon} \rho,$$

$$\mathrm{div}\vec{A} + \frac{\epsilon\mu}{c} \frac{\partial \Phi}{\partial t} = 0.$$

We apply the gauge transformation

$$\vec{A} \to \vec{A}' = \vec{A} + \vec{\nabla}\chi, \ \Phi \to \Phi' = \Phi - \frac{1}{c}\dot{\chi}$$

to the vector potential (2.66) which plays the role of the photon wave function. We choose the generating function χ in the form

$$\chi = \alpha \exp\left[i(\vec{k}\vec{r} - \omega t)\right],$$

where α will be determined later. Thus,

$$\vec{A}' = (N\vec{e} + i\alpha\vec{k})\exp\left[i(\vec{k}\vec{r} - \omega t)\right], \quad \Phi' = \frac{i\omega\alpha}{c}\exp\left[i(\vec{k}\vec{r} - \omega t)\right],$$

where $\vec{e}$ is given by (2.67). We require the disappearance of the ρ component of $\vec{A}'$. This fixes α:

$$\alpha = \frac{N}{ik}\cot\theta_k.$$

The nonvanishing components of $\vec{A}'$ are given by

$$A'_z = \frac{N}{\sin\theta_k}\exp\left[i(\vec{k}\vec{r} - \omega t)\right], \quad A'_0 = \frac{N}{n}\cot\theta_k \exp\left[i(\vec{k}\vec{r} - \omega t)\right].$$

It is easy to see that $A'_z = \beta A'_0$. This completes the proof of (2.68).

Now we take into account that photons described by the wave function (2.68) are created by the axially symmetrical current of a moving charge. According to Glauber ([13], Lecture 3) to obtain the VP in the coordinate representation, one should form a superposition of the wave functions (2.68) by taking into account the relation (2.65) which tells us that the photon is emitted at the Cherenkov angle θ_k defined by (2.64). This superposition is given by

$$A_\mu(x) = iN' \int p_\mu \exp\left(ikx\right)\delta(pk)d^3k/\omega.$$

The factor $1/\omega$ is introduced using the analogy with the photon wave function in vacuum where it is needed for the relativistic covariance of A_μ. The expression $p_\mu\delta(pu)$ is (up to a factor) the Fourier transform of the classical current of the uniformly moving charge. This current creates photons in coherent states which are observed experimentally. In particular, they are manifested as a classical electromagnetic radiation. We rewrite A_μ in a slightly extended form

$$A_\mu = iN' \int p_\mu \exp\left[i(\vec{k}\vec{r} - \omega t)\right]\delta\left[\frac{\mathcal{E}\omega}{c^2}(1 - \beta_n\cos\theta)\right]$$

$$\times \frac{n^3}{c^3} d\phi \, d\cos\theta \, \omega \, d\omega. \tag{2.69}$$

Introducing the cylindrical coordinates ($\vec{r} = \rho\vec{n}_\rho + z\vec{n}_z$), we present $\vec{k}\vec{r}$ in the form

$$\vec{k}\vec{r} = \frac{\omega}{c_n}[\rho\sin\theta\cos(\phi - \phi_r) + z\cos\theta].$$

Inserting this into (2.69) we find

$$A_\mu(\vec{r}, t) = iN'' \int p_\mu \exp\left[i\omega\left(\frac{z}{c_n}\cos\theta_k - t\right)\right]$$

$$\times \exp\left[\frac{i\omega}{c_n}\rho\sin\theta_k\cos(\phi - \phi_r)\right]d\phi d\omega,$$

where N'' is the real modified normalization constant and ϕ_r is the azimuthal angle in the usual space. Integration over ϕ gives

$$A_0(\vec{r}, t) = A_z(\vec{r}, t)/\beta, \quad A_z(\vec{r}, t) = \int\limits_0^\infty \exp(-i\omega t)A_z(\vec{r}, \omega)d\omega,$$

where

$$A_z(\vec{r}, \omega) = \frac{2\pi i N''}{\sin\theta_k}\exp\left(\frac{i\omega}{c_n}\cos\theta_k z\right)J_0\left(\frac{\omega}{c_n}\rho\sin\theta_k\right). \tag{2.70}$$

We see that $A_z(\vec{r}, \omega)$ is the oscillating function of the frequency ω without a pronounced δ function maximum. In the $\vec{r}, t$ representation $A_z(\vec{r}, t)$ (and, therefore, photon's wave function) is singular on the Cherenkov cone $vt - z = \rho/\gamma_n$

$$\mathrm{Re}A_z = 2\pi N'' p_z \int \sin\omega(t - z/v)J_0\left(\frac{\omega\rho}{c_n}\sin\theta_k\right)d\omega$$

$$= 2\pi N'' p_z \frac{v}{[(z - vt)^2 - \rho^2/\gamma_n^2]^{1/2}}\Theta((z - vt)^2 - \rho^2/\gamma_n^2),$$

$$\mathrm{Im}A_z = 2\pi N'' p_z \int \cos\omega(t - z/v)J_0(\frac{\omega\rho}{c_n}\sin\theta_k)d\omega =$$

$$= 2\pi N'' p_z \frac{v}{[\rho^2/\gamma_n^2 - (z - vt)^2]^{1/2}}\Theta(\rho^2/\gamma_n^2 - (z - vt)^2)$$

Despite the fact that the wave function (2.69) satisfies the free wave equation and does not contain singular Neumann functions N_0 (needed to satisfy Maxwell equations with a moving charge current in their r.h.s.), its real part (which, roughly speaking, corresponds to the classic electromagnetic potential) properly describes the main features of the VC radiation.

So far, our conclusion on the absence of CSW in Eqs.(2.26) and (2.27) has been proved only for the dispersion-free case (as only in this case we have exact solution). At this time we are unable to prove the same result in the general case with dispersion. We see that the Tamm formula (2.29) describes evolution and interference of two generalized BS shock waves emitted at the beginning and at the end of the charge motion in the spatial region lying outside the plateau to which the CSW is confined. The Tamm formula does not describe the CSW originating from the charge uniform motion in the interval $(-z_0, z_0)$. On the other hand, the exact solution of the Tamm problem found in [7] contains both CSW and the BS shock wave and not in any way can be reduced to the superposition of two BS waves.

Now the paradoxical results of [8,14] in which the Tamm formula (2.29) was investigated numerically become understandable. Their authors associated the Tamm radiation intensity (2.29) with the interference of the BS shock waves emitted at the beginning and end of the charge motion. Without knowing that the CSW associated with the charge uniform motion in the interval $(-z_0, z_0)$ is absent in the approximate Tamm equations (2.26) they concluded that the CSW is a result of the interference of the above BS shock waves. We quote them:

> Summing up, one can say that radiation of a charge moving with a constant velocity along the limited section of its path (the Tamm problem) is the result of interference of two bremsstrahlungs produced in the beginning and at the end of motion. This is especially clear when the charge moves in vacuum where the laws of electrodynamics prohibit radiation of a charge moving with a constant velocity.
>
> In the Tamm problem the constant-velocity charge motion over the distance l between the charge acceleration and stopping instants in the beginning and at the end of the path only affects the result of interference but does not cause the radiation.
>
> As was shown by Tamm [1] and it follows from our paper the radiation emitted by the charge moving at a constant velocity over the finite section of the trajectory l has the same characteristics in the limit $l \to \infty$ as the VCR in the Tamm-Frank theory [6]. Since the Tamm-Frank theory is a limiting case of the Tamm theory, one can consider the same conclusion is valid for it as well.
>
> Noteworthy is that already in 1939 Vavilov [10] expressed his opinion that deceleration of the electrons is the most probable reason for the glow observed in Cerenkov's experiments.

(We have left the numeration of references in this citation the same as it was in [14]).

We agree with the authors of [8,14] that the Tamm approximate formulae (2.26),(2.29) and (2.31) can be interpreted as the interference between two BS waves if by them we understand the mixture of three shock waves mentioned above (the BS shock wave associated with the beginning and the end of the motion and BS shock waves arising when the charge velocity coincides with the velocity of light in the medium). The Tamm angular intensity (2.29) is valid everywhere except for the angular interval $\theta_2 < \theta < \theta_1$, where θ_1 and θ_2 are defined by (2.40). For the observational distances large compared with the interval of the motion $(r \gg z_0)$,

$$\theta_1 = \arccos \frac{1}{\beta_n} + \delta\theta \quad \text{and} \quad \theta_2 = \arccos \frac{1}{\beta_n} - \delta\theta,$$

where $\delta\theta = \epsilon_0/\beta_n\gamma_n$, $\epsilon_0 = z_0/r$. Although the angular region $2\delta\theta$ tends to zero for $r \gg z_0$, the length of the arc corresponding to it is finite: $\delta L = 2z_0/\beta_n\gamma_n$. On this part of the observational sphere the Tamm angular intensity (2.29) is not valid.

Equation (2.64) defining the position of the maxima of field strengths in the spectral representation is valid when the point charge moves with the velocity $v > c_n$ in the finite spatial interval small compared with the radius r of the observational sphere $(z_0 \ll r)$. When the value of z_0 is comparable or larger than r the pronounced maximum of the Fourier transforms of the field strengths at the angle $\cos\theta = 1/\beta_n$ disappears. Instead, many maxima of the same amplitude distributed over the finite region of space arise. In particular, for the charge unbounded motion the mentioned above Fourier transforms are highly oscillating functions of space variables distributed over the whole space. It follows from the present consideration that Eq. (2.64) (relating to the particular Fourier component) cannot be used for the identification of the charge velocity if the motion interval is comparable with the observational distance.

In the usual space-time representation the field strengths, in the absence of dispersion, are singular in the spatial region $\rho\gamma_n - z_0 \leq z \leq \rho\gamma_n + z_0$ shown in Fig. 2.4. When the dispersion is taken into account, many maxima in the angular distribution of field strengths (in the space-time representation) appear, but the main maximum is at the same position where the Cherenkov singularity lies in the absence of dispersion (see Chapter 4).

It should be noted that doubts about the validity of the Tamm formula (2.64) for the maximum of Fourier components were earlier pointed out by D.V. Skobeltzyne [15]. We mean the so-called Abragam-Minkowski controversy between the photon energy and its momentum.

2.4. Back to the original Tamm problem

In this section we reproduce the results of section (2.1) beginning with the spectral representation. This allows us to analyse the approximations involved.

2.4.1. EXACT SOLUTION

Let a charge be at rest at the point $z = -z_0$ up to an instant $t = -t_0$. In the time interval $-t_0 < t < t_0$ it moves with a constant velocity v. Finally, after the instant t_0 it is again at rest at the point $z = z_0$. The corresponding charge and current densities are

$$\rho(t) = e\delta(x)\delta(y)\times$$

$$[\delta(z + z_0)\Theta(-t - t_0) + \delta(z - z_0)\Theta(t - t_0) + \delta(z - vt)\Theta(t + t_0)\Theta(t_0 - t)],$$

$$\vec{j} = j\vec{n}_z, \quad j = v\delta(z - vt)\Theta(t + t_0)\Theta(t_0 - t), \quad t_0 = \frac{z_0}{v}.$$

Their Fourier transforms are

$$\rho(\omega) = \frac{1}{2\pi}\int \rho(t)\exp(-i\omega t)dt = \rho_1(\omega) + \rho_2(\omega) + \rho_3(\omega),$$

$$j(\omega) = v\rho_3(\omega), \tag{2.71}$$

where

$$\rho_1(\omega) = -\frac{e}{2\pi i\omega}\delta(z + z_0)\delta(x)\delta(y)[\exp(i\omega t_0) - \exp(i\omega T)],$$

$$\rho_2(\omega) = -\frac{e}{2\pi i\omega}\delta(z - z_0)\delta(x)\delta(y)[\exp(-i\omega T) - \exp(-i\omega t_0)],$$

$$\rho_3(\omega) = \frac{e}{2\pi v}\delta(x)\delta(y)\Theta(z + z_0)\Theta(z_0 - z)\exp(-i\omega z/v), \quad j = v\rho_3.$$

In (2.71) the integration over t is performed from $-T$ to T, where $T > t_0$. Later we take the limit $T \to \infty$.

The electromagnetic potentials are equal to

$$\Phi(\omega) = \Phi_1(\omega) + \Phi_2(\omega) + \Phi_3(\omega), \quad A(\omega) \equiv A_z(\omega) = \epsilon\mu\beta\Phi_3(\omega), \tag{2.72}$$

where

$$\Phi_1(\omega) = -\frac{e}{2\pi i\omega\epsilon}[\exp(i\omega t_0) - \exp(i\omega T)]\frac{\exp(-ik_n R_1)}{R_1},$$

$$\Phi_2(\omega) = -\frac{e}{2\pi i\omega\epsilon}[\exp(-i\omega T) - \exp(-i\omega t_0)]\frac{\exp(-ik_n R_2)}{R_2},$$

$$\Phi_3(\omega) = \frac{e}{2\pi v\epsilon}\int_{-z_0}^{z_0}\frac{dz'}{R}\exp(-\frac{i\omega z'}{v})\exp(-ik_n R).$$

Here $R_1 = [(z + z_0)^2 + \rho^2]^{1/2}$, $R_2 = [(z - z_0)^2 + \rho^2]^{1/2}$, $R = [(z - z')^2 + \rho^2]^{1/2}$, $k_n = \omega/c_n$, $c_n = c/n$ is the velocity of light in medium, n is its refractive index.

These potentials satisfy the gauge condition

$$\mathrm{div}\,\vec{A} + \frac{\epsilon\mu}{c}\frac{\partial\Phi}{\partial t} = 0,$$

whilst

$$\mathrm{div}\,\vec{A} + \frac{\epsilon\mu}{c}\frac{\partial\Phi_3}{\partial t} \neq 0.$$

Thus Φ_1 and Φ_2 should be taken into account. Another argument for this is to evaluate

$$E_r = -\frac{\partial\Phi}{\partial r} - \frac{i\omega}{c}A_r, \quad A_r = A\cos\theta.$$

It is easy to check that E_r decreases like $1/r^2$ for $r \to \infty$, whilst it decreases like $1/r$ if Φ is substituted by Φ_3. Thus Φ_1 and Φ_2 are needed to guarantee the correct asymptotic behaviour of electromagnetic field strengths (if we evaluate $\vec{E}$ according to $\vec{E} = -\nabla\Phi - i\omega\vec{A}/c$).

We are primarily interested in the radial energy flux $S_r \sim E_\theta H_\phi$. In the expression

$$E_\theta = -\frac{1}{r}\frac{\partial\Phi}{\partial\theta} - \frac{i\omega}{c}A_\theta, \quad A_\theta = -\sin\theta A$$

the first term in E_θ is the $1/kr$ part of the second term, and therefore it can be disregarded (since in realistic conditions kr is about 10^7). Thus obtained E_θ differs from the exact E_θ by terms of the order $1/kr$.

To make clear the physical meaning of electromagnetic potentials (2.72), we rewrite them in the time representation:

$$\Phi(t) = \int \exp(i\omega t)\Phi(\omega)d\omega, \quad \Phi(t) = \Phi_1(t) + \Phi_2(t) + \Phi_3(t),$$

$$\Phi_1(t) = \frac{e}{\epsilon r_1}\Theta[r_1 - c_n(t + t_0)], \quad \Phi_2(t) = \frac{e}{\epsilon r_2}\Theta[c_n(t - t_0) - r_2],$$

$$\Phi_3(t) = \frac{e}{\epsilon v}\int_{-z_0}^{z_0}\frac{dz'}{R}\delta(t - \frac{z'}{v} - k_n R), \quad R = [(z - z')^2 + \rho^2]^{1/2},$$

$$A(t) = \epsilon\mu\beta\Phi_3(t). \tag{2.73}$$

When evaluating $\Phi_1(t)$ and $\Phi_2(t)$ it was taken into account that

$$\int_{-\infty}^{\infty} \exp(i\omega x)d\omega/\omega = i\pi\mathrm{sign}(x).$$

The following notation will be useful: the spheres $r_1 \equiv [\rho^2 + (z + z_0)^2]^{1/2}$ and $r_2 \equiv [\rho^2 + (z - z_0)^2]^{1/2}$ will be denoted by S_1 and S_2. We say that a particular spatial point lies inside or outside S_1 if $r_1 < c_n(t + t_0)$ and $r_1 > c_n(t + t_0)$, respectively. And similarly for S_2.

We see that $\Phi_1(t)$ differs from zero outside the sphere S_1, i.e., at those points which are not reached by the information about the beginning of the motion. Furthermore, $\Phi_2(t)$ differs from zero inside the sphere S_2, i.e., at those points which are reached by the information about the termination of the motion. Or, in other words, Φ_1 and Φ_2 describe the electrostatic fields of a charge which is at rest at the point $z = -z_0$ up to an instant $t = -t_0$ (beginning of motion) and at the point $z = z_0$ after the instant $t = t_0$ (the termination of motion). In what follows, electrostatic fields associated with Φ_1 and Φ_2 will be denoted by E_1 and E_2, respectively. Obviously, Φ_1 and Φ_2 coincide with the first two terms in (2.13).

To evaluate $\Phi_3(t)$, we use the well-known relation

$$\delta[f(z)] = \sum_i \frac{\delta(z - z_i)}{|f'(z_i)|},$$

where the summation runs over all roots of the equation $f(z) = 0$ and

$$f'(z_i) = \frac{df(z')}{dz'}\Big|_{z=z_i}.$$

We should find the roots of the equation

$$t - \frac{z'}{v} = \frac{R}{c_n}, \quad R = [(z - z')^2 + \rho^2]^{1/2}. \tag{2.74}$$

Squaring this equation we obtain a quadratic equation relative to z' with the roots

$$z_1 = \gamma_n^2(vt - z\beta_n^2 - \beta_n r_m), \quad z_2 = \gamma_n^2(vt - z\beta_n^2 + \beta_n r_m),$$

$$r_m = [(z - vt)^2 + (1 - \beta_n^2)\rho^2]^{1/2}, \quad \gamma_n^2 = \frac{1}{1 - \beta_n^2}. \tag{2.75}$$

Charge's velocity is smaller than the velocity of light in medium
Consider first the case $\beta_n < 1$. Then, only the root z_1 satisfies (2.74) (the appearance of the second root in (2.75) is because the quadratic equation following from (2.74) can have roots which do not satisfy (2.74)). Now we impose the condition $-z_0 < z' < z_0$ which means that the motion takes

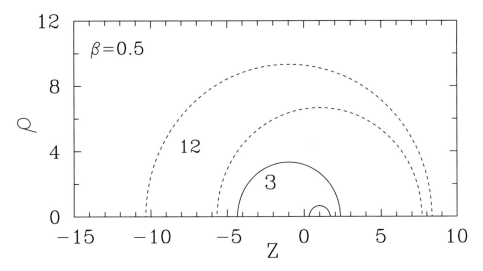

Figure 2.15. Positions of shock waves for $T = 3$ and $T = 12$ in the exact Tamm problem for the charge velocity ($\beta = 0.5$) smaller than the velocity of light in the medium. Here $T = ct/z_0$. The vector potential differs from zero between the solid lines for $T = 3$ and between the dotted lines for $T = 12$; ρ and z are in units of z_0. The interval of motion and refractive index are: $L = 0.5\,\text{cm}$ and $n = 1.5$, respectively.

place on the interval $(-z_0, z_0)$. It then follows from (2.74), that $\Phi_3(t) \neq 0$ for the spatial points lying inside S_1 and outside S_2:

$$\Phi_3(t) = \frac{e}{r_m\epsilon}\Theta[c_n(t + t_0) - r_1]\Theta[r_2 - c_n(t - t_0)],$$

$$A_z = \beta\epsilon\mu\Phi_3, \quad t_0 = \frac{z_0}{v}, \tag{2.76}$$

Physically, Φ_3 describes the EMF of a charge moving on the interval $(-z_0, z_0)$. It differs from zero at those spatial points which obtained information on the beginning of motion and did not obtain information on its termination. It is easy to see that for $\beta_n < 1$ the S_2 sphere lies entirely inside S_1, i.e., there are no intersections between them. The positions of S_1 and S_2 spheres for two different instants of time are shown in Fig. 2.15. The region where $\Phi_3 \neq 0$ is between S_1 and S_2 belonging to the same t. Static fields Φ_1 and Φ_2 lie outside S_1 and inside S_2, respectively.

Equation (2.76) coincides with the last term in Φ given by (2.13).

Charge's velocity is greater than the velocity of light in medium
Now let $\beta_n > 1$. Then Φ_1, Φ_2 and their physical meanings are the same as for $\beta_n < 1$. We now turn to Φ_3. It is easy to check that the two roots satisfy (2.74) if $z < vt$, and there are no roots if $z > vt$. We need further notation.

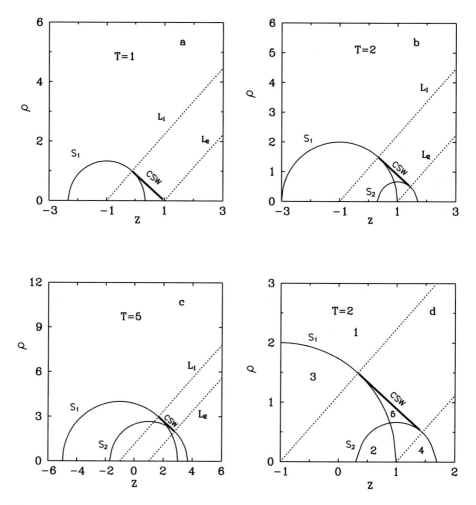

Figure 2.16. Time evolution of shock waves in the exact Tamm problem for the charge velocity ($\beta = 1$) greater than the velocity of light in medium. S_1 and S_2 are shock waves radiated at the beginning and termination of motion, respectively. CSW is the Cherenkov shock wave. The time $T = 1$ corresponds to the instant when the wave S_2 arises (a). For larger times the CSW is tangential both to S_1 and S_2 and is confined between the straight lines L_1 and L_2 (b,c). Part (d) of the figure is a magnified version of (b). The vector potential is zero in region 2 lying inside S_1 and S_2 and in region 2 lying outside S_1 and S_2 and above the CSW. Only one retarded time contributes in region 3 (lying inside S_1 and outside S_2) and in region 4 (lying inside S_2 and outside S_1). Two retarded times contribute to region 5 lying outside S_1 and S_2 and below the CSW. Other parameters are the same as in Fig. 2.15.

We denote by L_1 and L_2 the straight lines $z = -z_0 + \rho|\gamma_n|$ and $z = z_0 + \rho|\gamma_n|$, respectively (Fig. 2.16). We say that a particular point is to the left or right of L_1 if $z < -z_0 + \rho|\gamma_n|$ or $z > -z_0 + \rho|\gamma_n|$, respectively. And similarly for L_2. Correspondingly, a particular point lies between L_1 and L_2 if $-z_0 + \rho|\gamma_n| <$

$z < z_0 + \rho/|\gamma_n|$. The straight lines L_1 and L_2 are inclined towards the motion axis at the Cherenkov angle $\theta_{\mathrm{Ch}} = \arccos(1/\beta_n)$. The CSW is the straight line $z + \rho/|\gamma_n| = vt$, perpendicular both to L_1 and L_2 straight lines and enclosed between them. We observe that the denominators r_m vanish exactly for $z + \rho/|\gamma_n| = vt$, i.e., on the CSW. There are no other zeroes of r_m. We say also that a particular point lies under or above the CSW if $z + \rho/|\gamma_n| < vt$ or $z + \rho/|\gamma_n| > vt$, respectively.

We impose the condition for motion to be on the interval $(-z_0, z_0)$. Then, the first root exists in the following space-time domains (Fig. 2.16, d): i) To the right of L_2, it exists only outside S_1 and inside S_2; ii) Between L_1 and L_2, it exists outside S_1 and under the CSW. The contribution of the first root to Φ_3 is:

$$\Phi_3^{(1)} = \frac{e}{\epsilon r_m}\{\Theta(z + z_0 - \rho|\gamma_n|)\Theta(z_0 + \rho|\gamma_n| - z)\Theta(t - \frac{z + \rho/|\gamma_n|}{v})$$

$$+\Theta(z - z_0 - \rho|\gamma_n|)\Theta[c_n(t - t_0) - r_2)]\}\Theta[r_1 - c_n(t + t_0)]. \qquad (2.77)$$

The first term in (2.77) is singular on the CSW (since $r_m = 0$ on it) enclosed between the straight lines L_1 and L_2. The second term in (2.77) does not contain singularities except for the point where $S_2(=\mathrm{BS}_2)$ meets with L_2 and CSW.

Now we turn to the second root: i) To the left of the L_1, it exists only inside S_1 and outside S_2; ii) Between L_1 and L_2, it exists outside S_2 and under the CSW. Correspondingly, the contribution of the second root is

$$\Phi_3^{(2)} = \frac{e}{\epsilon r_m}\{\Theta(z + z_0 - \rho|\gamma_n|)\Theta(z_0 + \rho|\gamma_n| - z)\Theta(t - \frac{z + \rho/|\gamma_n|}{v})$$

$$+\Theta(\rho|\gamma_n| - z - z_0)\Theta[c_n(t + t_0) - r_1)]\}\Theta[r_2 - c_n(t - t_0)]. \qquad (2.78)$$

Again, the first term in this expression is singular on the same CSW. while the second term does not contain singularities except for the point where $S_1(=\mathrm{BS}_1)$ meets with L_1 and CSW.

The contribution of two roots to Φ_3 is

$$\Phi_3 = \Phi_3^{(1)} + \Phi_3^{(2)}, \quad A_z(t) = \beta\epsilon\mu\Phi_3(t). \qquad (2.79)$$

This Φ_3 coincides with Φ_m in (2.36). In Figs. 2.16 (a,b,c) there are shown positions of S_1, S_2 and CSW shock waves at various instants of time. In Fig. 2.16 (d), which is a magnified image of Fig. 2.16 (b), we see five regions in which the EMF differs from zero. The region 1 lies outside S_1 and S_2 and above the CSW. There is only the electrostatic field E_1 there. In the region 2 lying inside S_1 and S_2 there is only the electrostatic field E_2. In the region 3 lying inside S_1 and outside S_2 there is the EMF of a moving charge (only the second root contributes). In the region 4 lying inside S_2 and outside S_1, there is EMF of a moving charge (only the first root contributes) and electrostatic fields E_1 and E_2. Finally, in the region 5 lying outside S_1 and S_2 and below the CSW, there is the EMF of a moving charge (both roots contribute) and electrostatic field E_1.

So far we have suggested that for $t < -t_0$ and $t > t_0$ a charge is at rest at points $z = -z_0$ and $z - z_0$, respectively. However, usually, when dealing with the Tamm problem, one uses only the vector potential describing the charge motion on the interval $(-z_0 < z < z_0)$. It is given by $A = \mu \epsilon \beta \Phi_3$. One then evaluates the magnetic and electric fields using the relations $\mu \vec{H} = \mathrm{curl} \vec{A}$ and $\mathrm{curl} \vec{H} = i \epsilon k \omega \vec{E}$ valid in the spectral representation. In this case the terms Φ_1 and Φ_2 drop out of consideration. There are then nonzero electromagnetic potentials corresponding to the first root in region 4, the second root in region 3, and first and second roots in region 5. In other spatial regions potentials are zero. On the border of regions 3, 4 and 5 with regions 1 and 2 potentials exhibit jumps, and therefore field strengths have delta singularities.

Experimentalists insist that they measure $\vec{E}(\omega)$ and $\vec{H}(\omega)$ (in fact, they detect photons with a definite frequency). It is just the reason that enabled us to operate in preceding sections with the Fourier transforms $\vec{E}(\omega)$ and $\vec{H}(\omega)$.

2.4.2. RESTORING VECTOR POTENTIAL IN THE SPECTRAL REPRESENTATION

We turn now to the vector potential in the spectral representation given by (2.72):

$$A_z(\omega) = \frac{e\mu}{2\pi c} \int\limits_{-z_0}^{z_0} \frac{dz'}{R} \exp\left(-\frac{i\omega z'}{v}\right) \exp(-ik_n R).$$

This expression contains both the BS and Cherenkov radiation in an indivisible form. On the other hand, the vector potential in the time representation is

$$A_z(t) = \beta \epsilon \mu \Phi_3(t),$$

where $\Phi_3(t)$ is defined by (2.79). Equations (2.77)-(2.79) demonstrate that contributions of the BS and Cherenkov radiation are unambiguously separated. We now apply the inverse Fourier transformation to particular pieces of $A_z(t)$ and try to separate the above contributions in the spectral representation. But first, for pedagogical purposes we consider the case $\beta n < 1$. The corresponding VP, in the time representation, is given by (2.76):

$$A_z(t) = \frac{e\mu\beta}{r_m} \Theta[c_n(t+t_0) - r_1] \Theta[r_2 - c_n(t-t_0)].$$

In the spectral representation, one gets

$$A_z(\omega) = \frac{e\mu\beta}{2\pi} \int\limits_{-t_0+r_1/c_n}^{t_0+r_2/c_n} \frac{dt}{r_m} \exp(-i\omega t).$$

Making the change of the integration variable

$$t = \frac{z}{v} + \frac{\rho}{\beta|\gamma_n|} \sinh \chi$$

one has

$$A_z(\omega) = \frac{e\mu}{2\pi c} \exp\left(-\frac{ikz}{\beta}\right) \int\limits_{\chi_1}^{\chi_2} d\chi \exp\left(-\frac{ik\rho}{\beta|\gamma_n|} \sinh \chi\right),$$

where χ_1 and χ_2 are defined by

$$\sinh \chi_1 = \frac{r_1\beta_n - z - z_0}{\rho}|\gamma_n|, \quad \sinh \chi_2 = \frac{r_2\beta_n - z + z_0}{\rho}|\gamma_n|.$$

When the interval of motion is much larger than the observational distance,

$$\sinh \chi_1 \to -\frac{z_0(1 - \beta_n)}{\rho}|\gamma_n| \approx -\infty, \quad \sinh \chi_2 \to \frac{z_0(1 - \beta_n)}{\rho}|\gamma_n| \approx \infty$$

and

$$A_z(\omega) \to \frac{e\mu}{\pi c} \exp\left(-\frac{ikz}{\beta}\right) K_0\left(\frac{k\rho}{\beta|\gamma_n|}\right).$$

We now apply the quasi-classical method for the evaluation of $A_z(\omega)$. This gives

$$A_z(\omega) = \frac{ie\mu\beta|\gamma_n|}{2\pi ck\rho}(C_2 - C_1),$$

where

$$C_1 = \frac{1}{\cosh \chi_1} \exp\left[-\frac{ik}{\beta}(r_1\beta_n - z_0)\right],$$

$$C_2 = \frac{1}{\cosh \chi_2} \exp\left[-\frac{ik}{\beta}(r_2\beta_n + z_0)\right].$$

Now let $\beta n > 1$. Then according to (2.77)-(2.79) the VP consists of three pieces defined in the spatial regions lying to the left of L_1, to the right of L_2 and between L_1 and L_2 (Fig. 2.16):

$$A_z(\omega) = A_z^{(1)}(\omega) + A_z^{(2)}(\omega) + A_z^{(3)}(\omega),$$

where

$$A_z^{(1)}(\omega) = \Theta(z - z_0 - \rho|\gamma_n|)\frac{e\mu}{2\pi c}\exp\left(-\frac{ikz}{\beta}\right)\int_{\chi_2}^{\chi_1}\exp\left(-\frac{ik\rho}{\beta|\gamma_n|}\cosh\chi\right)d\chi,$$

$$A_z^{(2)}(\omega) = \Theta(\rho|\gamma_n| - z - z_0)\frac{e\mu}{2\pi c}\exp\left(-\frac{ikz}{\beta}\right)\int_{\chi_1}^{\chi_2}\exp\left(-\frac{ik\rho}{\beta|\gamma_n|}\cosh\chi\right)d\chi,$$

$$A_z^{(3)}(\omega) = \Theta(\rho|\gamma_n| - z + z_0)\Theta(z + z_0 - \rho|\gamma_n|)\frac{e\mu}{2\pi c}\exp\left(-\frac{ikz}{\beta}\right)$$

$$\times\left(\int_0^{\chi_1} + \int_0^{\chi_2}\right)\exp\left(-\frac{ik\rho}{\beta|\gamma_n|}\cosh\chi\right)d\chi,$$

where χ_1 and χ_2 are now defined as follows:

$$\cosh\chi_1 = \frac{\beta_n r_1 - z - z_0}{\rho}|\gamma_n|, \quad \cosh\chi_2 = \frac{\beta_n r_2 - z + z_0}{\rho}|\gamma_n|.$$

In the quasi-classical approximation, one gets

$$A_z^{(1)}(\omega) = -\Theta(z - z_0 - \rho|\gamma_n|)\frac{ie\mu\beta|\gamma_n|}{2\pi ck\rho}\exp\left(-\frac{ikz}{\beta}\right)(S_2 - S_1),$$

$$A_z^{(2)}(\omega) = \Theta(\rho|\gamma_n| - z_0 - z)\frac{ie\mu\beta|\gamma_n|}{2\pi ck\rho}\exp\left(-\frac{ikz}{\beta}\right)(S_2 - S_1),$$

$$A_z^{(3)}(\omega) = \Theta(\rho|\gamma_n| - z + z_0)\Theta(z + z_0 - \rho|\gamma_n|)\frac{e\mu}{2\pi c}\exp\left(-\frac{ikz}{\beta}\right)$$

$$\times\left[\frac{i\beta|\gamma_n|}{kr\sin\theta}(S_1 + S_2) + \exp\left(-i\frac{k\rho}{\beta|\gamma_n|}\right)\exp\left(-\frac{i\pi}{4}\right)\sqrt{\frac{2\pi\beta|\gamma_n|}{kr\sin\theta}}\right],$$

where

$$S_1 = \frac{1}{\sinh\chi_1}\exp\left(-i\frac{k\rho}{\beta|\gamma_n|}\cosh\chi_1\right),$$

$$S_2 = \frac{1}{\sinh\chi_2}\exp\left(-i\frac{k\rho}{\beta|\gamma_n|}\cosh\chi_2\right).$$

For the observational distances much larger than the interval of motion, one obtains ($\epsilon_0 = z_0/r$)

$$\cosh\chi_2 \approx \frac{|\gamma_n|}{\sin\theta}[\beta n - \cos\theta + \epsilon_0(1 - \beta n)],$$

$$\cosh\chi_1 \approx \frac{|\gamma n|}{\sin\theta}[\beta n - \cos\theta - \epsilon_0(1 - \beta n)],$$

$$\epsilon_0 = z_0/r, \quad \sinh\chi_1 \approx \sinh\chi_2 \approx \frac{|\gamma n|}{\sin\theta}|1 - \beta n\cos\theta|,$$

$$\Theta(z - z_0 - \rho|\gamma n|) \approx \Theta\left(\cos\theta - \frac{1}{\beta n} - \frac{\epsilon_0}{\beta_n^2|\gamma n|^2}\right),$$

$$\Theta(\rho|\gamma n| - z_0 - z) \approx \Theta\left(\frac{1}{\beta n} - \frac{\epsilon_0}{\beta_n^2|\gamma n|^2} - \cos\theta\right)$$

Under these approximations, $A_z^{(1)}(\omega)$ and $A_z^{(2)}(\omega)$ coincide with the Tamm VP (2.26), whilst $A_z^{(3)}(\omega)$ goes into

$$A_z^{(3)}(\omega) \approx \Theta\left(\frac{1}{\beta n} + \frac{\epsilon_0}{\beta_n^2|\gamma n|^2} - \cos\theta\right)\Theta\left(\cos\theta - \frac{1}{\beta n} + \frac{\epsilon_0}{\beta_n^2|\gamma n|^2}\right)$$

$$\times\{\frac{e\mu\beta}{\pi ckr}\exp(-iknr)\frac{\sin[kz_0(1 - \beta n\cos\theta)/\beta]}{1 - \beta n\cos\theta}$$

$$+\frac{e\mu}{2\pi c}\exp\left(-\frac{ikz}{\beta}\right)\exp\left(-\frac{i\pi}{4}\right)\sqrt{\frac{2\pi\beta|\gamma n|}{kr\sin\theta}}\exp\left(-i\frac{k\rho}{\beta|\gamma n|}\right)\}.$$

It is seen that the term $A_z^{(3)}(\omega)$ (which is absent in the Tamm vector potential (2.26)) differs from zero in a beam of width $2z_0/\beta_n\gamma_n$. Another important observation is that $A_z^{(1)}(\omega)$ and $A_z^{(2)}(\omega)$ decrease as $1/kr$ for $kr \to \infty$, whilst $A_z^{(3)}(\omega)$ decreases as $1/\sqrt{kr}$.

The same result is obtained if one applies the WKB approximation for the evaluation of A_z entering into (2.72). In fact, the integral (2.49) defining it has a stationary point $z' = z - \rho\gamma_n$ which lies within the interval $(-z_0, z_0)$ for $\theta_2 < \theta < \theta_1$, to the left of $(-z_0)$ for $\theta > \theta_1$ and to the right of (z_0) for $\theta < \theta_2$. Here

$$\cos\theta_1 = \frac{1}{\beta n}\sqrt{1 - \frac{\epsilon_0^2}{\beta_n^2|\gamma_n^2|}} - \frac{\epsilon_0}{\beta_n^2|\gamma_n^2|}, \quad \cos\theta_2 = \frac{1}{\beta n}\sqrt{1 - \frac{\epsilon_0^2}{\beta_n^2|\gamma_n^2|}} + \frac{\epsilon_0}{\beta_n^2|\gamma_n^2|}.$$

It is easy to check that in the angular regions $\theta > \theta_1$ and $\theta < \theta_2$ only the boundary points $\mp z_0$ of the interval of motion contribute to BS shock waves. On the other hand, in the angular region $\theta_2 < \theta < \theta_1$ the stationary point lying inside the interval of motion $-z_0 < z < z_0$ contributes to the Cherenkov shock wave, whilst the boundary points $(\pm z_0)$ contribute to the BS shock waves.

From the definition (2.49) of the magnetic vector potential in the spectral representation it follows that all the points z' of the interval of motion $(-z_0, z_0)$ contribute to it. In the time representation the factor $\delta(t - z'/v -$

$k_n R$) appears inside the integral. After integration over z', one obtains $A_z(t)$ given by (2.79) which differs from zero inside the spatial region bounded by the BS and Cherenkov shock waves. The electromagnetic field strengths have delta singularities on the borders of this region. Thus the integration in (2.49) over the interval of motion in the spectral representation language results in the appearance of BS and Cherenkov shock waves in the time representation.

2.4.3. THE TAMM APPROXIMATE SOLUTION

The Tamm vector potential in the spectral representation is

$$A_T(\omega) = \frac{e\mu}{\pi r n \omega (\cos\theta - 1/\beta_n)} \exp(-ik_n r) \sin[k_n z_0 (\cos\theta - 1/\beta_n)]. \quad (2.80)$$

It is obtained from $A_z(\omega)$ given by (2.72) when the conditions

$$z_0 \ll r, \quad kr \ll 1, \quad \text{and} \quad k z_0^2 / r \ll 1$$

are satisfied. Using (2.80) for the evaluation of field strengths and the radiation intensity, one gets the famous Tamm formula (2.29) for the radiation intensity. Going in (2.80) to the time representation, one gets

$$A_T(t) = \frac{e\mu}{rn|\cos\theta - 1/\beta_n|}[\Theta\left(\frac{1}{\beta_n} - \cos\theta\right) \cdot \Theta(r - R_1) \cdot \Theta(R_2 - r)$$

$$+ \Theta\left(\cos\theta - \frac{1}{\beta_n}\right) \cdot \Theta(r - R_2) \cdot \Theta(R_1 - r)]. \quad (2.81)$$

Here

$$R_1 = c_n t + z_0(\frac{1}{\beta_n} - \cos\theta), \quad \text{and} \quad R_2 = c_n t - z_0(\frac{1}{\beta_n} - \cos\theta).$$

Equation (2.81) is an extended version of (2.42). For $\beta_n < 1$, (2.81) is transformed into

$$A_T(t) = \frac{e\mu}{rn(1/\beta_n - \cos\theta)} \cdot \Theta(r - R_2)\Theta(R_1 - r), \quad (2.82)$$

that is, at a fixed instant of time the electromagnetic field differs from zero between two non-intersecting curves S_1 and S_2 defined by $r = R_1$ and $r = R_2$, respectively. (Fig. 2.17 (a)).

On the other hand, for $\beta_n > 1$

$$A_T(t) = \frac{e\mu}{rn(\cos\theta - 1/\beta_n)}\Theta(r - R_1) \cdot \Theta(R_2 - r) \quad (2.83)$$

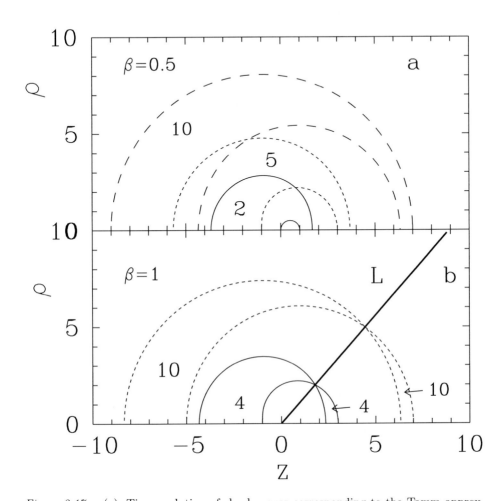

Figure 2.17.   (a): Time evolution of shock waves corresponding to the Tamm approximate vector potential (2.82) for the charge velocity smaller than the velocity of light
in the medium. The Tamm vector potential differs from zero between two solid lines for
$T = 2$, between two dotted lines for $T = 5$, and between two dashed lines for $T = 10$;
(b): The same as in (a), but for the charge velocity greater than the velocity of light in
medium. The Tamm vector potential (2.83) and (2.84) differing from zero between two
solid lines for $T = 4$ and between two dotted lines for $T = 10$, is singular at the intersection of lines with the same T. The straight line passing through these singular points is
shown by a thick line. The energy flux propagates mainly along this straight line. Probably, the absence of CSW in this approximate picture has given rise to associate above
singularities with an interference (intersection) of BS shock waves. Other parameters are
the same as in Fig. 2.15.

for $\cos\theta > 1/\beta_n$ and

$$A_T(t) = \frac{e\mu}{rn(1/\beta_n - \cos\theta)}\Theta(r - R_2)\cdot\Theta(R_1 - r) \qquad (2.84)$$

for $\cos\theta < 1/\beta_n$. For $\beta_n > 1$ the curves S_1 and S_2 are intersected at $\cos\theta = 1/\beta_n$.

The region in which $A_T(t) \neq 0$ lies between S_1 and S_2 (Fig.2.17 (b)). By comparing this figure with Fig. 2.16 we observe that the CSW shown in Fig. 2.16 by the thick line and enclosed between the straight lines L_1 and L_2 degenerates into a point coinciding with the intersection of curves 1 and 2. These intersection points at different instants of time lie on the same straight line L inclined towards the motion axis under the Cherenkov angle $\cos\theta_{Ch} = 1/\beta n$. The electromagnetic potentials and field strengths are infinite on this line at the distance $r = c_n t$ from the origin, and therefore, the major part of the energy flux propagates under the angle θ_{Ch} towards the motion axis (Fig. 2.17 (b)).

For $\beta_n > 1$ the curves S_1 and S_2 are always intersected at large distances (where the Tamm approximation holds). Probably this fact and the absence of the CSW gave rise to a number of attempts [8,14] to interpret the Tamm intensity (2.29) as the interference between BS shock waves emitted at the boundary $z = \pm z_0$ points. The standard approach [1,4] associates (2.80) and (2.81) with the radiation produced by a charge uniformly moving in medium, in a finite spatial interval, with a velocity $v > c_n$. We believe that this dilemma cannot be resolved in the framework of the Tamm approximate solution (2.80).

The question arises of at which stage the CSW has dropped from the vector potential (2.80)? We have seen above that it presents both in (2.73) and (2.79). But (2.73) is just the Fourier transform of $A(\omega)$ defined in (2.72). The Tamm vector potential (2.80) is obtained from the exact (2.72) by changing $R \to r$ in the denominator and $R \to r - z'\cos\theta$ in the exponent. The first approximation is not essential if the observational distance is much larger than the interval of motion. It is the second approximation that is responsible for the disappearance of the CSW. The condition for the validity of the second of these approximations is not valid in realistic cases. Exact analytical and numerical calculations show that an enormous broadening of the angular intensity spectrum takes place in the spectral representation (see Chapter 5). In the time representation this broadening leads to the appearance of the CSW enclosed between L_1 and L_2 straight lines shown in Fig. 2.16. Equations similar to (2.76)-(2.79) were obtained in section 2.1 but without use the spectral representation (2.72) as an intermediate step. The latter is needed to recover at what stage of approximations the CSW drops out from consideration and to make a choice between opposite interpretations of the Tamm formula for radiation intensity.

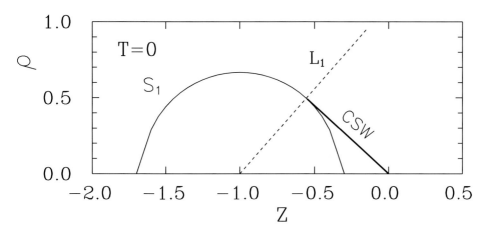

Figure 2.18. A counter-example showing that in the exact Tamm model the presence of two BS waves is not needed for the existence of the Cherenkov shock wave. In the time interval $-t_0 < t < t_0$ there is a shock wave S_1 arising at the beginning of motion and the CSW. The S_2 shock wave has still not appeared. Other parameters are the same as in Fig. 2.15.

2.4.4. CONCRETE EXAMPLE SHOWING THAT THE CSW IS NOT ALWAYS REDUCED TO THE INTERFERENCE OF BS SHOCK WAVES

In Fig. 2.18 there are shown positions of shock waves at the instant $t = 0$ lying inside the interval $-t_0 < t < t_0$. At this instant, the shock wave S_1 associated with the beginning of motion has arisen, but S_2 shock wave associated with the termination of motion has not still appeared. In this figure we see the part of a Cherenkov wave, enclosed between the motion axis and S_1, tangential to the latter and having a normal inclined at the angle $\theta_{Ch} = \arccos(1/\beta n)$ toward the motion axis. Since the shock wave S_2 is absent, the appearance of the CSW cannot be attributed to the interference of the waves S_1 and S_2.

Therefore in the time representation the existence of the shock wave S_2 is not needed for the appearance of the CSW. In some time interval the CSW is enclosed between the motion axis and the shock wave S_1. (Figs. 2.16 (a) and 2.18). As time advances, the shock wave S_2 arises. For large times the CSW is tangential to S_1 and S_2 and is enclosed between them (Fig. 2.16, (b),(c),(d)).

Since the frequency distribution of the radiation intensity $\sigma_r(\omega)$ involves integration over all times, all particular configurations shown in Fig. 2.16 contribute to $\sigma_r(\omega)$. Thus it is still possible to associate the Tamm problem with the interference of S_1 and S_2 shock waves (one may argue that, since all times contribute to the radiation intensity in the spectral representation, the large times, when S_1 and S_2 shock waves are intersected, also give a contribution to the frequency representation just mentioned). The contribution of CSW is confined to the region

$$\rho|\gamma_n| - z_0 < z < z_0 + \rho|\gamma_n|,$$

degenerating (if one drops z_0 in this expression) into the straight line inclined at the angle θ_c, $\cos\theta_c = 1/\beta n$ towards the motion axis.

2.5. Schwinger's approach to the Tamm problem

We begin with the continuity equation following from Maxwell equations

$$\mathrm{div}\vec{S} + \frac{\partial}{\partial t}\mathcal{E} = -\vec{j}\vec{E}. \qquad (2.85)$$

Here

$$\vec{S} = \frac{c}{4\pi}(\vec{E}\times\vec{H}), \quad \mathcal{E} = \frac{1}{8\pi}(\epsilon E^2 + \mu H^2).$$

Integrating this equation over the volume V of the sphere S of radius r surrounding a moving charge, one finds the following equation describing the energy conservation

$$\int S_r r^2 d\Omega + \frac{\partial}{\partial t}\int \mathcal{E}dV = -\int \vec{j}\vec{E}dV. \qquad (2.86)$$

Usual interpretation of this equation proceeds as follows (see, e.g., [16], pp.276-277):

> The first term on the left-hand side represents the electromagnetic energy flowing out of the volume V through the surface S_r, and the second term represents the time rate of change of the energy stored by the electromagnetic field within V.

And further:

> The right-hand side, on the other hand, represents the power supplied by the external forces that maintain the charges in dynamic equilibrium.

Schwinger [17] identifies energy losses of a moving charge with the integral in the r.h.s. of (2.85)

$$W_S = -\int \vec{j}\vec{E}dV. \qquad (2.87)$$

Substituting $\vec{E} = -\vec{\nabla}\Phi - \dot{\vec{A}}/c$ and integrating by parts one has

$$W_S = -\int \vec{j}\vec{E}dV = \int \vec{j}(\vec{\nabla}\Phi + \dot{\vec{A}}/c)dV = -\int (\mathrm{div}\vec{j} - \vec{j}\dot{\vec{A}}/c)dV$$

$$= \int (\rho\dot{\Phi} + \vec{j}\dot{\vec{A}}/c)dV = \frac{d}{dt}\int \rho\Phi dV - \int (\rho\dot{\Phi} - \vec{j}\dot{\vec{A}}/c)dV. \qquad (2.88)$$

By definition W_S is the energy lost by a moving charge per unit time. Schwinger discards the first term in the second line of (2.88) on the grounds that

it is of an accelerated energy type.

The retarded and advanced electromagnetic potentials corresponding to charge current densities ρ and $\vec{j}$ are given by

$$\Phi_{ret,adv} = \frac{1}{\epsilon}\int \frac{1}{R}\rho(\vec{r}',t')\delta(t' - t \pm R/c_n)dV'dt'$$

$$= \frac{1}{2\pi\epsilon}\int_{-\infty}^{\infty} d\omega \frac{1}{R}\rho(\vec{r}',t')\exp[i\omega(t' - t \pm R/c_n)]dV'dt',$$

$$\vec{A}_{ret,adv} = \frac{\mu}{c}\int \frac{1}{R}\vec{j}(\vec{r}',t')\delta(t' - t \pm R/c_n)dV'dt'$$

$$= \frac{\mu}{2\pi c}\int_{-\infty}^{\infty} d\omega \frac{1}{R}\vec{j}(\vec{r}',t')\exp[i\omega(t' - t \pm R/c_n)]d\omega dV'dt', \qquad (2.89)$$

where ϵ and μ are the electric and magnetic permittivities, respectively; $R = |\vec{r} - \vec{r}'|$ and $+$ and $-$ signs refer to retarded and advanced potentials, respectively. Furthermore, Schwinger represents retarded electromagnetic potentials in the form

$$\Phi_{ret} = \frac{1}{2}(\Phi_{ret} + \Phi_{adv}) + \frac{1}{2}(\Phi_{ret} - \Phi_{adv}),$$

$$\vec{A}_{ret} = \frac{1}{2}(\vec{A}_{ret} + \vec{A}_{adv}) + \frac{1}{2}(\vec{A}_{ret} - \vec{A}_{adv}) \qquad (2.90)$$

and discards the symmetrical part of these equations on the grounds that

the first part of (2.90), derived from the symmetrical combination of $\vec{E}_{ret}$ and $\vec{E}_{adv}$, changes sign on reversing the positive sense of time and therefore represents reactive power. It describes the rate at which the electron stores energy in the electromagnetic field, an inertial effect with which we are not concerned. However, the second part of (2.90), derived from the antisymmetrical combination of $\vec{E}_{ret}$ and $\vec{E}_{adv}$, remains unchanged on reversing the positive sense of time and therefore represents resistive power. Subject to one qualification, it describes the rate of irreversible energy transfer to the electromagnetic field, which is the desired rate of radiation.

Correspondingly, electromagnetic potentials are reduced to

$$\Phi = -\frac{1}{\pi\epsilon}\int\limits_0^\infty d\omega\,\frac{1}{R}\rho(\vec{r}',t')\sin[\omega(t'-t)]\sin(k_nR)dV'dt',$$

$$\vec{A} = -\frac{\mu}{\pi c}\int\limits_0^\infty d\omega\,\frac{1}{R}\vec{j}(\vec{r}',t')\sin[\omega(t'-t)]\sin(k_nR)dV'dt', \quad k_n = \frac{\omega}{c_n}. \quad (2.91)$$

Substituting this into (2.88) we obtain

$$W_S = \int\limits_0^\infty P(\omega,t)d\omega, \tag{2.92}$$

where

$$P(\omega,t) = \frac{d^2E}{dtd\omega} = -\frac{\omega}{\pi\epsilon}\int dV\,dV'dt'\frac{\sin k_nR}{R}\cos\omega(t-t')$$

$$\times\left[\rho(\vec{r},t)\rho(\vec{r}',t') - \frac{1}{c_n^2}\vec{j}(\vec{r},t)\vec{j}(\vec{r}',t')\right] \tag{2.93}$$

is the energy lost by a moving charge per unit time and per frequency unit. The angular distribution $P(\vec{n},\omega,t)$ is defined as

$$P(\omega,t) = \int P(\vec{n},\omega,t)d\Omega, \tag{2.94}$$

where

$$P(\vec{n},\omega,t) = \frac{d^3E}{dtd\omega d\Omega} = -\frac{n\omega^2}{4\pi^2ce}\int dV\,dV'dt'\cos\omega\left[(t'-t) + \frac{1}{c_n}\vec{n}(\vec{r}-\vec{r}')\right]$$

$$\times\left[\rho(\vec{r},t)\rho(\vec{r}',t') - \frac{1}{c_n^2}\vec{j}(\vec{r},t)\vec{j}(\vec{r}',t')\right] \tag{2.95}$$

is the energy lost by a moving charge per unit time, per frequency unit, and per unit solid angle. Here $\vec{n}$ is the vector defining the observational point.

Equations (2.93) and (2.95) were obtained by Schwinger [17]. We apply them to the Tamm problem. In what follows we limit ourselves to dielectric medium for which $\epsilon = n^2$.

2.5.1.   INSTANTANEOUS POWER FREQUENCY SPECTRUM

For the Tamm problem treated, charge and current densities are given by

$$j_z = ev\delta(x)\delta(y)\Theta(t+t_0)\Theta(t_0-t)\delta(z-vt), \ \rho(\vec{r},t) = e\delta(x)\delta(y)$$

$$\times[\Theta(-t-t_0)\delta(z+z_0)+\Theta(t+t_0)\Theta(t_0-t)\delta(z-vt)+\Theta(t-t_0)\delta(z-z_0)]. \quad (2.96)$$

Inserting these expressions into (2.93) and performing integrations, one gets

$$P(\omega,t) = -\frac{\omega e^2}{\pi\epsilon}[\Theta(-t-t_0)P_1+\Theta(t-t_0)P_2+\Theta(t+t_0)\Theta(t_0-t)P_3], \quad (2.97)$$

where

$$P_1 = -\frac{\sin\omega(t+t_0)}{c_n}+\frac{\sin 2\omega t_0\beta_n}{2\omega t_0 v}\sin\omega(t-t_0)$$

$$+\frac{1}{2v}\cos\omega(t+t_0)\{\text{si}[2t_0\omega(1+\beta_n)]-\text{si}[2t_0\omega(1-\beta_n)]\}$$

$$+\frac{1}{2v}\sin\omega(t+t_0)\left\{\frac{1}{2}\ln\left(\frac{1+\beta_n}{1-\beta_n}\right)^2+\text{ci}[2\omega t_0|1-\beta_n|]-\text{ci}[2\omega t_0(1+\beta_n)]\right\},$$

$$P_2 = \frac{\sin\omega(t-t_0)}{c_n}-\frac{\sin 2\omega t_0\beta_n}{2vt_0\omega}\sin\omega(t+t_0)$$

$$+\frac{1}{2v}\cos\omega(t-t_0)\{\text{si}[2t_0\omega(1+\beta_n)]-\text{si}[2t_0\omega(1-\beta_n)]\}$$

$$-\frac{1}{2v}\sin\omega(t-t_0)\left\{\frac{1}{2}\ln\left(\frac{1+\beta_n}{1-\beta_n}\right)^2+\text{ci}[2\omega t_0|1-\beta_n|]-\text{ci}[2\omega t_0\omega(1+\beta_n)]\right\},$$

$$P_3 = -\frac{\sin\omega\beta_n(t+t_0)}{v(t+t_0)}\frac{\sin\omega(t+t_0)}{\omega}+\frac{\sin\omega\beta_n(t-t_0)}{v(t-t_0)}\frac{\sin\omega(t-t_0)}{\omega}$$

$$-\frac{1-\beta_n^2}{2v}\{\text{si}[(1-\beta_n)\omega(t_0-t)]-\text{si}[(1+\beta_n)\omega(t_0-t)]$$

$$\text{si}[(1-\beta_n)\omega(t_0+t)]-\text{si}[(1+\beta_n)\omega(t_0+t)]\}. \quad (2.98)$$

Here $\text{si}(x)$ and $\text{ci}(x)$ are the integral sine and cosine. They are defined by the equations

$$\text{si}(x) = -\int_x^\infty\frac{\sin t}{t}dt = -\frac{\pi}{2}+\int_0^x\frac{\sin t}{t}dt = -\frac{\pi}{2}-\sum_{k=1}^\infty\frac{(-1)^k}{(2k-1)(2k-1)!}x^{2k-1},$$

$$\text{ci}(x) = -\int_x^\infty\frac{\cos t}{t}dt = C+\ln x-\int_0^x\frac{1-\cos t}{t}dt = C+\ln x+\sum_{k=1}^\infty\frac{(-1)^k}{2k(2k)!}x^{2k}.$$

Here $C\approx 0.577$ is Euler's constant. For large and small x, $\text{si}(x)$ and $\text{ci}(x)$ behave as

$$\text{si}(x)\to-\frac{\cos x}{x}-\frac{\sin x}{x^2}, \quad \text{ci}(x)\to\frac{\sin x}{x}-\frac{\cos x}{x^2} \quad \text{for}\quad x\to+\infty,$$

$$\text{si}(x) \to -\pi + \frac{\cos x}{|x|} + \frac{\sin |x|}{x^2} \quad \text{for} \quad x \to -\infty,$$

$$\text{si}(x) \to -\frac{\pi}{2} + x, \quad \text{ci}(x) \to C + \ln x - \frac{x^2}{4} \quad \text{for} \quad x \to 0.$$

The following relations

$$\int_0^x \frac{\sin^2 t}{t} dt = \frac{1}{2}C + \frac{1}{2}\ln 2|x| - \frac{1}{2}\text{ci}(2|x|), \quad \text{si}(x) + \text{si}(-x) = -\pi$$

will be also useful.

The nonvanishing of P_1 and P_2 terms in (2.97) is because the Fourier transforms of a static charge density corresponding to charge at rest prior to the beginning of the charge motion ($t < -t_0$) and after its termination ($t > t_0$) contribute to (2.93) and (2.95). To see this explicitly we write out the Fourier transform of charge density (2.96):

$$\rho(\vec{r}, \omega) = \frac{1}{2\pi} \int_{-\infty}^{\infty} \exp(-i\omega t) \rho(\vec{r}, t) dt =$$

$$= \frac{1}{2\pi} e\delta(x)\delta(y)[\delta(z + z_0) \int_{-\infty}^{-t_0} \exp(-i\omega t) dt + \delta(z - z_0) \int_{t_0}^{\infty} \exp(-i\omega t) dt$$

$$+ \frac{1}{v}\Theta(z + z_0)\Theta(z_0 - z) \exp(-i\omega z/v)].$$

The first term in the r.h.s. corresponds to the charge which is at rest at the point $z = -z_0$ up to an instant $t = -t_0$; the second term in the r.h.s. corresponds to the charge which is at rest at the point $z = z_0$ after the instant $t = t_0$. Finally, the third term corresponds to the charge moving between $-z_0$ and z_0 points in the time interval $-t_0 < t < t_0$. It should be noted that the first and second terms in this expression are Fourier densities of a charge which is not permanently at rest at the points $z = \pm z_0$, but up to a instant $-t_0$ and after the instant t_0, respectively. In fact, the Fourier density corresponding to charge which is permanently at rest at the point $z = z_0$ is

$$\frac{e}{2\pi}\delta(z - z_0) \int_{-\infty}^{\infty} \exp(i\omega t) dt = e\delta(z - z_0)\delta(\omega).$$

In the limit $\omega t_0 \to \infty$ Eqs (2.98) pass into

$$P_1 = -\frac{1}{c_n} \sin[\omega(t + t_0)] \left(1 - \frac{1}{2\beta_n} \ln \frac{1 + \beta_n}{1 - \beta_n}\right),$$

$$P_2 = +\frac{1}{c_n} \sin[\omega(t - t_0)] \left(1 - \frac{1}{2\beta_n} \ln \frac{1 + \beta_n}{1 - \beta_n}\right), \quad P_3 = 0$$

for $\beta_n < 1$ and

$$P_1 = -\frac{1}{c_n} \sin[\omega(t + t_0)] \left(1 - \frac{1}{2\beta_n} \ln \frac{1 + \beta_n}{\beta_n - 1}\right) + \frac{\pi}{2v} \cos \omega(t + \frac{z_0}{v}),$$

$$P_2 = \frac{1}{c_n} \sin[\omega(t - t_0)] \left(1 - \frac{1}{2\beta_n} \ln \frac{1 + \beta_n}{\beta_n - 1}\right) + \frac{\pi}{2v} \cos \omega(t - \frac{z_0}{v}),$$

$$P_3 = -\frac{\pi}{v}(\beta_n^2 - 1) \tag{2.99}$$

for $\beta_n > 1$. It is seen that the energy radiated during the time interval $-t_1 < t < t_1$, $t_1 < t_0$ is equal to zero for $\beta_n < 1$ and to $2\omega v e^2 t_1 (1 - 1/\beta_n^2)/c^2$ for $\beta_n > 1$. This coincides exactly with the VCR spectrum for the unbounded charge motion (see, e.g., Frank's book [1]). It should be noted that expressions for P_3 in (2.99) were obtained under the assumption that the arguments of si and ci entering into P_3 (see (2.98)) are sufficiently large, that is, there should be $\omega(t_0 - t) \gg 1$. This means that P_3 in (2.99) is valid if the observational instant t is not too close to t_0.

On the other hand, the terms P_1 and P_2 in (2.99) were obtained without this assumption. In particular, the term P_2 different from zero for $t > t_0$ shows how the bremsstrahlung (BS) and VCR behave for $t > t_0$, i.e., after termination of the charge motion. Since the part of P_2

$$\frac{1}{c_n} \sin\left[\omega\left(t - \frac{z_0}{v}\right)\right] \left(1 - \frac{1}{2\beta_n} \ln \frac{\beta_n + 1}{|\beta_n - 1|}\right)$$

is present both for $\beta_n < 1$ and $\beta_n > 1$, it may be associated with BS. On the other hand, the part of P_2

$$\frac{\pi}{2v} \cos\left[\omega\left(t - \frac{z_0}{v}\right)\right]$$

that differs from zero only for $\beta_n > 1$ may be conditionally attributed to the Cherenkov post-action.

We observe that for $t < -t_0$ and $t > t_0$ (P_1 and P_2 terms in (2.97)), the radiation intensity is a rapidly oscillating function of time t. The time average of this intensity is zero, so it could hardly be observed experimentally. Since, on the other hand, for $\beta_n > 1$ the term P_3 in the radiation intensity (2.97) does not depend on time in the time interval $-t_1 < t < t_1$ ($t_1 \ll t_0$), it contributes coherently to the radiated energy.

To obtain the energy radiated for a finite time interval, one should integrate (2.97) over t. However, the arising integrals involve integral sine and cosine functions. Since we did not succeed in evaluating these integrals in a closed form, we follow an indirect way in next sections. In subsection 2.5.2 we evaluate the instant angular-frequency distribution of the radiated energy. Integrating it over time we obtain (subsect. 2.5.3) the angular-frequency distribution of the energy radiated for a finite time interval. Finally, integrating the latter over angular variables we obtain a closed expression for the frequency distribution of the energy radiated for a finite time interval (Sect. 2.5.4).

2.5.2. INSTANTANEOUS ANGULAR-FREQUENCY DISTRIBUTION OF THE POWER SPECTRUM

Owing to the axial symmetry of the problem, $\vec{n}(\vec{r} - \vec{r}') = \cos\theta(z - z')$ in the integrand in (2.95), where θ is the inclination angle of $\vec{n}$ towards the motion axis. Integration over space-time variables in (2.95) gives

$$P(\vec{n}, \omega, t) = \frac{d^3\mathcal{E}}{dt\, d\omega\, d\Omega} = -\frac{\omega e^2 \beta}{2\pi^2 c} \frac{\sin[\omega t_0 (1 - \beta_n \cos\theta)]}{1 - \beta_n \cos\theta}$$

$$\times [\Theta(-t - t_0)P_{1n} + \Theta(t - t_0)P_{2n} + \Theta(t + t_0)\Theta(t_0 - t)P_{3n}]. \qquad (2.100)$$

Here

$$P_{1n} = \cos\theta \cos[\omega(t + t_0\beta_n \cos\theta)],$$

$$P_{2n} = \cos\theta \cos[\omega(t - t_0\beta_n \cos\theta)],$$

$$P_{3n} = (\cos\theta - \beta_n) \cos[\omega t(1 - \beta_n \cos\theta)].$$

2.5.3. ANGULAR-FREQUENCY DISTRIBUTION OF THE RADIATED ENERGY FOR A FINITE TIME INTERVAL

Integrating (2.100) over the observational time interval $-t_1 < t < t_1$, $t_1 < t_0$, one obtains the Fourier distribution of the energy detected for a time $2t_1$ radiated by a charge moving in the time interval $2t_0$ (it is suggested that the observational interval is smaller than the motion one):

$$\mathcal{E}(\vec{n}, \omega, t_1) = \int_{-t_1}^{t_1} P(\vec{n}, \omega, t)\, dt$$

$$= \frac{e^2 \beta}{\pi^2 c}(\beta_n - \cos\theta)\frac{\sin\omega t_0(1 - \beta_n \cos\theta)}{1 - \beta_n \cos\theta}\frac{\sin\omega t_1(1 - \beta_n \cos\theta)}{1 - \beta_n \cos\theta}. \qquad (2.101)$$

Let $\omega t_0 \to \infty$. Then

$$\mathcal{E}(\vec{n}, \omega, t_1) \to \frac{e^2 \beta \omega t_1}{\pi c}\left(1 - \frac{1}{\beta_n^2}\right)\delta\left(\cos\theta - \frac{1}{\beta_n}\right). \qquad (2.102)$$

This coincides with the angular-frequency distribution of the radiated energy in Tamm-Frank theory [11] describing the unbounded charge motion. For $\cos\theta = 1/\beta_n$ Eq. (2.101) reduces to

$$\mathcal{E}(\vec{n}, \omega, t_1) = \frac{e^2}{\pi n c}(\beta_n^2 - 1)\omega^2 t_0 t_1.$$

It vanishes for $\beta_n = 1$.

Let the observational time be greater than the charge motion interval $(t_1 > t_0)$. Then,

$$\mathcal{E}(\vec{n}, \omega, t_1) = \frac{e^2\beta}{\pi^2 c}\frac{\sin[\omega t_0(1 - \beta_n\cos\theta)]}{1 - \beta_n\cos\theta}$$

$$\times \left[\beta_n\sin^2\theta\frac{\sin\omega t_0(1 - \beta_n\cos\theta)}{1 - \beta_n\cos\theta} - \cos\theta\sin\omega(t_1 - t_0\beta_n\cos\theta)\right] \quad (2.103)$$

is the angular-frequency distribution of the energy detected for the time interval $2t_1 > 2t_0$. The first term in square brackets coincides with the Tamm angular distribution (2.29). The second term originating from integration of P_1 and P_2 terms in (2.100) describes the boundary effects. The physical reason for the appearance of the extra term in (2.103) (second term in square brackets) is owed to the following reason. The magnetic field $\vec{H}$ is defined as the curl of VP (2.83). Tamm obtained electric field from the Maxwell equation

$$\mathrm{curl}\vec{H} = \frac{\epsilon}{c}\frac{\partial\vec{E}}{\partial t}$$

valid outside the interval of motion. In the ω representation this equation looks like

$$\mathrm{curl}\vec{H}_\omega = \frac{i\omega\epsilon}{c}\vec{E}_\omega.$$

This equation suggests that contribution of static electric field existing before beginning of charge motion and after its termination has dropped from the Tamm formula (2.29) (because VP (2.25) and magnetic field (2.26) describe only the charge motion on the interval $(-z_0, z_0)$). On the other hand, Schwinger's equations (2.93) and (2.95) contain the static electric field contributions of a charge which is at rest up to the instant $t = -t_0$ and after the instant $t = t_0$. They are responsible for the appearance of extra term in (2.103). In the $\vec{r}, t$ representation, the contribution of the static electromagnetic field strengths is not essential in the wave zone. Taking into account that

$$\frac{\sin\alpha x}{x} \to \pi\delta(x) \quad \text{and} \quad \frac{1}{\alpha}\left(\frac{\sin\alpha x}{x}\right)^2 \to \pi\delta(x) \quad \text{for} \quad \alpha \to \infty, \quad (2.104)$$

one obtains from (2.103) for large ωt_0

$$\mathcal{E}(\vec{n}, \omega, t_1) = \frac{e^2}{\pi c n} \delta(1 - \beta_n \cos\theta)[\omega t_0(\beta_n^2 - 1) - \sin\omega(t_1 - \frac{z_0}{v})]. \quad (2.105)$$

For $\beta_n \neq 1$ the second term inside the square brackets may be discarded, and one obtains

$$\mathcal{E}(\vec{n}, \omega, t_1) = \frac{e^2}{\pi c n} \omega t_0(\beta_n^2 - 1)\delta(1 - \beta_n \cos\theta). \quad (2.106)$$

For $\cos\theta = 1/\beta_n$ Eq. (2.103) is reduced to

$$\mathcal{E}(\vec{n}, \omega, t_1) = \frac{e^2}{\pi n c}(\beta_n^2 - 1)\omega^2 t_0^2 - \frac{e^2}{\pi n c}\omega t_0 \sin\omega(t_1 - t_0).$$

It does not vanish at $\beta_n = 1$. Equations (2.101) and (2.103) generalize the Tamm angular-frequency distribution (2.29) for $t_1 \neq t_0$.

2.5.4. FREQUENCY DISTRIBUTION OF THE RADIATED ENERGY

Let $t_0 > t_1$ (i.e., the detection time is smaller than the motion time). Integrating (2.101) over the solid angle one finds the following expression for the frequency distribution of the radiated power:

$$\mathcal{E}(\omega, t_1) = \frac{e^2\beta}{\pi c}\left(1 - \frac{1}{\beta_n^2}\right)\{\frac{\cos(\omega(t_1 - t_0)(1 - \beta_n))}{1 - \beta_n} - \frac{\cos(\omega(t_0 - t_1)(1 + \beta_n))}{1 + \beta_n}$$

$$-\frac{\cos(\omega(t_1 + t_0)(1 - \beta_n))}{1 - \beta_n} + \frac{\cos(\omega(t_1 + t_0)(1 + \beta_n))}{1 + \beta_n}$$

$$+\omega(t_0 - t_1)[\mathrm{si}(\omega(t_0 - t_1)(1 - \beta_n)) - \mathrm{si}(\omega(t_0 - t_1)(1 + \beta_n))]$$

$$-\omega(t_0 + t_1)[\mathrm{si}(\omega(t_0 + t_1)(1 - \beta_n)) - \mathrm{si}(\omega(t_0 + t_1)(1 + \beta_n))]\}$$

$$-\frac{e^2}{\pi \epsilon v}[\mathrm{ci}(\omega(t_0 - t_1)|1 - \beta_n|) - \mathrm{ci}(\omega(t_0 - t_1)(1 + \beta_n))$$

$$-\mathrm{ci}(\omega(t_0 + t_1)|1 - \beta_n|) + \mathrm{ci}(\omega(t_0 + t_1)(1 + \beta_n))]. \quad (2.107)$$

Now let $t_1 > t_0$ (i.e., the detection time is greater than the motion time). Then,

$$\mathcal{E}(\omega, t_1) = \frac{2e^2\beta}{\pi c}(\beta_n I_1 - I_2), \quad (2.108)$$

where

$$I_1 = \int \sin^3\theta d\theta[\frac{\sin\omega t_0(1 - \beta_n \cos\theta)}{1 - \beta_n \cos\theta}]^2 = \frac{1}{\beta_n}\left(1 - \frac{1}{\beta_n^2}\right)$$

$$\times\{\frac{\sin^2 \omega t_0(1-\beta_n)}{1-\beta_n} - \frac{\sin^2 \omega t_0(1+\beta_n)}{1+\beta_n} - \omega t_0[\mathrm{si}(2\omega t_0(1-\beta_n)) - \mathrm{si}(2\omega t_0(1+\beta_n))]\}$$

$$-\frac{1}{\beta_n^3}\left[\ln\frac{|1-\beta_n|}{1+\beta_n} - \mathrm{ci}(2\omega t_0|1-\beta_n|) + \mathrm{ci}(2\omega t_0(1+\beta_n))\right]$$

$$-\frac{1}{\beta_n^2} - \frac{1}{4\beta_n^3\omega t_0}[\sin(2\omega t_0(1-\beta_n)) - \sin(2\omega t_0(1+\beta_n))],$$

$$I_2 = \int \sin\theta \cos\theta d\theta \frac{\sin \omega t_0(1-\beta_n \cos\theta)\sin \omega(t_1 - t_0\beta_n \cos\theta)}{1-\beta_n \cos\theta}$$

$$= -\frac{1}{4\beta_n^2\omega t_0}\sin \omega(t_1 - t_0)[\cos(2\omega t_0(1-\beta_n)) - \cos(2\omega t_0(1+\beta_n))]$$

$$-\frac{1}{\beta_n}\cos \omega(t_1 - t_0) - \frac{1}{4\beta_n^2\omega t_0}\cos \omega(t_1 - t_0)[\sin(2\omega t_0(1-\beta_n)) - \sin(2\omega t_0(1+\beta_n))]$$

$$-\frac{1}{2\beta_n^2}\sin \omega(t_1 - t_0)[\mathrm{si}(2\omega t_0(1-\beta_n)) - \mathrm{si}(2\omega t_0(1+\beta_n))]$$

$$-\frac{1}{2\beta_n^2}\cos \omega(t_1 - t_0)[\ln\frac{|1-\beta_n|}{1+\beta_n} - \mathrm{ci}(2\omega t_0|1-\beta_n|) + \mathrm{ci}(2\omega t_0(1+\beta_n))].$$

The typical dependence of $\mathcal{E}$ on t_0 for t_1 fixed is shown in Fig. 2.19.

For large ωt_0 and $\beta_n < 1$, it oscillates around zero. For large ωt_0 and $\beta_n > 1$, $\mathcal{E}$ oscillates around the value

$$\frac{2e^2\omega t_1\beta}{c}\left(1 - \frac{1}{\beta_n^2}\right),$$

given by the Tamm-Frank theory [1]. In both cases the amplitude of oscillations decreases like $1/\omega t_0$ for large t_0. The typical dependence of $\mathcal{E}$ on t_1 for t_0 fixed is shown in Fig. 2.20.

Since I_2 is a periodic function of t_1 and I_1 does not depend on t_1, $\mathcal{E}$ oscillates around the value $2e^2\beta^2 n I_1/\pi c$. Previously the frequency distribution of the radiated energy in the framework of the Tamm theory was given by Kobzev and Frank [18] and by Kobzev et al [19]. It is obtained by integrating the Tamm angular distribution (2.29) over the angular variables:

$$\frac{d\mathcal{E}}{d\omega} = \frac{2e^2\beta}{\pi c}(1 - \frac{1}{\beta_n^2})\{\frac{\sin^2 \omega t_0(1-\beta_n)}{1-\beta_n} - \frac{\sin^2 \omega t_0(1+\beta_n)}{1+\beta_n}$$

$$-\omega t_0[\mathrm{si}(2\omega t_0(1-\beta_n)) - \mathrm{si}(2\omega t_0(1+\beta_n))]\}$$

$$-\frac{2e^2}{\pi c n^2\beta}\left[\ln\frac{|1-\beta_n|}{1+\beta_n} - \mathrm{ci}(2\omega t_0|1-\beta_n|) + \mathrm{ci}(2\omega t_0(1+\beta_n))\right]$$

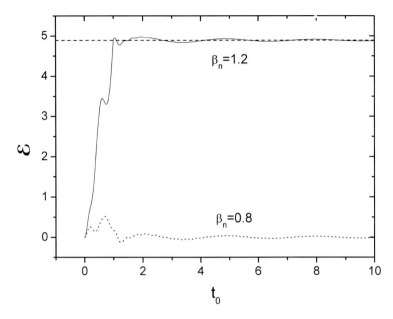

Figure 2.19.   Energy $\mathcal{E}$ detected in a fixed time interval t_1 as a function of the charge motion time t_0. For $\beta_n < 1$, $\mathcal{E}$ oscillates around zero. For $\beta_n > 1$ it oscillates around the finite value (2.31). The amplitude of oscillations decreases like $1/\omega t_0$ for a large time of motion t_0. $\mathcal{E}$ is given in units of e^2/c, t_0 in units of t_1.

$$-\frac{e^2}{\pi c n^2 \beta} \left\{ 2\beta_n + \frac{1}{2\omega t_0} \left[\sin 2\omega t_0 (1 - \beta_n)) - \sin 2\omega t_0 (1 + \beta_n)) \right] \right\}. \quad (2.109)$$

This expression coincides with the first term in (2.108) which involves I_1. For large ωt_0, (2.109) goes into the Tamm equations (2.29).

The frequency dependences of the energy radiated for the time t_1 and given by (2.108) are shown in Figs. 2.21 and 2.22. In Fig. 2.21 one sees the frequency dependence for the case when the observational time $2t_1$ is twice as small as the charge motion time $2t_0$. For $\beta_n < 1$, the radiated energy is concentrated near zero, while for $\beta_n > 1$ it rises linearly with frequency

$$\mathcal{E} \sim \frac{2e^2 \omega t_1 \beta}{c} \left(1 - \frac{1}{\beta_n^2} \right).$$

The frequency dependence for the case when the observational time $2t_1$ is twice as large as the charge motion time $2t_0$ is shown in Fig. 2.22.   For

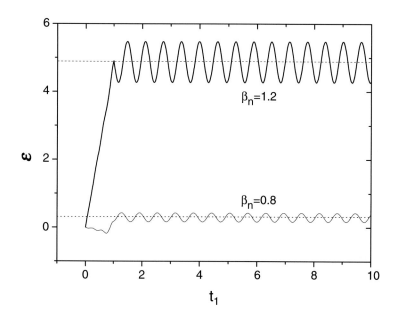

Figure 2.20. Energy $\mathcal{E}$ as a function of the detection time t_1 for the fixed time of motion t_0. The time interval of motion t_0 is fixed. For $\beta_n < 1$ and $\beta_n > 1$, $\mathcal{E}$ oscillates around the Tamm values (2.5) and (2.6), respectively. Contrary to the previous figure, there is no damping of oscillations. $\mathcal{E}$ is given in units of e^2/c; t_1, in units of t_0.

$\beta_n < 1$ the radiated energy oscillates around the Tamm value

$$\frac{2e^2}{\pi c \beta n^2}\left(\ln\frac{1+\beta_n}{1-\beta_n} - 2\beta_n\right),$$

whilst for $\beta_n > 1$ it again rises linearly but with a coefficient different from the case $t_1 < t_0$:

$$\mathcal{E} \sim \frac{2e^2\omega t_0\beta}{c}\left(1 - \frac{1}{\beta_n^2}\right).$$

It is interesting to compare the frequency distribution (2.109) obtained by integration the Tamm angular-frequency distribution over the solid angle with its approximate version (2.31) given by Tamm. Equation (2.31) has a singularity at $\beta = 1/n$, whilst (2.109) is not singular there. To see how they agree with each other we present them and their difference (Fig. 2.23) as a function of the velocity β for the parameters $L = 2z_0 = 0.1$ cm and

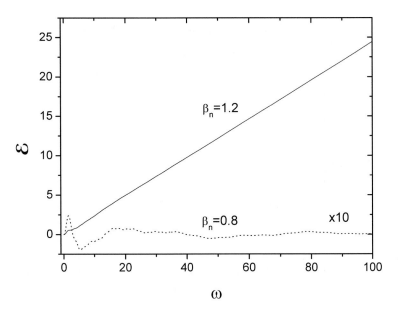

Figure 2.21.   Frequency dependence of the radiated energy for $t_1/t_0 = 0.5$. $\mathcal{E}$ is given in units of e^2/c; ω, in units of $1/t_0$.

$\lambda = 4 \cdot 10^{-5}$ cm used above. It is seen that they coincide with each other everywhere except for the closest vicinity of $\beta = 1/n$.

Large interval of motion
Let the observational time be less than the motion time $(t_1 < t_0)$. Then, for $\omega(t_0 - t_1) \gg 1$, $\mathcal{E}(\omega, t_1)$ is very small for $\beta_n < 1$. On the other hand, for $\beta_n > 1$,

$$\mathcal{E}(\omega, t_1) = \frac{2\omega t_1 e^2 \beta}{c}\left(1 - \frac{1}{\beta_n^2}\right). \tag{2.110}$$

This coincides with the frequency distribution of the radiated energy during the whole charge motion in the Frank-Tamm theory.

Let now the observational time be greater than the motion time $(t_1 > t_0)$. Then, for $\omega t_0 \gg 1$ (but $t_1 > t_0$) one finds

$$\mathcal{E}(\omega, t_1) \approx -\frac{2e^2}{\pi c n}[2 - \cos\omega(t_1 - t_0)]\left(1 + \frac{1}{2\beta_n}\ln\frac{1 - \beta_n}{1 + \beta_n}\right) \tag{2.111}$$

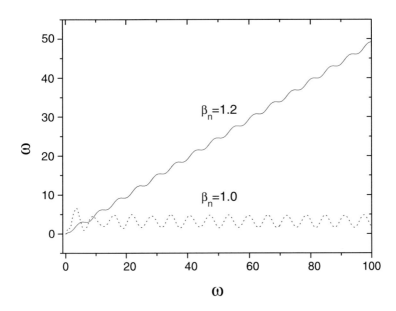

Figure 2.22.   Frequency dependence of the radiated energy for $t_1/t_0 = 2$. $\mathcal{E}$ is given in units of e^2/c; ω, in units of $1/t_0$.

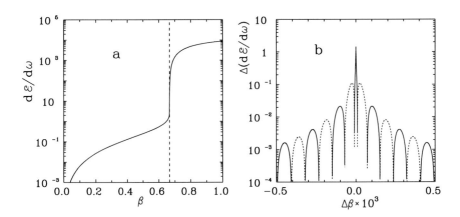

Figure 2.23.   (a) Frequency distributions of the radiated energy (in e^2/c units) given by (2.109) and its simplified version (2.31) as functions of the charge velocity. They are indistinguishable in this scale; (b) the difference between (2.31) and (2.109). The regions where this difference is negative are shown by dotted lines; $\Delta\beta$ means $\beta - 1/n$.

for $\beta_n < 1$ and

$$\mathcal{E}(\omega, t_1) \approx \frac{2e^2\beta}{\pi c}\{\pi\omega t_0(1 - \frac{1}{\beta_n^2})$$

$$-\frac{1}{\beta_n}[2 - \cos\omega(t_1 - t_0)]\left(1 + \frac{1}{2\beta_n}\ln\frac{\beta_n - 1}{1 + \beta_n}\right) - \frac{\pi}{2\beta_n^2}\sin\omega(t_1 - t_0)\} \quad (2.112)$$

for $\beta_n > 1$.

Non-oscillating parts of these expressions coincide with Eqs. (2.31) given by Tamm. According to his own words, Eqs. (2.31)

are obtained by neglecting the fast-oscillating terms of the form $\sin\omega t_0$ (Tamm gives only Eqs.(2.31) without deriving them). On the other hand, Eq.(2.109) obtained in [18,19] gives, in the limit $\omega t_0 \to \infty$, the Tamm expressions (2.31) with additional oscillating terms decreasing like $1/\omega t_0$.

Since some terms in (2.107) and (2.108) depend on the parameters $(1 - \beta_n)(t_0 - t_1)$ and $(1 - \beta_n)(t_0 + t_1)$, Eqs.(2.110)-(2.112) are not valid for $\beta_n \sim 1$ (this corresponds to Cherenkov's threshold).

Frequency distribution at the Cherenkov threshold
Thus, the case $\beta_n = 1$ needs a special consideration. One obtains

$$\mathcal{E}(\omega, t_1) = -\frac{e^2}{\pi nc}\left[\ln\frac{t_0 - t_1}{t_0 + t_1} - \text{ci}(2\omega(t_0 - t_1)) + \text{ci}(2\omega(t_0 + t_1))\right] \quad (2.113)$$

for $t_1 < t_0$. This expression tends to zero for t_1 fixed and $t_0 \to \infty$.
On the other hand, for $t_1 > t_0$

$$\mathcal{E}(\omega, t_1) = \frac{2e^2}{\pi nc}\{\left[1 - \frac{1}{2}\cos\omega(t_1 - t_0)\right][C + \ln(4\omega t_0) - \text{ci}(4\omega t_0)]$$

$$-[1 - \cos\omega(t_1 - t_0)]\left[1 - \frac{\sin(4\omega t_0)}{4\omega t_0}\right] + \sin\omega(t_1 - t_0)$$

$$\times\left[\frac{1 - \cos(4\omega t_0)}{4\omega t_0} - \frac{\pi}{4} - \frac{1}{2}\sin(4\omega t_0)\right]\}. \quad (2.114)$$

The non-oscillating part of this expression coincides with that given by Tamm [1]:

$$\mathcal{E}_{\text{T}} = \frac{2e^2}{\pi nc}[C + \ln(4\omega t_0) - 1].$$

On the other hand, Eq.(2.111) obtained by Kobzev and Frank for $\beta_n = 1$ goes into

$$\mathcal{E}_{\text{KF}} = \frac{2e^2}{\pi nc}\left[C + \ln(4\omega t_0) - 1 - \text{ci}(4\omega t_0) + \frac{\sin(4\omega t_0)}{4\omega t_0}\right].$$

For $(t_1 - t_0)$ fixed and $t_0 \to \infty$, Eq.(2.114) is reduced to

$$\mathcal{E}(\omega, t_1) \to \frac{2e^2}{\pi n c}\left\{\left[1 - \frac{1}{2}\cos\omega(t_1 - t_0)\right][C + \ln(4\omega t_0)] - 1 + \right.$$

$$\left. \cos\omega(t_1 - t_0) - \sin\omega(t_1 - t_0)\left[\frac{\pi}{4} + \frac{1}{2}\sin(4\omega t_0)\right]\right\}. \tag{2.115}$$

In the limit $t_0 \to \infty$, $\mathcal{E}_{KF}$ goes into $\mathcal{E}_T$ plus oscillating terms decreasing like $1/\omega t_0$.

The main result of this consideration is that the Schwinger approach incorporates both Tamm-Frank and Tamm problems. The Tamm-Frank results are obtained when the observational time t_1 is smaller than the charge motion time t_0 and $t_0 \to \infty$. In particular, there is no radiation when the charge velocity is smaller than the velocity of light in medium. The radiated energy rises in direct proportion to the observational time t_1 for $\beta_n > 1$. The Tamm problem is obtained when $t_1 > t_0$ and t_0 (and therefore t_1) tends to ∞. The intensity oscillates around the Tamm value for $\beta_n < 1$ and rises in proportion to the time of charge motion t_0 for $\beta_n > 1$.

2.6. The Tamm problem in the spherical basis

2.6.1. EXPANSION OF THE TAMM PROBLEM IN TERMS OF THE LEGENDRE POLYNOMIALS

We need the expansion of the Green function

$$G = \exp(ik_n R)/R, \quad R = |\vec{r} - \vec{r}'|$$

in spherical coordinates. It is given by

$$G = 2\sum_{m \geq 0} \epsilon_m(2l + 1)\frac{(l - m)!}{(l + m)!}\cos m(\phi - \phi')$$

$$\times G_l(r, r')P_l^m(\cos\theta)P_l^m(\cos\theta'), \tag{2.116}$$

where

$$G_l(r, r') = ik_n j_l(k_n r_<)h_l(k_n r_>),$$

$$j_l(x) = \sqrt{\frac{\pi}{2x}}J_{l+1/2}(x) \quad \text{and} \quad h_l(x) = \sqrt{\frac{\pi}{2x}}H_{l+1/2}^{(1)}(x)$$

are the spherical Bessel and Hankel functions; $P_l^m(x)$ is the adjoint Legendre polynomial.

Let a charge move in medium in a finite interval $(-z_0, z_0)$ (this corresponds to the so-called Tamm problem). The current density corresponding to the Tamm problem, in cartesian coordinates is given by

$$j_z(\omega) = \frac{e}{2\pi} \exp(i\omega z/v)\delta(x)\delta(y)\Theta(z+z_0)\Theta(z_0-z).$$

We rewrite this in spherical coordinates:

$$j_z(\omega) = \frac{e}{4\pi^2 r^2 \sin\theta} \left[\delta(\theta)\exp(\frac{ikr}{\beta}) + \delta(\theta-\pi)\exp(-\frac{ikr}{\beta})\right]\Theta(z_0-r).$$

Then on the sphere of the radius $r > z_0$ one obtains

$$A_z(\omega) = \frac{ie\mu kn}{2\pi c}\sum_l (2l+1)P_l h_l(knr)J_l(0, z_0),$$

$$H_\phi(\omega) = -\frac{iek^2 n^2}{2\pi c}\sum_l P_l^1 h_l(knr)\tilde{J}_l(0, z_0),$$

$$E_\theta(\omega) = -\frac{ek^2 \mu n}{2\pi c}\sum_l P_l^1 H_l(knr)\tilde{J}_l(0, z_0). \qquad (2.117)$$

Here

$$J_l(0, z_0) = \int_0^{z_0} j_l(knr')f_l(r')dr', \quad \tilde{J}_l(0, z_0) = J_{l-1}(0, z_0) + J_{l+1}(0, z_0),$$

$$f_l(r') = \exp(\frac{ikr'}{\beta}) + (-1)^l \exp(\frac{ikr'}{\beta}),$$

$$H_l(x) = \dot{h}_l(x) + \frac{h_l(x)}{x} = \frac{1}{2l+1}[(l+1)h_{l-1}(x) - lh_{l+1}(x)].$$

In (2.117) and further on, we omit the arguments of the Legendre polynomials if they are equal to $\cos\theta$ (θ is the observational angle). At large distances ($kr \gg 1$)

$$A_z \sim \frac{e\mu}{2\pi cr}\exp(iknr)\sum_l (2l+1)i^{-l}P_l J_l(0, z_0),$$

$$H_\phi \sim -\frac{ekn}{2\pi cr}\exp(iknr)\sum_l i^{-l}P_l^1 \tilde{J}_l(0, z_0),$$

$$E_\theta \sim -\frac{ek\mu}{2\pi cr}\exp(iknr)\sum_l i^{-l}P_l^1 \tilde{J}_l(0, z_0).$$

The distribution of the radiation intensity on the sphere of the radius r

$$\frac{d^2\mathcal{E}}{d\omega d\Omega} = \frac{1}{2}cr^2(E_\theta H_\phi^* + \text{c.c.}) = \frac{e^2k^2n\mu}{4\pi^2 c}|\sum i^{-1}P_1^1\tilde{J}_1(0, z_0)|^2$$

$$= \frac{e^2k^2n\mu\sin^2\theta}{4\pi^2 c}|\sum (2l+1)i^{-l}P_lJ_l(0, z_0)|^2. \tag{2.118}$$

Or, in a manifest form,

$$\frac{d^2\mathcal{E}}{d\omega d\Omega} = \frac{e^2n\mu}{\pi^2 c}\sin^2\theta(S_1 + S_2)^2. \tag{2.119}$$

where

$$S_1 = \sum_{l=0}^\infty (-1)^l(4l+1)P_{2l}(\cos\theta)I_{2l}^c, \quad S_2 = \sum_{l=0}^\infty (-1)^l(4l+3)P_{2l+1}(\cos\theta)I_{2l+1}^s,$$

$$I_{2l}^c = \int_0^{kz_0} j_{2l}(nx)\cos(\frac{x}{\beta})dx, \quad I_{2l+1}^s = \int_0^{kz_0} j_{2l+1}(nx)\sin(\frac{x}{\beta})dx. \tag{2.120}$$

Integrating over the solid angle, one obtains the frequency distribution of the radiation:

$$\frac{d\mathcal{E}}{d\omega} = \frac{e^2k^2n\mu}{\pi c}\sum \frac{l(l+1)}{2l+1}|\tilde{J}_l(0, z_0)|^2$$

$$= \frac{8e^2n\mu}{\pi c}(I_c + I_s), \tag{2.121}$$

where

$$I_c = \sum \frac{(l+1)(2l+1)}{4l+3}(I_{2l}^c + I_{2l+2}^c)^2$$

and

$$I_s = \sum \frac{l(2l+1)}{4l+1}(I_{2l+1}^s + I_{2l-1}^s)^2.$$

These equations are valid if the radius r of the observational sphere is larger than z_0. Eqs. (2.121) and (2.109) should coincide since the same approximations were involved in their derivation. Numerical calculations support this claim.

We concentrate now on the vector potential. For this we rewrite it as

$$A_\omega = \frac{ie\mu n}{\pi c}\sum_{l=0}^\infty (4l+1)h_{2l}(knr)P_{2l}(\cos\theta)I_{2l}^c$$

$$-\frac{e\mu n}{\pi c}\sum_{l=0}^{\infty}(4l+3)h_{2l+1}(knr)P_{2l+1}(\cos\theta)I_{2l+1}^s \qquad (2.122)$$

Usually observations are made at large distances ($kr \gg 1$). For example, for $\lambda = 4 \times 10^{-5}$ cm and $r = 1$ m, $kr = 2\pi r/\lambda \sim 10^7$. Changing the Hankel functions by their asymptotic values, one finds

$$A_\omega = \frac{e\mu}{kr\pi c}\exp(iknr)(S_1 + S_2). \qquad (2.123)$$

Obviously vector potentials (2.123) and (2.26) are the same (since the same approximations are involved in their derivation). Equating them one has

$$S_1 + S_2 = \frac{1}{n}\frac{\sin[kz_0 n(\cos\theta - 1/\beta n)}{\cos\theta - 1/\beta n}. \qquad (2.124)$$

Now we consider the coefficients I_{2l}^c and I_{2l+1}^s. In the limit $kz_0 \to \infty$ the integrals defining I_c and I_s are:

$$I_{2l}^c = \int_0^{\infty}j_{2l}(nx)\cos(\frac{x}{\beta})dx, \; I_{2l+1}^s = \int_0^{\infty}j_{2l+1}(nx)\sin(\frac{x}{\beta})dx.$$

These integrals can be evaluated in a closed form (see, e.g., [20]). They are given 0 for $\beta n < 1$ and

$$I_{2l}^c = \frac{\pi}{2n}(-1)^l P_{2l}(1/\beta n), \; I_{2l+1}^s = \frac{\pi}{2n}(-1)^l P_{2l+1}(1/\beta n)$$

for $\beta n > 1$. Substituting them into (2.122) one obtains

$$A_\omega = \frac{e\mu}{2nkrc}\exp(iknr)$$

$$\times\left[\sum_{l=0}^{\infty}(4l+1)P_{2l}(\cos\theta)P_{2l}(\frac{1}{\beta n}) + \sum_{l=0}^{\infty}(4l+3)P_{2l+1}(\cos\theta)P_{2l+1}(\frac{1}{\beta n})\right]$$

$$=\frac{e\mu}{2nkrc}\exp(iknr)\sum_{l=0}^{\infty}(2l+1)P_l(\cos\theta)P_l\left(\frac{1}{\beta n}\right)$$

$$=\frac{e\mu}{nkrc}\exp(iknr)\delta\left(\cos\theta - \frac{1}{\beta n}\right). \qquad (2.125)$$

In deriving this, we used the relation

$$\sum_{l=0}^{\infty}(l+1/2)P_l(x)P_l(x') = \delta(x - x').$$

Vector potential (2.125) coincides with the one entering (2.26').

2.7. Short résumé of this chapter

What can we learn from this chapter?

1. The approximate Tamm formula (2.29) for the energy radiated by a moving charge in a finite interval $(-z_0, z_0)$ describes the interference of two BS shock waves arising at the beginning and termination of motion and does not describe the CSW properly. However, some reservation is needed. In the next chapter the instantaneous velocity jumps of the original Tamm problem will be replaced by the velocity linearly rising (or decreasing) with time. It will be shown there that, in addition to the BS shock wave arising at the beginning of the motion, two new shock waves arise at the instant when the charge velocity coincides with the velocity of light in medium. Owing to the instantaneous jump in velocity in the original Tamm problem, the above three shock waves are created simultaneously. When discussing the BS shock waves throughout this chapter, we implied the mixture of these three shock waves.

2. The exact solution of the Tamm problem contains the Cherenkov shock wave in addition to the BS shock waves. This Cherenkov shock wave propagates between two straight lines L_1 and L_2 originating from the boundary points $\pm z_0$ of the interval of motion and inclined at the angle θ_c, $\cos \theta_c = 1/\beta n$ towards the motion axis.

3. Applying the Schwinger approach to the solution of the Tamm problem, we have found that angular-frequency distributions of the energy radiated by a moving charge depend not only on the interval of motion but also on the observational time interval. This should be kept in mind when discussing the experimental results.

4. We have made an expansion of the electromagnetic field and radiation intensity corresponding to the Tamm problem in terms of Legendre polynomials. This will be used in Chapter 7.

References

1. Frank I.M. (1988) *Vavilov-Cherenkov Radiation*, Nauka, Moscow.
2. Sommerfeld A. (1905) Zur Elektronentheorie. III. Ueber Lichtgeschwindigkeits- und Ueberlichtgeschwindigkeits-Elektronen *Gotting. Nachricht.*, pp. 201-235.
3. Heaviside O. (1922) *Electromagnetic Theory*, Benn Brothers Ltd, London (Reprinted Edition).
4. Tamm I.E. (1939) Radiation emitted by Uniformly Moving Electrons, *J. Phys. USSR*, **1, No 5-6**, pp. 439-461.
5. Frank I.M. and Tamm I.E. (1937) Coherent Visible Radiation of Fast Electrons Passing through Matter *Dokl. Akad. Nauk SSSR*, **14**, pp. 107-113.
6. Volkoff G.M. (1963) Electric Field of a Charge Moving in Medium *Amer.J.Phys.*,**31**, pp.601-605.

7. Afanasiev G.N., Beshtoev Kh. and Stepanovsky Yu.P. (1996) Vavilov-Cherenkov Radiation in a Finite Region of Space *Helv. Phys. Acta*, **69**, pp. 111-129.
8. Zrelov V.P. and Ruzicka J. (1992) Optical Bremsstrahlung of Relativistic Particles in a Transparent Medium and its Relation to the Vavilov-Cherenkov Radiation *Czech. J. Phys.*, **42**, pp. 45-57.
9. Lawson J.D. (1954) On the Relation between Cherenkov Radiation and Bremsstrahlung *Phil. Mag.*, **45**, pp.748-750.
10. Lawson J.D. (1965) Cherenkov Radiation, "Physical" and "Unphysical", and its Relation to Radiation from an Accelerated Electron *Amer. J. Phys.*, **33**, pp. 1002-1005.
11. Ginzburg V.L. (1940) Quantum Theory of Radiation of Electron Uniformly Moving in Medium, *Zh. Eksp. Teor. Fiz.*, **10** pp. 589-600.
12. Akhiezer A.I. and Berestetzky V.B., 1981, *Quantum Electrodynamics*, Nauka, Moscow.
13. Glauber R., 1965, in *Quantum Optics and Electronics* (Lectures delivered at Les Houches, 1964, Eds.: C.DeWitt, A.Blandin and C.Cohen-Tannoudji), pp.93-279, Gordon and Breach, New York.
14. Zrelov V.P. and Ruzicka J. (1989) Analysis of Tamm's Problem on Charge Radiation at its Uniform Motion over a Finite Trajectory *Czech. J. Phys.*, **B 39**, pp. 368-383.
15. Skobeltzyne D.V. (1975) Sur l'impulsiom-énergie du photon et l'equilibr/'e thermodynamique du champ de radiation dans le milieu réfrigent *C.R. Acad. Sci.Paris*, **B 280**, pp.251-254;
    Skobeltzyne D.V. (1975) Sur les postulats d'Einstein concernant l'emission induite de la lumiére dans un milieu réfrigent et le tenseur d'impulsion-énergie du champ de radiation dans ce milieu *C.R. Acad. Sci.Paris*, **B 280**, pp.287-290;
    Skobeltzyne D.V. (1977) Quantum Theory Paradoxes of the Vavilov-Cherenkov and Doppler Effects *Usp. Fiz. Nauk*, **122**, pp. 295-324.
16. Fano R.M., Chu L.J. and Adler R.B. (1960) *Electromagnetic fields, energy and forces*, John Wiley, New York.
17. Schwinger J. (1949) On the Classical Radiation of Accelerated Electrons *Phys.Rev.*,**A 75**, pp. 1912-1925.
18. Kobzev A.P. and Frank I.M. (1981) Some Peculiarities of the Vavilov-Cherenkov Radiation due to the Finite Thickness of the Radiator *Yadernaja Fizika*, **34**, pp. 125-133.
19. Kobzev A.P., Krawczyk A. and Rutkowski J. (1988) Charged Particle Radiation along a Finite Trajectory in a Medium *Acta Physica Polonica*, **B 119**, pp.853-861.
20. Gradshteyn I.S. and Ryzik I.M. (1965) *Tables of Integrals, Series and Products*, Academic Press, New York.

CHAPTER 3

NON-UNIFORM CHARGE MOTION IN A DISPERSION-FREE MEDIUM

3.1. Introduction

Although the Vavilov-Cherenkov effect is a well established phenomenon widely used in physics and technology [1,2], many its aspects remain uninvestigated up to now. In particular, it is not clear how takes place a transition from the subluminal velocity régime to the superluminal régime. Some time ago [3,4], it was suggested that alongside with the usual Cherenkov and bremsstrahlung (BS) shock waves, the shock wave arises when the charge velocity coincides with the light velocity in medium. The consideration presented there was purely qualitative without any formulae and numerical results. It was grounded on the analogy with phenomena occurring in acoustics and hydrodynamics. It seems to us that this analogy is not complete. In fact, the electromagnetic waves are pure transversal, whilst acoustic and hydrodynamic waves contain longitudinal components. Furthermore, the analogy itself cannot be considered as a final proof. This fact and experimental ambiguity in distinguishing the Cherenkov radiation from the BS [5] make us consider effects arising from the overcoming the velocity of light barrier in the framework of the completely solvable model. To be precise, we consider the accelerated straight line motion of the point charge in medium and evaluate the arising electromagnetic field (EMF). We prove the existence of the shock wave arising at the moment when a charge overcomes the velocity of light barrier. This wave has essentially the same singularity as the Cherenkov shock wave. It is much stronger than the singularity of the bremsstrahlung shock wave. Formerly, the accelerated motion of a point charge in a vacuum was considered by Schott [6]. However, his qualitative consideration was purely geometrical, not allowing the numerical investigations. In the next sections the following definitions will be used:

1) BS shock wave. By it we mean a singular wave arising at the beginning or termination of a charge motion.

2) Shock wave originating when a charge velocity exceeds the velocity of light in medium. By it we mean a singular wave emitted when the charge velocity coincides the velocity of light in medium.

3) Cherenkov shock wave. By it we mean the Cherenkov shock wave attached to a moving charge.

Although these linear waves have some features typical of shock waves (finite or infinite jumps of certain quantities on their boundaries), they are not shock waves in the meaning used in hydrodynamics or gas dynamics where these waves are highly nonlinear formations. This is valid especially for the BS shock wave. However, for other two singular waves the linearity is illusory. We demonstrate this using the Cherenkov shock wave as an example. Consider a charge moving uniformly in vacuum with a velocity only slightly smaller than that of light. Its EMF is completely different from the Cherenkov radiation field. Now let this charge move with the same velocity in medium. The moving charge interacts with atoms of medium, excites and ionizes them. The EMFs arising from the electron transitions between atomic levels, from the acceleration of secondary knocked out electrons, all these fields being added give the Cherenkov radiation field. Obviously, this is a highly nonlinear phenomenon and this, in turn, justifies the term 'shock wave' used above. Usually, when considering the charge motion inside medium one disregards ionization phenomena and takes into account only excitations of atomic levels. The atomic electrons are treated as harmonic oscillators. For non-magnetized substances one finds the Lorentz-Lorenz formula in classical theory and the Kramers-Heisenberg dispersion formula in quantum theory.

In the present approach we take the refractive index to be independent of ω. This permits us to solve the problem under consideration explicitly. The cost of disregarding the dependence of ω is the divergence of integrals quadratic in Fourier transforms of field strengths (such as the total energy). Physically, these infinities are owed to the infinite self-energy of a point-like charge. To avoid divergences one should either make a cut-off procedure integrating up to some maximal frequency [1], or consider a charge of a finite size [7,8] (see also Chapter 7). Note that despite the infinite value of the radiated energy (in the absence of ω dispersion) for a uniformly moving charge with $v > c_n$, the usual theory correctly describes the position and propagation of the Cherenkov singularity. We believe that the approach adopted here is also adequate for the description of space-time distributions of EMF arising from accelerated motion of a charge.

3.2. Statement of the physical problem

Let a charged particle move inside the non-dispersive medium with polarizabilities ϵ and μ along the given trajectory $\vec{\xi}(t)$. Then its EMF at the observational point (ρ, z) is given by the Liénard-Wiechert potentials

$$\Phi(\vec{r}, t) = \frac{e}{\epsilon} \sum \frac{1}{|R_i|}, \quad \vec{A}(\vec{r}, t) = \frac{e\mu}{c} \sum \frac{\vec{v}_i}{|R_i|}, \quad \mathrm{div}\vec{A} + \frac{\epsilon\mu}{c}\dot{\Phi} = 0. \quad (3.1)$$

Here

$$\vec{v}_i = \left(\frac{d\vec{\xi}}{dt}\right)\Big|_{t=t_i}, \quad R_i = |\vec{r} - \vec{\xi}(t_i)| - \vec{v}_i(\vec{r} - \vec{\xi}(t_i))/c_n$$

and c_n is the velocity of light inside the medium $(c_n = c/\sqrt{\epsilon\mu})$. The summation in (3.1) is performed over all physical roots of the equation

$$c_n(t - t') = |\vec{r} - \vec{\xi}(t')|. \quad (3.2)$$

To preserve the causality, the time t' of the radiation should be smaller than the observational time t. Obviously, t' depends on the coordinates $\vec{r}, t$ of the point P at which the EMF is observed. With the account of (3.2) one finds for R_i

$$R_i = c_n(t - t_i) - \vec{v}_i(\vec{r} - \vec{\xi}(t_i))/c_n. \quad (3.3)$$

3.2.1. SIMPLEST ACCELERATED AND DECELERATED MOTIONS [9]

Consider the motion of the charged point-like particle inside the medium with a constant acceleration $2a$ (thus our acceleration is one half of the usual) along the Z axis:

$$\xi = at^2. \quad (3.4)$$

At first glance it seems that this equation describes the nonrelativistic motion. We analyze this question slightly later. The retarded times t' satisfy the following equation

$$c_n(t - t') = [\rho^2 + (z - at'^2)^2]^{1/2}. \quad (3.5)$$

It is convenient to introduce the dimensionless variables

$$\tilde{t} = at/c_n, \quad \tilde{z} = az/c_n^2, \quad \tilde{\rho} = a\rho/c_n^2. \quad (3.6)$$

Then

$$\tilde{t} - \tilde{t}' = [\tilde{\rho}^2 + (\tilde{z} - \tilde{t}'^2)^2]^{1/2}. \quad (3.7)$$

In order not to overload the exposition, we drop the tilda signs:

$$t - t' = [\rho^2 + (z - t'^2)^2]^{1/2} \quad (3.8)$$

For the case of treated one-dimensional motion the denominators R_i are given by:

$$R_i = \frac{c_n^2}{a}r_i, \quad r_i = (t - t_i) - 2t_i(z - t_i^2). \quad (3.9)$$

Eq. (3.8) can be reduced to the following equation of fourth degree

$$t'^4 + pt'^2 + qt' + R = 0. \tag{3.10}$$

Here $p = -2(z + 1/2)$, $q = 2t$, $R = r^2 - t^2$.

We consider the following two problems:

I. A charged particle is at rest at the origin up to a moment $t' = 0$. After that, it is uniformly accelerated in the positive direction of the Z axis. In this case only positive retarded times t' are nontrivial.

II. A charged particle is uniformly decelerated moving from $z = \infty$ to the origin. After the moment $t' = 0$ it is at rest there. Only negative retarded times are nontrivial in this case.

It is easy to check that the moving charge acquires the velocity of light c_n at the instants $t_c = \pm 1/2$ for the accelerated and decelerated motion, respectively. The position of a charge at those instants is $z_c = 1/4$.

It is our aim to investigate the space-time distribution of the EMF arising from such particle motions.

We intend to solve Eq. (3.10). It is obtained by squaring Eq. (3.8). As a result, extra false roots are possible. They are discarded on the following physical grounds:

1) physical roots should be real;

2) physical roots should preserve causality. For this the radiation time t' should be smaller than the observational time t;

3) the treated accelerated motion takes place for $t' > 0$. Negative values of $t' = t - r$ correspond to a charge at rest at the origin. If amongst the roots of (3.10) there occurs a negative one which does not coincide with $t' = t - r$, it should be discarded. Similarly, the treated decelerated motion takes place for $t' < 0$. Positive values of $t' = t - r$ correspond to a charge resting at the origin. So if amongst the roots of (3.10) there occurs a positive one not coinciding with $t' = t - r$, it should be discarded. Here $r = \sqrt{x^2 + y^2 + z^2}$.

These conditions define space-time domains in which the solutions of Eqs. (3.8) and (3.10) exist.

Accelerated motion

For the first of the problems treated (uniform acceleration of the charge which initially is at rest at the origin) the resulting configuration of the shock waves for the typical case corresponding to $t = 2$ is shown in Fig. 3.1. We see on it the Cherenkov shock wave $C_M^{(1)}$, the shock wave $C_L^{(1)}$ closing the Cherenkov-Mach cone and the sphere C_0 representing the spherical shock wave arising from the beginning of the charge motion. It turns out that the surface $C_L^{(1)}$ is approximated to a high accuracy by the part of the sphere $\rho^2 + (z - 1/4)^2 = (t - 1/2)^2$ (shown by the short dash curve C) which corresponds to the shock wave emitted from the point $z = 1/4$ at the

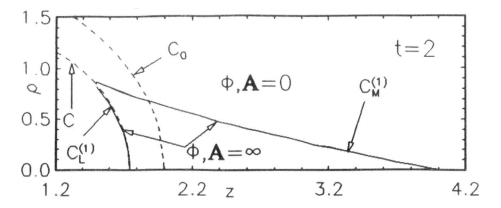

Figure 3.1. Distribution of the shock waves for a uniformly accelerated charge for $t = 2$. The short dash curve C represents the spherical wave emitted from the point $z = 1/4$ at the instant $t = 1/2$ when the accelerated charge overcomes the velocity of light barrier.

instant $t = 1/2$ when the velocity of the charged particle coincides with the velocity of light in the medium. On the internal sides of the surfaces $C_L^{(1)}$ and $C_M^{(1)}$ electromagnetic potentials acquire infinite values. On the external side of $C_M^{(1)}$ lying outside of C_0 the electromagnetic potentials are zero (as there are no solutions there). On the external sides of $C_L^{(1)}$ and on the part of the $C_M^{(1)}$ surface lying inside C_0 the electromagnetic potentials have finite values (owing to the presence of BS shock waves there).

Consider the time evolution of the arising shock waves for the accelerated motion of the charge beginning from the origin at the instant $t = 0$. It is shown in Figs. 3.2 and 3.3. All the Cherenkov (Mach) cones shown in Figs. 3.2 and 3.3 exist only for $t > 1/2$, $z > 1/4$. This means that the observer placed in the spatial region with $z < 1/4$ will not see either the Cherenkov shock wave or the shock wave originating from the overcoming the velocity of light barrier in any instant of time. Only the shock wave C_0 (not shown in these figures) associated with the beginning of the charge motion reaches him at the instant $c_n t = r$. Moreover, the aforementioned shock waves ($C_L^{(1)}$ and $C_M^{(1)}$) in the $z > 1/4$ region exist only if the distance ρ from the Z axis satisfies the equation

$$\rho < \rho_c, \ \rho_c = \frac{4}{3\sqrt{3}} \left(z - \frac{1}{4} \right)^{3/2}, \quad z > \frac{1}{4}. \tag{3.11}$$

Inside this region the observer sees at first the Cherenkov shock wave $C_M^{(1)}$. Later he detects the BS shock wave C_0 and the shock wave $C_L^{(1)}$ associ-

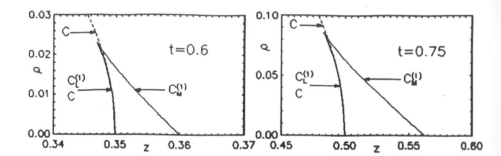

Figure 3.2. The positions of the Cherenkov shock wave $C_M^{(1)}$ and the shock wave $C_L^{(1)}$ arising from the charge exceeding the velocity of light barrier for the accelerated charge are shown for the instant $t = 0.6$ (left) and $t = 0.75$ (right). The short dash curve C represents the spherical wave emitted from the point $z = 1/4$ at the instant $t = 1/2$ when the accelerated charge overcomes the light barrier.

ated with the exceeding the velocity of light barrier at $z = 1/4$ at the time $t = 1/2$ when the charge velocity is equal to c_n. Outside the region defined by (3.11) the observer sees only the BS shock wave C_0 which reaches him at the instant $c_n t = r$. Furthermore, for $t < 1/2$ only one retarded solution (t_1) exists. It is confined to the sphere C_0 of the radius $r = c_n t$. Therefore the observer in this time interval will not detect either the Cherenkov shock wave or that of originating from the exceeding the velocity of light barrier. The dimensions of the Cherenkov cones shown in Figs. 3.2 and 3.3 are zero for $t = 1/2$ and continuously rise with time for $t > 1/2$. The physical reason for this behaviour is that the shock wave $C_L^{(1)}$ closing the Cherenkov cone propagates with the velocity of light c_n, while the head part of the Cherenkov cone (i.e., the Cherenkov shock wave $C_M^{(1)}$) attached to a moving charge propagates with the velocity $v > c_n$. In the gas dynamics the existence of at least two shock waves attached to the finite body moving with a supersonic velocity was proved on the very general grounds by Landau and Lifshitz [10], Chapter 13). In the present context we associate them with the shock waves $C_L^{(1)}$ and $C_M^{(1)}$.

Decelerated motion

Now we turn to the uniform deceleration of a charged particle. Let it move along the positive z semi-axis up to an instant $t = 0$, after which it is at rest at the origin. In this case only negative retarded times t_i have a physical meaning.

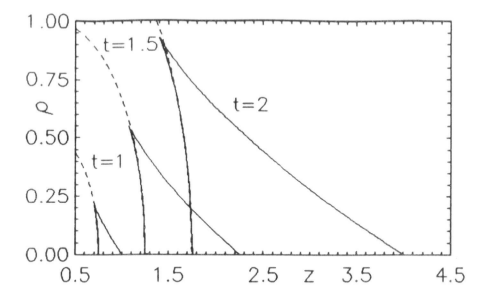

Figure 3.3.   The same as Fig. 3.2, but for $t = 1$, 1.5, and 2.

For the observational time $t = 2$ the resulting configuration of the shock waves is shown in Fig. 3.4. We see the BS shock wave C_0 arising from the termination of the charge motion and the blunt shock wave $C_L^{(2)}$. Its head part is described with a high accuracy by the sphere $\rho^2 + (z - 1/4)^2 = (t + 1/2)^2$ (shown by the short dash curve) corresponding to the shock wave emitted from the point $z = 1/4$ at the instant $t = -1/2$ when the velocity of the decelerated charged particle coincides with the velocity of light in the medium. The electromagnetic potentials vanish outside of $C_L^{(2)}$ (as no solutions exist there) and acquire infinite values on the internal part of $C_L^{(2)}$ (owing to the vanishing of their denominators R_1 and R_2). Therefore the surface $C_L^{(2)}$ represents the shock wave. As a result, for $t > 0$, $t' < 0$ one has the shock wave $C_L^{(2)}$ and the BS wave C_0 arising from the termination of the particle motion.

For the decelerated motion and the observational time $t < 0$ the physical solutions exist only inside the Cherenkov cone $C_M^{(2)}$ (Fig. 3.5). On its internal boundary the electromagnetic potentials acquire infinite values. On the external boundary the electromagnetic potentials are zero (as no solutions exist there). Thus for the case of decelerated motion and the observational time $t = -2$ the physical solutions are contained inside the Cherenkov cone $C_M^{(2)}$.

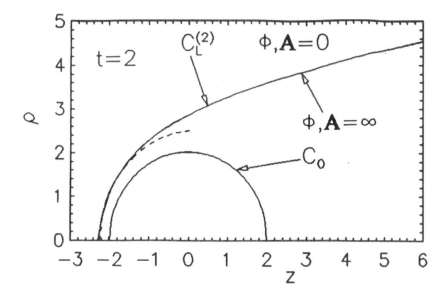

Figure 3.4. Distribution of the shock waves for a uniformly decelerated charge for $t = 2$. The short dash curve represents the spherical wave emitted from the point $z = 1/4$ at the instant $t = -1/2$ when the accelerated charge overcomes the light barrier.

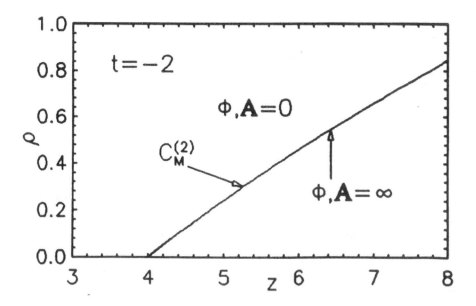

Figure 3.5. The same as Fig. 3.4, but for $t = -2$.

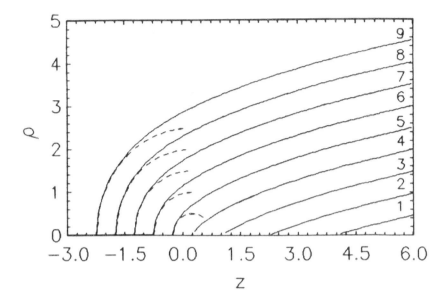

Figure 3.6. Continuous transformation of the Cherenkov shock wave (1) into the blunt shock wave (9) for the decelerated motion. The charge motion terminates at the point $z = 0$ at the instant $t = 0$. The numbers 1-9 refer to the instants of time $t = -2; -1.5;$ $-1; -0.5; 0; 0.5; 1; 1.5$ and 2, respectively. Short dash curves represent the positions of the spherical wave emitted from the point $z = 1/4$ at the instant $t = -1/2$ when the velocity of the decelerated charge coincides with the velocity of light in medium.

For the decelerated motion the time evolution of shock waves is shown in Fig. 3.6. The observer in the spatial region $z < 0$ detects the blunt shock wave $C_L^{(2)}$ first and the bremsstrahlung shock wave C_0 later. It turns out that the head part of this blunt wave coincides to a high accuracy with the sphere $\rho^2 + (z - 1/4)^2 = (t + 1/2)^2$ describing the spherical wave emitted from the point $z = 1/4$ at the instant $t = -1/2$ when the charge velocity coincides with c_n. The observer in the $z > 1/2$ region detects the Cherenkov shock wave $C_M^{(2)}$ first and the bremsstrahlung shock wave C_0 later. In order not to hamper the exposition, we have not mentioned in this section the continuous radiation which reaches the observer between the arrival of two shock waves or after the arrival of the last shock wave. It is easily restored from the above figures.

3.2.2. COMPLETELY RELATIVISTIC ACCELERATED AND DECELERATED MOTIONS [11]

To avoid troubles arising from the nonrelativistic nature of the motion law (3.4), we consider the motion of a point-like charge of rest mass m inside the medium according to the motion law [12]

$$z(t) = \sqrt{z_0^2 + c^2 t^2} + C.$$

It may be realized in a constant electric field E directed along the Z axis: $z_0 = |mc^2/eE| > 0$. Here C is an arbitrary constant. We choose it from the condition $z(t) = 0$ for $t = 0$. Therefore

$$z(t) = \sqrt{z_0^2 + c^2 t^2} - z_0. \tag{3.12}$$

This law of motion, being manifestly relativistic, corresponds to constant proper acceleration [12]. The charge velocity is given by

$$v = \frac{dz}{dt} = c^2 t (z_0^2 + c^2 t^2)^{-1/2}.$$

Clearly, it tends to the velocity of light in vacuum as $t \to \infty$. The retarded times t' satisfy the following equation:

$$c_n(t - t') = \left[\rho^2 + \left(z + z_0 - \sqrt{z_0^2 + c^2 t'^2} \right)^2 \right]^{1/2}. \tag{3.13}$$

It is convenient to introduce the dimensionless variables

$$\tilde{t} = ct/z_0, \quad \tilde{z} = z/z_0, \quad \tilde{\rho} = \rho/z_0. \tag{3.14}$$

Then

$$\alpha(\tilde{t} - \tilde{t}') = \left[\tilde{\rho}^2 + \left(\tilde{z} + 1 - \sqrt{1 + \tilde{t}'^2} \right)^2 \right]^{1/2}, \tag{3.15}$$

where $\alpha = c_n/c = 1/n$ is the ratio of the velocity of light in medium to that in vacuum. In order not to overload the exposition we drop the tilde signs

$$\alpha(t - t') = \left[\rho^2 + \left(z + 1 - \sqrt{1 + t'^2} \right)^2 \right]^{1/2}. \tag{3.16}$$

For the treated one-dimensional motion the denominators R_i entering into are (3.3) given by

$$R_i = \frac{z_0}{\alpha \sqrt{1 + t_i^2}} \left[\alpha^2 (t - t_i) \sqrt{1 + t_i^2} - t_i \left(z + 1 - \sqrt{1 + t_i^2} \right) \right] \tag{3.17}$$

It is easy to check that the moving charge acquires the velocity of light c_n in medium at the instants $t_l = \pm\alpha/\sqrt{1-\alpha^2}$ for the accelerated and decelerated motion, respectively. The position of a charge at those instants is $z_l = 1/\sqrt{1-\alpha^2} - 1$.

It is our aim to investigate the space-time distribution of EMF arising from such particle motions. For this we should solve Eq.(3.16). Taking its square we obtain the fourth degree algebraic equation relative to t'. Solving it we find space-time domains in which the EMF exists. It is just this way of finding the EMF which was adopted in [9]. It was shown in the same reference that there is another, much simpler, approach for recovering EMF singularities (which was extensively used by Schott [6]). We seek the zeros of the denominators R_i entering into the definition of the electromagnetic potentials (3.1). They are obtained from the equation

$$\alpha^2(t-t')\sqrt{1+t'^2} - t'(z+1-\sqrt{1+t'^2}) = 0. \qquad (3.18)$$

We rewrite (3.16) in the form

$$\rho^2 = \alpha^2(t-t')^2 - (z+1-\sqrt{1+t'^2})^2. \qquad (3.19)$$

Recovering t' from (3.18) and substituting it into (3.19) we find the surfaces $\rho(z,t)$ carrying the singularities of the electromagnetic potentials. They are just the shock waves which we seek. It turns out that BS shock waves (i.e., moving singularities arising from the beginning or termination of a charge motion) are not described by Eqs. (3.18) and (3.19). The physical reason for this is that on these surfaces the BS field strengths, not potentials, are singular [6]. The simplified procedure mentioned above for recovering moving EMF singularities is to find solutions of (3.18) and (3.19) and add to them 'by hand' the positions of BS shock waves defined by the equation $r = \alpha t$, $r = \sqrt{\rho^2 + z^2}$. The equivalence of this approach to the complete solution of (3.13) has been proved in [9] where the complete description of the EMF (not only its moving singularities as in the present approach) of a moving charge was given. It was shown there that the electromagnetic potentials exhibit infinite (for the Cherenkov and the shock waves under consideration) jumps when one crosses the above singular surfaces. Correspondingly, field strengths have the δ type singularities on these surfaces whilst the space-time propagation of these surfaces describes the propagation of the radiated energy flux.

In what follows we consider the typical case when the ratio α of the velocity of light in medium to that in vacuum is equal to 0.8.

Accelerated motion

For the uniform acceleration of the charge resting at the origin up to $t = 0$ only positive retarded times t_i have a physical meaning (negative t_i corre-

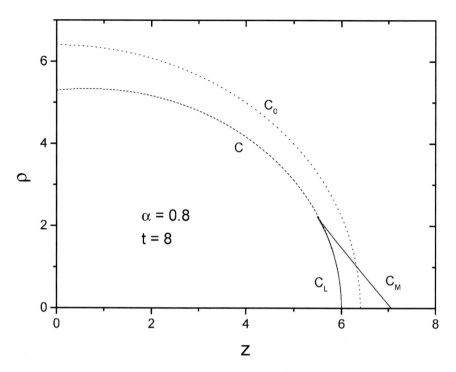

Figure 3.7.   Typical distribution of the shock waves emitted by an accelerated charge. C_M is the Cherenkov shock wave, C_L is the shock wave emitted from the point $z_l = (1 - \alpha^2)^{-1/2}$ at the instant $t_l = \alpha(1 - \alpha^2)^{-1/2}$ when the charge velocity coincides with the velocity of light in medium. Part of it is described to good accuracy by the fictitious spherical surface C $(\rho^2 + (z - z_l)^2 = (t - t_l)^2)$; C_0 is the bremsstrahlung shock wave originating from the beginning (at the instant $t = 0$) of the charge motion.

spond to a charge at rest at the origin). The resulting configuration of the shock waves for the typical observational time $t = 8$ is shown in Fig. 3.7. We see in it:

i) The Cherenkov shock wave C_M having the form of the Cherenkov cone;

ii) The shock wave C_L closing the Cherenkov cone and describing the shock wave emitted from the point $z_l = (1 - \alpha^2)^{-1/2} - 1$ at the instant $t_l = \alpha(1 - \alpha^2)^{-1/2}$ when the velocity of a charge coincides with the velocity of light in medium;

iii) The BS shock wave C_0 arising at the beginning of notion.

It turns out that the surface C_L is approximated to good accuracy by the spherical surface $\rho^2 + (z - z_l)^2 = (t - t_l)^2$ (shown by the short dash curve C). It should be noted that only the part of C coinciding with C_L has a physical meaning.

On the internal sides of the surfaces C_L and C_M electromagnetic po-

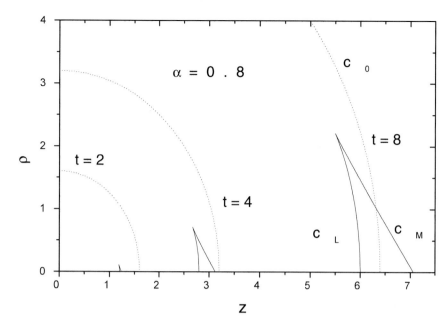

Figure 3.8.   Time evolution of shock waves emitted by an accelerated charge. C_M and C_L are respectively the usual Cherenkov shock wave and the shock wave arising at the instant when the charge velocity coincides with the velocity of light in medium. Pointed curves are bremsstrahlung shock waves.

tentials acquire infinite values. On the external side of C_M lying outside C_0 the magnetic vector potential is zero (as there are no solutions of Eqs. (3.18),(3.19) there), whilst the electric scalar potential coincides with that of the charge at rest. On the external sides of C_L and on the part of the surface C_M lying inside C_0 the electromagnetic potentials have finite values (as bremsstrahlung has reached these spatial regions).

In the negative z semi-space an experimentalist will detect only the BS shock wave. In the positive z semi-space, for the sufficiently large times $(t > 2\alpha/(1-\alpha^2))$, an observer close to the z axis will detect the Cherenkov shock wave C_M first, the BS shock wave C_0 later, and, finally, the shock wave C_L originating from the exceeding the velocity of light in medium. For the observer more remote from the z axis the BS shock wave C_0 arrives first, then C_M and finally C_L (Fig. 3.7). For the larger distances from the z axis the observer will see only the BS shock wave.

The positions of the shock waves for different observational times are shown in Fig. 3.8. The dimension of the Cherenkov cone is zero for $t \le t_l$ and continuously increases with time for $t > t_l$. The physical reason for this is that the C_L shock wave closing the Cherenkov cone propagates with the velocity of light c_n, whilst the head part of the Cherenkov cone C_M

attached to a moving charge propagates with a velocity $v > c_n$. It is seen that for small observational times ($t = 2$ and $t = 4$) the BS shock wave C_0 (pointed curve) precedes C_M. Later, C_M reaches (this happens at the instant $t = 2\alpha/(1 - \alpha^2)$) and partly passes BS shock wave C_0 ($t = 8$). However, the C_L shock wave is always behind C_0 (as both of them propagate with the velocity c_n, but C_L is born at the later instant $t = t_l$). A picture similar to the $t = 8$ case remains essentially the same for later times.

Decelerated motion

Now we turn to the second problem (uniform deceleration of the charged particle along the positive z semi-axis up to a instant $t = 0$ after which it is at rest at the origin). In this case only negative retarded times t_i have a physical meaning (positive t_i correspond to the charge at rest at the origin).

For an observational time $t > 0$ the resulting configuration of the shock waves is shown in Fig. 3.9 where one sees the BS shock wave C_0 arising from the termination of the charge motion (at the instant $t = 0$) and the blunt shock wave C_M into which the CSW transforms after the termination of the motion. The head part of C_M is described to good accuracy by the sphere $\rho^2 + (z - z_l)^2 = (t + t_l)^2$ corresponding to the fictitious shock wave C emitted from the point $z_l = (1 - \alpha^2)^{-1/2} - 1$ at the instant $t_l = -\alpha(1 - \alpha^2)^{-1/2}$ when the velocity of the decelerated charge coincides with the velocity of light in medium. Only the part of C coinciding with C_M has a physical meaning. The electromagnetic potentials vanish outside C_M (as no solutions exist there) and acquire infinite values on the internal part of C_M. Therefore the surface C_M represents the shock wave. As a result, for the decelerated motion after termination of the particle motion ($t > 0$) one has the shock wave C_M detached from a moving charge and the BS shock wave C_0 arising from the termination of the particle motion. After the C_0 shock wave reaches the observer, he will see the electrostatic field of a charge at rest and bremsstrahlung from remote parts of the charge trajectory.

The positions of shock waves at different times are shown in Fig. 3.10 where one sees how the acute CSW attached to the moving charge ($t = -2$) transforms into the blunt shock wave detached from it ($t = 8$). The pointed curves mean the BS shock waves described by the equation $r = \alpha t$ (in dimensional variables it has the form $r = c_n t$). For the decelerated motion and $t < 0$ (i.e., before termination of the charge motion) physical solutions exist only inside the Cherenkov cone C_M ($t = -2$ on Fig. 3.10). On the internal boundary of the Cherenkov cone the electromagnetic potentials acquire infinite values. On their external boundaries the electromagnetic potentials are zero (as no solutions exist there). When the charge velocity coincides with c_n the CSW leaves the charge and transforms into the C_M

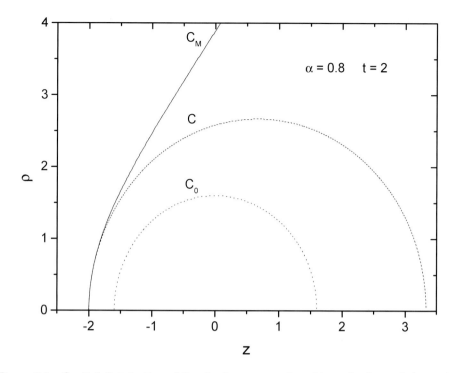

Figure 3.9.   Spatial distribution of the shock waves produced by a decelerated charge in an uniform electric field. C_M is the blunt shock wave into which the CSW transforms after the charge velocity coincides with the velocity of light in medium. Part of it is approximated to good accuracy by the fictitious spherical surface C. C_0 is the bremsstrahlung shock wave originating from the termination of the charge motion at $t = 0$.

shock wave which propagates with the velocity c_n ($t = 2$, 4 and 8 on Fig. 3.10). As has been mentioned, the blunt head parts of these waves are approximated to a good accuracy by the fictitious surface $\rho^2 + (z - z_l)^2 = (t + t_l)^2$ corresponding to the shock wave emitted at the instant when the charge velocity coincides with the velocity of light in the medium.

In the negative z half-space an experimentalist will detect the blunt shock wave first and BS shock wave (short dash curve) later.

In the positive z half-space, for the observational point close to the z axis the observer will see the CSW first and BS shock wave later. For larger distances from the z axis he will see at first the blunt shock C_M into which the CSW degenerates after the termination of the charge motion and the BS shock wave later (Fig. 3.10).

It should be mentioned about the continuous radiation which reaches the observer between the arrival of the above shock waves, about the continuous radiation and the electrostatic field of a charge at rest reaching the observer after the arrival of the last shock wave. They are easily restored from the

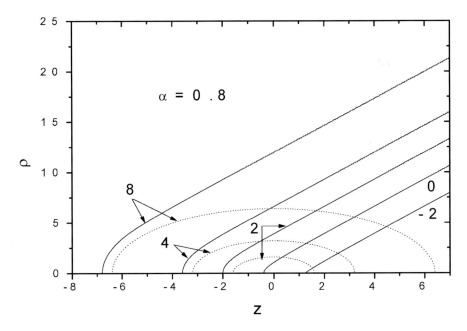

Figure 3.10.   Continuous transformation of the acute Cherenkov shock wave attached to a moving charge ($t = -2$) into the blunt shock wave detached from a charge ($t = 8$) for the decelerated motion. The numbers at the curves mean the observational times. Pointed curves are bremsstrahlung shock waves. Charge motion is terminated at $t = 0$.

complete exposition presented in [9] for the $z = at^2$ motion law.

We have investigated the space-time distribution of the electromagnetic field arising from the accelerated manifestly relativistic charge motion. This motion is maintained by the constant electric field. Probably this field is easier to create in gases (than in solids in which the screening effects are essential) where the Vavilov-Cherenkov effect is also observed. We have confirmed the intuitive predictions made by Tyapkin [3] and Zrelov et al. [4] concerning the existence of the new shock wave (in addition to the Cherenkov and bremsstrahlung shock waves) arising when the charge velocity coincides with the velocity of light in medium. For the accelerated motion this shock wave forms indivisible unity with Cherenkov's shock wave. It closes the Vavilov-Cherenkov radiation cone and propagates with the velocity of light in the medium. For the decelerated motion the above shock wave detaches from a moving charge when its velocity coincides with the velocity of light in medium.

The quantitative conclusions made in [9] for a less realistic external electric field maintaining the accelerated charge motion are also confirmed. We have specified under what conditions and in which space-time regions the above-mentioned new shock waves do exist. It would be interesting to observe these shock waves experimentally.

3.3. Smooth Tamm problem in the time representation

In 1939, Tamm [13] solved approximately the following problem: A point charge is at rest at a fixed point of medium up to some instant $t = -t_0$, after which it exhibits an instantaneous infinite acceleration and moves uniformly with a velocity greater than the velocity of light in that medium. At the instant $t = t_0$ the charge decelerates instantaneously and comes to a state of rest. Later this problem was qualitatively investigated by Aitken [14] and Lawson [15] and numerically by Ruzicka and Zrelov [5,16]. The analytic solution of this problem in the absence of dispersion was found in [17]. However, in all these studies the information concerning the transition effects was lost owing to the instantaneous charge acceleration. The main drawbacks of the original Tamm problem are instantaneous acceleration and deceleration of a moving charge.

On the other hand, effects arising from unbounded accelerated and decelerated motions of a charge were considered in a previous section. It was shown there that alongside with the bremsstrahlung and Cherenkov shock waves, a new shock wave arises when the charge velocity coincides with c_n.

The aim of this consideration is to avoid infinite acceleration and deceleration typical for the Tamm problem by applying methods developed in [9,17]. For this aim we consider the following charge motion: a charge is smoothly accelerated, then moves with a constant velocity, and, finally, is smoothly decelerated (Fig. 3.11).

3.3.1. MOVING SINGULARITIES OF ELECTROMAGNETIC FIELD

Let a point charge move inside the medium with permittivities ϵ and μ along the given trajectory $\vec{\xi}(t)$. Its EMF at the observational point (ρ, z) is then given by the Liénard-Wiechert potentials (3.1). Summation in (3.1) runs over all physical roots of the equation (3.2). Obviously, t' depends on the coordinates $\vec{r}, t$ of the observational point P.

To investigate the space-time distribution of the EMF of a moving charge one should find (for the given observational point $\vec{r}, t$) the retarded times from Eq.(3.2) and substitute them into (3.1).

There is another much simpler method (suggested by Schott [6]) for recovering EMF singularities. We seek zeros of the denominators R_i entering into the definition of electromagnetic potentials (3.1). They are obtained from the equation

$$c_n(t - t') = \frac{v(t')}{c_n}(z - \xi(t')), \qquad (3.20)$$

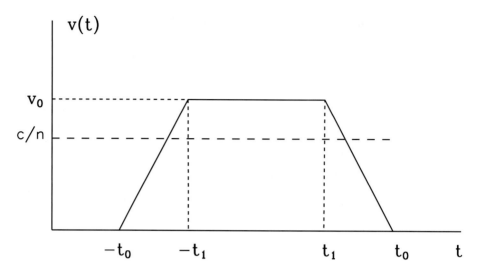

Figure 3.11. Schematic presentation of the smooth Tamm model. Charge accelerates, moves uniformly with a velocity v_0, and decelerates in the time intervals $(-t_0, -t_1)$, $(-t_1, t_1)$ and (t_1, t_0), respectively.

Combining (3.20) and (3.33) we find $\rho(t')$ and $z(t')$

$$z = \xi(t') + \frac{c^2}{n^2 v}(t - t'), \quad \rho = \frac{c^2(t - t')}{n^2 v \gamma_n}. \tag{3.21}$$

Here $\gamma_n = 1/\sqrt{\beta^2 n^2 - 1}$, $\beta = v/c$.

Our procedure reduces to the following one. For the fixed observation time t, we vary t' over the motion interval, evaluate $z(t')$ and $\rho(t')$ and draw the dependence $\rho(z)$ for the fixed t. Due to the axial symmetry of the problem, this curve is in fact the surface on which the electromagnetic potentials are singular. It follows from (3.21) that these singular surfaces exist only if $v > c/n$, that is if the charge velocity is greater than the light velocity in medium. There are other surfaces on which the EMF strengths are singular and which are not described by (3.21). For example, on the surfaces of the bremsstrahlung (BS) shock waves arising at the start or the end of motion, the electromagnetic potentials exhibit finite jumps. The corresponding EMF strengths have δ singularities on these surfaces.

Moving singularity of the original Tamm problem
In the time interval $-t_0 < t' < t_0$ ($t_0 = z_0/v_0$) where a charge moves uniformly with the velocity v_0 equations (3.21) look like

$$\rho = \frac{c_n^2}{v_0\gamma_{0n}}(t - t'), \quad z = v_0t' + \frac{c_n^2}{v_0}(t - t'). \tag{3.22}$$

Here $\gamma_{0n} = 1/\sqrt{v_0^2/c_n^2 - 1}$. Excluding t' from these equations one finds

$$\rho = (v_0 t - z)\gamma_{0n}, \tag{3.23}$$

where ρ and z are changed in the intervals

$$z_1^0 < z < v_0 t, \quad 0 < \rho < \frac{c_n^2}{v_0\gamma_{0n}}(t + t_0)$$

for $-t_0 < t < t_0$ and

$$z_1^0 < z < z_2^0, \quad \rho_2 < \rho < \rho_1$$

for $t > t_0$. Here

$$z_1^0 = \frac{c^2}{v_0 n^2}(t + t_0) - z_0, \quad \rho_1 = \frac{c^2}{v_0 n^2 \gamma_{0n}}(t + t_0),$$

$$z_2^0 = \frac{c_n^2}{v_0}(t - t_0) + z_0, \quad \rho_2 = \frac{c_n^2}{v_0\gamma_{0n}}(t - t_0).$$

We define the straight lines L_1 ($z = -z_0 + \rho\gamma_n$) and L_2 ($z = z_0 + \rho\gamma_n$) (Fig. 3.12 (a)). They originate from the $\mp z_0$ points and are inclined at the angle θ_c ($\cos\theta_c = 1/\beta_0 n$) towards the motion axis. It is seen that for each $t > t_0$ the singular segment (3.23) enclosed between the straight lines L_1 and L_2 is perpendicular to both of them and coincides with the CSW defined in Chapter 2. Its normal is inclined at the angle θ_c towards the motion axis. As time goes, it propagates between L_1 and L_2. For $-t_0 < t < t_0$ the CSW is enclosed between the moving charge and the straight line L_1.

Smooth Tamm problem
In the smooth Tamm problem (Fig. 3.11) a charge is at rest at the spatial point $z = -z_0$ up to an instant $t = -t_0$. In the space-time interval $-t_0 < t < -t_1$, $-z_0 < z < -z_1$ (we refer to this interval as to region 1) it moves with constant acceleration a

$$\xi(t') = -z_0 + \frac{1}{2}a(t' + t_0)^2, \quad v(t') = a(t' + t_0).$$

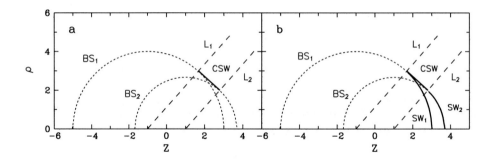

Figure 3.12. (a): The position of the shock waves in the original Tamm problem. BS_1 and BS_2 are the bremsstrahlung shock waves emitted at the beginning and the end of the charge motion; CSW (thick straight line) is the Cherenkov shock wave; (b): The position of the shock waves in the limiting case of the smooth Tamm problem (see Fig. 11) when the lengths of accelerated and decelerated parts of the charge trajectory tend to zero. The thick curves SW_1 and SW_2 are the shock waves arising at the accelerated and decelerated parts of the charge trajectory, respectively. Due to the instantaneous velocity jumps, SW_1 and SW_2 partly coincide with the BS_1 and BS_2 shock waves, respectively.

In the space-time interval $-t_1 < t < t_1$, $-z_1 < z < z_1$ (region 2) it moves with the constant velocity v_0

$$\xi(t') = v_0 t', \quad v(t') = v_0.$$

In the space-time interval $t_1 < t < t_0$, $z_1 < z < z_0$ (region 3) a charge moves with constant deceleration a down reaching the state of rest at $t = t_0$:

$$\xi(t') = z_0 - \frac{1}{2}a(t' - t_0)^2, \quad v(t') = a(t_0 - t').$$

The matching conditions of $\xi(t')$ and $v(t')$ at the $z = \pm z_1$ points define a, t_0 and t_1:

$$a = \frac{v_0^2}{2(z_0 - z_1)}, \quad t_0 = \frac{2z_0 - z_1}{v_0}, \quad t_1 = \frac{z_1}{v_0}.$$

Space region 1.   In the space region 1 equations (3.21) are

$$z = -z_0 + \frac{1}{2}a(t' + t_0)^2 + \frac{c^2}{n^2 v}(t - t'), \quad \rho = \frac{c^2(t - t')}{n^2 v \gamma_n}, \quad (3.24)$$

where $v = a(t' + t_0)$. It follows from this that the charge velocity coincides with the velocity of light in medium $c_n = c/n$ at $t' = -t_c$, $t_c \equiv t_0 - c_n/a$. At this instant

$$\rho_c = \rho(t' = -t_c) = 0,$$

$$z_c^{(1)} = z(t' = -t_c) = c_n t - (1 - \frac{1}{\beta_0 n})[z_0 - \frac{1}{n\beta_0}(z_0 - z_1)]. \tag{3.25}$$

For the observation time t smaller than the time $-t_1$ corresponding the right boundary of the motion interval 1, $\rho(t')$ has two zeroes (at $t' = t_c$ and $t' = t$). There is a maximum between them (Fig. 3.13 (a)) at

$$t' = t'_m \equiv -t_0 + (\frac{c_n}{a})^{2/3}(t + t_0)^{1/3}. \tag{3.26}$$

Obviously, $t_c < t'_m < t$. The corresponding ρ and z are equal to

$$\rho_m = \frac{c_n^2}{a}\{[\frac{a(t + t_0)}{c_n}]^{2/3} - 1\}^{3/2},$$

$$z_m = -z_0 + \frac{c_n^2}{a}\{\frac{3}{2}[\frac{a(t + t_0)}{c_n}]^{2/3} - 1\}. \tag{3.27}$$

This solution coincides with the analytical solution found in [9] for the semi-infinite motion beginning from the state of rest. The dependence $\rho(z)$ has a moon sickle-like form. This complex arises when the charge velocity coincides with the velocity of light c_n in medium. It consists of the curvilinear Cherenkov shock wave CSW attached to a moving charge and the shock wave closing the Cherenkov cone. As time goes, the dimensions of this complex rise (since a charge moves with the velocity v while SW_1 propagates with the velocity c_n).

For the observation time t greater than the time $-t_1$, ρ has only one zero. It has a maximum if $-t_1 < t < -t_0 + 2(z_0 - z_1)v_0/c_n^2$. The corresponding t'_m, ρ_m, and z_m are given by (3.26) and (3.27). In the interval $t'_m < t' < -t_1$, ρ decreases reaching the value

$$\rho_1 = \rho(t' = -t_1) = \frac{c^2}{v_0 n^2 \gamma_n}(t + t_1) \tag{3.28}$$

at the boundary point of the motion interval. The corresponding z is equal to

$$\tilde{z}_1 = z(t' = -t_1) = \frac{c^2}{v_0 n^2}(t + t_1) - z_1. \tag{3.29}$$

It is easy to check that z as a function of t' has a minimum at $t' = t'_m$: it decreases from $z_c^{(1)}$ at $t' = -t_c$ down to

$$z_m = -z_0 + \frac{c_n^2}{a}\{[\frac{a(t + t_0)}{c_n}]^{2/3} - 1\} \tag{3.30}$$

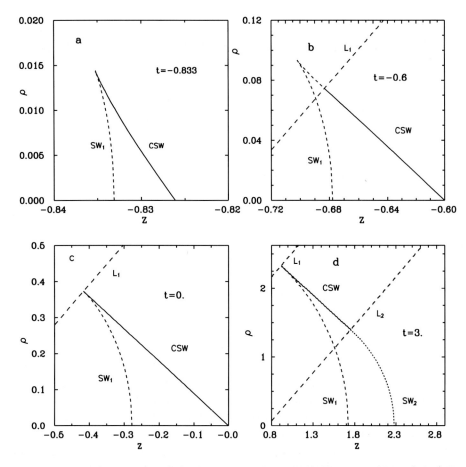

Figure 3.13.   The position of shock waves in the smooth Tamm problem. (a,b,c) For small and moderate observation times the singularity complex consists of the Cherenkov shock wave (CSW) attached to a moving charge and the shock wave SW_1 closing the Cherenkov cone and inclined at the right angle towards the motion axis; (d) For large observation times this complex detaches from a moving charge and propagates with the velocity of light c_n in medium. It consists of the CSW and the singular shock waves SW_1 and SW_2 perpendicular to the motion axis and arising at the accelerated and decelerated parts of the charge trajectory.

at $t' = t'_m$ and then increases up to $\tilde{z}_1$ for $t' = -t_1$ (Fig. 3.13(b), dotted line). For $t > -t_0 + 2(z_0 - z_1)v_0/c_n^2$ there is no maximum of $\rho(t')$ which rises steadily from 0 for $t' = t_c$ up to ρ_1 given by (5.5) for $t' = -t_1$ (Fig. 3.13(c), dotted line). In particular, $\rho_m = \rho_1$, $z_m = \tilde{z}_1$ for $t = -t_0 + 2(z_0 - z_1)v_0/c_n^2$.

Space region 2. In the time interval $-t_1 < t' < t_1$ $(t_1 = z_1/v_0)$ where a charge moves uniformly with the velocity v_0 equations (3.21) look like

$$\rho = \frac{c_n^2}{v_0\gamma_{0n}}(t - t'), \quad z = v_0 t' + \frac{c_n^2}{v_0}(t - t'). \tag{3.31}$$

Here $\gamma_{0n} = 1/\sqrt{v_0^2/c_n^2 - 1}$. Excluding t' from these equations one finds

$$\rho = (v_0 t - z)\gamma_{0n}, \tag{3.32}$$

where ρ and z change in the intervals

$$\tilde{z}_1 < z < v_0 t, \quad 0 < \rho < \frac{c_n^2}{v_0\gamma_{0n}}(t + t_1)$$

for $-t_1 < t < t_1$ and

$$\tilde{z}_1 < z < z_2, \quad \rho_2 < \rho < \rho_1$$

for $t > t_1$. Here $\tilde{z}_1$ and ρ_1 are the same as above, and

$$z_2 = \frac{c_n^2}{v_0}(t - t_1) + z_1, \quad \rho_2 = \frac{c_n^2}{v_0\gamma_{0n}}(t - t_1). \tag{3.33}$$

It is seen that for each $t > t_1$ the singular segment (3.33) is enclosed between the straight lines L_1 $(\rho = (z + z_1)/\gamma_{0n})$ and L_2 $(\rho = (z - z_1)/\gamma_{0n})$ originating from the boundary points of the interval 2 and inclined at the angle θ_c $(\cos\theta_c = 1/\beta_0 n)$ towards the motion axis (Fig. 3.13(d), solid line). The singular segment (3.32) is a piece of the Cherenkov shock wave which is enclosed between L_1 and L_2 and perpendicular to both of them. Its normal is inclined at the angle θ_c towards the motion axis. As time goes, it propagates between L_1 and L_2. For $-t_1 < t < t_1$ the singular segment (3.32) is enclosed between the moving charge and the straight line L_1 (Fig. 3.13 (c), solid line).

Space region 3. In the time interval $t_1 < t' < t_0$ where a charge moves with deceleration a equations (3.21) look like

$$z = z_0 - \frac{1}{2}a(t' - t_0)^2 + \frac{c^2}{n^2 v}(t - t'), \quad \rho = \frac{c^2(t - t')}{n^2 v\gamma_n}, \tag{3.34}$$

where $v = a(t_0 - t')$. The charge velocity changes steadily from v_0 at $t' = t_1$ down to 0 at $t = t_0$. The above singularity surfaces exist only if $c_n < v < v_0$. The charge velocity coincides with the velocity of light in medium $c_n = c/n$ at $t' = t_c$. At this instant

$$\rho_c = \rho(t' = t_c) = 0,$$

$$z_c^{(2)} = z(t' = t_c) = c_n t + (1 - \frac{1}{\beta_0 n})[z_0 - \frac{1}{\beta_0 n}(z_0 - z_1)]. \tag{3.35}$$

The radius $\rho(t')$ vanishes at the position of a moving charge $(t' = t)$ for $t < t_c$ and at $t' = t_c$ for $t > t_c$ (Fig 3.13(d)). It is maximal at the start of the third motion interval 3 $(t' = t_1)$ where

$$\rho(t' = t_1) = \rho_2, \quad z(t' = t_1) = z_2$$

(ρ_2 and z_2 are the same as in (3.33)).

A complete singular contour composed of its singular pieces defined in the regions 1,2 and 3 is always closed for the fixed observation time t. In fact, for $-t_c < t < -t_1$ the singular contour lies completely in the region 1. It begins at the point $z = z_c^{(1)}$, $\rho = 0$ and ends at the point $\rho = 0, z = -z_0 + a(t + t_0)^2/2$ coinciding with the current charge position (Fig. 3.13(a)). For $-t_1 < t < t_1$ the singular contour lies in the regions 1 and 2 (Figs. 3.13 (b,c)). Its branch lying in the region 1 begins at the point $z = z_c^{(1)}$, $\rho = 0$ and ends at the point $z = \tilde{z}_1$, $\rho = \rho_1$. Its branch lying in the region 2 begins at the point $z = \tilde{z}_1$, $\rho = \rho_1$ and ends at the point $z = v_0 t$, $\rho = 0$ coinciding with the current charge position. For $t > t_1$ the singular contour lies in the regions 1,2 and 3 (Fig. 3.13(d)). Its branch in region 1 is the same as above. Its branch lying in the region 2 begins at the point $z = \tilde{z}_1$, $\rho = \rho_1$ and ends at the point $z = z_2$, $\rho = \rho_2$. Its branch lying in region 3 begins at the point $z = z_2$, $\rho = \rho_2$ and ends at the point $z = z_c^{(2)}$, $\rho = 0$.

Transition to instantaneous velocity jumps
It is instructive to consider the limit $z_1 \to z_0$ corresponding to the instantaneous velocity jumps at the start and the end of the charge motion. Intuitively it is expected that the original Tamm problem should appear in this limit. Turning to (3.24) we observe that the second term entering into z vanishes. In fact, it is equal to

$$\frac{1}{2}a(t' + t_0)^2 = \frac{z_0 - z_1}{\beta^2 n^2}$$

at $t' = -t_c$ and

$$\frac{1}{2}a(t' + t_0)^2 = \frac{z_0 - z_1}{\beta^2 n^2}$$

at $t' = -t_1$. Therefore, in the limit $z_1 \to z_0$ it disappears at the boundaries of the charge motion interval and, therefore, inside this interval since the above term is a monotone function of t'. Then, (3.24) reduces to

$$z = -z_0 + \frac{2(z_0 - z_1)}{\beta_0^2 n^2}\frac{t - t_0}{t' + t_0},$$

$$\rho = \frac{2(z_0 - z_1)}{\beta_0^2 n^2} \frac{t - t_0}{t' + t_0} \left\{ \left[\frac{n\beta_0^2 c(t' + t_0)}{2(z_0 - z_1)} \right]^2 - 1 \right\}^{1/2}. \tag{3.36}$$

On the other hand, we cannot drop the terms with $(z_0 - z_1)$ in (3.36) since the denominator $(t' + t_0)$ is of the same order of smallness. It is seen that $z = z_c^{(1)}$, $\rho = \rho_c^{(1)} = 0$ at $t' = -t_c$ and $z = \tilde{z}_1$, $\rho = \rho_1$ at $t' = -t_1$ Here

$$z_c^{(1)} = c_n t - z_0(1 - \frac{1}{\beta_{0n}}), \quad \tilde{z}_1 = \frac{c_n^2}{v_0} t - z_0(1 - \frac{1}{\beta_{0n}^2}), \quad \rho_1 = \frac{c_n^2}{v_0 \gamma_{0n}}(t + t_0).$$

It follows from (3.36) that

$$\rho^2 + (z + z_0)^2 = c_n^2(t + t_0)^2 \tag{3.37}$$

that coincides with the equation of the BS shock wave arising at the beginning of the charge motion (BS$_1$, for short). This singular contour (SW$_1$ in Fig. 3.12 (b)) begins at the point $z = z_c^{(1)}$, $\rho = \rho_c^{(1)} = 0$ and ends at the point $z = \tilde{z}_1$, $\rho = \rho_1$. It represents the shock wave arising when the charge velocity coincides with the velocity of light in medium at the accelerated part of the charge trajectory. The fact that SW$_1$ and BS$_1$ are described by the same equation (3.37) is physically understandable since both these waves, due to the instantaneous velocity jump, are created at the same instant $t = -t_0$, at the same space point $z = -z_0$, and propagate with the same velocity c_n. It should be noted that the BS$_1$ shock wave is distributed over the whole sphere (3.37) while the singular shock wave SW$_1$ fills only its part.

The second part of the singular contour is the Cherenkov shock wave (CSW in Fig. 3.12 (b)) extending from the point $z = \tilde{z}_1$, $\rho = \rho_1$ to the point $z = z_2$, $\rho = \rho_2$. Here

$$z_2 = \frac{c_n^2}{v_0} t + z_0(1 - \frac{1}{\beta_{0n}^2}), \quad \rho_2 = \frac{c_n^2}{v_0 \gamma_{0n}}(t - t_0),$$

The third part of the singularity contour (SW$_2$ in Fig. 3.12 (b)) begins at the point $z = z_2$, $\rho = \rho_2$ and ends at $z = z_c^{(2)}$, $\rho = \rho_c^{(2)} = 0$. Here

$$z_c^{(2)} = c_n t + z_0(1 - \frac{1}{\beta_{0n}}).$$

This part of the singularity contour represents the shock wave arising at the decelerated part of the charge trajectory. It is described by the equation

$$\rho^2 + (z - z_0)^2 = c_n^2(t - t_0)^2 \tag{3.38}$$

coinciding with the equation of the BS$_2$ shock wave emitted at the end $(t = t_0, z = z_0)$ of a charge motion. Again, the singularity fills only part of the sphere (3.38).

Now we discuss why the configuration of the shock waves in the limiting case of the smooth Tamm problem (Fig. 3.12(b)) does not coincide with that of the original Tamm problem (Fig. 3.12(a)). It was shown in [18,19] that in the spectral representation the radiation intensity (for the fixed observation wavelength) of the smooth Tamm problem transforms into the radiation intensity of the original Tamm problem when the length of the trajectory along which a charge moves nonuniformly tends to zero. However, Figs. 3.12 ((a),(b)) describe the position of the EMF singularities at the fixed moment of the observational time (or, in other words, Figs. 3.12 ((a),(b)) correspond to the time representation). The time and spectral representations of the EMF are related by the Fourier transformation. For an arbitrary small but finite length l of the charge nonuniform motion in the smooth Tamm problem, the contribution of the non-uniform motion to the radiation intensity becomes essential and comparable with the contribution of the uniform motion for high frequencies. This was clearly shown in [18, 20]. Thus, the appearance of additional shock waves in Fig. 3.12 (b) is due to the contribution of high frequencies.

3.4. Concluding remarks for this chapter

What can we learn from this chapter?

1. For an accelerated charge motion beginning from a state of rest, the bremsstrahlung shock wave arises at the start of the motion. When the charge velocity coincides with the velocity of light c_n in medium, the complex arises consisting from two shock waves. One of them is the Cherenkov shock wave inclined at the angle θ_c, ($\cos\theta_c = 1/\beta n$, β is the current charge velocity) towards the motion axis. The other shock wave, closing the Cherenkov cone behind it, is perpendicular to the motion axis. As time advances, the dimensions of this complex grow.

2. For a decelerated motion terminating with the state of rest, the initial Cherenkov shock wave is transformed into a blunt shock wave when the charge velocity coincides with c_n. This blunt shock wave detaches from a charge and propagates with the velocity c_n.

3. For the smooth Tamm problem consisting of accelerated, decelerated and uniform motions, the bremsstrahlung shock wave arises at the beginning of the motion. At the instant when the charge velocity coincides with c_n the above complex consisting of the Cherenkov shock wave and the shock wave closing the Cherenkov cone appears. At the uniform part of the charge motion this complex moves without changing its form (only its dimensions grow). At the decelerated part of a charge trajectory the slope of the Cherenkov shock wave towards the motion axis tends to $\pi/2$, as the charge velocity approaches c_n. At this instant the above complex

detaches from the charge and propagates with the velocity c_n. When the charge motion terminates, the bremsstrahlung shock wave arises.

References

1. Frank I.M. (1988) *Vavilov-Cherenkov Radiation*, Nauka, Moscow.
2. Zrelov V.P. (1970) *Vavilov-Cherenkov Radiation in High-Energy Physics*, vols. 1 and 2, Israel Program for Scientific Translations.
3. Tyapkin A.A. (1993) On the Induced Radiation Caused by a Charged Relativistic Particle Below Cherenkov Threshold in a Gas *JINR Rapid Communications*, **No 3**, pp. 26-31.
4. Zrelov V.P., Ruzicka J. and Tyapkin A.A. (1998) Pre-Cherenkov Radiation as a Phenomenon of "Light Barrier" *JINR Rapid Communications* **No1[87]-98**, pp.23-25.
5. Zrelov V.P. and Ruzicka J. (1989) Analysis of Tamm's Problem on Charge Radiation at its Uniform Motion over a Finite Trajectory *Czech. J. Phys.*, **B 39**, pp. 368-383.
6. Schott G.A. (1912) *Electromagnetic Radiation*, Cambridge Univ. Press, Cambridge.
7. Smith G.S. (1993) Cherenkov Radiation from a Charge of Finite Size or a Bunch of Charges *Amer. J. Phys.*, **61**, pp. 147-155.
8. Afanasiev G.N., Shilov V.M. and Stepanovsky Yu.P. (2003) Questions concerning observation of the Vavilov-Cherenkov radiation *J. Phys. D* **36**, pp. 88-102.
9. Afanasiev G.N., Eliseev S.M. and Stepanovsky Yu.P. (1998) Transition of the Light Velocity in the Vavilov-Cherenkov Effect *Proc. Roy. Soc. London*, **A 454**, pp. 1049-1072.
10. Landau L.D. and Lifshitz E.M. (1962), *Fluid Mechanics*, Addison-Wesley, Reading.
11. Afanasiev G.N. and Kartavenko V.G. (1999) Cherenkov-like shock waves associated with surpassing the light velocity barrier *Canadian J. Phys.*, **77**, pp. 561-569.
12. Landau L.D. and Lifshitz E.M. (1971) *The Classical Theory of Fields*, Reading, Massachusetts, Pergamon, Oxford and Addison-Wesley.
13. Tamm I.E. (1939) Radiation Induced by Uniformly Moving Electrons, *J. Phys. USSR*, **1, No 5-6**, pp. 439-461.
14. Aitken D.K. et al. (1963) Transition Radiation in Cherenkov Detectors *Proc. Phys. Soc.*, **83**, pp. 710-722.
15. Lawson J.D. (1954) On the Relation between Cherenkov Radiation and Bremsstrahlung *Phil. Mag.*, **45**, pp.748-750;
    Lawson J.D. (1965) Cherenkov Radiation, "Physical" and "Unphysical", and its Relation to Radiation from an Accelerated Electron *Amer. J. Phys.*, **33**, pp. 1002-1005.
16. Zrelov V.P. and Ruzicka J. (1992) Optical Bremsstrahlung of Relativistic Particles in a Transparent Medium and its Relation to the Vavilov-Cherenkov Radiation *Czech. J. Phys.*, **42**, pp. 45-57.
17. Afanasiev G.N., Beshtoev Kh. and Stepanovsky Yu.P. (1996) Vavilov-Cherenkov Radiation in a Finite Region of Space *Helv. Phys. Acta*, **69**, pp. 111-129.
    Afanasiev G.N., Kartavenko V.G. and Stepanovsky Yu.P. (1999) On Tamm's Problem in the Vavilov-Cherenkov Radiation Theory *J.Phys. D: Applied Physics*, **32**, pp. 2029-2043.
18. Afanasiev G.N. and Shilov V.M. (2002) Cherenkov Radiation versus Bremsstrahlung in the Tamm Problem *J. Phys. D*, **35**, pp. 854-866.
19. Afanasiev G.N., Kartavenko V.G. and Stepanovsky Yu.P. (2003) Vavilov-Cherenkov and Transition Radiations on the dielectric and Metallic Spheres, *J. Math. Phys.* **44**, pp. 4026-4056.
20. Afanasiev G.N., Shilov V.M. and Stepanovsky Yu.P (2003) Numerical and Analytical Treatment of the Smoothed Tamm Problem *Ann. Phys. (Leipzig)* **12**, pp. 51-79.

CHAPTER 4

CHERENKOV RADIATION
IN A DISPERSIVE MEDIUM

4.1. Introduction

The radiation produced by fast electrons moving in medium was observed by P.A. Cherenkov in 1934 [1]. Tamm and Frank [2] considered the motion of a point charge in dispersive medium. They showed that a charge should radiate when its velocity exceeds the velocity of light in medium c_n. For the frequency independent electric permittivity, the electromagnetic strengths have δ-like singularities on the surface of the so-called Cherenkov (or Mach) cone [3]-[6]. This leads to the divergence of the quantities involving the product of electromagnetic strengths. In particular, this is true for the flux of the EMF. There are some ways of overcoming this difficulty. Tamm and Frank operated in the Fourier transformation. They integrated the energy flux up to some maximal frequency ω_0. The other way [7], widely used in quantum electrodynamics, is to represent the square of the δ function as a product of two factors: one is a δ function and other is the integral from the exponent taken over the interval $(-T, T)$ with a subsequent transition to the $T \rightarrow \infty$ limit. Owing to the δ function, the second integral reduces to $2T$. Dividing both parts of the equation (in which the product of two δ functions appears) by $2T$, one obtains, e.g., the energy flux per unit time.

The goal of this consideration is to evaluate the electromagnetic field (EMF) arising from the uniform motion of a charge in a non-magnetic medium described by the frequency dependent one-pole electric permittivity

$$\epsilon(\omega) = 1 + \frac{\omega_L^2}{\omega_0^2 - \omega^2}. \tag{4.1}$$

Equation (4.1) is a standard parametrization describing a lot of optical phenomena [8]. It is valid when the wavelength of the electromagnetic field is much larger than the distance between the particles of a medium on which the light scatters. The typical atomic dimensions are of the order $a \approx \hbar/mc\alpha$, $\alpha = e^2/\hbar c$, and m is the electron mass. This gives $\lambda = c/\omega \gg a$ or $\omega \ll mc^2\alpha/\hbar \approx 5 \times 10^{18}\mathrm{s}^{-1}$. The typical atomic frequencies are of the order $\omega_0 \approx mc^2/\hbar\alpha^2 \approx 10^{16}$ s^{-1}. Thus the integration region extends well beyond ω_0 [9]. For $\omega \gg \omega_0$, $\epsilon(\omega) \approx 1$, that is, atomic electrons have no enough time to be excited. Following the book [10] and review [11], we extrapolate the parametrization (4.1) to all ω. This means that we disregard the excitation of nuclear levels and discrete structure of scatterers. According to Brillouin ([10], p. 20):

> Also, we use the formulas of the dispersion theory in a somewhat more general way than can be justified physically. Namely, we extend these formulas to infinitesimally small wavelengths, while their derivation is justified only for wavelengths large compared with the distance between dispersing particles.

Sometimes in physical literature another representation of the dielectric permittivity is used (known as the Lorentz-Lorenz or Clausius-Mossotti formula (see, e.g., [9,10]):

$$\epsilon' = \frac{1 + 2\alpha(\omega)/3}{1 - \alpha(\omega)/3} = 1 + \frac{\omega_L^2}{\omega_0'^2 - \omega^2}, \quad \alpha(\omega) = \frac{\omega_L^2}{\omega_0^2 - \omega^2},$$

$$\omega_0'^2 = \omega_0^2 - \omega_L^2/3. \tag{4.2}$$

It is generally believed that $\epsilon(\omega)$ given by (4.1) describes optical properties of media for which $\epsilon(\omega)$ differs only slightly from unity (e.g., gases), whereas $\epsilon'(\omega)$ describes more general media (liquids, solids, etc.). We see that the qualitative behaviour of ϵ and ϵ' is almost the same if we identify ω_0 and ω_L with ω_0' and ω_L, respectively. This permits us to limit ourselves to the ϵ representation in the form (4.1).

So we intend to consider the effects arising from the charge motion in medium with $\epsilon(\omega)$ given by (4.1). This was partly done by E. Fermi in 1940 [12]. He showed that a charged particle moving uniformly in medium with permittivity (4.1) should radiate at every velocity. He also showed that energy losses as a function of the charge velocity are less than those predicted by the Bohr theory [13]. However, Fermi did not evaluate the electromagnetic strengths for various charge velocities and did not show how the transition takes place from the subluminal regime to the superluminal. The Fermi theory was extended to the case of many poles case by Sternheimer [14] who obtained satisfactory agreement with experimental data. Another development of the Fermi theory is its quantum generalization [15]-[17].

In this consideration we restrict ourselves to the classical theory of the Vavilov-Cherenkov radiation with electric permittivity given by (4.1) and its complex analog. It is suggested that the uniform motion of a charge is maintained by some external force the origin of which is not of interest for us.

There are experimental indications [18]-[20] that a uniformly moving charge radiates even if its velocity is less than the velocity of light in medium. It seems that the present consideration supports this claim.

4.2. Mathematical preliminaries

Consider a point charge e moving uniformly in a non-magnetic medium with a velocity v directed along the z axis. Its charge and current densities are given by

$$\rho(\vec{r}, t) = e\delta(x)\delta(y)\delta(z - vt), \quad j_z = v\rho.$$

Their Fourier transforms are

$$\rho(\vec{k}, \omega) = \int \rho(\vec{r}, t) \exp[i(\vec{k}\vec{r} - \omega t)]d^3\vec{r}dt = 2\pi e\delta(\omega - \vec{k}\vec{v}),$$

$$j_z(\vec{k}, \omega) = v\rho(\vec{k}, \omega). \tag{4.3}$$

In the $(\vec{k}, \omega)$ space the electromagnetic potentials are given by (see, e.g., [21])

$$\Phi(\vec{k}, \omega) = \frac{4\pi}{\epsilon} \frac{\rho(\vec{k}, \omega)}{k^2 - \frac{\omega^2}{c^2}\epsilon}, \quad A_z(\vec{k}, \omega) = 4\pi\beta \frac{\rho(\vec{k}, \omega)}{k^2 - \frac{\omega^2}{c^2}\epsilon}, \quad \beta = v/c. \tag{4.4}$$

Here $\epsilon(\omega)$ is the electric permittivity of medium. Its frequency dependence is chosen in a standard form (4.1). In the usual interpretation ω_L and ω_0 are the plasma frequency $\omega_L^2 = 4\pi N_e e^2/m$ (N_e is the number of electrons per unit volume, m is the electron mass) and some resonance frequency, respectively. Quantum mechanically, it can be associated with the energy excitation of the lowest atomic level. Our subsequent exposition does not depend on this particular interpretation of ω_L and ω_0. The static limit of $\epsilon(\omega)$ is

$$\epsilon_0 = \epsilon(\omega = 0) = 1 + \frac{\omega_L^2}{\omega_0^2}.$$

$\epsilon(\omega)$ has poles at $\omega = \pm\omega_0$. Being positive for $\omega^2 < \omega_0^2$ it jumps from $+\infty$ to $-\infty$ when one crosses the point $\omega^2 = \omega_0^2$; $\epsilon(\omega)$ has zero at $\omega^2 = \omega_3^2 = \omega_0^2 + \omega_L^2$ and tends to unity for $\omega \to \infty$. In Eq. (4.1) $\epsilon(\omega)$ is negative for $\omega_0^2 < \omega^2 < \omega_3^2$ (Fig. 4.1,a). For the free electromagnetic wave this leads to its damping in this ω region even for real $\epsilon(\omega)$ (see, e.g., [10,22]). It is seen that

$$\epsilon^{-1}(\omega) = 1 - \frac{\omega_L^2}{\omega_3^2 - \omega^2}$$

has a zero at $\omega^2 = \omega_0^2$ and a pole at $\omega^2 = \omega_3^2$.

For the EMF radiated by a point charge moving uniformly in a dielectric medium, the conditions for the damping are modified. It turns out that the damping takes place for $1 - \beta^2\epsilon > 0$. Otherwise ($1 - \beta^2\epsilon < 0$) there is no damping. This corresponds to the Tamm-Frank radiation condition. We now define domains where $1 - \beta^2\epsilon > 0$ and $1 - \beta^2\epsilon < 0$.

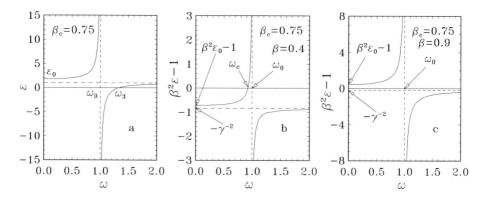

Figure 4.1. (a): For a free electromagnetic wave propagating in medium the damping region where $\epsilon < 0$ corresponds to $\omega_0^2 < \omega^2 < \omega_3^2 = \omega_0^2 + \omega_L^2$; (b): For the electromagnetic field radiated by a charge moving uniformly in medium with velocity $v < v_c$, the damping region where $1 - \beta^2\epsilon > 0$ lies within the intervals $0 < \omega < \omega_c$ and $\omega_0 < \omega < \infty$; (c): For the electromagnetic field radiated by a charge uniformly moving in medium with velocity $v > v_c$, the damping region where $1 - \beta^2\epsilon > 0$ extends from $\omega = \omega_0$ to $\omega = \infty$.

For $\beta < \beta_c$ one has $1 - \beta^2\epsilon > 0$ for $\omega^2 < \omega_c^2$ and $\omega^2 > \omega_0^2$ and $1 - \beta^2\epsilon < 0$ for $\omega_c^2 < \omega^2 < \omega_0^2$ (Fig. 1,b). For $\beta > \beta_c$ one obtains $1 - \beta^2\epsilon > 0$ for $\omega^2 > \omega_0^2$ and $1 - \beta^2\epsilon < 0$ for $0 < \omega^2 < \omega_0^2$ (Fig. 1 c) . Here

$$\beta_c = \epsilon_0^{-1/2} = 1/\sqrt{1 + \omega_L^2/\omega_0^2}, \quad \omega_c = \omega_0\sqrt{1 - \tilde\epsilon},$$

$$\tilde\epsilon = \beta^2\gamma^2/\beta_c^2\gamma_c^2, \quad \gamma^2 = (1 - \beta^2)^{-1}, \quad \gamma_c^2 = (1 - \beta_c^2)^{-1}.$$

In what follows, β_c, despite its formal appearance and independence of ω, will play an important role for the analysis of the EMF induced by a charge moving in medium with a frequency dependent permittivity. We apply Eq. (4.1) to the medium with $\beta_c = 0.75$, $n = \sqrt{\epsilon_0} = 1/\beta_c = 1.333$. The optical properties of this medium are close to those of water for which $n = 1.334$. It is seen that β_c changes from $\beta_c = 0$ for $N \gg 1$ up to $\beta_c = 1$ for $N = 0$. We refer to these limit cases as to optically dense and rarefied media, respectively.

In the $\vec{r}, t$ representation $\Phi(\vec{r}, t)$ and $\vec{A}(\vec{r}, t)$ are given by

$$\Phi(\vec{r}, t) = \frac{e}{\pi v}\int\frac{d\omega}{\epsilon}e^{i\omega(t - z/v)}\frac{kdk}{k^2 + (\omega^2/v^2)(1 - \beta^2\epsilon)}J_0(k\rho).$$

$$A_z(\vec{r}, t) = \frac{e}{\pi c}\int d\omega e^{i\omega(t - z/v)}\frac{kdk}{k^2 + (\omega^2/v^2)(1 - \beta^2\epsilon)}J_0(k\rho). \qquad (4.5)$$

The usual way of handling with these integrals is to integrate them first over k. For this we use the Table integral (see, e.g., [23])

$$\int_0^\infty \frac{kdk}{k^2 + q^2} J_0(k\rho) = K_0(\rho q), \qquad (4.6)$$

where in the right hand side the value of square root $\sqrt{q^2}$ corresponding to its positive real part should be taken.

4.3. Electromagnetic potentials and field strengths

As was shown in [11], the inclusion of the ω dependencies in ϵ and ϵ' effectively takes into account the retardation effects. The very fact that the velocity of light in medium c_n is less than the velocity of light in vacuum c means that oscillators of medium react to the initial electromagnetic field with some delay. The deviation of c_n from c is owed to the deviation of ϵ from unity. For the incoming plane wave and frequency independent ω this was clearly demonstrated in [24]-[26]. At first glance it seems that c_n will be greater than c for $\epsilon < 1$. However, a more accurate analysis shows [10] that the group velocity of light in medium is always less than c.

To evaluate integrals entering into (4.5) one should satisfy the condition $Re\sqrt{1 - \beta^2\epsilon} > 0$. It is satisfied automatically if $1 - \beta^2\epsilon > 0$. In this case the argument of the K_0 function is $(|\omega|\rho/v)\sqrt{1 - \beta^2\epsilon}$ where the square root means its arithmetic value. Now let $1 - \beta^2\epsilon < 0$. First, we consider the case when ϵ has the imaginary part:

$$\epsilon(\omega) = 1 + \frac{\omega_L^2}{\omega_0^2 - \omega^2 + ip\omega}, \qquad p > 0. \qquad (4.7)$$

The positivity of p leads to poles of $\epsilon(\omega)$ lying only in the upper complex ω half-plane. This is required to satisfy the causality condition (for details see [27]). Sometimes in physical literature [22] it is stated that the causality condition is satisfied if the poles of $\epsilon(\omega)$ lie in the lower ω half-plane. This is because of a different definition of the Fourier transforms corresponding to different signs of ω inside the exponentials occurring in (4.3). We are now able to write out explicit expressions for electromagnetic potentials and field strengths. In the cylindrical coordinates they are given by

$$\Phi = \frac{e}{\pi v} \int_{-\infty}^\infty \frac{d\omega}{\epsilon} e^{i\alpha} K_0(k\rho), \quad A_z = \frac{e}{\pi c} \int_{-\infty}^\infty d\omega e^{i\alpha} K_0(k\rho),$$

$$H_\phi = \beta D_\rho = \frac{e}{\pi c} \int_{-\infty}^\infty d\omega e^{i\alpha} k K_1(k\rho), \quad E_\rho = \frac{e}{\pi v} \int_{-\infty}^\infty \frac{d\omega}{\epsilon} e^{i\alpha} k K_1(k\rho),$$

$$E_z = -\frac{ie}{\pi c^2} \int_{-\infty}^{\infty} d\omega \omega (1 - \frac{1}{\beta^2 \epsilon}) e^{i\alpha} K_0(k\rho),$$

$$D_z = \frac{ie}{\pi v^2} \int_{-\infty}^{\infty} d\omega \omega (1 - \beta^2 \epsilon) e^{i\alpha} K_0(k\rho). \qquad (4.8)$$

Here $\alpha = \omega(t - z/v)$, $k^2 = (1 - \beta^2\epsilon)\omega^2/v^2$. Again, k in Eq.(4.8) means the value of $\sqrt{k^2}$ corresponding to $\mathrm{Re}\, k > 0$.

These expressions were obtained by Fermi [12]. Their drawback is that modified Bessel functions K are complex even for real ϵ (when $1 - \beta^2\epsilon < 0$). We intend now to present Eqs. (4.8) in a manifestly real form. This greatly simplifies calculations. We write $1 - \beta^2\epsilon$ in the form

$$1 - \beta^2 \epsilon = a + ib = \sqrt{a^2 + b^2}(\cos\phi + i\sin\phi) \qquad (4.9)$$

where

$$a = 1 - \beta^2 - \beta^2\omega_L^2 \frac{\omega_0^2 - \omega^2}{(\omega_0^2 - \omega^2)^2 + p^2\omega^2}, \qquad b = \beta^2\omega_L^2 \frac{\omega p}{(\omega_0^2 - \omega^2)^2 + p^2\omega^2},$$

$$\cos\phi = \frac{a}{\sqrt{a^2 + b^2}}, \qquad \sin\phi = \frac{b}{\sqrt{a^2 + b^2}}.$$

Now we take the square root of $1 - \beta^2\epsilon$. The positivity of $\mathrm{Re}\sqrt{1 - \beta^2\epsilon}$ defines it uniquely:

$$\sqrt{1 - \beta^2\epsilon} = (a^2 + b^2)^{1/4}(\cos\frac{\phi}{2} + i\sin\frac{\phi}{2}),$$

$$\cos\frac{\phi}{2} = \frac{1}{\sqrt{2}}(1 + \frac{a}{\sqrt{a^2 + b^2}})^{1/2}, \qquad \sin\frac{\phi}{2} = \frac{1}{\sqrt{2}}\frac{b}{|b|}(1 - \frac{a}{\sqrt{a^2 + b^2}})^{1/2}. \quad (4.10)$$

Thus the argument of K functions entering into (4.8) is

$$\rho \frac{|\omega|}{v}(a^2 + b^2)^{1/4}\left(\cos\frac{\phi}{2} + i\sin\frac{\phi}{2}\right). \qquad (4.11)$$

Although the integrands in (4.8) are complex, the integrals defining electromagnetic potentials and strengths are real. This is due to the fact that $\epsilon(-\omega) = \epsilon^*(\omega)$.

We now take the limit $p \to 0+$. Let $1 - \beta^2\epsilon > 0$ in this limit. Then $a > 0$, $b \to 0$, $\cos(\phi/2) \to 1$, $\sin(\phi/2) \to 0$, and $\sqrt{1 - \beta^2\epsilon}$ coincides with its arithmetic value. Now let $1 - \beta^2\epsilon < 0$. Then, $a < 0$, $b \to 0$, $\cos(\phi/2) \to 0$, $\sin(\phi/2) \to b/|b|$ and $\sqrt{1 - \beta^2\epsilon} = i\sqrt{|1 - \beta^2\epsilon|}\,\mathrm{sign}(\omega)$. (it has been taken into account that $p > 0$). This shows that the functions K entering into the right hand side of Eq. (4.8) reduce to

$$
K_0\left(i\rho\frac{|\omega|}{v}\sqrt{|1-\beta^2\epsilon|}\right) = -\frac{i\pi}{2}H_0^{(2)}\left(\rho\frac{|\omega|}{v}\sqrt{|1-\beta^2\epsilon|}\right),
$$

$$
K_1\left(i\rho\frac{|\omega|}{v}\sqrt{|1-\beta^2\epsilon|}\right) = -\frac{\pi}{2}H_1^{(2)}\left(\rho\frac{|\omega|}{v}\sqrt{|1-\beta^2\epsilon|}\right)
$$

for $\omega > 0$ and

$$
K_0\left(-i\rho\frac{|\omega|}{v}\sqrt{|1-\beta^2\epsilon|}\right) = \frac{i\pi}{2}H_0^{(1)}\left(\rho\frac{|\omega|}{v}\sqrt{|1-\beta^2\epsilon|}\right),
$$

$$
K_1\left(-i\rho\frac{|\omega|}{v}\sqrt{|1-\beta^2\epsilon|}\right) = -\frac{\pi}{2}H_1^{(1)}\left(\rho\frac{|\omega|}{v}\sqrt{|1-\beta^2\epsilon|}\right)
$$

for $\omega < 0$. Now we are able to write out electromagnetic potentials and field strengths in a manifestly real form. For $\beta < \beta_c$ one finds

$$
\Phi(\vec{r},t) = \frac{2e}{\pi v}\left(\int_0^{\omega_c}+\int_{\omega_0}^\infty\right)\frac{d\omega}{\epsilon}\cos\alpha K_0 + \frac{e}{v}\int_{\omega_c}^{\omega_0}\frac{d\omega}{\epsilon}\left(\sin\alpha J_0 - \cos\alpha N_0\right),
$$

$$
A_z(\vec{r},t) = \frac{2e}{\pi c}\left(\int_0^{\omega_c}+\int_{\omega_0}^\infty\right)d\omega\cos\alpha K_0 + \frac{e}{c}\int_{\omega_c}^{\omega_0}d\omega\left(\sin\alpha J_0 - \cos\alpha N_0\right), \quad (4.12)
$$

$$
H_\phi(\vec{r},t) = \frac{2e}{\pi cv}\left(\int_0^{\omega_c}+\int_{\omega_0}^\infty\right)\omega d\omega\sqrt{|1-\beta^2\epsilon|}\cos\alpha K_1
$$

$$
+\frac{e}{cv}\int_{\omega_c}^{\omega_0}\omega d\omega\sqrt{|1-\beta^2\epsilon|}\left(\sin\alpha J_1 - \cos\alpha N_1\right),
$$

$$
E_z = \frac{2e}{\pi c^2}\left(\int_0^{\omega_c}+\int_{\omega_0}^\infty\right)\left(1-\frac{1}{\epsilon\beta^2}\right)\omega d\omega\sin\alpha K_0
$$

$$
-\frac{e}{c^2}\int_{\omega_c}^{\omega_0}\left(1-\frac{1}{\epsilon\beta^2}\right)\omega d\omega(N_0\sin\alpha + J_0\cos\alpha),
$$

$$
E_\rho = \frac{2e}{\pi v^2}\left(\int_0^{\omega_c}+\int_{\omega_0}^\infty\right)d\omega\frac{\omega}{\epsilon}\sqrt{|1-\beta^2\epsilon|}\cos\alpha K_1
$$

$$+\frac{e}{v^2}\int\limits_{\omega_c}^{\omega_0} d\omega\frac{\omega}{\epsilon}\sqrt{|1-\beta^2\epsilon|}(\sin\alpha J_1-\cos\alpha N_1).$$

On the other hand, for $\beta > \beta_c$

$$\Phi(\vec{r},t)=\frac{2e}{\pi v}\int\limits_{\omega_0}^{\infty}\frac{d\omega}{\epsilon}\cos\alpha K_0+\frac{e}{v}\int\limits_{0}^{\omega_0}\frac{d\omega}{\epsilon}(\sin\alpha J_0-\cos\alpha N_0),$$

$$A_z(\vec{r},t)=\frac{2e}{\pi c}\int\limits_{\omega_0}^{\infty} d\omega\cos\alpha K_0+\frac{e}{c}\int\limits_{0}^{\omega_0} d\omega(\sin\alpha J_0-\cos\alpha N_0), \qquad (4.13)$$

$$H_\phi(\vec{r},t)=\frac{2e}{\pi c v}\int\limits_{\omega_0}^{\infty}\omega d\omega\sqrt{|1-\beta^2\epsilon|}\cos\alpha K_1+$$

$$+\frac{e}{c v}\int\limits_{0}^{\omega_0}\omega d\omega\sqrt{|1-\beta^2\epsilon|}(\sin\alpha J_1-\cos\alpha N_1),$$

$$E_z=\frac{2e}{\pi c^2}\int\limits_{\omega_0}^{\infty}\left(1-\frac{1}{\epsilon\beta^2}\right)\omega d\omega\sin\alpha K_0$$

$$-\frac{e}{c^2}\int\limits_{0}^{\omega_0}\left(1-\frac{1}{\epsilon\beta^2}\right)\omega d\omega(N_0\sin\alpha+J_0\cos\alpha),$$

$$E_\rho=\frac{2e}{\pi v^2}\int\limits_{\omega_0}^{\infty} d\omega\frac{\omega}{\epsilon}\sqrt{|1-\beta^2\epsilon|}\cos\alpha K_1$$

$$+\frac{e}{v^2}\int\limits_{0}^{\omega_0} d\omega\frac{\omega}{\epsilon}\sqrt{|1-\beta^2\epsilon|}\left(\sin\alpha J_1-\cos\alpha N_1\right).$$

Here $\alpha=\omega(t-z/v)$. The argument of all the Bessel functions is

$$\sqrt{|1-\beta^2\epsilon|}\rho\omega/v.$$

We observe that integrals containing usual $(J, \ N)$ and modified (K) Bessel functions are taken over spatial regions where $1-\beta^2\epsilon < 0$ and $1-\beta^2\epsilon > 0$, respectively. Consider particular cases of these expressions.

For $\omega_L \to 0$ we obtain: $\epsilon \to 1, \beta_c \to 1, \omega_c \to \omega_0$,

$$\Phi=\frac{2e}{\pi v}\int\limits_{0}^{\infty} d\omega\cos\alpha K_0(\frac{\rho\omega}{v\gamma})=\frac{e}{[(z-vt)^2+\rho^2/\gamma^2]^{1/2}},$$

$$A_z = \beta\Phi, \quad \gamma = 1/\sqrt{1-\beta^2}$$

i.e., we obtain the field of a charge moving uniformly in vacuum.

Let $v \to 0$. Then $\omega_c = \omega_0$ and

$$\Phi = \frac{2e}{\pi\epsilon_0}\int_0^\infty d\omega \cos\left(\frac{\omega z}{c}\right)K_0\left(\frac{\rho\omega}{c}\right) = \frac{e}{\epsilon_0}\frac{1}{\sqrt{\rho^2+z^2}}, \quad A_z = 0$$

i.e., we obtain the field of a charge resting in medium.

Let $\omega_0 \to \infty$, $\omega_L \to \infty$,, but ω_L/ω_0 is finite. Then

$$\omega_c = \omega_0\sqrt{1-\beta^2\gamma^2\omega_L^2/\omega_0^2} \to \infty, \quad \epsilon(\omega) \to \epsilon_0$$

and

$$\Phi = \frac{2e}{\pi v\epsilon_0}\int_0^\infty d\omega \cos\alpha K_0\left(\frac{\rho\omega}{v}\sqrt{1-\beta^2\epsilon_0}\right)$$

$$= \frac{e}{\epsilon_0}\frac{1}{[(z-vt)^2+\rho^2/\gamma_n^2]^{1/2}}, \quad A_z = \beta\epsilon_0\Phi$$

for $\beta < \beta_c$ and

$$\Phi = \frac{e}{v\epsilon_0}\int_0^\infty d\omega(\sin\alpha J_0 - \cos\alpha N_0)$$

$$= \frac{2e}{\epsilon_0}\frac{1}{[(z-vt)^2-\rho^2/\gamma_n^2]^{1/2}}\Theta(vt-z-\rho/\gamma_n), \quad A_z = \beta\epsilon_0\Phi$$

for $\beta > \beta_c$. Here $\gamma_n = 1/\sqrt{|1-\beta_n^2|}$, $\beta_n = v/c_n$, $c_n = c/\sqrt{\epsilon_0}$. Thus, we arrive at a charge motion in a medium with a constant electric permittivity $\epsilon = \epsilon_0$. It is seen that the EMF has the form of an oblate ellipsoid for $\beta < \beta_C$ and the Mach (or Cherenkov) cone with its vertex at the charge current position for $\beta > \beta_c$ (Fig. 4.2). Electromagnetic potentials are zero outside the Cherenkov cone ($z > vt - \rho/\gamma_n$), singular at its surface ($z = vt-\rho/\gamma_n$), and decrease as $1/r$ inside the Cherenkov cone ($z < vt - \rho/\gamma_n$). It should be stressed that the integration over the whole range of ω is required for obtaining correct limit expressions and for guaranteeing the reversibility of the Fourier transformation.

The distributions of the magnetic field strength H_ϕ as a function of z on the surface of a cylinder C_ρ of the radius ρ are shown in Figs. 4.3-4.5 for ϵ given by (4.1). If the dependence ϵ of ω were neglected ($\epsilon(\omega) = \epsilon_0$), then for $\beta > \beta_c = 1/n$ the electromagnetic field would be confined to the interior of the Cherenkov cone with the solution angle $2\theta_c$, $\sin\theta_c = \beta_c/\beta = 1/\beta n$ (Fig. 4.2). This means that on the surface of C_ρ the electromagnetic field would be zero for $-z_c < z < \infty$, $z_c = \rho\cot\theta_c = \rho\sqrt{\beta^2 n^2 - 1}$.

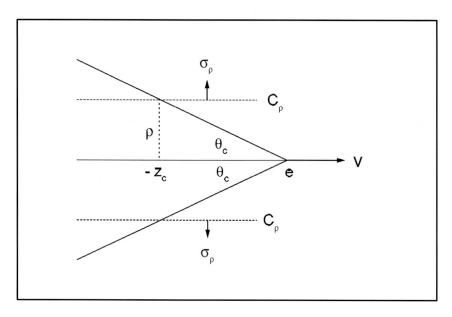

Figure 4.2. Schematic presentation of the Cherenkov cone attached to a charge moving in a dispersion-free medium. The radiation field is confined to the surface of the cone, the field inside the cone does not contribute to the radiation. On the surface of the cylinder C_ρ the electromagnetic field is zero for $z > -z_c$; σ_ρ means the radial energy flux through the cylinder surface.

What can we learn from figures 4.3-4.5 ?.

For a small charge velocity ($\beta \leq 0.4$) the magnetic field coincides with that of a charge moving inside medium with the constant $\epsilon = \epsilon_0$. For β slightly less than β_c ($\beta \approx 0.6$) oscillations appear for negative values of z. Their amplitude grows as β increases. For $\beta \approx \beta_c$ we see a number of peaks in the neighborhood of $z = 0$ with the amplitude slowly decreasing in the $z < 0$ region. For $\beta = \beta_c$ there is a large maximum at $z = -z_c$ and smaller ones in the region $z < -z_c$. The period of these oscillations approximately coincides with that of the medium polarization $T_z \approx 2\pi v \beta_c / \omega_0$.

Figures 4.3-4.5 demonstrate how the EMF is distributed over the surface of the cylinder C_ρ at a fixed instant of time t. Since all electromagnetic strengths depend on z and t via $z - vt$, the periodic dependence on time (with the period $2\pi \beta_c / \omega_0$) should be observed at a fixed spatial point.

It is seen that despite the ω dependence of ϵ, the critical velocity $\beta_c = 1/\sqrt{\epsilon_0}$ still has a physical meaning. Indeed, for $\beta > \beta_c$ the magnetic vector potential and field strength are very small outside the Mach cone ($z > -z_c$) exhibiting oscillations inside it ($z < -z_c$). For $\beta < \beta_c$ the Mach cone

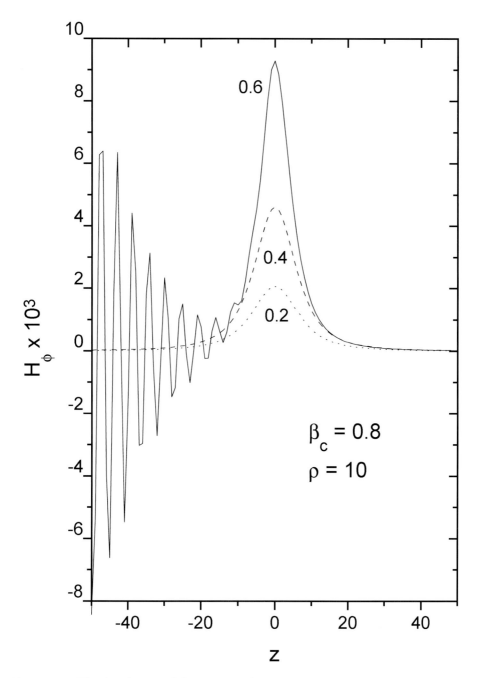

Figure 4.3. The distribution of the magnetic field strength on the surface of the cylinder C_ρ. The number of a particular curve means $\beta = v/c$; z and ρ are in units of c/ω_0; H_ϕ is in units of $e\omega_0^2/c^2$.

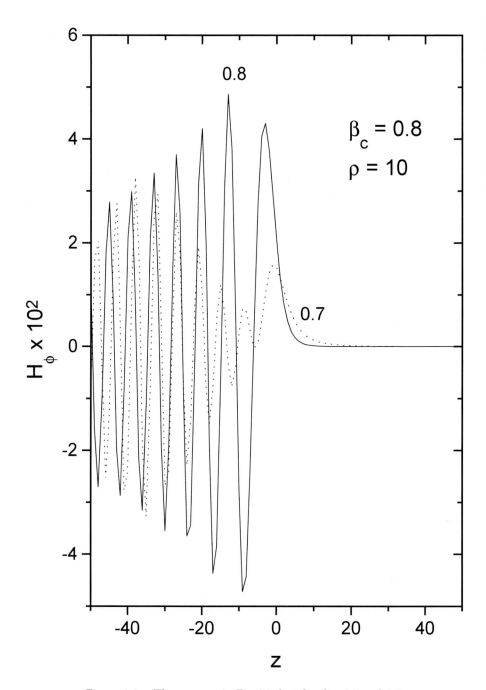

Figure 4.4.   The same as in Fig.4.3, but for $\beta = 0.7$ and 0.8.

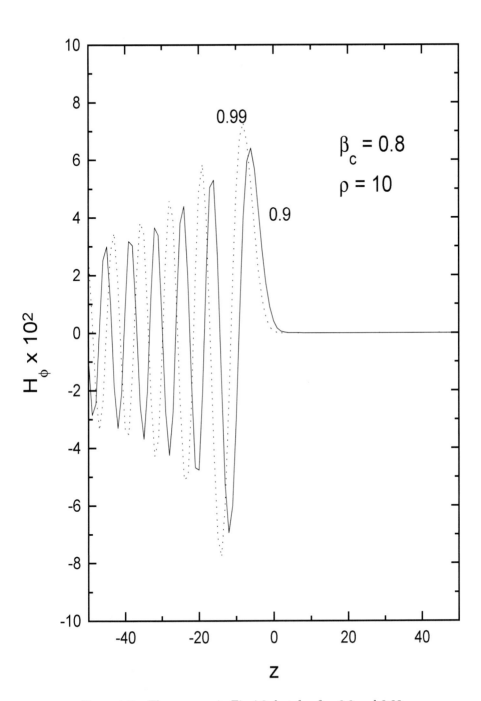

Figure 4.5. The same as in Fig.4.3, but for $\beta = 0.9$ and 0.99.

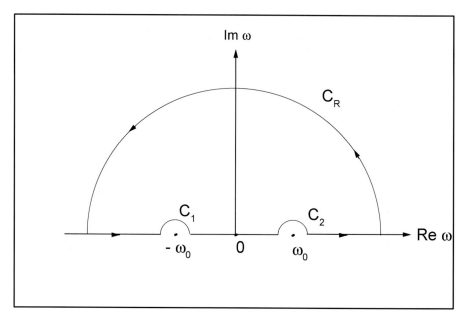

Figure 4.6.   The integration contour discussed in the text

disappears. The EMF being relatively small differs from zero everywhere. The magnetic field presented in Figs. 4.3-4.5 can be compared with its non-oscillating behaviour for the frequency-independent $\epsilon = \epsilon_0$:

$$H_\phi = \frac{e\beta\rho(\beta^2 n^2 - 1)}{[(z - vt)^2 - \rho^2(\beta^2 n^2 - 1)]^{3/2}}\Theta(vt - z - \rho/\gamma_n)$$

$$+ \frac{e\beta}{\gamma_n}\frac{\delta(vt - z - \rho/\gamma_n)}{[(z - vt)^2 - \rho^2(\beta^2 n_0^2 - 1)]^{1/2}}.$$

We turn again to Eqs. (4.12) and (4.13). The Fourier components of Φ and $\vec{E}$ have a pole at $\omega = \omega_3 = \sqrt{\omega_0^2 + \omega_L^2}$. This leads to the divergence of integrals defining Φ and $\vec{E}$. It would be tempting to approximate these integrals by their principal values. We illustrate this using Φ as an example (see Eq.(4.8)). Consider a closed contour C consisting of three real intervals $((-\infty, -\omega_0-\delta), \quad (-\omega_0+\delta, \omega_0-\delta), \quad (\omega_0+\delta, \infty))$, of two semi-circles C_1 and C_2 of the radius δ with their centers at $z = -\omega_0$ and $z = \omega_0$, respectively, and of a semi-circle C_R of the large radius R (Fig. 4.6). All semi-circles C_1, C_2 and C_R lie in the upper half-plane. The integral

$$\int \frac{d\omega}{\epsilon} e^{i\alpha} K_0(k\rho)$$

taken over the closed contour C equals zero if the function K_0 has no singularities inside C. The same integral taken over C_R is also 0 for $t - z/v > 0$ due to the exponential factor $e^{i\alpha}$. Therefore,

$$\left(\int_{-\infty}^{-\omega_0-\delta} + \int_{-\omega_0+\delta}^{\omega_0-\delta} + \int_{\omega_0+\delta}^{\infty} + \int_{C_1} + \int_{C_2} \right) \frac{d\omega}{\epsilon} e^{i\alpha} K_0(k\rho) = 0.$$

In the limit $\delta \to 0$ one obtains

$$\text{V.P.} \int_{-\infty}^{\infty} \frac{d\omega}{\epsilon} e^{i\alpha} K_0(k\rho) = -\left(\int_{C_1} + \int_{C_2} \right) \frac{d\omega}{\epsilon} e^{i\alpha} K_0(k\rho)$$

$$= -2\pi \frac{\omega_L^2}{\omega_3} \Theta(t - z/v) \sin \omega_3(t - z/v) K_0\left(\rho \frac{|\omega_3|}{v} \right).$$

For the electric potential one then finds

$$\Phi = -2 \frac{e}{v} \frac{\omega_L^2}{\omega_3} \Theta(t - z/v) \sin \omega_3(t - z/v) K_0\left(\rho \frac{|\omega_3|}{v} \right). \qquad (4.14)$$

We see that the principal value of the integral treated does not describe the Cherenkov cone. Probably, this is owing to singularities (poles and branch points) of the modified Bessel function in the upper ω half-plane. When evaluating (4.14) we did not take them into account.

4.4. Time-dependent polarization of the medium

Another, more physical, way to obtain EMF of a charge uniformly moving in medium is to start with the Maxwell equations

$$\text{div}\vec{D} = 4\pi\rho, \quad \text{div}\vec{B} = 0, \quad \text{curl}\vec{E} = -\frac{1}{c}\dot{\vec{B}}, \quad \text{curl}\vec{H} = \frac{1}{c}\dot{\vec{D}} + \frac{4\pi}{c}\vec{j}. \quad (4.15)$$

As the medium is non-magnetic, $\vec{B} = \vec{H}$. The second and third Maxwell equations are satisfied if we put

$$\vec{H} = \vec{\nabla} \times \vec{A}, \quad \vec{E} = -\vec{\nabla}\Phi - \frac{1}{c}\dot{\vec{A}}.$$

We rewrite Maxwell equations in the ω representation:

$$H_\phi^\omega = -\frac{\partial}{\partial\rho} A_z^\omega, \quad E_z^\omega = \frac{i\omega}{v}(\Phi^\omega - \beta A_z^\omega),$$

$$\frac{1}{\rho}\frac{\partial}{\partial\rho}\rho(E_\rho^\omega + 4\pi P_\rho^\omega) - \frac{i\omega}{v}(E_z^\omega + 4\pi P_z^\omega) = 4\pi\rho^\omega,$$

$$H_\phi^\omega = \beta(E_\rho^\omega + 4\pi P_\rho^\omega), \quad \frac{i\omega}{v} E_\rho^\omega + \frac{\partial E_z^\omega}{\partial \rho} = \frac{i\omega}{c} H_\phi^\omega. \qquad (4.16)$$

The last equation is satisfied trivially if we express electromagnetic strengths through the electromagnetic potentials:

$$E_\rho^\omega = -\frac{\partial \Phi^\omega}{\partial \rho}, \quad E_z^\omega = \frac{i\omega}{v} \Phi^\omega - \frac{i\omega}{c} A_z^\omega, \quad H_\phi^\omega = -\frac{\partial A_z^\omega}{\partial \rho}.$$

In deriving these equations we have taken into account that the z and t dependencies of the electromagnetic potentials, field strengths, polarization, charge and current densities enter through the factor $\exp[i\omega(t - z/v)]$ in their Fourier transforms.

The electric field $\vec{E}$ of a moving charge induces the polarization $\vec{P}(\vec{r}, t)$ which, being added to $\vec{E}$, gives the electric induction $\vec{D} = \vec{E} + 4\pi\vec{P}$. Usually it is believed (see, e.g., [8]-[11], [22], [27] that the ω components of $\vec{P}$ and $\vec{E}$

$$\vec{P}_\omega = \frac{1}{2\pi} \int e^{-i\omega t} \vec{P}(\vec{r}, t) dt, \quad \vec{E}_\omega = \frac{1}{2\pi} \int e^{-i\omega t} \vec{E}(\vec{r}, t) dt$$

are related by the formula

$$4\pi \vec{P}_\omega = \frac{\omega_L^2}{\omega_0^2 - \omega^2 + ip\omega} \vec{E}_\omega. \qquad (4.17)$$

Using this fact and expressing electromagnetic strengths in Eq.(4.16) through the potentials we obtain (taking into account that the last equation (4.16) is satisfied trivially):

$$\Delta_2 \Phi^\omega - \frac{\omega^2}{v^2} \Phi^\omega + \frac{i\omega}{c} \mathrm{div}\,\vec{A}^\omega = -\frac{1}{\epsilon} 4\pi \rho^\omega,$$

$$\Delta_2 A_z^\omega + \frac{\omega^2}{c^2} \epsilon A_z^\omega - \frac{\omega^2}{cv} \epsilon \Phi^\omega = -\frac{4\pi}{c} j_z^\omega, \quad \frac{\partial A_z^\omega}{\partial \rho} = \beta\epsilon \frac{\partial \Phi^\omega}{\partial \rho}. \qquad (4.18)$$

Here

$$\rho^\omega = \frac{e}{v}\delta(x)\delta(y)\exp(-i\omega z/v), \quad j_z^\omega = e\delta(x)\delta(y)\exp(-i\omega z/v),$$

$$\Delta_2 = \frac{1}{\rho}\frac{\partial}{\partial \rho}\left(\rho\frac{\partial}{\partial \rho}\right).$$

The last equation (4.18) is satisfied if we choose

$$A_z^\omega = \beta\epsilon(\omega)\Phi^\omega. \qquad (4.19)$$

whilst two others coincide after this substitution. The solutions of these equations are

$$\Phi^\omega = \frac{2e}{v\epsilon} K_0(\frac{\rho|\omega|}{v}\sqrt{1-\beta^2\epsilon}), \quad A_z^\omega = \frac{2e}{c} K_0(\frac{\rho|\omega|}{v}\sqrt{1-\beta^2\epsilon})$$

(again, a square root means its value with a positive real part).
Now we rewrite Eq.(4.17) in the $(\vec{r}, t)$ representation:

$$\vec{P}(t) = \frac{1}{8\pi^2} \int\limits_{-\infty}^{\infty} G(t-t')\vec{E}(t'),$$

where

$$G(t-t') = \omega_L^2 \int\limits_{-\infty}^{+\infty} \frac{d\omega}{\omega_0^2 - \omega^2 + ip\omega} e^{i\omega(t-t')}, \tag{4.20}$$

Taking into account the positivity of p one finds:
    a) for $p < \omega_0$:
$G(t-t') = 0$ for $t' > t$ and

$$G(t-t') = \frac{2\pi\omega_L^2}{\sqrt{\omega_0^2 - p^2/4}} \exp\left[-p(t-t')/2\right] \sin\left[\sqrt{\omega_0^2 - p^2/4}(t-t')\right]$$

for $t' < t$.
    b) for $p > \omega_0$ (this case is unrealistic because usually $p \ll \omega_0$):
$G(t-t') = 0$ for $t' > t$ and

$$G(t-t') = \frac{2\pi\omega_L^2}{\sqrt{\omega_0^2 - p^2/4}} \exp\left[-p(t-t')/2\right] \sinh\left[\sqrt{p^2/4 - \omega_0^2}(t-t')\right]$$

for $t' < t$.
    As a result of the positivity of p, the value of the polarization $\vec{P}$ at the instant t is defined by the values of the electric field $\vec{E}$ in preceding times (causality principle). The source of polarization is distributed along the z axis:

$$\text{div}\vec{P} = -\frac{e}{v}\delta(x)\delta(y)\frac{\omega_L^2}{\sqrt{\omega_0^2 + \omega_L^2 - p^2/4}} \exp\left[-p(t-z/v)/2\right]$$

$$\times \sin[\sqrt{\omega_0^2 + \omega_L^2 - p^2/4}(t-z/v)]$$

for $z < vt$ and $\text{div}\vec{P} = 0$ for $z > vt$ (this equation is related to the $\omega_0^2 + \omega_L^2 - p^2/4 > 0$ case). The origin of oscillations of the potentials and field strengths behind the Cherenkov cone now becomes understandable. A moving charge gives rise to a time-dependent polarization source which, in the absence of damping, oscillates with a frequency $\sqrt{\omega_0^2 + \omega_L^2}$. The oscillations of polarization, being added, lead to the appearance of the smoothed Cherenkov cones enclosed in each other. On the surface of the cylindrical surface C_ρ they are manifested as maxima of the potentials, field strengths, and intensities. The position of the first maximum approximately coincides with the position of the singular Cherenkov cone in the absence of dispersion. The latter case is obtained if we neglect the ω dependence in the denominator of the integral in (4.20):

$$G(t - t') = 2\pi \frac{\omega_L^2}{\omega_0^2} \delta(t - t').$$

Obviously this can be realized for large values of ω_0. The introduction of damping should lead to the decreasing of secondary maxima. To verify this we have evaluated the magnetic vector potential for various values of the parameter p (in units of ω_0) defining the imaginary part of $\epsilon(\omega)$. We see (Fig. 4.7) that for $p \geq 1$ the secondary oscillations disappear. Although the polarization formalism leads to the same expressions (4.12) and (4.13) for the electromagnetic potentials and field strengths, it presents another, more physical, point of view on the nature of the Vavilov-Cherenkov radiation.

4.4.1. ANOTHER CHOICE OF POLARIZATION

So far we have dealt with the gauge condition of the form $A_z^\omega = \beta\epsilon(\omega)\Phi^\omega$. It looks highly non-local in the $(\vec{r}, t)$ representation. There is another interesting possibility. We substitute

$$\vec{E} = -\vec{\nabla}\Phi - \frac{1}{c}\frac{\partial \vec{A}}{\partial t}, \quad \vec{H} = \vec{\nabla} \times \vec{A}$$

into the first and fourth Maxwell equations (4.15) (second and third equations are satisfied automatically) and obtain

$$\Delta\Phi + \frac{1}{c}\text{div}\vec{A} = -4\pi\rho + 4\pi\text{div}\vec{P},$$

$$\Delta\vec{A} - \frac{1}{c^2}\ddot{\vec{A}} = \vec{\nabla}(\text{div}\vec{A} + \frac{1}{c}\dot{\Phi}) - \frac{4\pi}{c}(\vec{j} + \dot{\vec{P}}).$$

We try to separate equations for Φ and $\vec{A}$ by imposing on them the Lorentz condition

$$\text{div}\vec{A} + \frac{1}{c}\dot{\Phi} = 0 \tag{4.21}.$$

This equation is satisfied automatically if we put

$$A_x = A_y = 0, \quad A_z = \beta\Phi \tag{4.22}$$

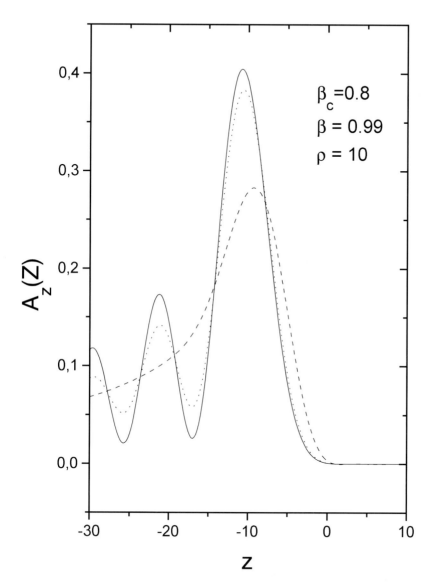

$\beta_c = 0.8$

$\beta = 0.99$

$\rho = 10$

Figure 4.7. Shows how switching on the imaginary part p of the dielectric permittivity affects the magnetic vector potential; z and A_z are in units of c/ω_0 and $e\omega_0/c$, respectively. The solid, point-like, and short dashed curves refer to $p = 0$, $p = 0.1$ and $p = 1$ (p is in ω_0 units) , respectively. It is seen that secondary maxima are damped for $p = 1$ more strongly than the main maximum.

(it has been taken into account that for the problem treated all the electromagnetic quantities depend on z and t through the combination $(z - vt)$). Thus we obtain

$$\Delta\Phi - \frac{1}{c^2}\ddot{\Phi} = -4\pi\rho + 4\pi\mathrm{div}\dot{\vec{P}},$$

$$\Delta\vec{A} - \frac{1}{c^2}\ddot{\vec{A}} = -4\frac{\pi}{c}\vec{j} - 4\frac{\pi}{c}\dot{\vec{P}}.$$

It follows from this that only the z component of $\vec{P}$ differs from zero in the chosen gauge (as only the z components of $\vec{A}$ and $\vec{j}$ differ from zero). We rewrite these equations in the ω representation

$$\Delta_2\Phi^\omega + \omega^2(\frac{1}{c^2} - \frac{1}{v^2})\Phi^\omega = -4\pi\rho^\omega - 4\pi\frac{i\omega}{v}P_\omega,$$

$$\Delta_2 A_z^\omega + \omega^2(\frac{1}{c^2} - \frac{1}{v^2})A_z^\omega = -\frac{4\pi}{c}j_z^\omega - 4\pi\frac{i\omega}{c}P_\omega \qquad (4.23)$$

As the medium treated is non-magnetic it is natural to require the coincidence of equations (4.18) and (4.23) for vector potentials satisfying different gauge conditions. This takes place if P_ω is chosen to be proportional to A_z^ω:

$$P_\omega = -\frac{i\omega}{4\pi c}(\epsilon - 1)A_z^\omega. \qquad (4.24)$$

One then obtains

$$\Delta_2\Phi^\omega + \omega^2(\frac{\epsilon}{c^2} - \frac{1}{v^2})\Phi^\omega = -4\pi\rho^\omega,$$

$$\Delta_2 A_z^\omega + \omega^2(\frac{\epsilon}{c^2} - \frac{1}{v^2})A_z^\omega = -\frac{4\pi}{c}j_z^\omega.$$

The solutions of these equations are

$$\Phi^\omega = \frac{2e}{v}K_0, \quad A_z^\omega = \frac{2e}{c}K_0,$$

$$H_\phi^\omega = \frac{2e|\omega|}{cv}\sqrt{1 - \beta^2\epsilon}K_1, \quad E_\rho^\omega = D_\rho^\omega = H_\phi^\omega/\beta,$$

$$E_z = \frac{2ie\omega}{v^2}(1 - \beta^2)K_0, \quad D_z = \frac{2ie\omega}{v^2}(1 - \beta^2\epsilon)K_0, \qquad (4.25)$$

where all K functions depend on the argument $(\rho\omega/v)\sqrt{1 - \beta^2\epsilon}$ in which the value of $\sqrt{1 - \beta^2\epsilon}$ corresponding to its positive real part should be taken. Obviously there is no proportionality between $\vec{D}$ and $\vec{E}$ for the chosen gauge. In the $(\vec{r}, t)$ representations the magnetic vector potential and field strength coincide with those in Eqs.(4.12) and (4.13), whilst for Φ, E_z, and E_ρ one has

$$\Phi(\vec{r},t) = \frac{2e}{\pi v}\left(\int_0^{\omega_c} + \int_{\omega_0}^{\infty}\right) d\omega \cos\alpha K_0 + \frac{e}{v}\int_{\omega_c}^{\omega_0} d\omega(\sin\alpha J_0 - \cos\alpha N_0),$$

$$E_z = \frac{2e}{\pi c^2}\left(1 - \frac{1}{\beta^2}\right)$$
$$\times \left[\left(\int_0^{\omega_c} + \int_{\omega_0}^{\infty}\right)\omega d\omega \sin\alpha K_0 - \frac{e}{c^2}\int_{\omega_c}^{\omega_0}\omega d\omega(N_0 \sin\alpha + J_0 \cos\alpha)\right],$$

$$E_\rho = \frac{2e}{\pi v^2}\left(\int_0^{\omega_c} + \int_{\omega_0}^{\infty}\right) d\omega\omega\sqrt{|1-\beta^2\epsilon|}\cos\alpha K_1$$
$$+\frac{e}{v^2}\int_{\omega_c}^{\omega_0} d\omega\omega\sqrt{|1-\beta^2\epsilon|}(\sin\alpha J_1 - \cos\alpha N_1).$$

for $\beta < \beta_c$ and

$$\Phi(\vec{r},t) = \frac{2e}{\pi v}\int_{\omega_0}^{\infty} d\omega \cos\alpha K_0 + \frac{e}{v}\int_0^{\omega_0} d\omega(\sin\alpha J_0 - \cos\alpha N_0),$$

$$E_z = \frac{2e}{\pi c^2}\left(1 - \frac{1}{\beta^2}\right)\left[\int_{\omega_0}^{\infty}\omega d\omega \sin\alpha K_0 - \frac{e}{c^2}\int_0^{\omega_0}\omega d\omega(N_0 \sin\alpha + J_0 \cos\alpha)\right],$$

$$E_\rho = \frac{2e}{\pi v^2}\int_{\omega_0}^{\infty} d\omega\omega\sqrt{|1-\beta^2\epsilon|}\cos\alpha K_1$$
$$+\frac{e}{v^2}\int_0^{\omega_0} d\omega\omega\sqrt{|1-\beta^2\epsilon|}(\sin\alpha J_1 - \cos\alpha N_1).$$

for $\beta > \beta_c$. These expressions satisfy the Maxwell equations but with the polarization different from that used earlier. We observe that the electric induction $\vec{D}$ is the same as above, but the electric strength differs. As the integrands defining Φ and $\vec{E}$ are finite for any value of ω, the corresponding integrals are convergent and can be evaluated numerically. We observe that $E_z \to 0$ for $\beta \to 1$. This means that for this choice of polarization and $v \approx c$ the energy flux in the transverse direction disappears, that is, for $v \approx c$ all the energy is radiated in the direction of the charge motion.

It is surprising that the choice (4.21) of the Lorentz condition almost inevitably leads to a solution with vanishing ρ component of polarization. But the physics cannot depend on the gauge choice. Checking all steps (4.21)-(4.25) in deriving field strengths we observe that the sole weak point in this chain is Eq. (4.22), which is the simplest realization of the gauge condition (4.21). Obviously, Eq. (4.22) can be realized in a variety ways. In particular, it can be realized with two non-vanishing components (A_z and A_ρ) of $\vec{A}$ ($A_\phi = 0$ owing to the axial symmetry of the treated problem). In this case we obtain the polarization and field strengths given in section 3 but with different electromagnetic potentials.

We conclude that different definitions (4.17) and (4.24) of the induced polarization proportional to the electric strength $\vec{E}$ and magnetic vector potential $\vec{A}$, respectively, lead to different physical consequences.

4.5. On the Krönig-Kramers dispersion relations

Up to now we have considered the case when the imaginary part of the dielectric penetrability was chosen to be zero. Can this be reconciled with the Krönig-Kramers dispersion relations? Since for the chosen form of the Fourier integrals the poles of $\epsilon(\omega)$ lie in the upper ω half-plane, one has (see, e.g.,[22]):

$$\int\limits_{-\infty}^{+\infty} \frac{\epsilon(x) - 1}{\omega - x} d\omega + i\pi[\epsilon(x) - 1] = 0.$$

Or, separating real and imaginary parts

$$\int\limits_{-\infty}^{\infty} \frac{\epsilon_r - 1}{\omega - x} d\omega = \pi\epsilon_i(x), \qquad \int\limits_{-\infty}^{\infty} \frac{\epsilon_i}{\omega - x} d\omega = -\pi[\epsilon_r(x) - 1] \qquad (4.26)$$

(by the integrals we mean their principal values obtained by closing the integration contour in the lower ω half-plane). Here ϵ_r and ϵ_i are the real and imaginary parts of ω:

$$\epsilon_r = 1 + \frac{\omega_L^2(\omega_0^2 - \omega^2)}{(\omega_0^2 - \omega^2)^2 + p^2\omega^2}, \qquad \epsilon_i = -\frac{p\omega\omega_L^2}{(\omega_0^2 - \omega^2)^2 + p^2\omega^2}. \qquad (4.27)$$

At first glance it seems that the relations (4.26) cannot be valid. Take, e.g., the second of them. For $\epsilon_i = 0$ its left hand side disappears, which is not valid for its right hand side. However, we cannot put $\epsilon_i = 0$ 'by hand'. The value of imaginary part of ϵ is determined by the parameter p in (4.27). Thus we should substitute ϵ_i given by (4.27) into (4.26) and then let p go to zero. For the integral entering into the left hand side of (4.26) one finds

$$\int_{-\infty}^{\infty} \frac{\epsilon_i}{\omega - x} d\omega = -p\omega_L^2 \int \frac{\omega d\omega}{\omega - x} \frac{1}{(\omega_0^2 - \omega^2)^2 + p^2\omega^2}. \qquad (4.28)$$

A detailed consideration shows that the integral in the right hand side of this equation is equal to

$$-\frac{\pi}{p} \frac{x^2 - \omega_0^2}{(x^2 - \omega_0^2)^2 + p^2 x^2}. \qquad (4.29)$$

The factor p of the integral in (4.28) compensates the factor $1/p$ in (4.29). Thus

$$\int_{-\infty}^{\infty} \frac{\epsilon_i}{\omega - x} d\omega = \pi \omega_L^2 \frac{x^2 - \omega_0^2}{(x^2 - \omega_0^2)^2 + p^2 x^2},$$

that coincides exactly with the right hand side of the second relation (4.26). The same reasoning proves the validity of the first relation (4.28). Thus, the Krönig-Kramers relations are valid for any small $p > 0$. The positivity of p defines how the integration contour should be closed, which in turn leads to the validity of the causality condition.

4.6. The energy flux and the number of photons

We evaluate now the energy flux per unit length through the surface of a cylinder C_ρ (Fig.4.2) coaxial with the z axis for the total time of motion. It is given by

$$W_\rho = 2\pi\rho \int_{-\infty}^{+\infty} S_\rho dt = \frac{2\pi\rho}{v} \int_{-\infty}^{+\infty} S_\rho dz,$$

$$S_\rho = \frac{c}{4\pi} (\vec{E} \times \vec{H})_\rho = -\frac{c}{4\pi} E_z H_\phi. \qquad (4.30)$$

Substituting E_z and H_ϕ from (4.12) and (4.13) and taking into account that

$$\int_{-\infty}^{\infty} dt \sin \omega t \cos \omega' t = 0, \quad \int_{-\infty}^{\infty} dt \sin \omega t \sin \omega' t = \pi[\delta(\omega - \omega') - \delta(\omega + \omega')],$$

$$\int_{-\infty}^{\infty} dt \cos \omega t \cos \omega' t = \pi[\delta(\omega - \omega') + \delta(\omega + \omega')],$$

we obtain for energy losses per unit length

$$W_\rho = \frac{e^2}{c^2} \int\limits_{\beta^2\epsilon>1} \omega d\omega \left(1 - \frac{1}{\epsilon\beta^2}\right). \tag{4.31}$$

This expression was obtained by Tamm and Frank [2]. Inserting $\epsilon(\omega)$ given by (4.1) into (4.31) we find

$$W_\rho = \frac{e^2}{c^2} \int\limits_{\omega_c}^{\omega_0} \omega d\omega \left(1 - \frac{1}{\epsilon\beta^2}\right) = -\frac{e^2\omega_0^2}{2c^2\beta_c^2\gamma_c^2}\left[1 + \frac{1}{\beta^2}\ln(1-\beta^2)\right] \tag{4.32}$$

for $\beta < \beta_c$ and

$$W_\rho = \frac{e^2}{c^2} \int\limits_0^{\omega_0} \omega d\omega \left(1 - \frac{1}{\epsilon\beta^2}\right) = \frac{e^2\omega_0^2}{2c^2}\left[-\frac{1}{\beta^2\gamma^2} + \frac{1}{\beta^2\beta_c^2\gamma_c^2}\ln(\gamma_c^2)\right] \tag{4.33}$$

for $\beta > \beta_c$.

Similar expressions were obtained by Fermi [12]. The validity of (4.33) is also confirmed by the results obtained by Sternheimer [14] (whose equations reduce to (4.33) in the limit $p \to 0$) and Ginzburg [28].

We observe that only those terms in (4.12) and (4.13) contribute to the radial energy flux for the total time of motion which contain the usual Bessel functions (J_μ and N_μ) and correspond to the $1 - \beta^2\epsilon < 0$ region without damping. This permits us to avoid difficulties connected with the above-mentioned pole of ϵ^{-1} (at $\omega = \omega_3$) which appears only in terms containing modified Bessel functions in the damping region where $1 - \beta^2\epsilon > 0$.

For $\beta \to 0$ the energy losses W_ρ tend to 0, whilst for $\beta \to 1$ (only this limit was considered by Tamm and Frank [29]) they tend to the finite value

$$\frac{e^2\omega_0^2}{2c^2\beta_c^2\gamma_c^2}\ln(\gamma_c^2).$$

In Fig. 4.8 we present the dimensionless quantity $F = W_\rho/(e^2\omega_0^2/c^2)$ as a function of the charge velocity β. The numbers on the curves mean β_c. The vertical lines with arrows divide each curve into two parts corresponding to the energy losses with velocities $\beta < \beta_c$ and $\beta > \beta_c$ and lying to the left and right of vertical lines, respectively. We see that a charge moving uniformly in a medium with dispersion law (4.1) radiates at every velocity.

Exactly the same Eqs. (4.31)-(4.33) are obtained if one starts from the complex $\epsilon(\omega)$ given by (4.7), evaluates electromagnetic strengths and radial energy flux, and then takes the limit $p \to 0$ in them. This will be shown below.

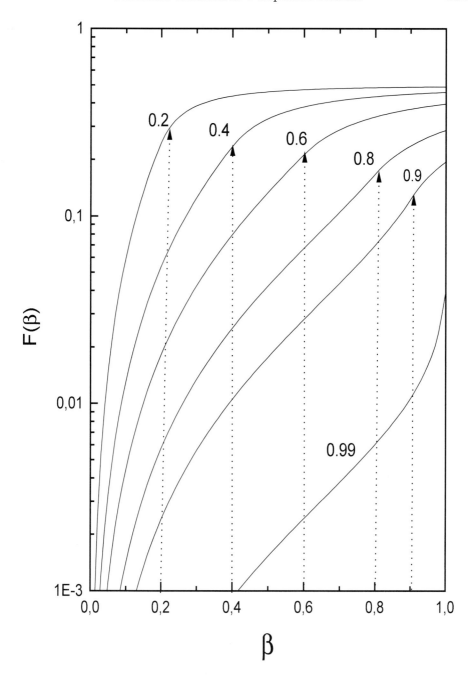

Figure 4.8. The radial energy losses per unit length (in units of $e^2\omega_0^2/c^2$) as a function of $\beta = v/c$. The number on a particular curve means the critical velocity β_c.

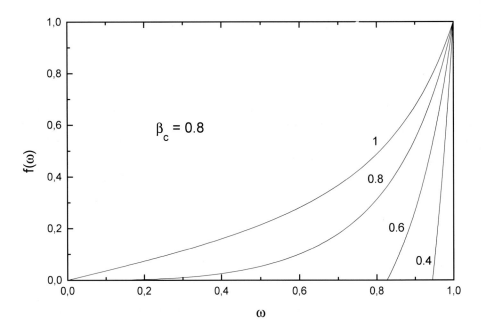

Figure 4.9. Spectral distribution of the energy losses (in units of $e^2\omega_0/c^2$); ω is in units of ω_0. The number on a particular curve refers to $\beta = v/c$.

The dimensionless spectral distributions $f(\omega) = w(\omega)/(e^2\omega_0/c^2)$ of the energy loss $W_\rho = \int\limits_0^\infty w(\omega)d\omega$ are shown in Fig. 4.9. The numbers on particular curves mean β. It is seen that for $\beta > \beta_c$ all ω from the interval $0 < \omega < \omega_0$ contribute to the energy losses. For $\beta < \beta_c$ the interval of permissible ω ($\omega_c < \omega < \omega_0$) diminishes.

The total number of photons emitted per unit length is given by

$$N = \frac{e^2}{\hbar c^2} \int\limits_{\omega_c}^{\omega_0} d\omega \left(1 - \frac{1}{\epsilon\beta^2} \right)$$

$$= \frac{e^2}{\hbar c^2} \left[\frac{\omega_c - \omega_0}{\beta^2\gamma^2} + \frac{\omega_L^2}{2\beta^2\omega_3} \ln\left(\frac{\omega_3 + \omega_0}{\omega_3 - \omega_0} \frac{\omega_3 - \omega_c}{\omega_3 + \omega_c} \right) \right]$$

for $\beta < \beta_c$ and

$$N = \frac{e^2}{\hbar c^2} \int\limits_0^{\omega_0} d\omega \left(1 - \frac{1}{\epsilon\beta^2} \right) = \frac{e^2}{\hbar c^2} \left[-\frac{\omega_0}{\beta^2\gamma^2} + \frac{\omega_L^2}{2\beta^2\omega_3} \ln\left(\frac{\omega_3 + \omega_0}{\omega_3 - \omega_0} \right) \right]$$

for $\beta > \beta_c$. It is seen that N grows from 0 for $\beta = 0$ up to

$$N = \frac{e^2}{\hbar c^2} \frac{\omega_L^2}{2\beta^2 \omega_3} \ln\left(\frac{\omega_3 + \omega_0}{\omega_3 - \omega_0}\right)$$

for $\beta = 1$. In Fig. 4.10 we present the dimensionless quantity $N/(e^2\omega_0/\hbar c^2)$ as a function of the particle velocity β. The numbers on the curves mean β_c. The vertical lines with arrows divide each curve into two parts corresponding to the photon numbers emitted by the charge with velocities $\beta < \beta_c$ and $\beta > \beta_c$ and lying to the left and right of vertical lines, respectively. We see that an uniformly moving charge emits photons at every velocity. The spectral distribution $n(\omega)$ of the photon number emitted per unit length and per unit frequency defined as $N = \int\limits_0^\infty n(\omega)d\omega$ is given by

$$n(\omega) = \frac{e^2}{\hbar c^2}\left(1 - \frac{1}{\epsilon\beta^2}\right).$$

For $\beta < \beta_c$, $n(\omega)$ changes from 0 at $\omega = \omega_c$ up to $n(\omega) = e^2/\hbar c^2$ at $\omega = \omega_0$. For $\beta > \beta_c$, $n(\omega)$ changes from $(e^2/\hbar c^2)(1 - 1/(\epsilon_0\beta^2))$ at $\omega = \omega_c$ up to $e^2/\hbar c^2$ at $\omega = \omega_0$. The dimensionless spectral distributions $n(\omega)/(e^2/\hbar c^2)$ of the photon number are shown in Fig. 4.11. The numbers of a particular curve mean β. It is seen that for $\beta > \beta_c$ all ω from the interval $0 < \omega < \omega_0$ contribute to the number of emitted photons. For $\beta < \beta_c$ the interval of permissible ω ($\omega_c < \omega < \omega_0$) diminishes, i.e., only high-energy photons contribute.

So far we have evaluated the total energy losses (i.e., for the whole time of the charge motion) per unit length. The question arises of how the radiated flux is distributed in space at a fixed instant of time. The distributions of the radial energy flux $\sigma_\rho = 2\pi\rho S_\rho$ on the surface of the cylinder C_ρ of the radius $\rho = 10$ (in units of c/ω_0) are shown in Figs. 4.12 and 4.13 for $\beta_c = 0.8$ and various charge velocities β. It is seen that despite the ω dependence of ϵ the critical velocity $\beta_c = 1/\sqrt{\epsilon_0}$ has still a physical meaning. Indeed, for $\beta > \beta_c$ the electromagnetic energy flux is very small outside the Cherenkov cone, exhibiting oscillations in its neighbourhood. For $\beta < \beta_c$ the radial flux diminishes and becomes negligible for $\beta \le 0.4$ (Fig. 4.13). This disagrees with Fig. 4.8, where for $\beta_c = 0.8$ one sees the finite value of energy losses for $\beta = 0.4$. In the next section we remove this inconsistency.

We have considered the distribution of the EMF on the surface of C_ρ at the fixed instant of time t. Since all electromagnetic strengths depend on z and t via the combination $z - vt$, the periodic dependence of time should be observed at a fixed spatial point.

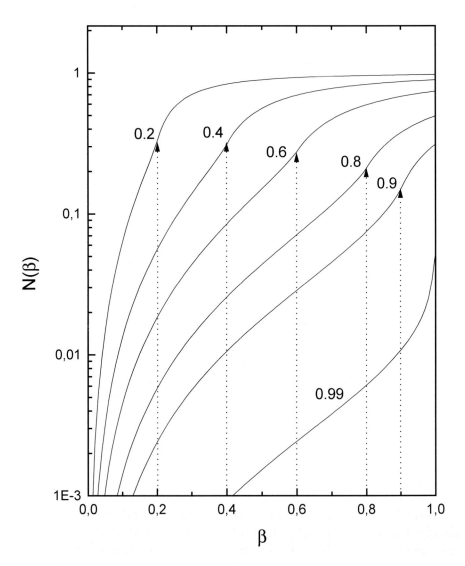

Figure 4.10. The number of emitted quanta in the radial ρ direction per unit length (in units of $e^2\omega_0/\hbar c^2$) as a function of $\beta = v/c$. The number on a particular curve is the critical velocity β_c.

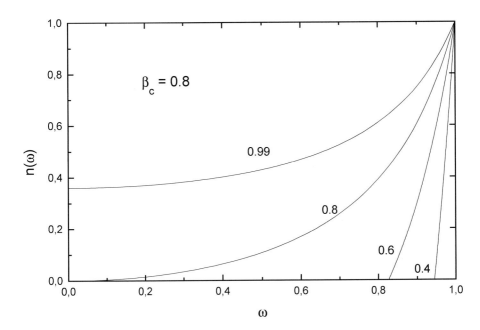

Figure 4.11. Spectral distribution of the emitted quanta (in units of $e^2/\hbar c^2$); ω is in units of ω_0. The number of a particular curve is $\beta = v/c$.

For the frequency-independent $\epsilon = \epsilon_0$ the energy flux is confined to the surface of the Cherenkov cone. Electromagnetic strengths inside the Cherenkov cone fall as r^{-2} at large distances, and therefore do not contribute to the radial flux.

4.7. WKB estimates

The radiation field (described by the integrals in (4.12) and (4.13) containing usual Bessel functions) can be handled by the WKB method. We follow closely Tamm's paper [30] (see also the review [31] and the book [32]). For this we replace the functions J_ν and N_ν by their asymptotic values:

$$J_\nu(x) \sim \sqrt{\frac{2}{\pi x}} \cos\left(x - \frac{\nu\pi}{2} - \frac{\pi}{4}\right), \qquad N_\nu(x) \sim \sqrt{\frac{2}{\pi x}} \sin\left(x - \frac{\nu\pi}{2} - \frac{\pi}{4}\right).$$

Then,

$$H_\phi = \frac{e}{c}\sqrt{\frac{2}{\pi v \rho}} \int d\omega \sqrt{\omega}(\beta^2\epsilon - 1)^{1/4} \cos\left(f + \frac{\pi}{4}\right),$$

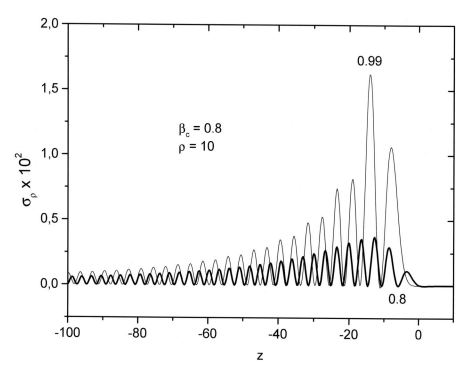

Figure 4.12.   The distribution of the radial energy flux (in units of $e^2\omega_0^3/c^3$) on the surface of the cylinder C_ρ, z is in units of c/ω_0. The number on a particular curve is $\beta = v/c$.

$$E_\rho = \frac{e}{v}\sqrt{\frac{2}{\pi v \rho}} \int d\omega \frac{1}{\epsilon}\sqrt{\omega}(\beta^2\epsilon - 1)^{1/4} \cos\left(f + \frac{\pi}{4}\right),$$

$$E_z = -\frac{e}{v}\sqrt{\frac{2}{\pi v \rho}} \int d\omega \frac{1}{\epsilon}\sqrt{\omega}(\beta^2\epsilon - 1)^{3/4} \cos\left(f + \frac{\pi}{4}\right). \qquad (4.34)$$

Here $f = \omega(t - z/v) - \sqrt{\beta^2\epsilon - 1}\rho\omega/v$. The argument of the cosine is a rapidly oscillating function of ω. The main contribution to the integrals comes from stationary points at which $df/d\omega = 0$. Or, explicitly,

$$(vt - z)\sqrt{\beta^2\epsilon - 1} = \rho\left[\beta^2 - 1 + \frac{\beta^2\omega_0^2\omega_L^2}{(\omega^2 - \omega_0^2)^2}\right]. \qquad (4.35)$$

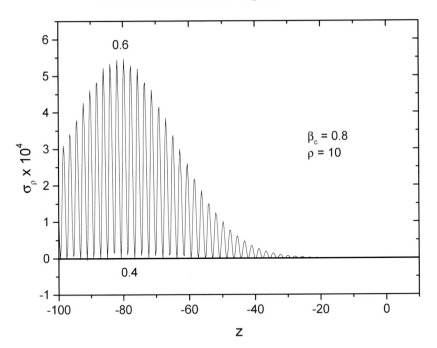

Figure 4.13. The same as in Fig. 4.12, but for $\beta < \beta_c$.

This equation defines ω as a function of ρ, z. Let this ω be $\omega_1(\rho, z)$. Then the WKB method gives

$$H_\phi = -\frac{2e}{c}\sqrt{\frac{\omega_1}{v\rho|\ddot{f}_1|}}(\beta^2\epsilon_1 - 1)^{1/4}\sin f_1,$$

$$E_\rho = -\frac{2e}{v\epsilon_1}\sqrt{\frac{\omega_1}{v\rho|\ddot{f}_1|}}(\beta^2\epsilon_1 - 1)^{1/4}\sin f_1,$$

$$E_z = \frac{2e}{v\epsilon_1}\sqrt{\frac{\omega_1}{v\rho|\ddot{f}_1|}}(\beta^2\epsilon_1 - 1)^{3/4}\sin f_1 \qquad (4.36)$$

for $\ddot{f}_1 > 0$ and

$$H_\phi = \frac{2e}{c}\sqrt{\frac{\omega_1}{v\rho|\ddot{f}_1|}}(\beta^2\epsilon_1 - 1)^{1/4}\cos f_1,$$

$$E_\rho = \frac{2e}{v\epsilon_1}\sqrt{\frac{\omega_1}{v\rho|\ddot{f}_1|}}(\beta^2\epsilon_1 - 1)^{1/4}\cos f_1,$$

$$E_z = -\frac{2e}{v\epsilon_1}\sqrt{\frac{\omega_1}{v\rho|\ddot{f}_1|}}(\beta^2\epsilon_1 - 1)^{3/4}\cos f_1 \qquad (4.37)$$

for $\ddot{f}_1 < 0$. Here

$$f_1 = f(\omega_1), \quad \epsilon_1 = \epsilon(\omega_1), \quad \ddot{f}_1 = \left(\frac{d^2 f}{d\omega^2}\right)_{\omega=\omega_1}.$$

The electromagnetic strengths are maximal if

$$\omega_1(vt - z) - \rho\omega_1\sqrt{\beta^2\epsilon_1 - 1} = (m - 1/2)\pi v \qquad (4.38)$$

for $\ddot{f}_1 > 0$ and

$$\omega_1(vt - z) - \rho\omega_1\sqrt{\beta^2\epsilon_1 - 1} = m\pi v \qquad (4.39)$$

for $\ddot{f}_1 < 0$. Here $m = 1, 2$, etc..

The combined solution of (4.35) and (4.38),(4.39) defines the set of surfaces on which the electromagnetic strengths and the Poynting vector are maximal. Due to the axial symmetry, these surfaces in the ρ, z coordinates look like lines. We refer to these lines as trajectories.

Equations (4.35)-(4.39) were obtained by Tamm [30]. We apply them to the particular $\epsilon(\omega)$ given by Eq.(4.1).

The electromagnetic field strengths and radial (i.e., in the ρ direction) energy flow have sharp maxima on some spatial surfaces. In the ρ, z coordinates these surfaces can be drawn (owing to the axial symmetry of the problem) by the lines. We refer to them as trajectories. Different trajectories are labelled by the integer numbers m. For the electric penetrability taken in the form (4.1), m runs from 1 to ∞. We make the notation $x_c^2 = 1 - \tilde{\epsilon}$, $\tilde{\epsilon} = \beta^2\gamma^2/\beta_c^2\gamma_c^2$. The trajectories can be parametrized by the equation

$$vt - z = \frac{m\pi c\beta}{\omega_0\tilde{\epsilon}x^3}[\tilde{\epsilon} - (x^2 - 1)^2], \quad \rho = \frac{m\pi c\beta\gamma}{\omega_0\tilde{\epsilon}x^3}(1 - x^2)^{3/2}(x^2 - x_c^2)^{1/2}. \quad (4.40)$$

To obtain the trajectory equation one should find x from the first of these equations and substitute it into the second one. Instead we prefer to vary x and compare ρ and $vt - z$ entering into (4.40) and corresponding to the same parameter x.

We consider cases $\beta > \beta_c$ and $\beta < \beta_c$ separately.

4.7.1. CHARGE VELOCITY EXCEEDS THE CRITICAL VELOCITY

It turns out that $x_c^2 < 0$ for $\beta > \beta_c$. In this case x runs in the interval $0 < x < 1$. The particular trajectory begins at the point $x = 1$ where $vt - z = m\pi c/\omega_0$ and $\rho = 0$. The slope of the trajectory is

$$\tan\theta = \gamma\frac{(1 - x^2)^{3/2}(x^2 - x_c^2)^{1/2}}{\tilde{\epsilon} - (x^2 - 1)^2}.$$

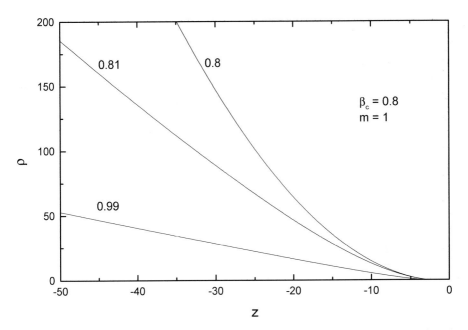

Figure 4.14. Spatial distribution of the $m = 1$ trajectory for charge velocities $\beta \geq \beta_c$. The slope of the trajectory increases as β approaches β_c.

When x decreases both $vt - z$ and ρ increase. For very small x

$$vt - z \sim \frac{m\pi c\beta}{\omega_0 \tilde{\epsilon} x^3}(\tilde{\epsilon} - 1), \quad \rho \sim \frac{m\pi c\beta\gamma}{\omega_0 \tilde{\epsilon} x^3}\sqrt{\tilde{\epsilon} - 1}.$$

The asymptotic slope of the trajectory is

$$\tan \theta = \frac{\rho}{vt - z} \sim (\frac{\beta^2}{\beta_c^2} - 1)^{-1/2}.$$

It is seen that the trajectory slope increases when β approaches β_c (Fig. 4.14). Let $v = c$, i.e., the charge moves with the velocity of light in vacuum. Then

$$vt - z = \frac{m\pi c}{\omega_0 x^3}, \quad \rho = \frac{m\pi c}{x^3 \omega_0}\beta_c\gamma_c(1 - x^2)^{3/2}.$$

Eliminating x one obtains

$$\rho = \beta_c\gamma_c(ct - z)\left[1 - \left(\frac{m\pi c}{\omega_0(ct - z)}\right)^{2/3}\right]^{3/2}.$$

For large $ct - z$ the trajectory is linear $(\rho = \beta_c \gamma_c (ct - z))$. For $\beta_c \to 0$ the trajectory approaches the motion axis. Let β be slightly greater than β_c,

$$\tilde{\epsilon} = 1 + \delta,\ 0 < \delta \ll 1,$$

i.e., charge moves almost with the velocity of light in medium. Then in the limit $\delta \to 0$,

$$vt - z = \frac{m\pi v}{\omega_0 x}(2 - x^2), \quad \rho = \frac{m\pi v \gamma}{\omega_0 x^2}(1 - x^2)^{3/2}. \tag{4.41}$$

Excluding x we obtain

$$\rho = \frac{m\pi c \beta_c \gamma_c}{\omega_0} \frac{[y\sqrt{2 + y^2/4} - 1 - y^2/2]^{3/2}}{y^2 + 2 - y\sqrt{2 + y^2/4}}.$$

Here $y = \omega_0(vt - z)/m\pi c \beta_c$. At large distances one has

$$\rho \sim \frac{\omega_0 \gamma_c}{4m\pi c \beta_c}(vt - z)^2.$$

That is, ρ increases quadratically with the rise of $vt - z$.

4.7.2. CHARGE VELOCITY IS SMALLER THAN THE CRITICAL VELOCITY

For $\beta < \beta_c$ one has $\tilde{\epsilon} < 1$ and $x_c^2 > 0$. The trajectory parametrization coincides with (4.40) when x lies within the interval $\sqrt{4 - 3\tilde{\epsilon}} - 1 < x^2 < 1$. We refer to this part of the trajectory as to branch 1. For $\beta < \beta_c$ and $1 - \tilde{\epsilon} < x^2 < \sqrt{4 - 3\tilde{\epsilon}} - 1$ the parametrization is given by Eq.(4.40) in which m should be replaced by $m - 1/2$. This part of the trajectory is denoted branch 2. These branches are marked by the numbers 1 and 2 in Fig. 4.15. It is seen that ρ vanishes for $x = x_c$ and $x = 1$. The corresponding $vt - z$ lie on the branches 1 and 2, respectively. As the values of $vt - z$ are finite for $\rho = 0$, the trajectories are closed for $\beta < \beta_c$.

Let β be slightly less than β_c, that is

$$\tilde{\epsilon} = 1 - \delta, \quad 0 < \delta \ll 1,$$

i.e., charge moves with a velocity slightly less than the velocity of light in medium. The parametrizations of $vt - z$ and ρ are then still given by (4.40), in which x changes in the interval $3\delta/2 < x^2 < 1$ for the first branch and in the interval $\delta/2 < x^2 < 3\delta/2$ for the second branch. This means that the first branch of the m trajectory for $\beta = \beta_c - \delta$ continuously passes into the corresponding m trajectory for $\beta = \beta_c + \delta$ for $\delta \to 0$.

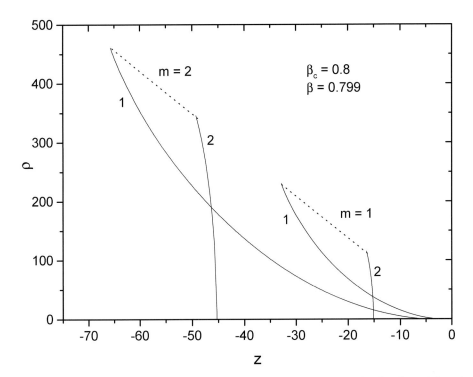

Figure 4.15. Space distribution of the $m = 1$ and $m = 2$ trajectories for $\beta_c = 0.8$ and $\beta = 0.799$. The trajectories for $\beta < \beta_c$ are closed (in contrast with the $\beta \geq \beta_c$ case shown in Fig. 4.14). Numbers 1 and 2 mean the branches of a particular trajectory.

As to the second branch, in the limit $\delta \to 0$ it degenerates into the almost vertical line. It begins at $z = (m - 1/2)\pi c\beta/\omega_0\sqrt{\delta}$, where $\rho = 0$, and terminates at $z = (m - 1/2)\pi c\beta 4\sqrt{2}/(3\sqrt{3}\omega_0\sqrt{\delta})$, where $\rho = 2(m - 1/2)\pi c\beta\gamma/(3\sqrt{3}\omega_0\delta)$ (see Fig. 4.15).

Let $\tilde{\epsilon} \to 0$. This may happen when the charge velocity is much less than the velocity of light in medium. However, this condition also takes place when $\beta \approx \beta_c \approx 1$, but β_c is much closer to 1 than β. This is possible because of the γ factors in the definition of $\tilde{\epsilon}$. In both cases one has

$$vt - z \to \frac{m\pi v}{\omega_0}, \quad \rho \to 0.$$

This means that the radiation flux is concentrated behind the charge on the motion axis.

The WKB approximation breaks at the neighbourhood of $x = x_m = (\sqrt{4 - 3\tilde{\epsilon}} - 1)^{1/2}$. This value can be reached only for $\beta < \beta_c$. The values of z and ρ at those points are

$$(vt - z)_1 \sim \frac{4m\pi c\beta}{\omega_0\tilde{\epsilon}} \frac{\tilde{\epsilon} + \sqrt{4 - 3\tilde{\epsilon}} - 2}{(\sqrt{4 - 3\tilde{\epsilon}} - 1)^{3/2}},$$

$$\rho_1 \sim \frac{m\pi c\beta\gamma}{\omega_0\tilde{\epsilon}} \frac{(\tilde{\epsilon} + \sqrt{4 - 3\tilde{\epsilon}} - 2)^{1/2}(2 - \sqrt{4 - 3\tilde{\epsilon}})^{3/2}}{(\sqrt{4 - 3\tilde{\epsilon}} - 1)^{3/2}}$$

for the branch 1. For the branch 2, m should be replaced by $m - 1/2$. The slope of the line C_m (strictly speaking, it is a cone rather than a line, but in the (ρ, z) plane it looks like a straight line (Figs. 4.16 and 4.17)) passing through the discontinuity points is given by

$$\tan\theta = \frac{\gamma}{4} \frac{(2 - \sqrt{4 - 3\tilde{\epsilon}})^{3/2}}{(\sqrt{4 - 3\tilde{\epsilon}} + \tilde{\epsilon} - 2)^{1/2}}.$$

In particular,

$$\tan\theta \sim \frac{3\sqrt{3}}{16}\gamma\tilde{\epsilon} \quad \text{for} \quad \tilde{\epsilon} \to 0$$

and

$$\tan\theta \sim \frac{1}{2\sqrt{2}}\frac{\gamma}{\sqrt{\delta}} \quad \text{for} \quad \tilde{\epsilon} \to 1 \quad (\tilde{\epsilon} = 1 - \delta, \quad \delta << 1).$$

That is, the slope of the line C_m tends to zero for the small charge velocity and becomes large as β approaches β_c. The meaning of this line is that on a particular trajectory (which itself is the line where field strengths are maximal) the field strengths become infinite as one approaches the point at which the WKB method breaks down.

On the surface of the cylinder C_ρ (see Fig. 4.2) the field strengths have maxima at those points in which C_ρ is intersected by the trajectories. Among these maxima the most pronounced (i.e., of the greatest amplitude) are expected to be those which lie near the point at which C_ρ is intersected by C_m (despite the WKB approximation breaking on it). In what follows we shall use this result as a tool for the rough estimation of the position where the radiation intensity is maximal. This will be confirmed by exact calculations).

Some of the trajectories corresponding to $\beta_c = 0.8$, $\beta = 0.4$ are shown in Figs. 4.16 and 4.17. It follows from them that there are no trajectories intersecting the surface of the cylinder C_ρ of the radius $\rho = 10$ in the interval $-100 < z < 0$ treated in Fig. 4.13. This means that there should be no radial energy flux there. The inspection of Fig. 4.17 tells us that for $\rho = 10$ the energy flux begins to penetrate the C_ρ surface at the distances $z \le -200$.

4.8. Numerical results

To verify WKB estimates we evaluated for $\beta = 0.4$ the distribution of the energy losses σ_ρ on the surface of C_ρ (Fig. 4.18). It is seen that the main contribution comes from the region in the neighbourhood $z \sim -300$. This

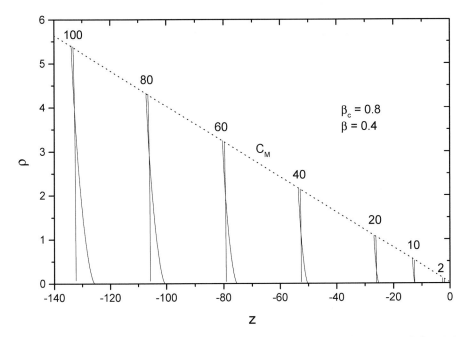

Figure 4.16.   Spatial distribution of the selected trajectories for $\beta_c = 0.8$ and $\beta = 0.4$.

σ_ρ distribution consists, in fact, of many peaks. Its fine structure in the small interval of z is shown in Fig. 4.19. The question arises of how the trajectories behave for other charge velocities β. It follows from Fig. 4.14 that for $\beta \geq \beta_c$ the trajectories are not closed, i.e., they go to infinity as z tends to $-\infty$. The slope of the trajectories increases as β approaches β_c. This means that for $\beta = \beta_c$ the EMF of a charge moving uniformly in a non-dispersive medium differs from zero only in the infinitely thin layer normal to the charge velocity [33].

Since for $\beta > \beta_c$ the trajectories intersect the surface C_ρ at small values of z, one should expect the appearance of the energy flux there.

In Figs. 4.20 and 4.21 we present the results of exact (i.e., not WKB) calculations of the intensity distribution for $\beta = 0.99$ and 0.8, respectively. We observe that for $\beta > \beta_c$ the main intensity maximum lies approximately at $z = -z_c$, $z_c = \rho\sqrt{\beta^2 n^2 - 1}$, i.e., at the place, where in the absence of the ω dispersion ($\epsilon = \epsilon_0 = \epsilon(0)$, $\beta_c^2 = 1/\epsilon_0$), the Cherenkov singular cone intersects C_ρ.

For $\beta < \beta_c$ the trajectories are closed (Figs. 4.15-4.17, and 4.22). As β decreases, the trajectories approach the motion axis. In this case the C_ρ surface is intersected by the trajectories with large m at larger values of

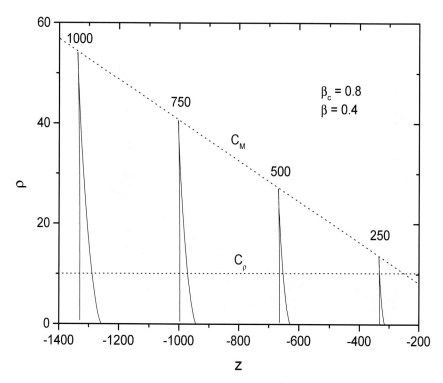

Figure 4.17. The same as in Fig. 4.16 but for a different z interval.

negative z (compared to the $\beta > \beta_c$ case) and the maxima of intensity should also be shifted to a large negative z. This is illustrated by Figs. 4.18 and 4.23 where the intensity spectra are shown for $\beta = 0.4$ and 0.6, respectively.

Consider now the distribution of the radiation flux on the surface of the sphere S (instead of on the cylinder surface, as we have done up to now).

From Figs. 4.16 and 4.17 based on the WKB estimates and numerical results presented in Fig. 4.18 it follows that for $\beta < \beta_c$ the radial radiation flux is confined to the narrow cone adjusted to the negative z semi-axis (Fig. 4.24). Its solution angle θ_c is approximately 5 degrees for $\beta_c = 0.8$ and $\beta = 0.4$.

We conclude that despite the ω dependence of ϵ, the critical velocity $\beta_c = 1/\sqrt{\epsilon_0}$ still conserves its physical meaning, thus separating closed ($\beta < \beta_c$) and unclosed ($\beta > \beta_c$) trajectories.

4.8.1. ESTIMATION OF NON-RADIATION TERMS

Up to now, when evaluating σ_ρ we have taken into account only those terms in $\vec{E}$ and $\vec{H}$ which contribute to the energy losses, i.e., to the W given by Eq. (4.30). They correspond to the terms of $\vec{E}$ and $\vec{H}$ containing the usual

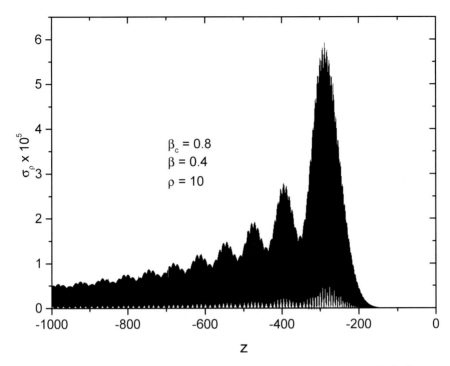

Figure 4.18. The distribution of the radial energy flux (in units of $e^2\omega_0^3/c^3$) on the surface of the cylinder C_ρ for $\beta = 0.4$; z is in units c/ω_0. It is seen that the main contribution comes from large negative z.

(non-modified) Bessel functions (see Eqs. (4.12) and (4.13)). However, we cannot use Eqs.(4.12) and (4.13) to evaluate terms with modified Bessel functions as their contribution to $\vec{E}$ is divergent. Instead, the following trick is used. We find $\vec{E}$ and $\vec{H}$ for the complex electric permittivity (4.7). They are finite for the non-zero value of parameter p defining the imaginary part of $\epsilon(\omega)$. The corresponding formulae are collected in Refs. [34,35] and in section 4.9. Then we tend the parameter p defining the imaginary part of ϵ to zero. We expect that for sufficiently small p we obtain the values of $\vec{E}$ and $\vec{H}$ which adequately describe the contribution of the terms with modified Bessel functions. There is also another approach (see [36] and section 4.11) in which the electric strength $\vec{E}$ is not singular (except for the charge motion axis) even for real ϵ. It turns out that electromagnetic strengths evaluated according to the formulae of section 4.9 are indistinguishable from those of [36] when the parameter p is of an order of 10^{-5}-10^{-4} in units of ω_0. In what follows, by the words 'terms with modified Bessel functions are taken into account' we mean that the calculations are made by means of formulae presented in section 4.9 for $p = 10^{-4}$.

When the terms with modified Bessel functions are taken into consid-

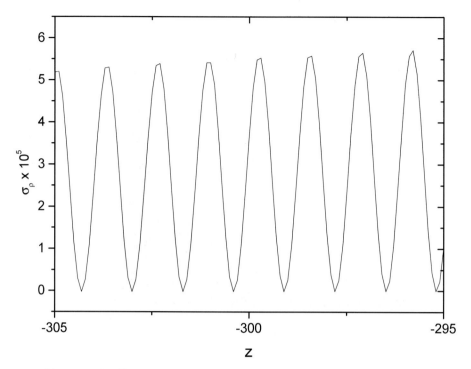

Figure 4.19.   Fine structure of the radial energy flux shown in Fig. 4.18.

eration, the characteristic oscillation of σ_ρ appears in the neigbourhood $z = 0$ (Figs. 4.25 and 4.26). For $\beta < \beta_c$ it is described approximately by the following expression:

$$\sigma_\rho^1 = -\frac{c\beta e^2}{2\epsilon_0}(1 - \beta^2/\beta_c^2)^2 \frac{\rho^2 z}{[z^2 + \rho^2(1 - \beta^2/\beta_c^2)]^3} \tag{4.42}$$

corresponding to the energy flux carried by the uniformly moving charge with the velocity $\beta < \beta_c$ in medium with a constant $\epsilon = \epsilon_0$. As we have mentioned, the terms in (4.12) and (4.13) containing modified Bessel functions do not contribute to the total energy losses (4.32). In particular, this is valid for σ_ρ^1 given by (4.42):

$$\int_{-\infty}^{\infty} \sigma_\rho^1 dz = 0$$

(owing to the antisymmetry of σ_ρ). For $z \gg \rho$ and $\rho \gg z$, σ_ρ^1 falls as ρ^2/z^5 and z/ρ^4, respectively.

For $\beta = 0.4$ we estimate the value of the term (4.42) in the region $z = -300$ where σ_ρ has a maximum (see Fig. 4.18). It turns out that

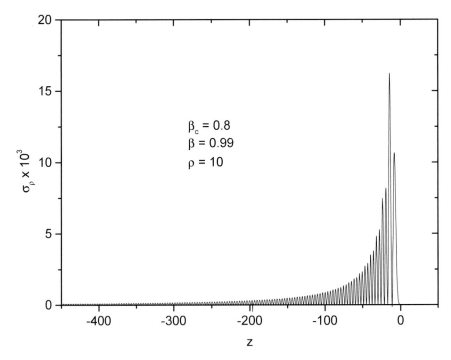

Figure 4.20. The distribution of the radial energy flux (in units of $e^2\omega_0^3/c^3$) on the surface of the cylinder C_ρ for $\beta = 0.99$; z is in units of c/ω_0 It is seen that the main contribution comes from the small negative values of z.

$\sigma_\rho \approx 6 \times 10^{-5}$ and $\sigma_\rho^1 \approx 5 \times 10^{-12}$ there, i.e., the contribution of σ_ρ^1 relative to σ_ρ is of an order of 10^{-7}, and therefore it is negligible.

For $\beta = 0.6$ we see in Fig. 4.23 the σ_ρ distribution evaluated via Eqs. (4.12) and (4.13) in which the terms with modified Bessel functions are omitted. Comparing Fig. 4.18 with 4.25 and Fig. 4.23 with 4.26 we conclude that they coincide everywhere except for the $z = 0$ region where the term (4.42) is essential.

For $\beta \geq \beta_c$ the contribution of the terms involving modified Bessel functions in (4.12) and (4.13) is very small. This illustrates Fig. 4.27 where two distributions σ_ρ with and without inclusion of the above-mentioned terms are shown for $\beta = 0.8$. They are indistinguishable on this figure and look like one curve. The same is valid for larger charge velocities.

4.9. The influence of the imaginary part of ϵ

So far we have evaluated the total energy losses per unit length (W) and their distribution along the z axis (σ_ρ) for the pure real electric permittivity given by (4.1). Equation (4.7) is a standard parametrization of the complex

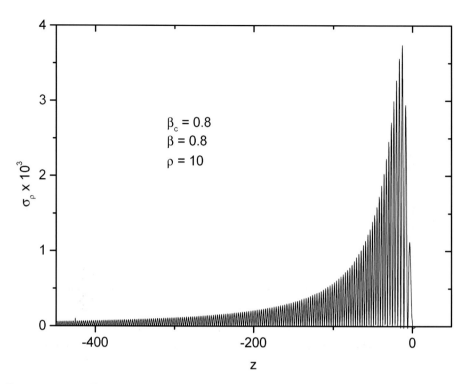

Figure 4.21.   The same as in Fig. 4.20 but for $\beta = 0.8$. The radial energy flux is distributed in a greater z interval.

electric permittivity [22,27]. For the chosen definition (4.5) of the Fourier transform the causality principle requires p to be positive.

We write out electromagnetic potentials and field strengths for a finite value of the parameter p defining the imaginary part of ϵ. Since $\epsilon(-\omega) = \epsilon^*(\omega)$, the EMF can be written in a manifestly real form

$$\Phi = \frac{2e}{\pi v} \int\limits_0^\infty [(\epsilon_r^{-1} \cos\alpha - \epsilon_i^{-1} \sin\alpha) K_{0r} - (\epsilon_i^{-1} \cos\alpha + \epsilon_r^{-1} \sin\alpha) K_{0i}] d\omega,$$

$$A_z = \frac{2e}{\pi c} \int\limits_0^\infty d\omega (\cos\alpha K_{0r} - \sin\alpha K_{0i}),$$

$$H_\phi = \frac{2e}{\pi vc} \int\limits_0^\infty \omega d\omega (a^2 + b^2)^{1/4} \left[\cos\left(\frac{\phi}{2} + \alpha\right) K_{1r} - \sin\left(\frac{\phi}{2} + \alpha\right) K_{1i}\right],$$

$$E_z = -\frac{2}{\pi v^2} \int\limits_0^\infty \omega d\omega \{[\cos\alpha(\epsilon_r^{-1} - \beta^2) - \sin\alpha\epsilon_i^{-1}] K_{0i}$$

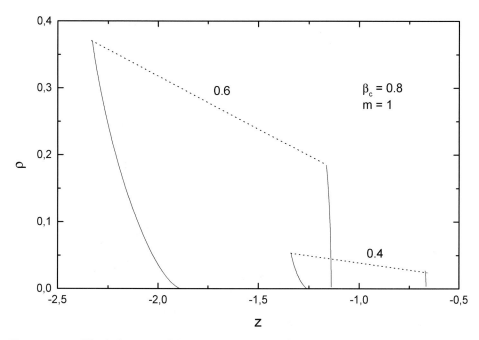

Figure 4.22. The behaviour of the $m = 1$ trajectory for $\beta = 0.4$ and $\beta = 0.6$. For $\beta < \beta_c$ the trajectories are grouped near the z axis. This shifts the maximum of the energy flux distribution to larger negative z.

$$+[\sin\alpha(\epsilon_r^{-1} - \beta^2) + \cos\alpha\epsilon_i^{-1}]K_{0r}\},$$

$$E_\rho = \frac{2}{\pi v^2} \int_0^\infty \omega d\omega (a^2 + b^2)^{1/4} [(\epsilon_r^{-1}\cos\alpha - \epsilon_i^{-1}\sin\alpha)$$

$$\times(\cos(\phi/2)K_{1r} - \sin(\phi/2)K_{1i})$$

$$-(\epsilon_i^{-1}\cos\alpha + \epsilon_r^{-1}\sin\alpha)(\sin(\phi/2)K_{1r} + \cos(\phi/2)K_{1i})]. \qquad (4.43)$$

Here we put

$$K_{0r} = \operatorname{Re}K_0\left(\frac{\rho\omega}{v}\sqrt{1 - \beta^2\epsilon}\right), \quad K_{0i} = \operatorname{Im}K_0\left(\frac{\rho\omega}{v}\sqrt{1 - \beta^2\epsilon}\right),$$

$$K_{1r} = \operatorname{Re}K_1\left(\frac{\rho\omega}{v}\sqrt{1 - \beta^2\epsilon}\right), \quad K_{1i} = \operatorname{Im}K_1\left(\frac{\rho\omega}{v}\sqrt{1 - \beta^2\epsilon}\right).$$

Furthermore, ϵ_r and ϵ_i are the real and imaginary parts of ω

$$\epsilon_r = 1 + \frac{\omega_L^2(\omega_0^2 - \omega^2)}{(\omega_0^2 - \omega^2)^2 + p^2\omega^2}, \quad \epsilon_i = -\frac{p\omega\omega_L^2}{(\omega_0^2 - \omega^2)^2 + p^2\omega^2},$$

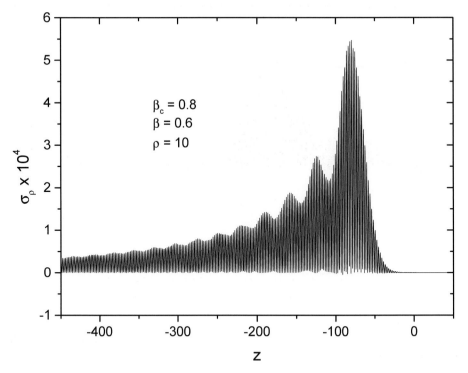

Figure 4.23. The same as in Fig. 4.18, but for the charge velocity $\beta = 0.6$.

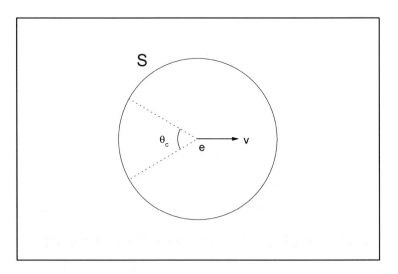

Figure 4.24. For the charge velocity β below some critical β_c the radial energy flux is confined to the narrow cone attached to the moving charge. For $\beta_c = 0.8$ and $\beta = 0.4$ the solution angle $\theta_c \approx 5°$.

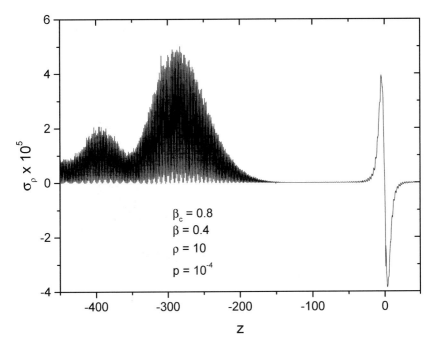

Figure 4.25. The same as in Fig. 4.18, but with the inclusion of the non-radiating term corresponding to the electromagnetic field carried by a moving charge.

$\epsilon_r^{-1} = \epsilon_r/(\epsilon_r^2 + \epsilon_i^2)$, $\quad \epsilon_i^{-1} = -\epsilon_i/(\epsilon_r^2 + \epsilon_i^2)$; $\alpha = \omega(t - z/v)$; a, b and ϕ are the same as in (4.9)-(4.11). The energy flux per unit length through the surface of a cylinder of the radius ρ coaxial with the z axis for the whole time of charge motion is defined by Eq.(4.30). Substituting E_z and H_ϕ given by (4.43) into it one finds

$$W = \int_0^\infty f(\omega)d\omega,$$

where

$$f(\omega) = -\frac{2e^2\rho}{\pi v^3}\omega^2(a^2 + b^2)^{1/4}$$

$$\times \{(K_{0r}K_{1r} + K_{0i}K_{1i})[(\epsilon_r^{-1} - \beta^2)\sin(\phi/2) - \epsilon_i^{-1}\cos(\phi/2)]$$

$$-(K_{0i}K_{1r} - K_{0r}K_{1i})[(\epsilon_r^{-1} - \beta^2)\cos(\phi/2) + \epsilon_i^{-1}\sin(\phi/2)]\}. \qquad (4.44)$$

It is surprising that $f(\omega)$ given by (4.44) differs from zero for all ω. That is, the Tamm-Frank radiation condition (stating that a charge moving uniformly in the dielectric medium radiates if the condition $\beta^2\epsilon > 1$ is satisfied) fails if $p \neq 0$. It restores in the limit $p \to 0$.

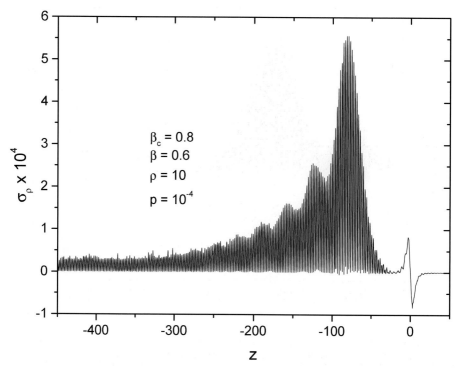

Figure 4.26. The same as in Fig. 4.25, but for the charge velocity $\beta = 0.6$.

Let $1 - \beta^2\epsilon > 0$ in this limit, then

$$\sin\frac{\phi}{2} \to 0, \quad \cos\frac{\phi}{2} \to 1, \quad \epsilon_i \to 0, \quad \epsilon_i^{-1} \to 0, \quad K_{0i} \to 0, \quad K_{1i} \to 0$$

and therefore $f(\omega) \to 0$ whilst electromagnetic potentials and field strengths coincide with those terms in (4.12) and (4.13) which contain modified Bessel functions. On the other hand, if in this limit $1 - \beta^2\epsilon < 0$, then

$$\sin\frac{\phi}{2} \to 1 \quad (\text{for } p > 0), \quad \cos\frac{\phi}{2} \to 0, \quad \epsilon_i \to 0, \quad \epsilon_i^{-1} \to 0,$$

$$K_{0r} \to -\frac{\pi}{2}N_0, \quad K_{0i} \to -\frac{\pi}{2}J_0, \quad K_{1r} \to -\frac{\pi}{2}J_1, \quad K_{1i} \to \frac{\pi}{2}N_1,$$

where the argument of the Bessel functions is $(\rho|\omega|/v)\sqrt{|1 - \beta^2\epsilon|}$. Substituting this into (4.44) and using the relation

$$J_\nu(x)N_{\nu+1}(x) - N_\nu(x)J_{\nu+1}(x) = -\frac{2}{\pi x}$$

one arrives at

$$f(\omega) = \frac{e^2\omega}{c^2}\left(1 - \frac{1}{\epsilon\beta^2}\right).$$

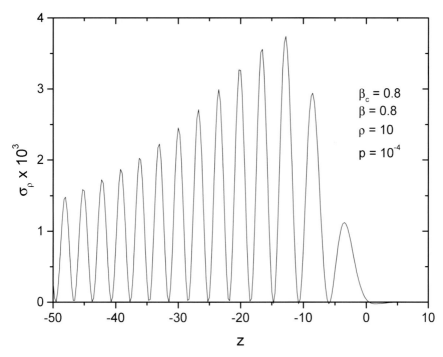

Figure 4.27. For $\beta = \beta_c$ the energy flux distributions with and without a non-radiating term are practically the same: they are indistinguishable in this figure. The same holds for $\beta > \beta_c$.

This in turn leads to W coinciding exactly with (4.31)-(4.33). Electromagnetic potentials and field strengths (4.43) coincide with the terms in (4.12) and (4.13) containing the ordinary Bessel functions.

Now we intend to clarify how the value of the parameter p affects the radiated electromagnetic field. For this we have evaluated σ_ρ for $\beta = 0.4$ on the surface of cylinder C_ρ, $\rho = 10$ for three different values of parameter p (in units ω_0): $p = 10^{-3}$ (Fig. 4.28), $p = 10^{-2}$, and $p = 0.1$ (Fig. 4.29). We observe that for $p = 10^{-3}$ the intensity amplitude is approximately two times less than for $p = 10^{-4}$ (Fig. 4.25). For $p = 10^{-2}$ and $p = 0.1$ all oscillations of σ_ρ on the negative z semi-axis practically disappear whilst the value of the term corresponding to the modified Bessel functions in (4.12) and (4.13) remains almost the same. In Figs. 4.30 and 4.31 there are given distributions of the radiated energy on the surface of σ_ρ for $\beta = 0.8$ and $\beta = 0.99$ for three different values of $p = 10^{-3}$, 0.1 and 1. We note that with a rise of p the oscillations for $\beta < \beta_c$ are damped much more strongly than for $\beta \geq \beta_c$. For example, for $p = 10^{-2}$ and $\beta = 0.99$ the values of the main maxima reduce only slightly (Fig. 4.31) whilst for $\beta = 0.4$ and the same p the oscillations of the radiation intensity completely disappear

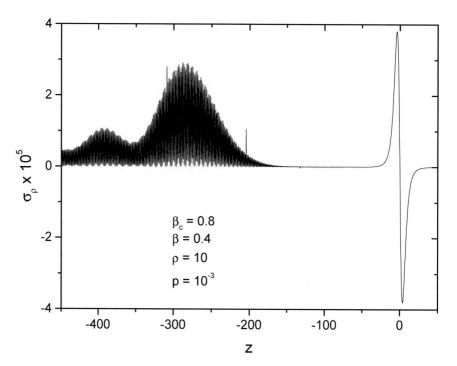

Figure 4.28.   The switching on the imaginary part of ϵ ($p = 10^{-3}$) reduces the oscillation amplitude by a factor of approximately 2 compared to that for $p = 10^{-4}$ (see Fig. 4.25). The non-radiating term is practically the same as in Fig. 4.25.

(Fig. 4.29).   Another observation is that secondary maxima are damped much more stronger than the main maximum. This is easily realized within the polarization formalism. In it a moving charge creates a time-dependent polarization source which, in the absence of damping, oscillates with a frequency $\sqrt{\omega_0^2 + \omega_L^2}$. The oscillating polarization results in the appearance of secondary electromagnetic waves, which being added are manifested as maxima of the potentials, field strengths, and intensities. The distribution of the polarization source for the electric permittivity (4.7) is given by [34,35]

$$\mathrm{div}\,\vec{P} = \frac{e}{v}\delta(x)\delta(y)\frac{\omega_L^2}{\sqrt{\omega_0^2 + \omega_L^2 - p^2/4}}$$

$$\times \exp\left[-p(t - z/v)/2\right] \times \sin[\sqrt{\omega_0^2 + \omega_L^2 - p^2/4}(t - z/v)]$$

for $z < vt$ and $\mathrm{div}\vec{P} = 0$ for $z > vt$ (this equation is related to the case $\omega_0^2 + \omega_L^2 - p^2/4 > 0$). As a result of the positivity of p the value of the polarization $\vec{P}$ at the instant t is defined by the values of the electric field $\vec{E}$ at preceding times (the causality principle). It follows that for large negative values of z

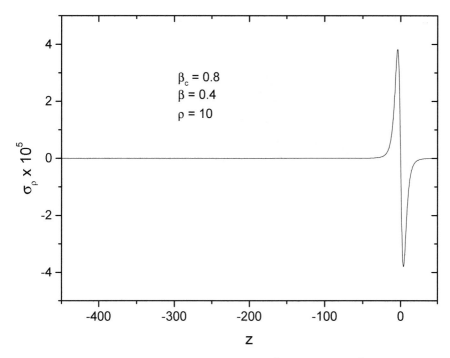

$\beta_c = 0.8$
$\beta = 0.4$
$p = 10$

Figure 4.29. The radial energy flux for $p = 10^{-2}$ and $p = 10^{-1}$. The oscillations completely disappeared, but the value of the non-radiating term remains practically the same.

the polarization source is suppressed much more strongly than for z values close to the current charge position. The position of the first maximum approximately coincides with the position of the singular Cherenkov cone in the absence of dispersion.

The total energy losses per unit length W (in units of $e^2\omega_0^2/c^2$) and the total number of emitted photons N (in units of $e^2\omega_0/\hbar c^2$) as a function of the charge velocity $\beta = v/c$ for $\beta_c = 0.8$ and different values of p are shown in Figs. 4.32 and 4.33. In most the cases W and N decrease with the rising of p. The sole exception, the origin of which remains unclear for us, is the intersection of $N(\beta)$ curves corresponding to $p = 0.1$ and $p = 1$ (Fig. 4.33). The corresponding ω densities $f(\omega)$ and $n(\omega)$ (entering $W = \int f(\omega)d\omega$ and

$N = \int n(\omega)d\omega$) are shown in Figs. 4.34 and 4.35.

4.10.   Application to concrete substances

We analyse two particular substances for which the parametrization of ϵ is known.

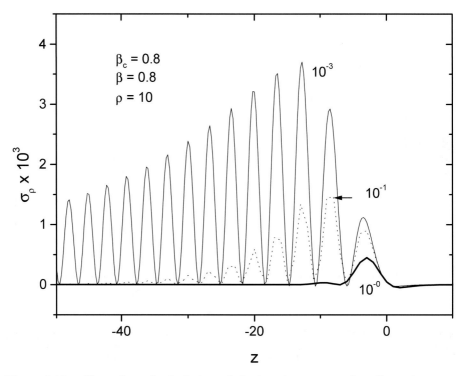

Figure 4.30. Shows how the inclusion of the imaginary part of ϵ affects the energy flux distribution. The number of a particular curve means the parameter p. The charge velocity is $\beta = 0.8$.

The first substance is iodine for which the parametrization of ϵ in the form (4.7) may be found in the Brillouin book [10]: Its resonance frequency lies in a far ultra-violet region and ϵ tends to 1 as $\omega \to \infty$. In this case, there is a critical velocity below and above which the properties of radiation differ appreciably. This parametrization is broadly used for the description of optical phenomena.

The following parametrization of ϵ

$$\epsilon = \epsilon_\infty + \frac{\omega_L^2}{\omega_0^2 - \omega^2 + ip\omega} \tag{4.45}$$

with $p = 0$ was found in [37] for ZnSe. Its resonance frequency lies in a far infrared region and ϵ tends to a constant value when $\omega \to \infty$. There are two critical velocities for this case. The behaviour of radiation is essentially different above the large critical velocity, between smaller and larger critical velocities and below the smaller critical velocity. Despite that the parametrizations (4.7) and (4.45) are valid in a quite narrow frequency

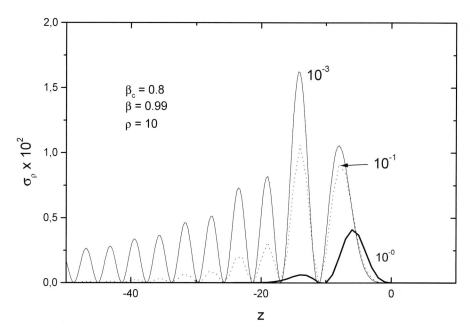

Figure 4.31.   The same as in Fig. 4.30 but for the charge velocity $\beta = 0.99$. Comparing this figure with Figs. 4.29-4.30, we observe that switching on the imaginary part of ϵ affects radiation intensities less for larger β.

region, we apply them to the whole ω semi-axis. Since we will deal with frequency distributions of radiation we can, at any step, limit consideration to the suitable frequency region.

The energy flux in the radial direction through the cylinder surface of the radius ρ is given by

$$\frac{d^3\mathcal{E}}{\rho d\phi dz dt} = -\frac{c}{4\pi} E_z(t) H_\phi(t).$$

Integrating this expression over the whole duration of the charge motion and over the azimuthal angle ϕ, and multiplying it by ρ, one obtains the energy radiated for the whole charge motion per unit length of the cylinder surface

$$\frac{d\mathcal{E}}{dz} = -\frac{c\rho}{2} \int E_z H_\phi dt.$$

Substituting here, instead of E_z and H_ϕ, their Fourier transforms and performing the time integration, one finds

$$\frac{d\mathcal{E}}{dz} = \int_0^\infty d\omega \sigma_\rho(\omega),$$

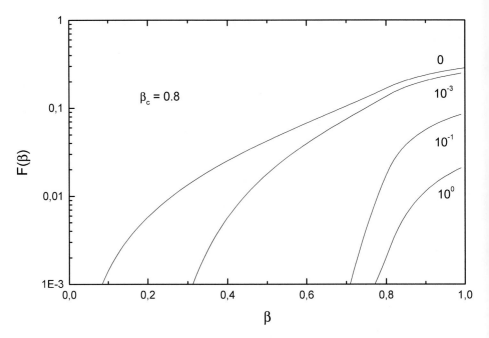

Figure 4.32. Shows how the inclusion of the imaginary part of ϵ affects the total energy losses W per unit length. The number on a particular curve is the parameter p; W and p are in units of $e^2\omega_0^2/c^2$ and ω_0, respectively.

where

$$\sigma_\rho(\omega) = \frac{d^2\mathcal{E}}{dzd\omega} = -\pi\rho c E_z(\omega) H_\phi^*(\omega) + \text{c.c.}$$

is the energy radiated in the radial direction per unit frequency and per unit length of the observational cylinder. The identification of the energy flux with σ_ρ is typical in the Tamm-Frank theory [29] describing the unbounded charge motion in medium. Finding electromagnetic field strengths from the Maxwell equations, one obtains

$$\sigma_\rho(\omega) = \frac{2ie^2\omega}{c^2}(1 - \frac{1}{\beta^2\epsilon})x^* K_0(x)[K_1(x)]^* + \text{c.c.}. \qquad (4.46)$$

Here $x = \sqrt{1 - \beta^2\epsilon} \cdot (\rho\omega/v)$. The sign of the square root should be chosen in such a way as to guarantee the positivity of its real part. In this case the modified Bessel functions decrease as $\rho \to \infty$. Equation (4.46), after reducing to the real form, was used for the evaluation of radiation intensities in [34,35]. In the limit $p \to 0$ it passes into the Tamm-Frank formula (2.32). For large $k\rho$ (k is the wave number, ρ is the radius of the observational

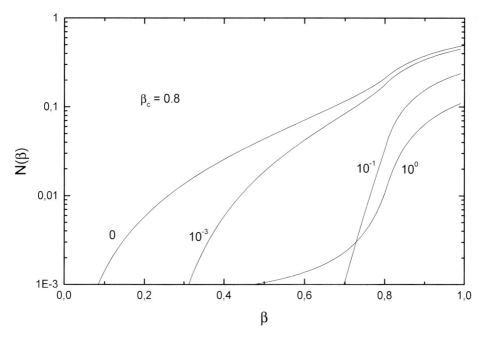

Figure 4.33.   The number of quanta emitted in the radial direction per unit length (in units of $e^2\omega_0/\hbar c^2$) as a function of the charge velocity β for different values of the parameter p.

cylinder C), the radiation intensity (4.46) goes into [35]

$$\sigma_\rho(\omega) = \frac{e^2\omega}{c^2}[(1-\frac{\tilde{\epsilon}_r}{\beta^2})\sin\frac{\phi}{2}+\tilde{\epsilon}_i\cos\frac{\phi}{2}]\exp[-\frac{2\rho\omega}{v}(a^2+b^2)^{1/4}\cos\frac{\phi}{2}], \quad (4.47)$$

where $\tilde{\epsilon}_r = \epsilon_r/(\epsilon_r^2 + \epsilon_i^2)$, $\quad \tilde{\epsilon}_i = -\epsilon_i/(\epsilon_r^2 + \epsilon_i^2)$; ϵ_r and ϵ_i (real and imaginary parts of ϵ), a, b and the angle ϕ were defined in (4.9)-(4.11). Usually, the condition $k\rho \gg 1$ is satisfied with great accuracy. For example, for a wavelength $\lambda = 4 \times 10^{-5}$ cm and $\rho = 10$ cm, one gets $k\rho \approx 10^6$. Equation (4.47) is valid for arbitrary dielectric permittivity. We apply it to (4.7) and (4.45).

4.10.1. DIELECTRIC PERMITTIVITY (4.7)

Dispersive medium without damping
For the sake of clarity we consider first the case of zero damping ($p = 0$). From (4.46) or (4.47) one then easily obtains the Tamm-Frank formula (4.31). According to Tamm and Frank [29], the total radiated energy is obtained by integrating $\mathcal{E}_{TF}(\omega)$ over the frequency region satisfying $\beta n > 1$. It is easy to check that for $\beta > \beta_c = 1/\sqrt{1 + \omega_L^2/\omega_0^2}$ this condition is satisfied

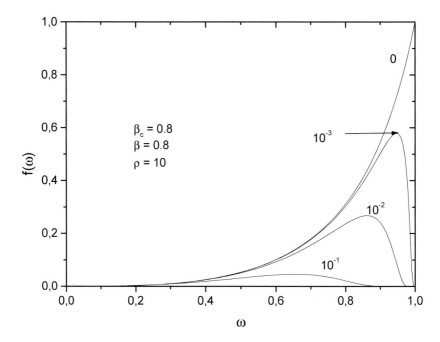

Figure 4.34. Spectral distribution of the energy losses (in units of $e^2\omega_0/c^2$); ω is in units of ω_0. The number of a particular curve means the parameter p.

for $0 < \omega < \omega_0$. For $\beta < \beta_c$ this condition is satisfied for $\omega_c < \omega < \omega_0$, where $\omega_c = \omega_0\sqrt{1 - \beta^2\gamma^2/\beta_c^2\gamma_c^2}$. This frequency window narrows as β diminishes. For $\beta \to 0$ the frequency spectrum is concentrated near the ω_0 frequency. The total energy radiated per unit length of the observational cylinder is equal to

$$\frac{d\mathcal{E}}{dz} = \int_0^\infty S_\rho(\omega)d\omega = \frac{e^2\omega_0^2}{2c^2}[1 - 1/\beta^2 - \frac{1}{\beta^2\beta_c^2\gamma_c^2}\ln(1 - \beta_c^2)] \qquad (4.48)$$

for $\beta > \beta_c$ and

$$\frac{d\mathcal{E}}{dz} = -\frac{e^2\omega_L^2}{2c^2}[1 + \frac{1}{\beta^2}\ln(1 - \beta^2)] \qquad (4.49)$$

for $\beta < \beta_c$.

Dispersive medium with damping

Obviously, the non-damping behaviour of EMF is possible when the index of the exponent in (4.47) is small. This takes place if $\cos\phi/2 \approx 0$. This, in turn implies that $a = 1 - \beta^2\epsilon_r < 0$, and $b \ll |a|$. We need, therefore, the frequency regions where $1 - \beta^2\epsilon_r < 0$.

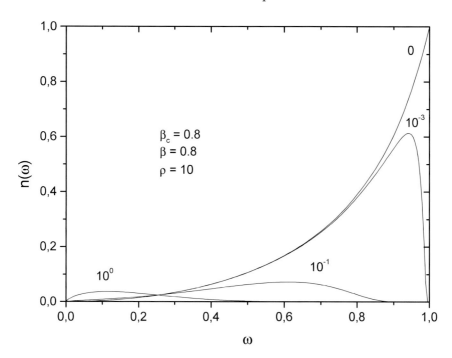

$n(\omega)$

ω

Figure 4.35. Spectral distribution of the emitted quanta (in units of $e^2/\hbar c^2$); ω is in units of ω_0. The number of a particular curve means the parameter p.

Let
$$\beta_c < \beta < 1, \quad \beta_c = 1/\sqrt{\epsilon_0}, \quad \epsilon_0 = \epsilon(0) = 1 + \omega_L^2/\omega_0^2.$$
Then $1 - \beta^2\epsilon_r < 0$ for $0 < \omega^2 < \omega_1^2$,
where

$$\omega_{1,2}^2 = \omega_0^2 \pm \Omega_0 - \frac{1}{2}(p^2 + \beta^2\gamma^2\omega_L^2), \quad \Omega_0 = \left[\frac{1}{4}(p^2 + \beta^2\gamma^2\omega_L^2)^2 - \omega_0^2p^2\right]^{1/2}.$$

In particular, $\omega_1 = \omega_0$ for $\beta = 1$ and $\omega_1 = \sqrt{\omega_0^2 - p^2}$, $\omega_2 = 0$ for $\beta = \beta_c$.
Let $\beta_p^2 < \beta^2 < \beta_c^2$, where

$$\beta_p^2 = \frac{2p\omega_0 - p^2}{\omega_L^2 + 2p\omega_0 - p^2}$$

(it is therefore suggested that p is sufficiently small to guarantee the positivity of β_p^2. This always takes place for transparent media in which the Cherenkov radiation is observed). Then $1 - \beta^2\epsilon_r < 0$ for $\omega_2 < \omega < \omega_1$. In particular, $\omega_1 = \omega_2 = \omega_0\sqrt{1 - p/\omega_0}$ for $\beta = \beta_p$.
    Finally, for $0 < \beta < \beta_p$ there is no room for $1 - \beta^2\epsilon_r < 0$.

We see that for $\beta > \beta_c$ the frequency distribution of the radiation differs from zero for $0 < \omega < \omega_1$, whilst for $\beta_p < \beta < \beta_c$ it is confined to the frequency window $\omega_2 < \omega < \omega_1$. Further decrease in β leads to the window narrowing. The window width disappears for $\beta = \beta_p$ when $\omega_1 = \omega_2 = \omega_0\sqrt{1 - p/\omega_0}$. Now the non-damping behaviour of the EMF strengths in addition to $1 - \beta^2\epsilon_r < 0$ requires also that $b \ll |a|$. This gives

$$\omega_L^2 \frac{\omega p - \omega_0^2 + \omega^2}{(\omega_0^2 - \omega^2)^2 + p^2\omega^2} \ll 1 - \frac{1}{\beta^2}$$

(it has been taken into account that $1 - \beta^2\epsilon_r < 0$). Since the r.h.s. of this inequality is smaller than 0 its l.h.s. should also be smaller than 0. This takes place if

$$\omega < \sqrt{\omega_0^2 + p^2/4} - p/2.$$

For small damping this reduces to $\omega < \omega_0 - p/2$.

Application to iodine

As an example we consider a dielectric medium with $\epsilon_0 = 1 + \omega_L^2/\omega_0^2 \approx 2.24$. The parameters of this medium are close to those given by Brillouin ([10], p. 56) for iodine. As to ω_0, Brillouin recommends $\omega_0 \approx 4 \cdot 10^{16}\mathrm{s}^{-1}$. This value of ω_0 is approximately 10 times larger than the average frequency of the visible region. However, since all formulae used for calculations depend only on the ratios ω_L/ω_0 and p/ω_0, we prefer to fix ω_0 only at the final stage.

To illustrate analytic results obtained above we present in Fig. 4.36 dimensionless spectral distributions $\sigma_\rho(\omega) = f(\omega)/(e^2\omega_0/c^2)$ for a number of charge velocities β and damping parameters p as a function of ω/ω_0. For $p = 0$ (Fig. 4.36 (a)), radiation intensities behave in the same way, as it was explained above. The switching on the damping parameter p affects radiation intensities for $\beta < \beta_c$ more strongly than for $\beta > \beta_c$. For example, the radiation intensity corresponding to $\beta = 0.4$ (smaller than $\beta_c \approx 0.668$) is very small even for $p/\omega_0 = 10^{-8}$ (Fig. 4.36(b)). For larger p the radiation intensity is so small that it cannot be depicted in the scale used For instance, for $\beta = \beta_c$ the maximal value of the radiation intensity equals 2×10^{-10} for $p/\omega_0 = 10^{-4}$ (Fig. 4.36(c)) and 3×10^{-14} for $p = 10^{-2}$ (Fig. 4.36(d)). With the rising of p the maximum of the frequency distribution shifts toward the smaller frequencies. This is owed to the large value of the index under the sign of exponent in (4.47) (and, especially, to the large value of $\rho\omega/v$).

So far we have not specified the resonance frequency ω_0. If, following Brillouin, we choose $\omega_0 = 4 \times 10^{16}\mathrm{s}^{-1}$ (which is approximately 10 times larger than average frequency of the visible light), then it follows from Fig. 4.36 (d) that for $p/\omega_0 = 10^{-2}$ (Brillouin recommends $p = 0.15$), frequency

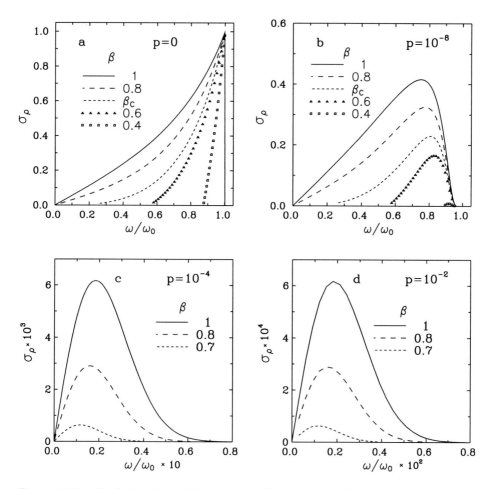

Figure 4.36. Radiation intensities corresponding to the dielectric permittivity (4.7) for a number of velocities and damping parameters p (in ω_0 units). The radius of the observational cylinder $\rho = 10$ cm. Other medium parameters are the same as suggested by Brilluoin for iodine. It is seen that the radiation spectrum shifts towards low frequencies with the rising of p.

distributions are practically zero inside the region of the visible light corresponding to $\omega \approx \omega_0/10$. This means, in particular, that space-time distributions of the radiated energy corresponding to realistic p are formed mainly by photons lying in the far infrared region, and therefore there is no chance of observing them in the region of visible light.

Up to now we have considered the radiation intensities on the surface of the cylinder C of the radius $\rho = 10$ cm. It is interesting to see how they look for smaller ρ. To be concrete, consider the radiation intensities corresponding to $p/\omega_0 = 10^{-2}$. From Fig. 4.36(d) we observe that the maximum

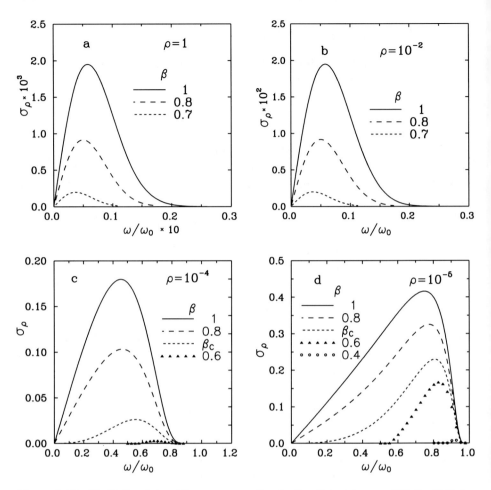

Figure 4.37. Radiation intensities corresponding to the dielectric permittivity (4.7) for $p/\omega_0 = 10^{-2}$ and for a number of velocities and observational cylinder radii ρ (in cm). It is seen that the frequency distribution of the radiation crucially depends on the radius ρ. This leads to the ambiguity in the interpretation of experimental data. The ρ dependence disappears in the absence of damping.

of σ_ρ is at $\omega/\omega_0 = 2 \times 10^{-3}$ for $\beta = 1$ and $\rho = 10$ cm. For $\rho = 1$ cm (Fig. 4.37(a)) the maximum of the same radiation intensity is at $\omega/\omega_0 \approx 6 \times 10^{-3}$. This means that all frequency distributions shown in this figure are shifted towards the larger ω/ω_0. This tendency is supported by Figs. 4.37(b,c,d) where the radiation intensities for $\rho = 10^{-2}$ cm, $\rho = 10^{-4}$ cm and $\rho = 10^{-5}$ cm are presented.

4.10.2. DIELECTRIC PERMITTIVITY (4.45)

There is an important difference between the parametrizations (4.7) and (4.45). It is seen that $\epsilon(\omega)$ given by (4.7) tends to unity for $\omega \to \infty$. This means that the medium oscillators do not have enough time to be excited in this limit. On the other hand, $\epsilon(\omega)$, given by (4.45), tends to ϵ_∞ in the same limit. This leads to the appearance of two critical velocities $\beta_\infty = 1/\sqrt{\epsilon_\infty}$ and $\beta_0 = 1/\sqrt{\epsilon_0}$, where $\epsilon_\infty = \epsilon(\omega = \infty)$ and $\epsilon_0 = \epsilon(\omega = 0) = \epsilon_\infty + \omega_L^2/\omega_0^2$. Now we evaluate the frequency distribution of the energy radiated by a point-like charge moving uniformly in ZnSe with the same parameters as in [37]. But first we make the preliminary estimates. For the parametrizations (4.45) with $p = 0$ the radiation $(1 - \beta^2\epsilon < 0)$ condition takes place in the following ω domains:

For a charge velocity greater than the larger critical velocity $(\beta > \beta_\infty)$ the radiation condition $1 - \beta^2\epsilon < 0$ holds if $0 < \omega < \omega_0$ and $\omega > \omega_1$. Here $\omega_1^2 = \omega_0^2(\beta^2\epsilon_0 - 1)/(\beta^2\epsilon_\infty - 1)$. At first glance it seems that for the parametrization (4.45) the frequency spectrum of the radiation extends to infinite frequencies. Fortunately this is not so. According to Chapter 7 the finite dimensions of a moving charge lead to the cut-off of the frequency spectrum at approximately $\omega_c = c/a$, where a is the charge dimension. If for a we take the classical electron radius (e^2/mc^2), then $\omega_c \sim 10^{23}\mathrm{s}^{-1}$, which is far above the frequency of the visible light $(\omega \sim 10^{15}\mathrm{s}^{-1})$. For $\beta \to \beta_\infty$, $\omega_1 \to \infty$, and only the low frequency part of the radiation spectrum survives.

For the charge velocity between two critical velocities $(\beta_0 < \beta < \beta_\infty)$ the radiation condition $1 - \beta^2\epsilon < 0$ takes place if $0 < \omega < \omega_0$.

Finally, for the charge velocity smaller than the minor critical velocity $(0 < \beta < \beta_0)$, the radiation condition $1 - \beta^2\epsilon < 0$ is realized in the frequency window $\omega_1 < \omega < \omega_0$. There is no radiation outside it. When $\beta \to 0$, $\omega_1 \to \omega_0$ and the frequency window becomes narrower.

Application to ZnSe
In [37] the following parameters of a dielectric permittivity (4.45) with $p = 0$ were found:

$$\epsilon_\infty = 5.79, \quad \epsilon_0 = 8.64, \quad \nu_0 = 6.3 \times 10^{12}\mathrm{Hz}, \quad \omega_0 = 2\pi\nu_0 \approx 4 \cdot 10^{13}\mathrm{s}^{-1}.$$

The corresponding critical velocities are given by $\beta_\infty = 0.416$ and $\beta_0 = 0.34$.

For $\beta > \beta_\infty$ the frequency distribution is confined to the following ω regions: $0 < \omega < \omega_0$ and $\omega > \omega_1$. At $p = 0$ the radiation intensities behave in accordance with above predictions (Fig. 4.38).

Let $p \neq 0$. For $\beta > \beta_\infty$ the radiation intensities corresponding to the high frequency branch $(\omega > \omega_1)$ vary quite slowly as p increases (Figs.

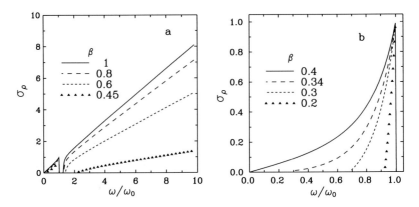

Figure 4.38.   Radiation intensities corresponding to the dielectric permittivity (4.45) for $p = 0$ and a number of charge velocities. The medium parameters are the same as for ZnSE. There are two critical velocities: $\beta_\infty \approx 0.416$ and $\beta_0 \approx 0.34$. (a): For $\beta > \beta_\infty$ there are two frequency regions ($0 < \omega < \omega_0$ and $\omega_1 < \omega < \infty$) to which frequency distributions are confined. For $\beta \to \beta_\infty$, $\omega_1 \to \infty$; (b): For $\beta_0 < \beta < \beta_\infty$ the radiation is confined to the frequency region $0 < \omega < \omega_0$ ($\beta = 0.4$ and 0.34). For $0 < \beta < \beta_0$, the radiation is confined to the frequency region $\omega_1 < \omega < \omega_0$. For $\beta \to 0$, $\omega_1 \to \omega_0$ and the frequency window becomes narrower ($\beta = 0.3$ and 0.2).

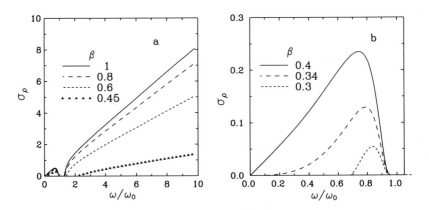

Figure 4.39.   The same as in Fig.4.38, but for a nonzero $p/\omega_0 = 10^{-8}$. (a): It is seen that for $\beta > \beta_\infty$, the high-frequency branch of the spectrum is almost the same as in the absence of damping. Radiation intensities in the low-frequency part of the spectrum are two times smaller than for $p = 0$; (b): For $\beta < \beta_\infty$, the frequency spectrum is more sensitive to the change of p. Its position is shifted towards the smaller ω. For $\beta < \beta_0$ the radiation intensities are very small. For example, for $\beta = 0.2$ the maximal value of the radiation intensity is $\approx 5 \times 10^{-6}$. The cylinder radius $\rho = 10$ cm.

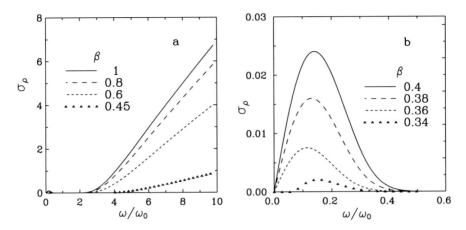

Figure 4.40. The same as in Fig. 4.38, but for a larger $p/\omega_0 = 10^{-6}$. (a): For $\beta > \beta_\infty$ the low-frequency part of the spectrum practically disappears. (b): For $\beta_0 < \beta < \beta_\infty$, the frequency spectrum is shifted towards the smaller ω. The radiation intensities are approximately ten times smaller than those in Fig. 4.39 (b). The radiation intensity corresponding to $\beta = \beta_0 = 0.34$ is multiplied by 100 (that is, the curve shown should be decreased in 100 times). For $\beta < \beta_0$ the radiation intensities are small and cannot be presented on this scale. Comparing this figure with Figs. 4.38 and 4.40, we observe that the position of the maximum of the frequency spectrum depends crucially on the damping parameter.

4.38(a) and 4.39 (a)). On the other hand, the low-energy branch of the radiation intensity ($0 < \omega < \omega_0$) is more sensitive to the damping increase: it is practically invisible even for a quite small value of $p/\omega_0 = 10^{-6}$ (Fig. 4.40 (a)).

Let $\beta_0 < \beta < \beta_\infty$. At $p/\omega_0 = 10^{-8}$ and $p/\omega_0 = 10^{-6}$ the maximal values of radiation intensities are, respectively, four and forty times smaller than for $p = 0$ (Figs. 4.38(b) and 4.39 (b)). In addition they are shifted towards the smaller ω. The radiation intensities decrease still more rapidly with rising p for $\beta < \beta_0$. For example, for $\beta = 0.2$ and $p/\omega_0 = 10^{-6}$ the maximal value of the radiation intensity is $\approx 5 \times 10^{-6}$.

The main result of this consideration is that, in absorptive media both the value and position of the maximum of the frequency distribution crucially depend on the distance at which observations are made. The diminishing of the radiation intensity is physically clear since only part of the radiated energy flux reaches the observer if $p \neq 0$. Does the frequency shift of the maximum of the radiation intensity mean that any discussion of the frequency distribution of the radiation intensity should be supplemented by an indication of the observational distance? In the absence of absorption ($p = 0$) the index of the exponent in (4.47) is zero and the dependence on the cylindrical radius ρ drops out. At first glance it is possible to associate the ρ independent frequency distribution of the radiation intensity with the pre-exponential factor in (4.47) which is the $\rho = 0$ limit of (4.47). But (4.47) is not valid at small distances. Instead, the exact Eq. (4.46) should be used there which is infinite at $\rho = 0$ (since a charge moves along the z axis).

4.11. Cherenkov radiation without use of the spectral representation

In the $\vec{r}, t$ representation $\Phi(\vec{r}, t)$ and $\vec{A}(\vec{r}, t)$ are given by

$$\Phi(\vec{r}, t) = \frac{e}{\pi v} \int \frac{d\omega}{\epsilon} e^{i\omega(t-z/v)} \frac{kdk}{k^2 + (\omega^2/v^2)(1 - \beta^2\epsilon)} J_0(k\rho).$$

$$A_z(\vec{r}, t) = \frac{e}{\pi c} \int d\omega e^{i\omega(t-z/v)} \frac{kdk}{k^2 + (\omega^2/v^2)(1 - \beta^2\epsilon)} J_0(k\rho). \qquad (4.50)$$

The usual way to handle these integrals is to integrate them first over k. This was done above in a closed form. The remaining integrals over ω are interpreted as frequency distributions of EMF associated with the uniform motion of charge in medium.

In this approach we prefer to take the above integrals first over ω [36]. The advantage of this approach is that arising integrals can be treated analytically in various particular cases. These integration methods complement each other. The Maxwell equations (4.15) describing the EMF of a uniformly moving charge can be handled without any appeal to the ω representation. To prove this we rewrite Eq. (4.17) in the $\vec{r}, t$ representation:

$$\vec{P}(t) = \frac{1}{8\pi^2} \int\limits_{-\infty}^{\infty} G(t - t')\vec{E}(t')dt',$$

where

$$G(t - t') = \lim_{p \to 0+} \omega_L^2 \int\limits_{-\infty}^{+\infty} \frac{d\omega}{\omega_0^2 - \omega^2 + ip\omega} e^{i\omega(t-t')}.$$

A direct calculation shows that

$$G(t-t') = 0 \quad \text{for} \quad t' > t \quad \text{and} \quad G(t-t') = \frac{2\pi\omega_L^2}{\omega_0} \sin[\omega(t-t')] \quad \text{for} \quad t' < t.$$

Substituting $\vec{P}$ into the Maxwell equations (4.15) one obtains the system of integro-differential equations which depend only on the charge velocity and the medium parameters and which do not contain the frequency ω.

We represent the denominator entering in (4.50) in the form

$$\frac{1}{k^2 + \omega^2(1 - \beta^2\epsilon)/v^2} = \frac{v^2}{1 - \beta^2}\frac{\omega^2 - \omega_0^2}{(\omega^2 - \omega_1^2)(\omega^2 + \omega_2^2)}$$

$$= \frac{v^2}{1 - \beta^2}\frac{\omega^2 - \omega_0^2}{\omega_1^2 + \omega_2^2}$$

$$\times \left[\frac{1}{2\omega_1}\left(\frac{1}{\omega - \omega_1} - \frac{1}{\omega + \omega_1}\right) - \frac{1}{2i\omega_2}\left(\frac{1}{\omega - i\omega_2} - \frac{1}{\omega + i\omega_2}\right)\right],$$

Here

$$k = \omega/c, \quad \omega_3^2 = \omega_0^2 + \omega_L^2, \quad \omega_1^2 = \omega_0^2 - \Omega + (\Omega^2 - \beta^2\gamma^2\omega_0^2\omega_L^2)^{1/2},$$

$$\omega_2^2 = -\omega_0^2 + \Omega + (\Omega^2 - \beta^2\gamma^2\omega_0^2\omega_L^2)^{1/2}, \quad \Omega = \frac{1}{2}[\omega_0^2 + \beta^2\gamma^2(k^2c^2 + \omega_L^2)].$$

Inserting these expressions into (4.50) and performing the ω integration we get for the electromagnetic potentials and field strengths

$$A_z = A_z^{(1)} + A_z^{(2)},$$

$$A_z^{(1)} = \frac{ev^2\gamma^2}{c}\int_0^\infty kdk\, J_0(k\rho)F_A^{(1)}, \quad A_z^{(2)} = \frac{ev^2\gamma^2}{c}\int_0^\infty kdk\, J_0(k\rho)F_A^{(2)}$$

$$\Phi = ev\gamma^2\int_0^\infty kdk\, J_0(k\rho)F_\phi - \frac{2e\omega_L^2}{v\omega_3}\sin[\omega_3(t - z/v)]\Theta(t - z/v)K_0(\rho\omega_3/v),$$

$$H_\phi = -\frac{\partial A_z}{\partial\rho} = e\beta^2 c\gamma^2\int_0^\infty k^2 dk\, J_1(k\rho)F_A, \quad D_\rho = H_\phi/\beta, \qquad (4.51)$$

$$E_\rho = e\gamma^2 v\int_0^\infty k^2 dk\, J_1(k\rho)F_\phi - \frac{2e\omega_L^2}{v^2}\sin\omega_3(t - z/v)\Theta(t - z/v)K_1(\rho\omega_3/v),$$

$$E_z = e\gamma^2\int_0^\infty kdk\, J_0(k\rho)[2\left(\beta^2 - \frac{\omega_1^2 - \omega_0^2}{\omega_1^2 - \omega_3^2}\right)$$

$$\times\frac{\omega_1^2 - \omega_0^2}{\omega_1^2 + \omega_2^2}\Theta(t - z/v)\cos\omega_1(t - z/v)$$

$$-\left(\beta^2 - \frac{\omega_0^2 + \omega_2^2}{\omega_2^2 + \omega_3^2}\right)\frac{\omega_0^2 + \omega_2^2}{\omega_2^2 + \omega_1^2}\cdot\text{sign}(z - vt)\exp\left(-\omega_2|t - z/v|\right)]$$

$$-\frac{2e\omega_L^2}{v^2}\cos\omega_3(t - z/v)\Theta(t - z/v)K_0(\rho\omega_3/v),$$

$$D_z = -2e \int_0^\infty kdk J_0(k\rho)\frac{\omega_1^2 - \omega_0^2 + \beta^2\gamma^2\omega_L^2}{\omega_1^2 + \omega_2^2}\Theta(t - z/v)\cos\omega_1(t - z/v)$$

$$-e \int_0^\infty kdk J_0(k\rho)\frac{\omega_2^2 + \omega_0^2 - \beta^2\gamma^2\omega_L^2}{\omega_1^2 + \omega_2^2}\exp(-\omega_2|t - z/v|)\cdot\operatorname{sign}(t - z/v).$$

Here we put:

$$F_A = F_A^{(1)} + F_A^{(2)}, \quad F_\phi = F_\phi^{(1)} + F_\phi^{(2)},$$

$$F_A^{(1)} = -\frac{\omega_1^2 - \omega_0^2}{\omega_1^2 + \omega_2^2}\frac{2}{\omega_1}\Theta(t - z/v)\sin\omega_1(t - z/v),$$

$$F_A^{(2)} = \frac{1}{\omega_2}\frac{\omega_2^2 + \omega_0^2}{\omega_1^2 + \omega_2^2}\exp\left(-\omega_2|t - z/v|\right),$$

$$F_\phi^{(1)} = -\frac{(\omega_1^2 - \omega_0^2)^2}{(\omega_1^2 + \omega_2^2)(\omega_1^2 - \omega_3^2)}\Theta(t - z/v)\frac{2}{\omega_1}\sin\omega_1(t - z/v),$$

$$F_\phi^{(2)} = \frac{(\omega_0^2 + \omega_2^2)^2}{\omega_2(\omega_1^2 + \omega_2^2)(\omega_3^2 + \omega_2^2)}\exp\left(-\omega_2|t - z/v|\right). \qquad (4.52)$$

The separation of F_A and F_ϕ into two parts is justified physically. It turns out (see the next section) that $F_A^{(1)}$, $F_\phi^{(1)}$ and $F_A^{(2)}$, $F_\phi^{(2)}$ describe correspondingly the radiation field and EMF carried by a uniformly moving charge. They originate from the ω poles lying in non-damping and damping regions, respectively.

When evaluating electromagnetic potentials and field strengths we have taken into account that $\epsilon(\omega)$ given by (4.1) is a limiting expression (as $p \to 0$) of

$$\epsilon(\omega) = 1 + \frac{\omega_L^2}{\omega_0^2 - \omega^2 + ip\omega}$$

having a pole in the upper ω half-plane (for the Fourier transform chosen in the form (4.3)). This in turn results in an infinitely small positive imaginary part in ω_1 and in factor 2 in the first terms in F_A and F_ϕ. The position of poles of $\epsilon(\omega)$ in the upper complex ω half-plane is needed to satisfy the causality condition. It is seen that Φ, E_ρ, and E_z are singular on the motion axis behind the moving charge. These singularities are due to the modified Bessel functions K outside the integrals in (4.51). For a fixed observational point z on the cylinder surface these singularities as functions of time oscillate with the frequency $\omega_3 = \omega_0/\beta_c$. For the fixed observational time t these singularities as functions of the observational point z oscillate with the frequency $\omega_0/\beta_c v$. Since the electric induction $\vec{D}$ is not singular on the motion axis, the electric polarization $\vec{P} = (\vec{D} - \vec{E})/4\pi$ has the same singularity as $\vec{E}$. As to the magnetic field $\vec{H}$, it tends to zero when one approaches the motion axis:

$$H_\phi \to \frac{e\omega_L^2\omega_0}{c^3}\Theta(t - z/v)\sin[\omega_0(t - z/v)]\rho K_0(\rho\omega_0/c) \quad \text{for} \quad \rho \to 0.$$

4.11.1. PARTICULAR CASES

Consider the limiting cases. In most cases we present analytic results for the magnetic vector potential (and, rarely, for the electric potential). The behaviour of EMF strengths is restored by the differentiation of potentials.

1) Let $v \to 0$. Then, $\omega_1 \to \omega_0$, $\omega_2 \to v\gamma k$, $A_z \to 0$, and

$$\Phi \to \frac{e\gamma}{1 + \omega_L^2/\omega_0^2} \int_0^\infty dk\, J_0(k\rho) \exp\left(-\beta\gamma kc|t - z/v|\right)$$

$$= \frac{e}{\epsilon_0} \frac{1}{[z^2 + \rho^2]^{1/2}}. \tag{4.53}$$

i.e., we obtain the field of a charge to be at rest in the medium. It turns out that only the second term in F_ϕ contributes to Φ.

2) Let $\omega_L \to 0$. This corresponds to the zero electron density, at which the moving charge exhibits scattering. Then, $\epsilon \to 1$, $\beta_c \to 1$, $\omega_1 \to 0$, $\omega_2 \to \gamma kv$,

$$A_z \to e\beta\gamma \int_0^\infty dk\, J_0(k\rho) \exp\left(-k\gamma|z - vt|\right) = \frac{e\beta}{[(z - vt)^2 + \rho^2/\gamma^2]^{1/2}},$$

$$\Phi \to \frac{e}{[(z - vt)^2 + \rho^2/\gamma^2]^{1/2}}, \tag{4.54}$$

i.e., we obtain the field of a charge moving uniformly in vacuum. Again, only second terms in F_ϕ and F_A contribute to Φ and A_z, respectively.

3) Let $\omega_L \to \infty$. This corresponds to an optically dense medium. Then,

$$\omega_1^2 \to \frac{\omega_0^2}{\omega_L^2}k^2c^2, \quad \omega_2^2 \to \beta^2\gamma^2(\omega_L^2 + k^2c^2) - \omega_0^2 + \frac{\omega_0^2}{\omega_L^2}k^2c^2,$$

$$F_A^{(1)} \to \frac{2\omega_0\omega_L}{\beta^2\gamma^2(\omega_L^2 + k^2c^2)kc}\Theta(t - z/v)\sin\frac{\omega_0 kc(t - z/v)}{\omega_L},$$

$$F_A^{(2)} \to \frac{e}{\beta\gamma\sqrt{\omega_L^2 + k^2c^2}}\exp(-\beta\gamma\sqrt{\omega_L^2 + k^2c^2}|t - z/v|).$$

$A_z^{(2)}$ can be evaluated in a closed form:

$$A_z^{(2)} \to \frac{e\beta}{R} \exp(-\gamma\omega_L R/c), \quad R = [(z - vt)^2 + \beta_c^2\gamma_c^2\rho^2/\gamma^2]^{1/2}. \quad (4.55)$$

whilst the analytic form of $A_z^{(1)}$ is available only for $\rho \ge \omega_0 c(t - z/v)/\omega_L$:

$$A_z^{(1)} \to \frac{2e\omega_0}{c} \Theta(t - z/v) \sinh[\omega_0(t - z/v)]K_0(\omega_L\rho/c). \quad (4.56)$$

(it is seen that $A_z^{(1)}$ decreases exponentially when ρ grows and increases exponentially with increasing of $t - z/v$), and on the motion axis:

$$A_z^{(1)} = \frac{e\omega_0}{c} \Theta(t - z/v)[\exp(-\omega_0(t - z/v))E_i(\omega_0(t - z/v))$$

$$- \exp(\omega_0(t - z/v))E_i(-\omega_0(t - z/v))].$$

Here $E_i(x)$ is an integral exponent. For small and large values of $\omega_0(t - z/v)$ this gives:

$$A_z^{(1)} \approx -2\frac{e\omega_0}{c} \Theta(t - z/v) \sin(\omega_0(t - z/v))[C + \ln(\omega_0(t - z/v))]$$

for $\omega_0(t - z/v) \ll 1$ and

$$A_z^{(1)} \approx \frac{2e}{c(t - z/v)}$$

for $\omega_0(t - z/v) \gg 1$. Here C is the Euler constant. Thus damped oscillations of the EMF should be observed on the motion axis behind the charge.

4) Let $\omega_0 \to \infty$, i.e., the resonance level lies very high. Then

$$\omega_1^2 \to \omega_0^2 - \beta^2\gamma^2\omega_L^2, \quad \omega_2^2 \to \beta^2\gamma^2k^2c^2,$$

$$F_A^{(1)} \to \frac{\omega_L^2\beta^2\gamma^2}{\omega_0^2 - \omega_L^2\beta^2\gamma^2 + \beta^2\gamma^2k^2c^2} \frac{2}{\sqrt{\omega_0^2 - \beta^2\gamma^2\omega_L^2}}$$

$$\times\Theta(t - z/v) \sin[\sqrt{\omega_0^2 - \omega_L^2\beta^2\gamma^2}(t - z/v)],$$

$$F_A^{(2)} \to \frac{1}{\beta\gamma kc} \exp(-\beta\gamma kc|t - z/v|,$$

$$A_z^{(1)} \to \frac{2e\omega_L^2\beta^2\gamma^2}{c\omega_0} \Theta(t - z/v) \sin[\omega_0(t - z/v)]K_0(\rho\omega_0/\beta\gamma c), \quad (4.57)$$

$$A_z^{(2)} \to \frac{e\beta}{[(z - vt)^2 + \rho^2/\gamma^2]^{1/2}}, \quad (4.58)$$

We see that a complete VP consists of the term $A_z^{(2)}$ describing the charge motion in vacuum and oscillating perturbation $A_z^{(1)}$ on the axis of the charge motion.

5) Let $\omega_0 \to 0$, i.e., the resonance level lies very low. Then,

$$\omega_1^2 \to \omega_0^2 \frac{k^2 c^2}{k^2 c^2 + \omega_L^2}, \quad \omega_2^2 \to \beta^2 \gamma^2 (k^2 c^2 + \omega_L^2) - \frac{\omega_L^2 \omega_0^2}{k^2 c^2 + \omega_L^2},$$

$$F_A^{(1)} \approx \frac{2 \omega_0^2 \omega_L^2}{\beta^2 \gamma^2 ck} \frac{1}{(k^2 c^2 + \omega_L^2)^{3/2}} \Theta(t - z/v) \sin\left[\frac{\omega_0 kc(t - z/v)}{\sqrt{k^2 c^2 + \omega_L^2}}\right],$$

$$F_A^{(2)} \approx \frac{1}{\beta\gamma} \frac{1}{\sqrt{k^2 c^2 + \omega_L^2}} \exp\left[-\beta\gamma\sqrt{k^2 c^2 + \omega_L^2}(t - z/v)\right],$$

$$A_z^{(2)} \approx \frac{e\beta}{R} \exp(-\gamma \omega_L R/c), \quad R = [(vt - z)^2 + \beta_c^2 \gamma_c^2 \rho^2/\gamma^2]^{1/2}. \quad (4.59)$$

We succeeded in evaluating $A_z^{(1)}$ in a closed form in two cases. For $\omega_0 \rho/c \ll 1$ the VP slowly oscillates behind the moving charge:

$$A_z^{(1)} \approx 2e\Theta(t - z/v)\frac{1 - \cos\omega_0(t - z/v)}{c(t - z/v)}, \quad (4.60)$$

On the other hand, for $\omega_0(t - z/v) \ll 1$

$$A_z^{(1)} \approx \frac{e\omega_0^2 \omega_L}{c^3} \Theta(t - z/v)\rho c(t - z/v) K_1(\rho \omega_L/c).$$

i.e., there are VP oscillations in the half-space behind the moving charge decreasing exponentially with increasing ρ.

6) Let $\omega_0 \to \infty$, $\omega_L \to \infty$, but ω_L/ω_0, and therefore β_c is finite. One then finds

$$\omega_1^2 \to \omega_0^2(1 - \tilde{\epsilon}) + x^2 \omega_0^2 \frac{\tilde{\epsilon}}{1 - \tilde{\epsilon}}, \quad \omega_2^2 = x^2 \omega_0^2 \frac{1}{1 - \tilde{\epsilon}}.$$

$$A_z \to \frac{2ec\beta^2\gamma^2\tilde{\epsilon}}{\omega_0(1 - \tilde{\epsilon})^{3/2}} \frac{\delta(\rho)}{\rho} \sin[\sqrt{1 - \tilde{\epsilon}}\omega_0(t - z/v)]$$

$$+ \frac{e\beta}{[(z - vt)^2 + \rho^2(1 - \beta^2\epsilon_0)]^{1/2}},$$

for $\beta < \beta_c$. Here $x = \beta\gamma kc/\omega_0$, $\tilde{\epsilon} = \beta^2\gamma^2/\beta_c^2\gamma_c^2$.

For $\beta > \beta_c$ one has

$$\omega_1^2 = \frac{\omega_0^2 x^2}{\tilde{\epsilon} - 1}, \quad \omega_2^2 = \omega_0^2(\tilde{\epsilon} - 1) + x^2 \omega_0^2 \frac{\tilde{\epsilon}}{\tilde{\epsilon} - 1},$$

$$A_z \rightarrow \frac{ec\beta^2\gamma^2\tilde{\epsilon}}{\omega_0(\tilde{\epsilon}-1)^{3/2}} \frac{\delta(\rho)}{\rho} \exp[-\sqrt{\tilde{\epsilon}-1}\,\omega_0(t-z/v)]$$

$$+\frac{2e\beta}{[(z-vt)^2 - \rho^2(\beta^2\epsilon_0-1)]^{1/2}},$$

(ϵ_0 is the same as above). The origin of the first and second terms in A_z and Φ is owed to the second and first terms in F_A and F_ϕ, respectively. Thus one obtains the EMF of a charge moving in a medium with a constant electric permittivity $\tilde{\epsilon} = \epsilon_0$ and the singular EMF on the motion axis.

7) Let the dimensionless quantity $\tilde{\epsilon} = \beta^2\gamma^2/\beta_c^2\gamma_c^2 \gg 1$. Then,

$$\omega_1^2 = \frac{\omega_0^2 x_c^2}{1+x_c^2}, \quad \omega_2^2 = \tilde{\epsilon}(1+x_c^2) - \frac{1}{1+x_c^2}, \quad x_c = \beta_c\gamma_c kc/\omega_0,$$

$$F_A = F_A^{(1)} + F_A^{(2)},$$

$$F_A^{(1)} = \frac{2}{\omega_0\tilde{\epsilon}x_c(1+x_c^2)^{3/2}}\Theta(t-z/v)\sin\left[\omega_0(t-z/v)\frac{x_c}{\sqrt{1+x_c^2}}\right],$$

$$F_A^{(2)} = \frac{1}{\omega_0\sqrt{\tilde{\epsilon}}}\frac{1}{\sqrt{1+x_c^2}}\exp(-\sqrt{\tilde{\epsilon}}\sqrt{1+x_c^2}\,\omega_0|t-z/v|). \qquad (4.61)$$

Correspondingly,

$$A_z = A_z^{(1)} + A_z^{(2)},$$

where

$$A_z^{(1)} = \frac{2e\omega_0}{c}\Theta(t-z/v)\int_0^\infty \frac{dx}{(1+x^2)^{3/2}}J_0\left(\frac{\rho\omega_0 x}{\beta_c\gamma_c c}\right)\sin\left[\frac{\omega_0(t-z/v)x}{\sqrt{1+x^2}}\right],$$

$$A_z^{(2)} = \frac{e\beta}{R}\exp\left(-\frac{\omega_0 R\gamma}{\beta_c\gamma_c c}\right), \quad R = [(z-vt)^2 + \rho^2/\gamma^2]^{1/2}.$$

We did not succeed in evaluating $A_z^{(1)}$ in a closed form. Instead, we consider particular cases when the condition $\tilde{\epsilon} \gg 1$ can be realized.

Let β be finite and $\beta_c \rightarrow 0$. This corresponds to an optically dense medium. Then $A_z^{(2)}$ is exponentially small whereas

$$A_z^{(1)} = \frac{2e\beta_c\gamma_c}{(\beta_c^2\gamma_c^2 c^2(t-z/v)^2 - \rho^2)^{1/2}}\Theta(t-z/v)\Theta[\beta_c\gamma_c c(t-z/v) - \rho]. \quad (4.62)$$

is confined to an infinitely narrow cone lying behind the moving charge. This equation is obtained by neglecting x_c^2 in the square roots in (4.61).

Let $\beta \to 1$, $\beta_c \to 1$ under the condition $\tilde{\epsilon} \gg 1$ (that is, β is much closer to unity than β_c). This inequality is possible because of the γ factors in the definition of $\tilde{\epsilon}$. Then

$$A_z^{(1)} = 2e\Theta(t - z/v)\frac{1 - \cos[\omega_0(t - z/v)]}{ct - z} \tag{4.63}$$

for small values of ρ. It is seen that the VP exhibits oscillations in a half-space behind the moving charge. More accurately, the condition under which Eq. (4.63) is valid looks like $\rho\omega_0/\beta_c\gamma_c c \ll 1$. This means that for β_c fixed in the interval $0 < \beta_c < 1$, $A_z^{(1)}$ oscillates for $\rho \ll \beta_c\gamma_c c/\omega_0$.

8) Let $\tilde{\epsilon} \ll 1$. Then

$$\omega_1^2 = \omega_0^2\left(1 - \frac{\tilde{\epsilon}}{1 + x^2}\right), \quad \omega_2^2 = \omega_0^2 x^2\left(1 + \frac{\tilde{\epsilon}}{1 + x^2}\right), \quad x = \beta\gamma kc/\omega_0,$$

$$F_A^{(1)} = \frac{2\tilde{\epsilon}}{\omega_0}\frac{1}{(1 + x^2)^2}\Theta(t - z/v)\sin\left[\omega_0(t - z/v)\left(1 - \frac{\tilde{\epsilon}}{2}\frac{1}{1 + x^2}\right)\right],$$

$$F_A^{(2)} = \frac{1}{\omega_0 x}\exp(-\omega_0 x|t - z/v|).$$

It turns out that $A_z^{(2)}$ coincides with the VP of a charge moving in a vacuum:

$$A_z^{(2)} = \frac{ev}{[(z - vt)^2 + \rho^2/\gamma^2]^{1/2}}.$$

As to $A_z^{(1)}$, it can be taken in an analytic form for $(t - z/v)\omega_0\tilde{\epsilon} \ll 1$:

$$A_z^{(1)} = \frac{e\rho\omega_0^2\beta\gamma}{c^2\beta_c^2\gamma_c^2}\Theta(t - z/v)\sin[\omega_0(t - z/v)]K_1(\rho\omega_0/\beta\gamma c). \tag{4.64}$$

The condition $\tilde{\epsilon} \ll 1$ can be realized in two ways. First, β_c can be finite but $\beta \ll 1$. In this case $A_z^{(1)}$ is confined to a narrow beam behind the moving charge:

$$A_z^{(1)} = e\left(\frac{\pi\rho\omega_0^3\beta^3\gamma^3}{2c^3}\right)^{1/2}\frac{1}{\beta_c^2\gamma_c^2}\Theta(t - z/v)\sin[\omega_0(t - z/v)]\exp(-\frac{\rho\omega_0}{\beta\gamma c}). \tag{4.65}$$

On the other hand, the condition $\tilde{\epsilon} \ll 1$ can be satisfied when β is close to 1, but β_c is much closer to it. Then,

$$A_z^{(1)} = \frac{e\omega_0\tilde{\epsilon}}{c}\Theta(t - z/v)\sin[\omega_0(t - z/v)]. \tag{4.66}$$

Thus $A_z^{(1)}$ is small (owing to the $\tilde{\epsilon}$ factor), but not exponentially small. This means that one should observe oscillations in the half-space behind the moving charge. Physically, $\beta_c \approx 1$, $\beta \approx 1$, $\tilde{\epsilon} \ll 1$ corresponds to the motion in an optically rarefied medium (e.g., gas) with a charge velocity slightly smaller than the velocity of light in medium.

We observe the a noticeable distinction between the cases $\beta \approx 1$, $\beta_c \approx 1$ corresponding to $\tilde{\epsilon} \gg 1$ and $\tilde{\epsilon} \ll 1$. In both cases $A_z^{(1)}$ oscillates in the half-space behind the moving charge, but the amplitude of oscillations is considerably smaller for $\beta < \beta_c$ (owing to the $\tilde{\epsilon}$ factor in (4.66)).

More precisely, the condition under which Eq. (4.66) is valid is $\rho\omega_0/\beta\gamma c \ll 1$. This means that for β fixed, the VP oscillations should take place for small values of ρ.

9) Let the charge velocity exactly coincide with the velocity of light in medium: $\beta = \beta_c$, $\tilde{\epsilon} = 1$. Then

$$\frac{\omega_1^2}{\omega_0^2} = -\frac{x^2}{2} + x(1 + x^2/4)^{1/2}, \qquad \frac{\omega_2^2}{\omega_0^2} = \frac{x^2}{2} + x(1 + x^2/4)^{1/2}.$$

Let $\beta = \beta_c \approx 1$. This corresponds to a fast charged particle moving in a rarefied medium. Then

$$A_z^{(1)} = \frac{4}{3}\frac{e\omega_0}{c}\Theta(t - z/v)\sin[\omega_0(t - z/v)]\left[K_0\left(\frac{\omega_0\rho}{\sqrt{2}\beta\gamma c}\right) - K_0\left(\sqrt{2}\frac{\omega_0\rho}{\beta\gamma c}\right)\right],$$

$$A_z^{(2)} = \frac{e\beta}{R}\exp(-\omega_0 R/v), \quad R = [(z - vt)^2 + \rho^2/\gamma^2]^{1/2}.$$

Thus $A_z^{(2)}$ differs from zero in a neighbourhood of the current charge position, whereas $A_z^{(1)}$ describes the oscillations in the half-plane behind the moving charge. As γ is very large, $A_z^{(1)}$ as a function of ρ diminishes rather slowly: it decreases essentially when the radius $\rho \approx \sqrt{2}c\gamma/\omega_0$.

4.11.2. NUMERICAL RESULTS.

In this section we present the results of numerical calculations. We intend to consider the EMF distribution on the surface of the cylinder C_ρ of the radius ρ (Fig. 4.2). This is a usual procedure in the consideration of VC effect (see, e.g., [29]).

For a frequency-independent electric permittivity ($\epsilon = \epsilon_0$) there is no radiation for $\beta < \beta_c = \epsilon_0^{-1/2}$. For $\beta > \beta_c$ the energy flux is infinite on the surface of the Cherenkov cone. On the surface of C_ρ it is equal to zero for $z > -z_c$, ($z_c = \rho\sqrt{\beta^2 n^2 - 1}$), and acquires an infinite value at $z = -z_c$ where C_ρ intersects the above cone. Inside the Cherenkov cone the electromagnetic strengths fall as r^{-2} at large distances, and therefore do not contribute to the radial flux.

In what follows, the results of numerical calculations will be presented in dimensionless variables. In particular, lengths will be expressed in units of c/ω_0, time in units of ω_0^{-1}, electromagnetic strengths in units of $e\omega_0^2/c^2$, the Poynting vector $\vec{P} = (c/4\pi)(\vec{E} \times \vec{H})$ in units of $e^2\omega_0^4/c^3$, etc.. The advantage of using dimensionless variables is that Cherenkov radiation can be considered at arbitrary distances.

In Fig. 4.8 we presented the dimensionless quantity $F = W_\rho/(e^2\omega_0^2/c^2)$ as a function of the particle velocity β. The numbers on curves are β_c. Vertical lines with arrows divide a curve into two parts corresponding to the energy losses with velocities $\beta < \beta_c$ and $\beta > \beta_c$ and lying to the left and right of vertical lines, respectively. We see that the charge uniformly moving in medium radiates at every velocity.

How is this flux distributed over the surface of C_ρ? For definiteness we take $\beta_c = 0.75$ to which corresponds the refractive index $n = 1/\beta_c = 1.333$. This is close to the refractive index of water ($n = 1.334$). The value of ρ is chosen to be $\rho = 10$ (in units of c/ω_0). In Fig. 4.41 it is shown how the quantity $\sigma_\rho = 2\pi S_\rho$ is distributed over the surface of C_ρ for $\beta = 0.3$. It is seen that the EMF (corresponding to the $A_z^{(1)}$ term in A_z) differs from zero only at large distances behind the moving charge. The isolated oscillation in the neighbourhood of $z = 0$ corresponds to the EMF carried by the moving charge. We refer to this part of EMF as the non-radiation EMF. Being originated from the $A_z^{(2)}$ term in A_z (see Eq.(4.51)), it is approximately equal to

$$\sigma_\rho^{(2)} = -\frac{c\beta e^2}{2\epsilon_0}(1 - \beta^2/\beta_c^2)^2\frac{\rho(z - vt)}{[z^2 + \rho^2(1 - \beta^2/\beta_c^2)]^3}. \qquad (4.67)$$

As we have mentioned, this corresponds to the radial energy flux carried by a uniformly moving charge with the velocity $\beta < \beta_c$ in medium with a constant $\epsilon = \epsilon_0$. Owing to its antisymmetry w.r.t. $z - vt$ the integral of it taken over either z or t is equal to zero.

If the distribution of the radiation flux on the surface of the sphere S (instead of on the cylinder surface C_ρ, as we have done up to now) were considered, the radial radiation flux S_ρ would be confined to the narrow cone adjusted to the negative z semi-axis. As follows from Fig. 4.41a the solution angle θ_c of this cone is equal to approximately 3 degrees for $\beta_c = 0.75$ and $\beta = 0.3$, i.e., the radiation is concentrated behind the moving charge near the motion axis.

When β grows, the relative contribution of the radiation term also increases. This is clearly demonstrated in Fig. 4.41(b) and 4.41(c) where the distributions of σ_ρ are presented for $\beta = 0.5$ and $\beta = 0.99$, respectively. The energy flux distributions presented in Figs. 4.41 (a,b,c) consist in fact of many oscillations. This is shown in Fig. 4.41(d) where the magnified image of σ_ρ for $\beta = 0.99$ is presented. It turns out that the first maximum of the radiation intensity is in the same place $z = -\rho\sqrt{\beta^2 n^2 - 1}$ where in

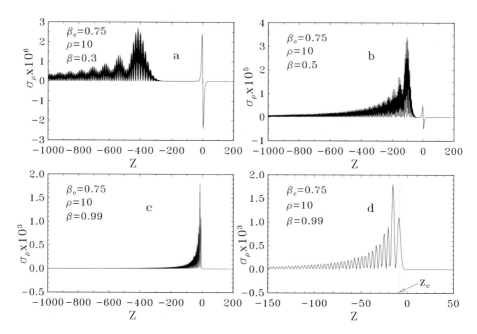

Figure 4.41. (a): Distribution of the radial energy flux on the surface of C_ρ for $\beta_c = 0.75$; and $\beta = 0.3$. The isolated oscillation in the neighbourhood of the plane $z = 0$ corresponds to the non-radiation field carried by a charge. The radiation and non-radiation terms are of the same order; (b): $\beta = 0.5$. The contribution of the non-radiation term relative to the radiation term is much smaller than for $\beta = 0.3$; (c): $\beta = 0.99$. The contribution of the non-radiation term relative to the radiation term is negligible; (d): Fine structure of the case $\beta = 0.99$. It is seen that a seemingly continuous distribution of (c) consists, in fact, of many peaks.

the absence of dispersion the singular Cherenkov cone intersects the surface of C_ρ. To detect the S_ρ component of $\vec{S}$, one should have a detector imbedded into a thin collimator and directed towards the charge motion axis. The collimator should be impenetrable for the γ quanta with directions different from the radial direction. It follows from Fig. 4.41 that in a particular detector (placed in the plane $z = \text{const}$), rapid oscillations of the radiation intensity as a function of time should be observed (since all the physical quantities and, in particular, S_ρ depend on t and z through the combination $z - vt$). It should be asked why so far nobody has observed these oscillations? From the $\beta = 0.99$, $\beta_c = 0.75$ case presented in Fig. 4.41 d it follows that the diffraction picture differs essentially from zero on the interval $-150 < z - vt < 0$, where z is expressed in units of c/ω_0. The typical ω_0 value taken from the Frank book [29] is $\omega_0 \approx 6 \times 10^{15} \text{s}^{-1}$. This gives $c/\omega_0 \approx 5 \times 10^{-6}$ cm. We see that the above interval is of the order 10^{-3} cm. The rapidly moving charge ($v \approx c$) traverses this distance for the time $10^{-3}c^{-1} \approx 3 \cdot 10^{-14}$s. It follows from Fig. 4.41 d that there are many oscillations in this time interval. Because of this, they can hardly be resolved experimentally.

Now we turn to experiments discussed recently in [18,19,20]. In them, for an electron moving in a gas with a fixed high energy $\beta \approx 1$), the radiation intensity was measured as a function of the gas pressure P. Let P_c corresponds $\beta = \beta_c$. For gas pressures below P_c (in this case $\beta < \beta_c$) the standard Tamm-Frank theory (see, e.g., [29]) predicts zero radiation intensity. A sharp reduction of the radiation intensity was observed in [18,19,20] for a gas pressure $P \approx P_c/100$. To this gas pressure there corresponds $\tilde{\epsilon} \ll 1$ despite the fact that $\beta \approx \beta_c \approx 1$ (this is possible because of the γ factors in the definition of $\tilde{\epsilon}$).

To clarify the nature of this phenomenon we turn to Eqs. (4.32) and (4.33) which for a fixed β define energy losses per unit length as a function of β_c. Typical curves are shown in Fig. 4.42 a,b. The numbers on curves are the charge velocity. It follows from Fig. 4.42 (b) that for $\beta = 0.99$ the radiation intensity diminishes approximately 60 fold when β_c changes from 0.9 to 0.999. The corresponding distributions of the energy flux on the surface of C_ρ are shown in Figs. 4 (c) and 4 (d). It is seen that the intensity at maxima is almost 1000 times smaller for $\beta_c = 0.999$ than for $\beta_c = 0.9$. The intensity distribution is very sharp for $\beta_c = 0.9$ and quite broad for $\beta_c = 0.999$. The physical reason for the sharp reduction of intensity lies in the increase for $\beta_c > \beta$ of the region in which the electromagnetic waves are damped. The sharp reduction of the radiation intensity when the gas pressure drops below P_c agrees with qualitative estimates of section 4.8.

So far we have evaluated S_ρ, the radial component of the Poynting vector. The integral $2\pi\rho \int S_\rho dz$ taken over the cylinder surface C_ρ is the same for any ρ. It is equal to vW_ρ, where v is the charge velocity while the quantity W_ρ independent of ρ is defined by Eqs. (4.31)-(4.33).

The Poynting vector $\vec{P}$ has another component, S_z. Both of them define the direction in which the radiation propagates. The distributions of $\sigma_z = 2\pi S_z$ on the surface of C_ρ are shown in Figs. 4.43 (a-d). for the charge velocities $\beta = 0.3$, 0.5, 0.75 and 0.99, respectively. The isolated peak in the neighbourhood of $z = 0$ plane corresponds to the EMF carried by the moving charge with itself. Being originated from the second term in A_z (see (4.51)) it is approximately equal to (for $\beta < \beta_c$)

$$\sigma_z^{(2)} \approx \frac{c\beta e^2 \rho}{2\epsilon_0 \gamma_n^4} \frac{1}{[(z - vt)^2 + \rho^2/\gamma_n^2]^3}, \quad \gamma_n^2 = (1 - \beta_n^2)^{-1}, \quad \beta_n = \beta/\beta_c.$$

It is seen that the qualitative behaviour of S_z is almost the same as S_ρ; however, the maxima of S_z are approximately twice of those of S_ρ. This

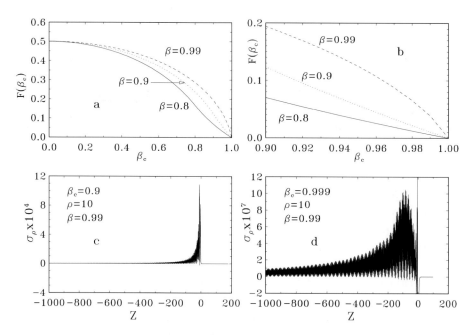

Figure 4.42. (a): Radial energy losses as a function of the critical velocity characterizing the medium properties. Values of β_c close to 1 and 0 correspond to optically rarefied and dense media, respectively. Numbers on curves are the charge velocity β; (b): The same as in (a), but for a smaller β_c interval; (c): Distribution of the radial energy flux on the surface of the cylinder C_ρ for a critical velocity ($\beta_c = 0.9$) slightly smaller than the charge velocity ($\beta = 0.99$) which in turn is slightly smaller than the velocity of light in vacuum. The intensity of radiation is concentrated near the plane $z = -z_c$; (d): The same as in (c), but for a critical velocity ($\beta_c = 0.999$) slightly greater than the charge velocity ($\beta = 0.99$). The distribution of the radiation intensity is very broad and by three orders smaller than in (c).

means that more radiation is emitted in the forward direction than in the transverse direction. To observe S_z one should orient the collimator (with a detector inside it) along the z axis. The collimator should be impenetrable for the γ quanta having directions non-parallel to the axis of motion. Again, the oscillations of intensity as a function of time should be detected during the charge motion.

To determine the major direction of the radiation, one should find surfaces on which the Poynting vector is maximal . Owing to the axial symmetry these surfaces look like lines in ρ, z variables. We shall refer to these lines as trajectories (see section 4.7). The behaviour of these trajectories is quite different depending on whether $\beta > \beta_c$ or $\beta < \beta_c$. For $\beta > \beta_c$ the trajectories are not closed. When $z \to \infty$, ρ also tends to ∞. For $\beta < \beta_c$

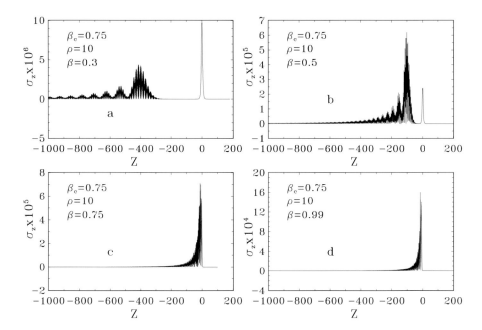

Figure 4.43. Distribution of the z component of the energy flux σ_z axis on the surface of C_ρ for $\beta_c = 0.75$; (a): $\beta = 0.3$. The isolated peak in the neighbourhood of $z = 0$ corresponds to the non-radiation field carried by a charge. The radiation and non-radiation terms are of the same order; (b): $\beta = 0.5$. The contribution of the non-radiation term relative to the radiation term is much smaller than for $\beta = 0.3$; (c): $\beta = 0.75$. The contribution of the non-radiation term relative to the radiation term is negligible. The radiation is concentrated near the plane $z = 0$; (d): $\beta = 0.99$. The contribution of the non-radiation term relative to the radiation term is negligible . The radiation is concentrated near the plane $z = -z_c$.

the trajectories are closed. In the WKB approach, on a particular one of the surfaces mentioned, the inclination of the Poynting vector towards the motion axis is given by [35,36]

$$\cos\theta_P = \frac{S_z}{\sqrt{S_\rho^2 + S_z^2}} = \frac{1}{\beta\sqrt{\epsilon(x)}}.$$

Here x is a parameter, $\epsilon(x) = 1 + [\beta_c^2\gamma_c^2(1 - x^2)]^{-1}$, $S_\rho = -cE_zH_\phi/4\pi$ and $S_z = cE_\rho H_\phi/4\pi$. For $\beta > \beta_c$, x changes from $x = 1$ for which ρ is zero, z is finite and $\theta_P = \pi/2$ up to $x = 0$ for which both ρ and z are infinite whilst $\cos\theta_P$ has the same value β_c/β as in the absence of dispersion.

For $\beta < \beta_c$ a particular trajectory intersects the motion axis two times: at $x = 1$ where z is finite and $\theta_P = \pi/2$ and at $x = \sqrt{1 - \tilde\epsilon}$ where z is finite and greater in absolute value than for $x = 1$, while $\theta_P = 0$ there. At the point of the trajectory where ρ is maximal the inclination of the Poynting vector towards the motion axis acquires the intermediate value

$$\cos\theta_P = \frac{1}{\beta}\left[1 + \frac{1}{\beta_c^2\gamma_c^2(2 - \sqrt{4 - 3\tilde{\epsilon}})}\right]^{-1/2}, \quad \tilde{\epsilon} = \beta^2\gamma^2/\beta_c^2\gamma_c^2.$$

Consider now the energy flux per unit time through the entire plane $z = const$. It is given by

$$W_z = \int S_z\rho d\rho d\phi = \frac{c}{2}\int E_\rho H_\phi \rho d\rho.$$

Substituting E_ρ and H_ϕ from (4.51) and using the well-known orthogonality relation between Bessel functions

$$\int\limits_0^\infty \rho d\rho J_m(k\rho)J_m(k'\rho) = \frac{1}{k}\delta(k - k'),$$

one obtains

$$W_z = \frac{1}{2}e^2 v^3\gamma^2 \int k^2 dk F_A(k, z - vt)$$

$$\times\{\gamma^2 F_\phi(k, z - vt) - \frac{2}{v^2}\frac{\omega_0}{\gamma_c^2\beta_c}\frac{1}{k^2 + \omega_0^2/v^2\beta_c^2}\sin[\omega_0(t - z/v)/\beta_c]\},$$

where F_A and F_ϕ are given by Eqs. (4.52).

It is not evident that W_z is positive-definite. In Fig. 4.44 (a) we present W_z as a function of z for $\beta_c = 0.75$ and $\beta = 0.99$. It is seen that W_z is almost constant in a very broad range of z except for the neighbourhood of the $z = const$ plane passing through the current charge position. The positivity of $W_z = 2\pi \int S_z\rho d\rho$ means that the energy flow of radiation follows for the moving charge and does not mean that S_z is also positive. This is illustrated in Fig. 6(b) where $\sigma_z = 2\pi S_z$ as a function of ρ is presented for a particular plane $z = -800$. It is seen that S_z contains both positive and negative parts. This may be understood within the polarization formalism [34,35,36]. In it the moving charge induces the time-dependent polarization of the medium. This in turn leads to the appearance of the radiation characterized by the Poynting vector $\vec{S}$. The positivity of S_z means that the part of the induced radiation flux follows for the moving charge. This fact has no relation to the well-known difficulty occurring for the radiation of the accelerated charge moving in a vacuum where the solutions of the Maxwell equations corresponding to the energy flux directed inward the moving charge are regarded as unphysical.

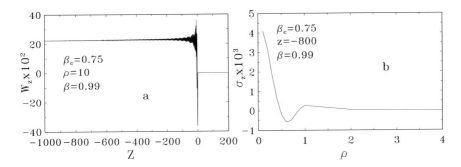

Figure 4.44. (a): The total integral energy flux W_z through the plane normal to the motion axis as a function of this plane position for $\beta = 0.99$ and $\beta_c = 0.75$; (b): The distribution of the energy flux in a particular ($z = -800$) plane normal to the motion axis as a function of the radial distance ρ. Positive and negative signs of σ_z correspond to the energy flow directed inwards the moving charge and outwards it, respectively.

The appearance of medium density oscillations behind the charge moving in a plasma was predicted in 1952 by Bohm and Pines [38]. The corresponding electric potential has been called the wake potential [39]. The electric field arising from such oscillations has been evaluated by Yu, Stenflo and Shukla [40]. For a charge moving in a metal the Cherenkov shock waves arise when the charge velocity exceeds the Fermi velocity of the solid [41]. The Cherenkov shock waves should be also induced by heavy ions moving in an electron plasma with the velocity greater the Fermi velocity of the electrons in the plasma [42]. However, in all these publications only the electric field has been evaluated, no attention has been paid to the magnetic field arising and to the Poynting vector defining the propagation of the electromagnetic field energy. The latter is the main goal of this investigation.

Recently, we were aware of an experiment performed by Stevens et al. [43] which seems to support the theoretical predictions of this Chapter. The experiment was performed on a single ZnSe crystal of the cubic form with a side of 5 mm. Its refractive index essentially differs from unity in the physically interesting frequency region. A laser pulse from an external source is injected into the sample. This laser pulse represents a wave packet centered around the frequency ω_L which may be varied in some interval. The injected pulse propagating with a group velocity defined by ω_L creates the distribution of electric dipoles following the laser pulse. The moving dipoles produce EMF, the properties of which depend on the dipole velocity v_d, which in its turn, is defined by ω_L. In particular, this velocity can be greater or smaller than c_0 ($c_0 = c/\epsilon_0$, $\epsilon_0 = \epsilon(\omega = 0)$). In the experiment treated the quantity measured was the electric field. The character of its time oscillations essentially depends on the fact whether $v_d > c_0$ or $v_d < c_0$. The observed time oscillations of electric field were in good agreement with the theoretical oscillations.

We believe that this experiment is a great achievement having both theoretical and technological meaning. However, Nature never resolutely says 'Yes'. We briefly enumerate the main reservations:

1) A bunch of electric dipoles is created at one side of the ZnSe cube and propagates towards the other. Such a motion corresponds to the so-called Tamm problem (see Chapter 2) describing the charge motion in a finite interval. Theory predicts that a charge uniformly moving in a finite dielectric slab radiates at each velocity even in the absence of dispersion. This assertion is not changed by the fact that the wavelength is much smaller than the motion interval (equal to the side of cube) in the experiment treated;

2) The switching of the imaginary part of dielectric permittivity leads to the damping of the EMF oscillations for $v < c_0$ and to their rather small attenuation for $v > c_0$. For realistic imaginary parts the oscillations for $v < c_0$ almost disappear (see this Chapter);

3) An important question is the distance at which the observations were made: oscillations of the EMF intensity sharply different from the Cherenkov ones appear at finite distances (see Chapters 5 and 9).

The experiment treated is so fundamental that any ambiguity in its interpretation should be excluded. Careful analysis of the influence of the above items on the experiment treated should be made.

4.12. Short résumé of this Chapter

We briefly summarize the main results discussed in this Chapter:

1. It is shown that a point charge moving uniformly in a dielectric medium with a standard choice (4.1) of electric permittivity should radiate at each velocity. The distributions of the radiated electromagnetic field differ drastically for the charge velocity v below and above some critical value v_c which depends on the medium properties and does not depend on the frequency (despite that the frequency dispersion is taken into account). For $v < v_c$ the radiation flux is concentrated behind the moving charge at a sufficiently remote distance from the charge.

2. The electromagnetic field radiated by a charge uniformly moving in a dielectric medium with $\epsilon(\omega)$ given by (4.1) consists of many oscillations which can be observed experimentally. We associate the appearance of these oscillations with the excitation of the lowest atomic level of the medium by a moving charge.

3. The results of recent experiments [18,19,20] and [43] dealing with the Vavilov-Cherenkov radiation and indicating on the existence of the radiation below the Cherenkov threshold seems to be supported by the present investigation. We associate this radiation with the frequency dependence of ϵ and the non-zero damping.

4. In an absorptive medium, both the value and position of the maximum of the frequency distribution depend crucially on the damping parameter and on the distance at which observations are made. The diminishing of the radiation intensity is physically clear since only part of the radiated energy flux reaches the observer for a non-zero damping parameter. Does the frequency shift of the maximum of the radiation intensity mean that any discussion of the frequency distribution of the radiation intensity should be supplemented by the indication of the observational distance and the damping parameter? In the absence of absorption the dependence on the observational cylindrical radius ρ disappears.

References

1. Cherenkov P.A. (1934) Visible luminescence of the pure fluids induced by γ rays *Dokl. Acad. Nauk SSSR*, **2**, pp. 451-454.
2. Frank I.M. and Tamm I.E. (1937) Coherent Visible Radiation of Fast Electrons Passing through Matter *Dokl. Akad. Nauk SSSR*, **14**, pp. 107-113.
3. Heaviside O. (1922) *Electromagnetic Theory*, **vol. 3**, Benn Brothers Ltd, London (Reprinted Edition).
4. Volkoff G.M. (1963) Electric Field of a Charge Moving in Medium *Amer.J.Phys.*,**31**, pp.601-605.
5. Zin G.N. (1961) General Theory of the Cherenkov Radiation *Nuovo Cimento*, **22**, pp. 706-778.
6. Afanasiev G.N., Beshtoev Kh. and Stepanovsky Yu.P. (1996) Vavilov-Cherenkov Radiation in a Finite Region of Space *Helv. Phys. Acta*, **69**, pp. 111-129; Afanasiev G.N., Eliseev S.M. and Stepanovsky Yu.P. (1998) Transition of the Light Velocity in the Vavilov-Cherenkov Effect *Proc. Roy. Soc. London*, **A 454**, pp. 1049-1072.
7. Brevik I. and Kolbenstvedt H. (1988) Quantum Detector Moving through a Dielectric Medium. 1. Constant Velocity, *Nuovo Cimento*, **102B**, pp.139-150.
8. Born M. and Wolf E. (1975), *Principles of Optics*, Pergamon, Oxford.
9. Ryazanov M.I. (1984) *Electrodynamics of Condensed Matter*, Nauka, Moscow, in Russian.
10. Brillouin L. (1960) *Wave Propagation and Group Velocity*, Academic Press, New York and London.
11. Lagendijk A. and Van Tiggelen B.A. (1996) Resonant Multiple Scattering of Light *Physics Reports*, **270**, pp. 143-216.
12. Fermi E. (1940) The Ionization Loss of Energy in Gases and in Condensed Materials, *Phys. Rev*, **57**, pp. 485-493.
13. Bohr N. (1913) On the Theory of the Decrease of Moving Electrified Particles on passing through Matter *Phil. Mag.*, **25**, pp. 10-31; Bohr N. (1915) On the Decrease of Velocity of Swiftly Moving Electrified Particles in passing through Matter *Phil. Mag.*, **30**, pp. 581-612;
14. Sternheimer R.M., 1953, The Energy Loss of a Fast Charged Particle by Cherenkov Radiation *Phys. Rev.*, **91**, pp.256-265.

15. Neamtan S.M. (1953) The Cherenkov Effect and the Dielectric Constant *Phys. Rev.*, **92**, pp.1362-1367.

16. Fano U. (1956) Atomic Theory of Electromagnetic Interactions in Dense Material *Phys. Rev.*, **103**, 1202-1218.

17. Tidman D.M. (1956,57) A Quantum Theory of Refractive Index, Cherenkov Radiation and the Energy Loss of a Fast Charged Particle *Nucl. Phys.*, **2**, pp. 289-346.

18. Ruzicka J.. and Zrelov V.P. (1992) Optical Transition Radiation in Transparent Medium and its Relation to the Vavilov-Cherenkov Radiation *JINR Preprint*,**P1-92-233**, Dubna.

19. Ruzicka J. (1993) *Doctor of Science Dissertation*, Dubna.

20. Zrelov V.P, Ruzicka J. and Tyapkin A.A. (1998) Pre-Cherenkov Radiation as a Phenomenon of 'Light Barrier, *JINR Rapid Communications*, **1[87]-98**, pp.23-25.

21. Akhiezer A.I. and Shulga N.F. (1993) *High Energy Electrodynamics in Medium*, Nauka, Moscow, In Russian.

22. Landau L.D. and Lifshitz E.M (1960), *Electrodynamics of Continuous Media*, Pergamon, Oxford.

23. Gradshteyn I.S. and Ryzik I.M. (1965) *Tables of Integrals, Series and Products*, Academic Press, New York.

24. James M.B. and Griffiths D.J. (1992) Why the Speed of Light is Reduced in a Transparent Medium *Am. J. Phys.*, **60**, pp. 309-313.

25. Diamond J.D. (1995) Comment on 'Why the Speed of Light is Reduced in a Transparent Medium' *Am. J. Phys.*, **63**, pp. 179-180.

26. Bart G. de Grooth (1997) Why is the propagation velocity of a photon in a transparent medium reduced? *Am. J. Phys.*, **65**, pp. 1156-1164.

27. Migdal A.B. (1975) *Qualitative Methods in Quantum Theory*, Nauka, Moscow, in Russian.

28. Ginzburg V.L. (1996) Radiation of Uniformly Moving Sources (Cherenkov Effect, Transition Radiation, and Other Phenomena *Usp. Fiz. Nauk*, **166**, pp. 1033-1042.

29. Frank I.M. (1988) *Vavilov-Cherenkov Radiation*, Nauka, Moscow, in Russian.

30. Tamm I.E. (1939) Radiation Induced by Uniformly Moving Electrons, *J. Phys. USSR*, **1, No 5-6**, pp. 439-461.

31. Bolotovsky B.M. (1957) Theory of the Vavilov-Cherenkov Effect *Usp. Fiz. Nauk*, **42**, pp. 201-350.

32. Zrelov V.P. (1970) *Vavilov-Cherenkov Radiation in High-Energy Physics*, Israel Program for Scientific Translations, Jerusalem.).

33. Landau L.D. and Lifshitz E.M. (1971) The Classical Theory of Fields Pergamon, Oxford and Addison-Wesley, Reading, Massachusetts.

34. Afanasiev G.N., Kartavenko V.G. and Magar E.N. (1999) Vavilov-Cherenkov Radiation in Dispersive Medium *Physica*, **B 269**, pp. 95-113.

35. Afanasiev G.N. and Kartavenko V.G. (1998) Radiation of a Point Charge Uniformly Moving in a Dielectric Medium *J. Phys. D: Applied Physics*, **31**, pp.2760-2776.

36. Afanasiev G.N., Eliseev S.M and Stepanovsky Yu.P. (1999) Semi-Analytic Treatment of the Vavilov-Cherenkov Radiation *Physica Scripta*, **60**, pp. 535-546.

37. Li H., 1984, Chem. Ref. Data, 13, 102; Gobel A. et al. (1999) Phonons and Fundamental Gap in ZnSe *Phys. Rev*B **59**, pp. 2749-2759; Hattori et al., 1973, Opt. Commun., 7, 229; Jensen B., Torabi A., 1983, Infrared Phys., 23, 359.

38. Pines D. and Bohm D. (1951) A Collective Description of Electron Interactions. I. Magnetic Interactions *Phys. Rev.*,**82**, pp. 625-634 (1951); Pines D. and Bohm D. (1952) A Collective Description of Electron Interactions. II. Collective vs Individual Particle Aspects of the Interactions *Phys. Rev.*,**85**, pp. 338-353.

39. Neelavathie V.N., Ritchie R.H. and Brandt R.H. (1974) Bound Electron States in the Wake of Swift Ions in Solids *Phys.Rev.Lett.* **33**, 302-305.

40. Yu M.Y., Stenflo L. and Shukla P.K., Radio Science,**7**, 1151 (1972).

41. Griepenkerl K., Schafer A. and Greiner W. (1995) Mach Shock Waves and Surface

Effects in Metals Ĵ.Phys.: Condensed Matter, **7**, pp. 9465-9473.

42. Schafer W., Stocker H., Muller B, and Greiner W. (1978) Mach Cones Induced by Fast Heavy Ions in Electron Plasma Z.Physik **A288**, pp. 349-352 (1978).

43. Stevens T.E., Wahlstrand J.K., Kuhl J. and Merlin R. (2001) Cherenkov Radiation at Speeds below the Light Threshold: Photon-Assisted Phase Matching *Science*, **291**, pp. 627-630.

CHAPTER 5

INFLUENCE OF FINITE OBSERVATIONAL DISTANCES AND CHARGE DECELERATION

5.1. Introduction

In Chapter 2 we analyzed frequency and angular distributions of the radiation in the so-called Tamm problem. The latter treats a point charge which is at rest in a medium at the spatial point $z = -z_0$ up to an instant $t = -t_0$. In the time interval $-t_0 < t < t_0$ the charge moves with a constant velocity v that can be smaller or greater than the velocity of light c_n in medium. After the instant $t = t_0$ the charge is again at rest at the point $z = z_0$. This problem was first considered by Tamm [1] in 1939. Later, it was analyzed qualitatively by Lawson [2,3] and numerically by Zrelov and Ruzicka [4,5]. In 1996 the exact solution of the Tamm problem was found for a non-dispersive medium [5.6]. A careful analysis of this solution given in [7] showed there that the Tamm formula does not always describe the VC radiation properly.

In the past, exact electromagnetic field (EMF) strengths and exact electromagnetic intensities of the Tamm problem were written out in [8]. It was shown there that the radiation intensity depends crucially on the observational sphere radius (the formula (2.29) given by Tamm corresponds to infinite observational distances). However, the calculations carried out there, were predominantly of a methodological character. The reason is that formulae obtained in [8] were not suitable for practical applications: EMF strengths were expressed through the integrals, the accurate evaluation of which for high frequencies, corresponding to visible light, required a great number of integration steps.

The goal of this consideration is to obtain more suitable practical formulae describing the radiation intensity of the Tamm problem at finite distances and having a greater range of applicability than the original Tamm formula. The original Tamm problem involves instantaneous jumps in velocity at the start and end of motion. To them correspond infinite acceleration and deceleration. There are no such jumps in reality. Our next goal is to study how a smooth transition from the state of rest to the uniform motion affects the radiation intensities.

The plan of our exposition is as follows. In Section 5.2.1 we reproduce the Tamm derivation of angular-frequency distributions of the radiation intensity produced by a point charge moving uniformly in a medium in a finite spatial interval. Criteria for the validity of the Tamm formula are given in the same section. Exact electromagnetic fields of the Tamm problem and radiation intensity are explicitly written out in Section 5.2.2. A closed expression for the radiation intensity which works at finite observational distances from a moving charge (the Tamm original formula corresponds to an infinite observational distance) is found in Section 5.2.3. This expression predicts the essential broadening of the angular Cherenkov spectrum if the measurements are made at realistic distances from a moving charge. The analytic formula taking into account both the deceleration of a moving charge owed to the energy losses and a finite distance of the observational point is presented in Section 5.2.4. It generalizes the formula found earlier in [9] that is valid only at infinite distances. In Section 5.2.5 we compare exact radiation intensities with approximate analytic intensities obtained in Sections 5.2.3 and 5.2.4. In all the cases corresponding to the real experimental situation, there is a perfect agreement between the exact radiation intensity and analytic formulae found in Sections 5.2.3 and 5.2.4. On the other hand, both of them sharply disagree with the Tamm radiation intensity. These formulae are applied to the description of the VC radiation observed in the Darmstadt experiments with heavy ions. The complications arising and the discussion of the results obtained are given. In the same section the experiment is proposed of testing the broadening of the radiation spectrum when it is measured at finite distances.

The analytic formulae obtained in Section 5.2.4 are valid for moderate accelerations when the loss of velocity is small compared to the velocity itself. The section 5.3 deals with arbitrary accelerations. Analytic formulae are obtained for the radiation intensity corresponding to a number of the smooth Tamm problem (when the transition from the state of versions rest to the uniform motion proceeds smoothly). These formulae are valid under the same approximations as the Tamm formula. Various analytic estimates are given and interesting limiting cases having numerous practical applications are considered.

5.2. Finite observational distances and small acceleration

5.2.1. THE ORIGINAL TAMM APPROACH

Tamm considered the following problem. A point charge is at rest at a point $z = -z_0$ of the z axis up to an instant $t = -t_0$ and at the point $z = z_0$ after the instant $t = t_0$. In the time interval $-t_0 < t < t_0$ it moves uniformly along the z axis with a velocity v greater than the velocity of light $c_n = c/n$ in medium. The non-vanishing z spectral component of the vector potential

(VP) is given by

$$A_z(x, y, z) = \frac{e\mu}{2\pi c} \int_{-z_0}^{z_0} \frac{dz'}{R} \exp(-i\psi),$$

$$R = [\rho^2 + (z - z')^2]^{1/2}, \quad \rho^2 = x^2 + y^2, \quad \psi = \omega\left(\frac{z'}{v} + \frac{R}{c_n}\right). \tag{5.1}$$

In what follows we limit ourselves to a dielectric medium ($\mu = 1$). At large distances from the moving charge where

$$R \gg z_0 \tag{5.2}$$

one obtains in the wave zone, where

$$k_n r \gg 1, \quad k_n = \omega/c_n \tag{5.3}$$

the following expression for the energy flux through a sphere of radius r for the whole time of observation

$$\mathcal{E} = r^2 \int S_r d\Omega dt = \int \frac{d^2\mathcal{E}}{d\Omega d\omega} d\Omega d\omega,$$

$$d\Omega = \sin\theta d\theta d\phi, \quad S_r = \frac{c}{4\pi} E_\theta H_\phi. \tag{5.4}$$

Here

$$\frac{d^2\mathcal{E}}{d\Omega d\omega} = \frac{e^2}{\pi^2 cn} [\sin\theta \frac{\sin\omega t_0(1 - \beta_n \cos\theta)}{\cos\theta - 1/\beta_n}]^2, \quad \beta_n = \frac{v}{c_n} \tag{5.5}$$

is the energy emitted into the solid angle $d\Omega$, in the frequency interval $d\omega$. This famous formula obtained by Tamm is frequently used by experimentalists (see, e.g., [10]-[13]) for the identification of the charge velocity.

The typical experimental situations described by the Tamm formula are:

i) β decay of a nucleus at one spatial point accompanied by a subsequent absorption of the emitted electron at another point;

ii) A high energy electron consequently moves in a vacuum, enters the dielectric slab, leaves the slab, and again propagates in the vacuum. Since an electron moving uniformly in a vacuum does not radiate (apart from the transition radiation arising at the boundaries of the dielectric slab), the experimentalists describe this situation by the Tamm formula, assuming that the electron is created on one side of the slab and is absorbed on the other.

In addition to the approximations (5.2) and (5.3), two other implicit assumptions are made when going from the exact VP (5.22) to the Tamm field strengths (5.4). The first of them

$$z_0 \ll 2r \left(\frac{1}{\beta_n} - \cos \theta \right) / \sin^2 \theta. \qquad (5.6)$$

means [7] that the second-order term in the expansion of ψ should be small as compared with the linear one (taken into account by Tamm). It is seen that the right hand side of this equation vanishes for $\cos \theta = 1/\beta_n$, i.e., at the angle where the VC radiation exists. Therefore in this angular region, the second-order terms may be important. The second of the conditions mentioned

$$\frac{z'^2 \omega \sin^2 \theta}{2rc_n} \ll \pi \qquad (5.7)$$

means that the second-order terms in the expansion of R should be small not only compared to the linear terms but also compared to π (since ψ is a phase in (5.1)). Or, taking for θ and z' their maximal values ($\theta = \pi/2$, $z' = z_0$), one obtains

$$\frac{nL^2}{8r\lambda} \ll 1, \quad L = 2z_0, \quad \lambda = \frac{2\pi c}{\omega}. \qquad (5.8)$$

This condition was mentioned by Frank on p. 59 of his book [10]. It should be noted that for gases these conditions are less restrictive than for solids and liquids. In fact, since for them $\beta_n \approx 1$, the angular spectrum is confined to the region $\theta \approx 0$ and conditions (5.6) and (5.7) are reduced to (5.2) and (5.3), respectively. As a result, for gases, the Tamm expression (5.5) for the radiated power works when Eqs. (5.6) and (5.7) are valid.

As an illustration, we turn to Ref. [14] where the angular distribution of the radiation ($\lambda \approx 4 \times 10^{-5}$ cm) arising from the passage of Au heavy ions ($\beta \approx 0.87$) through the LiF slab ($n \approx 1.39$) of width $L = 0.5$ cm was interpreted in terms of the Tamm formula. Substituting the parameters of [14] into (5.8) defining the validity of the Tamm formula (5.5), we find that the left hand side of (5.8) coincides with π for the observational sphere radius $r \approx 10m$. Obviously this value is unrealistic. Since a realistic r is about 10 cm, (5.8) is violated strongly. In this case the Tamm formula does not describe the experimental situation properly. Thus more accurate formulae are needed. In the next section, we present the exact EMF strengths of the Tamm problem.

5.2.2. EXACT ELECTROMAGNETIC FIELD STRENGTHS AND ANGULAR-FREQUENCY DISTRIBUTION OF THE RADIATED ENERGY

The energy flux through the unit solid angle of the sphere of the radius r for the whole time of a charge motion is given by

$$\frac{dW}{d\Omega} = \frac{c}{4\pi} r^2 \int_{-\infty}^{\infty} dt (\vec{E} \times \vec{H})_r. \tag{5.9}$$

Expressing $\vec{E}$ and $\vec{H}$ through their Fourier transforms

$$\vec{E} = \int \exp(i\omega t)\vec{E}_\omega d\omega, \quad \vec{H} = \int \exp(i\omega t)\vec{H}_\omega d\omega$$

and integrating over t one finds

$$\frac{dW}{d\Omega} = \frac{cr^2}{2} \int_{-\infty}^{\infty} (\vec{E}(\omega) \times \vec{H}(-\omega))_r d\omega = \int_{0}^{\infty} S(\omega) d\omega, \tag{5.10}$$

where

$$S(\omega, \theta) = \frac{d^2 W}{d\omega d\Omega} = cr^2 [\vec{E}_\theta^{(r)}(\omega) \vec{H}_\phi^{(r)}(\omega) + \vec{E}_\theta^{(i)}(\omega) \vec{H}_\phi^{(i)}(\omega)]. \tag{5.11}$$

This quantity shows how a particular Fourier component of the radiated energy is distributed over the sphere S. The superscripts (r) and (i) mean the real and imaginary parts of E_θ and H_ϕ. The exact field strengths obtained by differentiation of the exact vector potential (5.1) are given by

$$H_\phi^{(r)}(\omega) = \frac{ek_n r}{2\pi c} \sin\theta \int \frac{G}{R^2} dz', \quad H_\phi^{(i)}(\omega) = \frac{ek_n}{2\pi c} \sin\theta \int \frac{F}{R^2} dz',$$

$$E_\theta^{(r)}(\omega) = \frac{ek^2 r}{2\pi\omega} \sin\theta \left(\int \frac{r - z'\cos\theta}{R^3} F_1 dz' - \frac{2}{k_n r} \int \frac{F}{R^2} dz' \right),$$

$$E_\theta^{(i)}(\omega) = \frac{ek^2 r}{2\pi\omega} \sin\theta \left(\int \frac{r - z'\cos\theta}{R^3} G_1 dz' + \frac{2}{k_n r} \int \frac{G}{R^2} dz' \right), \tag{5.12}$$

where

$$F = \cos\psi - \frac{1}{k_n R}\sin\psi, \quad G = \sin\psi + \frac{1}{k_n R}\cos\psi,$$

$$F_1 = \sin\psi + 3\frac{\cos\psi}{k_n R} - 3\frac{\sin\psi}{k_n^2 R^2}, \quad G_1 = \cos\psi - 3\frac{\sin\psi}{k_n R} - 3\frac{\cos\psi}{k_n^2 R^2},$$

$$\psi = \frac{\omega z'}{v} + k_n R, \quad R = (r^2 - 2z'r\cos\theta + z'^2)^{1/2}, \quad \epsilon_0 = z_0/r. \quad (5.13)$$

The z' integration in (5.12) is performed over the interval $(-z_0, z_0)$. When Eqs. (5.2), (5.3), and (5.8) are satisfied, $S(\omega, \theta)$ given by (5.11) transforms into the Tamm formula (5.5).

Unfortunately, EMF strengths (5.12) given in [8] without derivation are not suitable for realistic cases corresponding to high frequencies. In fact, for visible light, $k = \omega/c$ is of the order $10^5 \mathrm{cm}^{-1}$. For an observational distance $r \sim 1$ m, one obtains $kr \sim 10^7$. A great number of steps of integration is needed to obtain the required accuracy. Therefore, some approximations are needed.

5.2.3. APPROXIMATIONS

In the wave zone where $k_n r \gg 1$, we omit the terms of the order $(k_n r)^{-1}$ and higher outside ψ and find

$$S(\omega, \theta) = \frac{e^2 k^2 r^4 n}{4\pi^2 c} \sin^2\theta \left[\int \frac{\sin\psi_1}{R^2} dz' \cdot \int \frac{\sin\psi_1}{R^3}(r - z'\cos\theta)dz' \right.$$

$$\left. + \int \frac{\cos\psi_1}{R^2} dz' \cdot \int \frac{\cos\psi_1}{R^3}(r - z'\cos\theta)dz' \right], \quad (5.14)$$

where

$$\psi_1 = \frac{kz'}{\beta} + k_n(R - r), \quad t_0 = z_0/v. \quad (5.15)$$

The condition $k_n r \gg 1$ in real experiments is satisfied to a great accuracy (we have seen kr is of the order 10^7 for $r = 1$ m).Therefore Eq.(5.14) is almost exact. Since ψ_1 in (5.15) contains $R - 1$, rather than R, its maximal value is of the order $k_n z_0$ rather than $k_n r$ as in Eq. (5.13). This makes numerical integration easier if $z_0 \ll r$ (the motion interval is much smaller than the observational distance). In the latter case one may disregard ϵ_0 outside ψ_1. Then

$$S(\omega, \theta) = \frac{e^2 k^2 n}{4\pi^2 c} \sin^2\theta \left[\left(\int \sin\psi_1 dz' \right)^2 + \left(\int \cos\psi_1 dz' \right)^2 \right]. \quad (5.16)$$

The expansion of ψ_1 up to the first order of ϵ_0 gives the Tamm formula (5.5) which does not always describe properly the real experimental situation. Therefore we expand R in ψ_1 up to the second order of ϵ_0

$$R = r - z'\cos\theta + \frac{z'^2}{2r}\sin^2\theta$$

and

$$\psi_1 = knz'(\frac{1}{\beta_n} - \cos\theta + \frac{z'z_0}{2r}\sin^2\theta).$$ (5.17)

With this ψ_1, $S(\omega, \theta)$ can be obtained in a closed form

$$S(\omega, \theta) = \frac{e^2 kr}{4\pi c}\{[S(z_+) - S(z_-)]^2 + [C(z_+) - C(z_-)]^2\},$$ (5.18)

where

$$z_\pm = \sqrt{\frac{\epsilon_0 k_n z_0}{2}}\sin\theta\left(\frac{1 - \beta_n\cos\theta}{\epsilon_0\beta_n\sin^2\theta} \pm 1\right),$$

$$S(x) = \sqrt{\frac{2}{\pi}}\int_0^x dt\,\sin t^2 \quad \text{and} \quad C(x) = \sqrt{\frac{2}{\pi}}\int_0^x dt\,\cos t^2$$

are the Fresnel integrals. For small and large arguments they behave as

$$S(x) \to \sqrt{\frac{2}{\pi}}\frac{x^3}{3}, \quad C(x) \to \sqrt{\frac{2}{\pi}}x - \frac{1}{\sqrt{2\pi}}\frac{x^5}{5} \quad \text{for} \quad x \to 0,$$

$$S(x) \to \frac{1}{2} - \frac{1}{\sqrt{2\pi}}\frac{\cos x^2}{x}, \quad C(x) \to \frac{1}{2} + \frac{1}{\sqrt{2\pi}}\frac{\sin x^2}{x} \quad \text{for} \quad x \to \infty.$$

It is instructive to see how a transition to the Tamm formula takes place. For this we present z_+ and z_- in the form

$$z_\pm = \frac{1 - \beta_n\cos\theta}{\beta_n\sin\theta}\sqrt{\frac{k_n r}{2}} \pm \sin\theta\sqrt{\frac{\epsilon_0 k_n z_0}{2}}.$$

Equation (5.18) was obtained under the assumptions $k_n r \gg 1$ and $r \gg z_0$. The first term in $z_\pm$ is then much larger than the second term everywhere except for $\cos\theta$ close to $1/\beta_n$. Therefore if $\cos\theta \neq 1/\beta_n$ then

$$C(z_+) - C(z_-) \approx \frac{1}{\sqrt{2\pi}}\left(\frac{\sin z_+^2}{z_+} - \frac{\sin z_-^2}{z_-}\right),$$

$$S(z_+) - S(z_-) \approx -\frac{1}{\sqrt{2\pi}}\left(\frac{\cos z_+^2}{z_+} - \frac{\cos z_-^2}{z_-}\right),$$

$$[C(z_+) - C(z_-)]^2 + [S(z_+) - S(z_-)]^2$$

$$\approx \frac{1}{2\pi}\left\{\frac{1}{z_+^2} + \frac{1}{z_-^2} - \frac{2}{z_+z_-}\cos\left[\frac{2k_n z_0}{\beta_n}(1 - \beta_n\cos\theta)\right]\right\}$$

$$\approx \frac{2}{\pi z^2}\sin^2\left[\frac{k_n z_0}{\beta_n}(1 - \beta_n\cos\theta)\right],$$

where we put $z_+^2 = z_-^2 = z^2 = k_n r(1 - \beta_n \cos\theta)/2\beta_n \sin\theta$ outside the sin and cos. Substituting this into (5.18), we get the Tamm formula (5.5).

It remains to consider the case $\cos\theta \approx 1/\beta_n$. Then

$$z_\pm \approx \pm z_0 \sin\theta \sqrt{\frac{k_n}{2r}},$$

$$[C(z_+) - C(z_-)]^2 + [S(z_+) - S(z_-)]^2$$
$$\approx 4C^2(z_0 \sin\theta\sqrt{k_n/2r}) + 4S^2(z_0 \sin\theta\sqrt{k_n/2r}).$$

The Tamm formula is valid if $\epsilon_0 k_n z_0/2 \ll 1$ which is equivalent to (5.8). Then

$$[C(z_+) - C(z_-)]^2 + [S(z_+) - S(z_-)]^2 \approx \frac{4}{\pi}\epsilon_0 k_n z_0 \sin^2\theta$$

and

$$S(\omega,\theta) \approx \frac{e^2 k^2 z_0^2 n}{\pi^2 c}\sin^2\theta.$$

This coincides with the limit $\cos\theta \to 1/\beta_n$ of the Tamm formula.

Equation (5.18) is valid if the third-order terms in the expansion (5.17) of ψ_1 are small compared to π:

$$\frac{1}{2}k_n r\epsilon_0^3 z'^3 \cos\theta \sin^2\theta \ll \pi \tag{5.19}$$

(π appears since ψ_1 is the phase). If we take for z' and $\cos\theta \sin^2\theta$ their maximal values one finds

$$\frac{nL^3}{8\lambda r^2} \ll 1. \tag{5.20}$$

We collect all approximations involved in derivation of (5.18)

$$k_n r \gg 1, \quad z_0 \ll r, \quad \frac{nL^3}{8\lambda r^2} \ll 1. \tag{5.21}$$

5.2.4. DECELERATED CHARGE MOTION

Consider the following problem. Let a point charge be at rest at the point $z = -z_0$ up to an instant $t = -t_0$. At $t = -t_0$, the charge acquires the velocity v_1. In the time interval $(-t_0 < t < t_0)$ the charge decelerates according to the law

$$\frac{z}{z_0} = \frac{t}{t_0} + \frac{at_0^2}{2z_0}\left(1 - \frac{t^2}{t_0^2}\right), \qquad \frac{dz}{dt} = \frac{z_0}{t_0} - at. \tag{5.22}$$

After the instant $t = t_0$ the charge is again at rest at the point $z = z_0$. The initial and final velocities of charge are equal to

$$v_{i,f} = v_0 \pm at_0.$$

Here

$$v_0 = \frac{v_i + v_f}{2} = \frac{z_0}{t_0}$$

is the charge velocity at the instant $t = 0$ and $at_0 = (v_i - v_f)/2$. It turns out that the same equations (5.11)-(5.13) are valid for the treated decelerated charge motion with the exception that the function ψ should be changed by

$$\psi = \omega t_0 T + k_n R, \tag{5.23}$$

where

$$T = \frac{1}{\delta}[1 - (1 + \delta^2 - 2\delta z'/z_0)^{1/2}], \quad \delta = \frac{at_0}{v_0} = \frac{v_i - v_f}{v_i + v_f}.$$

In the wave zone the same equation (5.14) is valid if one puts

$$\psi_1 = \omega t_0 T + k_n(R - r). \tag{5.24}$$

Dropping ϵ_0 outside the sines and cosines in (5.14), one arrives at (5.16) with ψ_1 given by (5.24).

Expanding square roots entering into R and T up to a second-order of ϵ_0 and δ, respectively, we obtain

$$R - r = -z' \cos\theta + \frac{z'^2}{r} \sin^2\theta, \quad T = \frac{z'}{z_0} - \frac{1}{2}\delta(1 - \frac{z'^2}{z_0^2}),$$

$$\psi_1 \approx kz'(\frac{1}{\beta} - n\cos\theta + \frac{z'}{2r} \sin^2\theta) - \frac{kz_0\delta}{2\beta}(1 - \frac{z'^2}{z_0^2}), \quad \beta_n = v_0/c_n. \tag{5.25}$$

With such ψ_1 integrals entering into (5.16) can be taken analytically, and one finds for $S(\omega, \theta)$

$$S(\omega, \theta) = \frac{e^2 kr}{4\pi c} \frac{\epsilon_0 \beta_n \sin^2\theta}{\epsilon_0 \beta_n \sin^2\theta + \delta}\{[S(z_+) - S(z_-)]^2 + [C(z_+) - C(z_-)]^2\}, \tag{5.26}$$

where

$$z_\pm = [\frac{\omega t_0}{2}(\delta + \beta_n \epsilon_0 \sin^2\theta)]^{1/2}[\frac{1 - \beta_n \cos\theta}{\delta + \beta_n \epsilon_0 \sin^2\theta} \pm 1].$$

Equation (5.26) works if, in addition to (5.21), the third-order term in the expansion of T entering into ψ_1 is small as compared with π:

$$\frac{kz'}{2\beta}\delta^2(1 - \frac{z'^2}{z_0^2}) \ll \pi \tag{5.27}$$

(again, π arises because ψ_1 is a phase). Taking for $z'(1 - z'^2)$ its maximal value ($\sim 2/5$), we obtain

$$\frac{\omega t_0}{5}\delta^2 \ll \pi, \quad \text{or} \quad \frac{L\delta^2}{5\beta\lambda} \ll 1, \quad \beta = v_0/c. \qquad (5.28)$$

This condition is satisfied for relatively small accelerations. In the limit $\delta \to 0$ (zero acceleration), Eq.(5.26) is reduced to (5.18).

For $\epsilon_0 \to 0$ (large radius of the observational sphere), one has

$$S(\omega, \theta) = \frac{e^2 k z_0 \beta_n \sin^2 \theta}{4\pi c \delta}\{[S(z_+) - S(z_-)]^2 + [C(z_+) - C(z_-)]^2\}, \quad (5.29)$$

where

$$z_\pm = \sqrt{\frac{k z_0 \delta}{2\beta}}(\frac{1 - \beta_n \cos\theta}{\delta} \pm 1).$$

An equation similar to (5.29) was obtained earlier in Ref. [9], but with the motion law different from (5.22).

Frequently the angular intensity is measured not on the sphere surface, but in the plane perpendicular to the motion axis (in the plane $z = const$ for the case treated). For $k_n z \gg 1$, the energy flux in the z direction is

$$S_z = \int S_z(\omega, \rho, z)d\omega d\phi \rho d\rho,$$

where

$$S_z(\omega, \rho, z) = \frac{d^3\mathcal{E}}{d\omega d\phi \rho d\rho} = c[E_\rho^r(\omega)H_\phi^r(\omega) + E_\rho^i(\omega)H_\phi^i(\omega)]$$

$$= \frac{e^2 k^2 \rho^2 n}{4\pi^2 c}(I_s I_s' + I_c I_c'), \qquad (5.30)$$

$$I_s = \int dz'\frac{\sin\psi_1}{R^2}, \quad I_s' = \int dz'(z - z')\frac{\sin\psi_1}{R^3}, \quad I_c = \int dz'\frac{\cos\psi_1}{R^2},$$

$$I_c' = \int dz'(z - z')\frac{\cos\psi_1}{R^3}, \quad r^2 = \rho^2 + z^2, \quad R^2 = \rho^2 + (z - z')^2,$$

z defines the plane in which the measurements are performed and ρ is the distance from the symmetry axis to the observational point. The integration over z' runs from $-z_0$ to z_0. If, in addition, $z_0 \ll z$ then

$$S_z(\omega, \rho, z) = \frac{e^2 k^2 \rho^2 n z}{4\pi^2 c r^5}\left[\left(\int dz' \sin\psi_1\right)^2 + \left(\int dz' \cos\psi_1\right)^2\right]. \qquad (5.31)$$

In the Fresnel approximation this reduces to

$$S_z(\omega, \rho, z) = \frac{e^2 k \beta_0 n z_0 z \rho^2}{4\pi c r^5 (\delta + \epsilon_0 \beta n \sin^2 \theta)}$$

$$\times \{[S(z_+) - S(z_-)]^2 + [C(z_+) - C(z_-)]^2\}, \tag{5.32}$$

where $z_\pm$ are obtained from $z_\pm$ entering (5.26) by setting in them $\sin \theta = \rho/r$, $\cos \theta = z/r$, $r = \sqrt{\rho^2 + z^2}$. The physical justification of this section considerations is as follows. When a charge enters into the dielectric slab it decelerates (owing to the VC radiation, ionization losses, etc.). For high-energy electrons these energy losses are negligible, and the uniform motion of the electron is a good approximation. However, for heavy ions for which the VC is also observed the energy losses are essential since they are proportional to the second-degree of heavy ion atomic number. Equations (5.14) and (5.16), with ψ_1 given by (5.24), are valid for arbitrary $\delta = (v_i - v_f)/(v_i + v_f)$. When conditions (5.21),(5.28) and $\delta \ll 1$ are satisfied, they are reduced to (5.29).

5.2.5. NUMERICAL RESULTS

With the parameters n, L, λ the same as in [14] (see Sect. 5.2.1) and $\beta = 0.868$, we have evaluated the almost exact radiation intensity (5.14) (because it was obtained from the exact intensity (5.11) by neglecting the terms of the order $1/k_n r$ and higher outside ψ) and the approximate Fresnel (5.18) angular distribution of the radiated energy on the spheres of the radii $r = 1$ cm (Fig. 5.1) $r = 10$ cm (Fig. 5.2), $r = 1$ m (Fig. 5.3) and $r = 10$ m (Fig. 5.4). It is seen that the radiation spectrum broadens enormously for small observational distances. For example, it occupies an angular region of approximately 20 degrees for $r = 1$ cm and 1.5 degrees for $r = 10$ cm. These figures demonstrate reasonable agreement between the Fresnel and exact intensities. In the case $r = 10$ cm, for which the condition (5.20) for the validity (5.18) is strongly violated (it looks like $14 \ll 1$), the agreement of (5.14) and (5.18) is quite satisfactory. Even for the case $r = 1cm$, for which the inequality (5.20) has the form $1400 \ll 1$, the Fresnel intensity although being shifted, qualitatively reproduces the exact radiation intensity (Fig. 5.1). In any case, the Fresnel intensity (5.18) can be used as a simple (although slightly rough) estimation of the position and the magnitude of the radiation intensity for realistic observational distances. On the other hand, both the Fresnel and exact intensities disagree sharply with the Tamm intensity (5.5). This demonstrates Fig. 5.5, where the exact (5.14) intensity on the sphere of radius $r = 10$ m is compared with the Tamm intensity (5.5) (which does not depend on r and which is obtained either from (5.14) or from (5.18) in the limit $r \to \infty$).

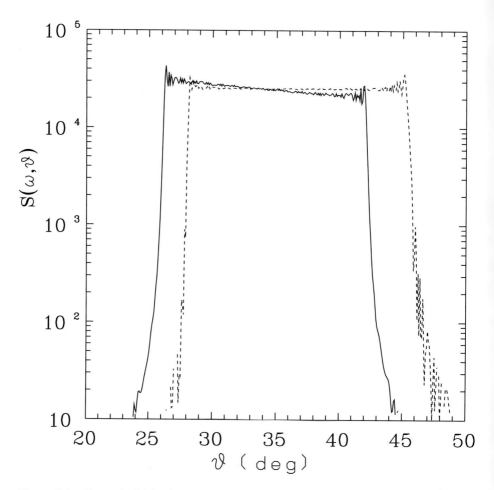

Figure 5.1.   Exact (solid line) and Fresnel (dotted line) intensities (in units of e^2/c) on the observational sphere of radius $r = 1$ cm. Parameters of the Tamm problem: the charge motion interval and velocity are $L = 0.5$ cm and $\beta = 0.868$, respectively.; wavelength $\lambda = 4 \cdot 10^{-5}$ cm; refractive index $n = 1.392$. It is seen that angular spectrum has a width approximately 20 degrees.

So far, we have investigated the influence of the radius of the observational sphere on the intensity distribution over this sphere. Now we analyze the influence of the charge deceleration on the radiation intensity on the sphere of infinite radius. The parameters n, L, λ and the initial velocity $\beta_i = 0.875$ are the same as in [14]. The radiation Fresnel intensities (5.29) for final velocities $\beta_f = 0.8\beta_i$ and $\beta_f = 0.2\beta_i$ are shown in Fig. 5.6. Their form remains practically the same for $\beta_f < 0.8\beta_i$. When β_f tends to β_i, the frequency spectrum tends to the Tamm intensity (5.5). This fact is illustrated in Fig. 5.7 (where the radiation intensities for $\beta_f = 0.9\beta_i$ and $\beta_f = 0.95\beta_i$ are shown) and in Fig. 5.8 (where intensities for $\beta_f = 0.99\beta_i$

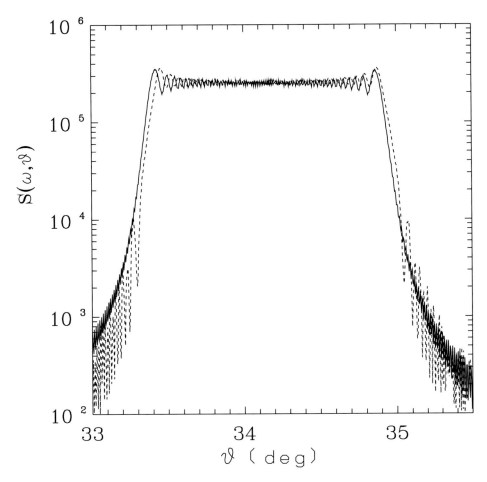

Figure 5.2.   The same as in Fig. 5.1, but for $r = 10$ cm. The width of the angular spectrum is about 1.5 degrees.

and $\beta_f = 0.999\beta_i$ are presented).

We turn now to experiments made recently in Darmstadt and discussed in [15]. In them, the beam of Au^{179}_{79} ions passed through the LiF slab creating the VC radiation. The initial energy of the ion beam (i.e., before entering the slab) was 905 MeV/n. One of the LiF slabs had the width $d = 0.5$ cm with the energy loss 73.3 MeV/n, while the other had width $d = 0.1$ cm with the energy loss 14.7 Mev/n. The authors of [15] compared the intensities for the slab widths $d = 0.5$ and 0.1 cm. In Figs. 5.9-5.12 we present Fresnel theoretical intensities (5.26) for $d = 0.1$ and $d = 0.5$ on the spheres of various radii. We observe that for observational distances larger than 1 m, the form of the radiation intensity practically does not depend on the distance, that is, the deceleration plays a major role at these distances. For

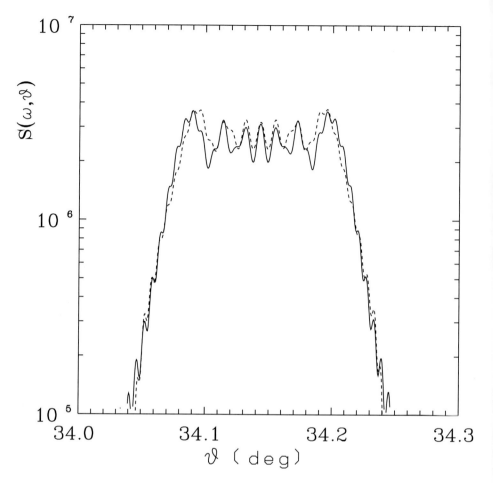

Figure 5.3. The same as in Fig.5.1, but for $r = 1$ m. The width of angular spectrum is about 0.15 degrees.

$r \geq 1$ m, theoretical intensities strongly resemble experimental radiation intensities measured in [15].

The relative radiation intensities were measured in [15] in the plane perpendicular to the motion axis. The position of this plane was not specified (as it was suggested to be irrelevant). According to one of authors of Ref.[15] (J. Ruzicka), it was approximately 3 cm. Dimensionless theoretical intensities $S_z(\omega, \rho, z)/(e^2/cz^2)$ (where $S_z(\omega, \rho, z)$ is given by (5.32) in the plane $z = 3$ cm, for $d = 0.1$ cm and $d = 0.5$ cm, are shown in Fig. 5.13. Although the positions of intensities maxima coincide with the experimental positions, their form differs appreciably. We now discuss the complications arising.

First, experimentalists claim (Zrelov V.P., private communication) that

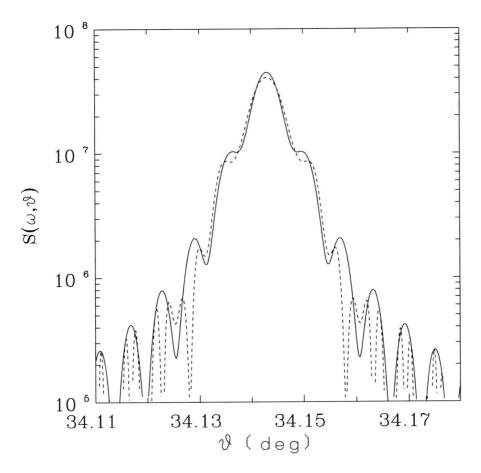

Figure 5.4.   The same as in Fig.5.1, but for $r = 10$ m.

the observed pronounced Cherenkov spectrum arising from the passage of relativistic protons through a transparent slab is described by the Tamm formula in the very neighbourhood of the slab. To resolve the inconsistency between the evaluated and observed angular spectra one may speculate (this is Zrelov's) that the Tamm picture (i.e., the charge particle propagation on a finite spatial interval) is displayed between spatial inhomogeneities of the medium. Since the distance between these inhomogeneities is much smaller than the length of the slab, the pronounced Cherenkov spectrum should be observed at arbitrary distance from the slab.

Dedrick [16] qualitatively showed that the angular spectrum broadens if the multiple scattering of a moving charge on the medium spatial inhomogeneities is taken into account. In this case the resulting interference picture is a superposition of Tamm's intensities from particular medium

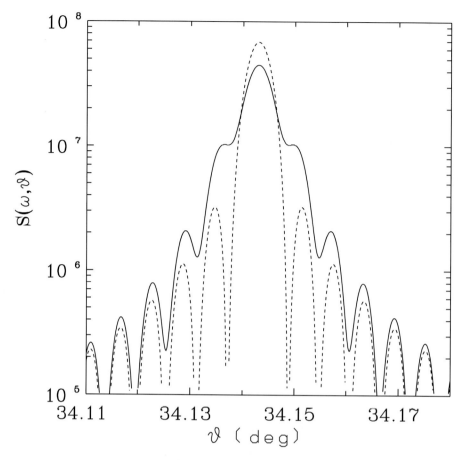

Figure 5.5. Exact angular intensity for $r = 10$ m (solid line) versus the Tamm intensity (dotted line) which does not depend on r. Their distinction is essential. Comparison of this figure with Figs. 5.1-5.4 demonstrates that for smaller r the exact radiation intensities differ drastically from that of Tamm.

inhomogeneities. Quantitatively this was confirmed in Refs. [17,18].

Another possible explanation of this phenomenon is owed to a rather specific measurement procedure used in experiments similar to [15]. In them the measurements were performed in the $z = const$ plane where the camera, with a photographic film inside it, was placed. The lens of this camera was focused on infinity. According to the authors of [15] (Ruzicka, Zrelov) this optical device effectively transforms the finite distance radiation spectrum into the infinite distance spectrum. We do not understand how this can be, but, if this really takes place, then in the $z = 3cm$ plane, the intensities should have a form corresponding to large distances. In passing, intensities shown in Figs 5.11 and 5.12, corresponding to large observational

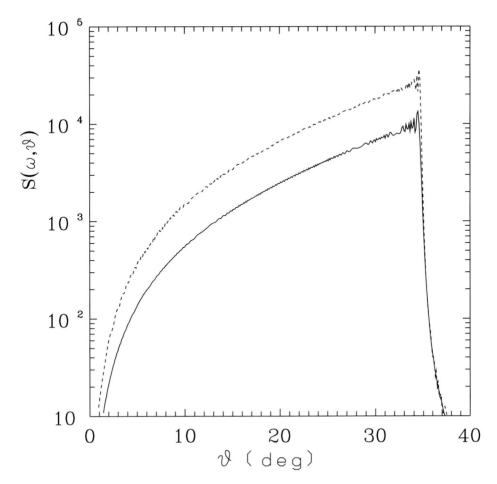

Figure 5.6. Exact angular intensities (in units of e^2/c) on the sphere of infinite radius arising from decelerated motion in a LiF slab ($n = 1.392$) of the width $L = 0.5$ cm; the observed wavelength $\lambda = 4 \times 10^{-5}$ cm; the initial velocity $\beta_i = 0.875$. Solid and dotted curves correspond to the final velocities $\beta_f = 0.2\beta_i$ and $\beta_f = 0.8\beta_i$. Qualitatively, the picture remains the same for smaller β_f.

distances, strongly resemble the experimental intensities. We feel that this question needs further consideration. It should be mentioned about the Schwinger approach [19] describing the radiation intensity of an arbitrary moving charge. The final formula contains only integrals of charge-current densities and does not depend on EMF strengths and the radius of the observational sphere. This formula was applied to the Tamm problem in [8] (see Chapter 2). It was shown there that the radiation intensity in the Schwinger approach strongly resembles that described by the Tamm formula. However, the Schwinger approach uses the half difference of the advanced and retarded potentials (this conflicts with causality) and *ad hoc*

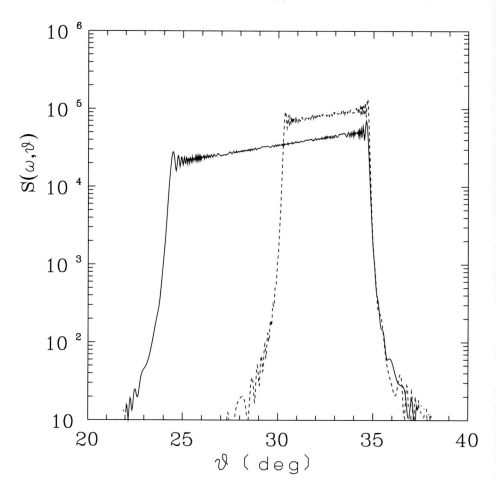

Figure 5.7.   The same as in Fig. 5.6 but for $\beta_f = 0.9\beta_i$ (solid line) and $\beta_f = 0.95\beta_i$ (dotted line).

neglects the terms with definite symmetry properties.

To observe the predicted broadening of the angular spectrum at finite distances, the measurement of the VC radiation produced by high-energy electrons (for which the energy losses are negligible) is needed at a distance from the target where the inequality (5.8), ensuring the validity of the Tamm formula, is violated. No optical devices distorting the radiation spectrum (in the sense defined above) should be used, if possible. Now, if the broadening of the angular spectrum will be observed at arbitrary distance from the dielectric slab then the multiple scattering mechanism suggested by Dedrick [16] takes place. On the other hand, the broadening of the angular spectrum in the immediate neighbourhood of the dielectric slab described by Eqs. (5.14),(5.18),(5.26) and (5.32) will support the

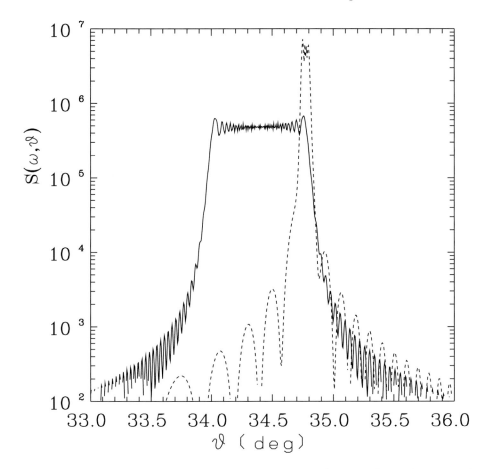

Figure 5.8.   The same as in Fig. 5.6 but for $\beta_f = 0.99\beta_i$ (solid line) and $\beta_f = 0.999\beta_i$ (dotted line). When β_f tends to β_i the angular spectrum tends to that of given by the Tamm formula (2.5).

validity of the original Tamm picture (with its modification for the finite observational distances). We hope, these formulae and considerations will be useful to experimentalists.

The frequency distribution of the radiated energy is defined as

$$S(\omega) = \int d\Omega S(\omega,\theta) = 2\pi \int \sin\theta d\theta S(\omega,\theta).$$

In Fig. 5.14, we present $S(\omega)$ in units of e^2/c evaluated for parameters the same as in [15] on the sphere of radius $r = 1$ m. For $S(\omega,\theta)$ we used its Fresnel approximation (5.26), which under the conditions of the Darmstadt experiments almost coincides with (5.1). In the same figures there are

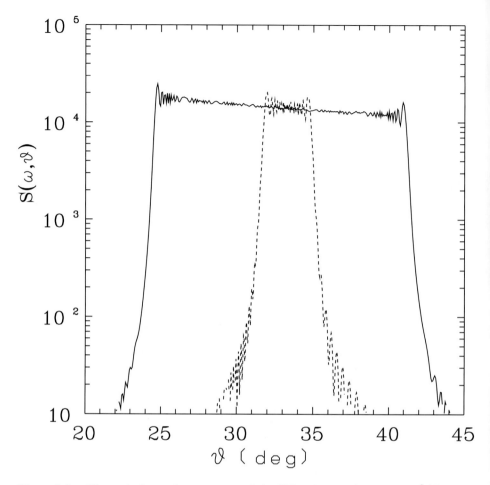

Figure 5.9. Theoretical angular spectrum of the VC radiation (in units of e^2/c) which should be observed in the Darmstadt experiments a the sphere of radius $r = 1$ cm. The solid and dotted curves refer to the widths of LiF ($n = 1.392$) slab $d = 0.5$ cm and $d = 0.1$ cm, respectively. The initial velocity $\beta_i \approx 0.86064$. The final velocity $\beta_f \approx 0.84781$ for $d = 0.5$ cm and $\beta_f \approx 0.8582$ for $d = 0.1$ cm. The observed wavelength $\lambda = 6.5 \times 10^{-5}$ cm.

shown the Tamm frequency distributions $S_T(\omega)$ obtained by integrating the Tamm formula (5.5) over a solid angle $d\Omega$:

$$S_T(\omega) = \frac{2e^2 k z_0}{c}\left(1 - \frac{1}{\beta_n^2}\right) + \frac{4e^2}{\pi c n}\left(\frac{1}{2\beta_n}\ln\frac{1+\beta_n}{\beta_n - 1} - 1\right) \qquad (5.33)$$

for $\beta_n > 1$ and

$$S_T(\omega) = \frac{4e^2}{\pi c n}\left(\frac{1}{2\beta_n}\ln\frac{1+\beta_n}{1-\beta_n} - 1\right)$$

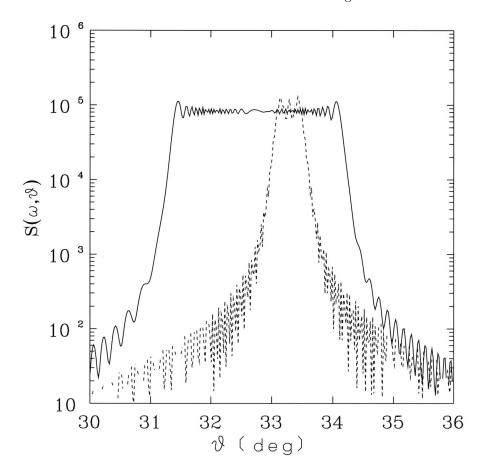

Figure 5.10.   The same as in Fig. 5.9, but for $r = 10$ cm.

for $\beta_n < 1$. Here $k = \omega/c$, $\beta_n = \beta n$, and $2z_0$ is the width of the slab. The value of β in (5.33) is chosen as the half-sum of the velocities of the Au ions before entering the LiF slab and after the passage this slab. We observe that the Tamm frequency intensities almost coincide with the Fresnel intensities despite the striking difference in corresponding angular-frequency distributions.

Up to now we have identified heavy ions with the point-like charged objects. Since the medium (a dielectric slab) is considered here as structureless (it is described by the refractive index depending only on the frequency), the condition for the validity of point-like approximation is the smallness of the heavy ion dimension R relative to the observed wavelength λ. If for R we take its radius $R = 1.5A^{1/3}$ fm and for λ we take the average wavelength of the optical region $\lambda \approx 6 \times 10^{-5}$ cm, then for $A \approx 200$ the above condition

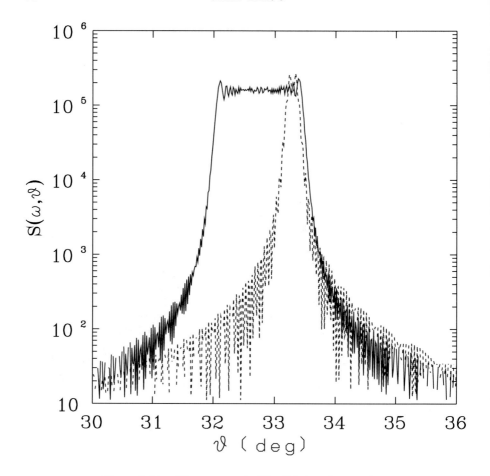

Figure 5.11. The same as in Fig. 5.9, but for $r = 1$ m.

is satisfied to a great accuracy: $R \approx 10^{-12}$ cm$\ll \lambda \approx 6 \times 10^{-5}$ cm.

Another estimation of the point-like approximation was made in an important paper [16] where the smallness of the wave packet dimension $\lambda_B = \hbar/(m_0 v \gamma)$ (coinciding with the de Broglie wavelength) relative to the motion length L (coinciding with the width of the dielectric slab) was postulated. Here $m_0 = m_N A$ is the rest mass of the heavy ion, v is its velocity, $\gamma = 1/\sqrt{1 - \beta^2}$, m_N is the mass of nucleon. For the case treated this condition is satisfied to a great accuracy: $\lambda_B \approx 5 \times 10^{-17}$ cm$\ll L \approx 0.1$ cm. In fact, λ_B is much smaller than the distance (10^{-8} cm) between the neighbouring atoms from which the dielectric slab is composed. This is essential for the multiple scattering of a charge on the medium spatial inhomogeneities considered in [16].

The influence of finite dimensions of a moving charge on the radiation of

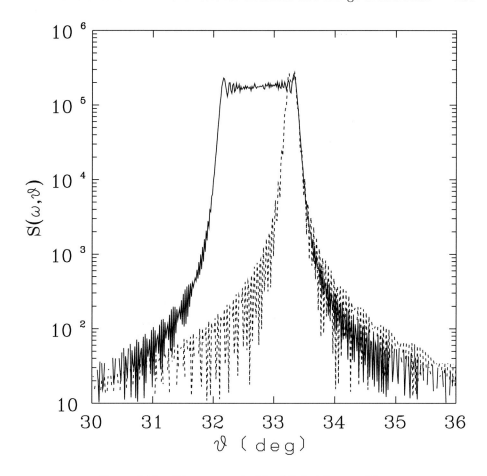

Figure 5.12.   The same as in Fig. 5.9, but for $r = 10$ m.

the EMF was studied in [20]. The moving charge density had the Gaussian form along the motion axis and zero dimensions in the directions perpendicular to it. It was shown there that the EMFs corresponding to a point-like and diffused charge densities were practically the same up to some critical frequency $\omega_c = c/a$, where c is the velocity of light and a is the parameter of the Gaussian distribution. If we identify a with the heavy ion radius R, then in the case treated, $\omega_c \approx 3 \cdot 10^{22} \mathrm{s}^{-1}$ which is far off the optical region ($10^{15} \mathrm{s}^{-1} < \omega < 10^{16} \mathrm{s}^{-1}$). Thus, a point-like approximation for heavy ions charge densities is satisfactory for the treated problem.

In the radiation intensities used in sections 4 and 5, e^2 should be changed to $Z^2 e^2$ if a propagation of heavy ion with an atomic number Z is considered. Alternatively one may think that Z^2 is included in e^2.

The moral of this section is that one should be very careful when ap-

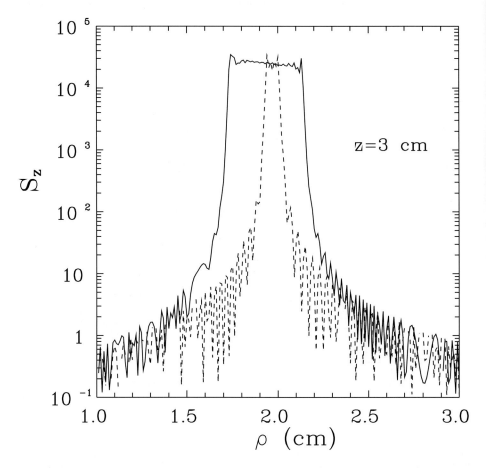

Figure 5.13.   Theoretical radial distribution of the VC radiation intensity (in units of e^2/cz^2) which should be observed in the Darmstadt experiments in the plane $z = 3$ cm. The solid and dotted curves refer to the widths of LiF slab $d = 0.5$ cm and $d = 0.1$ cm, respectively. Other parameters are the same as in Fig. 5.9.

plying the Tamm formula (5.5) to analyse experimental data. The validity of the conditions (5.2),(5.3), and (5.8) ensuring the validity of (5.5) should be verified. The almost exact energy flux (5.14) or the approximate expressions (5.18), (5.26), (5.29) or (5.32) should be used if these conditions are violated.

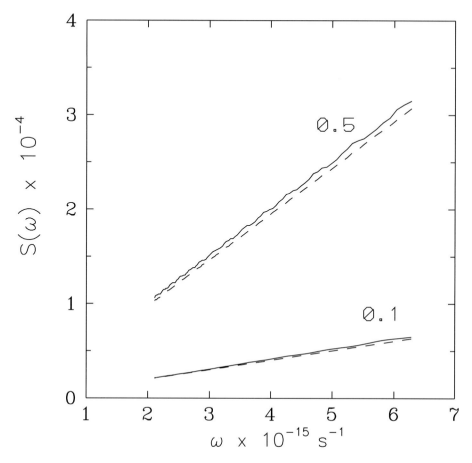

Figure 5.14.   Theoretical frequency spectrum (for the region of visible light) of VC radiation (in units of e^2/c) which should be observed in Darmstadt experiments on a sphere of radius $r = 1$ m for $d = 0.5$ cm and $d = 0.1$ cm (solid lines). Dotted lines correspond to the Tamm intensity (5.33).

5.3.   Motion in a finite spatial interval with arbitrary acceleration

5.3.1.   INTRODUCTION

In 1934-1937, the Russian physicist P.A. Cherenkov performed a series of experiments under the suggestion of his teacher S.I. Vavilov. In them photons emitted by Ra atoms passed through water. They induced the blue light observed visually. Applying an external magnetic field, Cherenkov recognized that this blue light was produced by secondary electrons knocked out by photons.

These experiments were explained by Tamm and Frank in 1937-1939 who attributed the blue light to the radiation of a charge uniformly moving in medium with a velocity greater than the velocity of light in medium.

Theoretically, when considering the VC radiation one usually treats either the unbounded charge motion with a constant velocity (this corresponds to the so-called Tamm-Frank problem [10]) or the charge motion in a finite interval with an instantaneous acceleration and deceleration of a charge at the beginning and at the end of its motion. This corresponds to the so-called Tamm problem [1]. The physical justification for the Tamm problem is as follows. A charge, moving initially uniformly in vacuum (where it does not radiate), penetrates into the transparent dielectric slab (where it radiates if the condition $\cos \theta_{\text{Ch}} = 1/\beta n$ for the Cherenkov angle is satisfied), and finally, after leaving the dielectric slab, moves again in vacuum without radiating (we disregard the transition radiation at the boundaries of the dielectric slab). The appearance of radiation at the instant when a charge enters the slab and its termination at the instant when it leaves the slab are usually interpreted in terms of the instantaneous charge acceleration at one side of the slab and its instantaneous deceleration at its other side. Since the Tamm problem is more physical than the Tamm-Frank problem, it is frequently used for the analysis of experimental data. Another possible application of the Tamm problem is the electron creation at some spatial point (nuclear β decay) with its subsequent absorption at another spatial point (nuclear β capture). Tamm obtained a remarkably simple analytic formula describing the intensity of radiation and interpreted it as the VC radiation in a finite interval.

Another viewpoint of the nature of the radiation observed by Cherenkov is owed to Vavilov [21]. According to him,

> We think that the most probable reason for the γ luminescence is the radiation arising from the deceleration of Compton electrons. The hardness and intensity of γ rays in the experiments of P.A. Cherenkov were very large. Therefore the number of Compton scattering events and the number of scattered electrons should be very considerable in fluids. The free electrons in a dense fluid should be decelerated at negligible distances. This should be followed by the radiation of the continuous spectrum. Thus, the weak visible radiation may arise, although the boundary of bremsstrahlung, and its maximum should be located somewhere in the Roentgen region. It follows from this that the energy distribution in the visible region should rise towards the violet part of spectrum, and the blue violet part of spectrum should be especially intense.

(our translation from Russian).

This Vavilov explanation of the Cherenkov effect has given rise to a number of attempts (see, e.g., [4,5]) in which the radiation described by the Tamm formula was attributed to the interference of bremsstrahlung (BS) arising at the start and end of motion.

On the other hand, the exact solution of the Tamm problem in a non-dispersive medium was found and analysed in [6,7]. It was shown there that the Cherenkov shock wave exists side by side with BS waves in no case can be reduced to them. Then, how can this fact be reconciled with the results of [4,5] which describe experimental data quite satisfactorily? The possible explanation of this controversy is that the exact solution obtained in [6,7] was written out in the space-time representation, while the authors of [4,5] operated with the Tamm formula related to the spectral representation. It might happen that the main contribution to the exact solution of describing the Cherenkov wave is owed to the integration over the frequency region lying outside the visible part of the intensity spectrum. Then, in principle, the radiation in the visible part of spectrum could be described by the Tamm formula frequently used for the interpretation of experimental data.

The aim of this consideration is to resolve this controversy. We shall operate simultaneously in the spectral representation as authors of [4,5] did and in the time representation used in [6,7]. Instead of the original Tamm problem in which a charge exhibits instantaneous acceleration and deceleration, we consider a charge motion with a finite acceleration and deceleration and uniform motion on the remaining part of a trajectory. This allows us to separate contributions from the uniform and non-uniform parts of a charge trajectory. In the past, analytic and numerical results for the motion with the change of velocity small compared with the charge velocity itself were obtained in [9,22]. Unfortunately, the analytic formulae obtained there do not work in the case treated, since the charge is accelerated from the state of rest up to a velocity close to that of light. Numerically, the smoothed Tamm problem with a large change of velocity was considered in [23], but their authors did not aim to resolve there the above controversy between Refs. [4,5] and [6,7].

5.3.2. MAIN MATHEMATICAL FORMULAE

Let a point charge move along the z axis with a trajectory $z = \xi(t)$ in a non-dispersive medium of refractive index n. Its charge and current densities then are equal to

$$\rho = e\delta(x)\delta(y)\delta(z - \xi(t)), \quad j_z = ev(t)\delta(x)\delta(y)\delta(z - \xi(t)), \quad v = \frac{d\xi}{dt}.$$

The Fourier transforms of these densities are equal to

$$\rho(\omega) = \frac{e}{2\pi} \int \exp(-i\omega t)\rho(t)dt = \frac{e}{2\pi}\delta(x)\delta(y)\int \exp(-i\omega t)\delta(z - \xi(t))dt$$

$$= \frac{e}{2\pi v}\delta(x)\delta(y)\exp(-i\omega\tau(z)),$$

$$j_z(\omega) = \frac{e}{2\pi}\delta(x)\delta(y)\exp(-i\omega\tau(z)), \tag{5.34}$$

where $\tau(z)$ is the root of the equation $z - \xi(t) = 0$. It was assumed here that $v > 0$, that is, a charge moves in the positive direction of the z axis.

The Fourier transform of the vector potential corresponding to these densities at the spatial point x, y, z is equal to

$$A_z(\omega) = \frac{e}{2\pi c}\int \frac{dz'}{R}\exp(-i\psi), \tag{5.35}$$

where $\psi = \omega\tau(z') + knR$ and $R = \sqrt{x^2 + y^2 + (z - z')^2}$ and $k = \omega/c$. The non-vanishing Fourier component of the magnetic field strength is

$$H_\phi(\omega) = \frac{iek_n r \sin\theta}{2\pi c}\int \frac{dz'}{R^2}\exp(-i\psi)(1 - \frac{i}{k_n R}). \tag{5.36}$$

Here $k_n = \omega/c_n$ and $c_n = c/n$ is the velocity of light in medium. Outside the motion axis, the electric field strengths are obtained from the Maxwell equation

$$\mathrm{curl}\vec{H}(\omega) = \frac{i\omega\epsilon}{c}\vec{E}(\omega). \tag{5.37}$$

The energy flux in the radial direction per unit time and per unit area of the observational sphere of the radius r is

$$S_r = \frac{d^2 W}{r^2 d\Omega dt} = \frac{c}{4\pi}E_\theta(t)H_\phi(t).$$

The energy radiated for the whole charge motion is

$$\int\limits_{-\infty}^{\infty} S_r dt = \frac{c}{4\pi}\int\limits_{-\infty}^{\infty} dt E_\theta(t)H_\phi(t)$$

$$= \frac{c}{2}\int\limits_{0}^{\infty} d\omega[E_\theta(\omega)H_\phi^*(\omega) + E_\theta^*(\omega)H_\phi^*(\omega)]. \tag{5.38}$$

Usually radial energy fluxes are related not to the unit area, but to the unit solid angle. For this one should multiply Eq. (5.38) by r^2 (r is the radius of the observational sphere). Then

$$r^2 \int_{-\infty}^{\infty} S_r dt = \int_0^{\infty} \sigma_r(\omega) d\omega,$$

where

$$\sigma_r(\omega, \theta) = \frac{d^2 W}{d\Omega d\omega} = \frac{c}{2} r^2 [E_\theta(\omega) H_\phi^*(\omega) + E_\theta^*(\omega) H_\phi(\omega)]. \tag{5.39}$$

Let the motion interval L be finite. Then under the conditions (5.2),(5.3) and (5.6)-(5.8) the radial radiation intensity is given by

$$\sigma_r(\omega, \theta) = \frac{e^2 k^2 n \sin^2 \theta}{4\pi^2 c} [(\int dz' \cos \psi_1)^2 + (\int dz' \sin \psi_1)^2] \tag{5.40}$$

with

$$\psi_1 = \omega\tau(z') - knz' \cos \theta. \tag{5.41}$$

For uniform rectilinear motion this approximation gives the famous Tamm formula

$$\sigma_T(\omega, \theta) = \frac{e^2}{\pi^2 cn} [\sin \theta \frac{\sin \omega t_0 (1 - \beta_n \cos \theta)}{\cos \theta - 1/\beta_n}]^2, \quad t_0 = \frac{z_0}{v} \quad \beta_n = \frac{v}{c_n}. \tag{5.42}$$

A question arises of why it is needed to use the approximate expression (5.40) even though the numerical integration is quite easy [8,22]. One of the reasons is the same as for the use of the Tamm formula which does not work at realistic distances [8,10]. Despite this and owing to its remarkable simplicity, the Tamm formula is extensively used by experimentalists for the planning and interpretation of experiments. Analytic formulae of the next section are also transparent. Since acceleration effects are treated in them exactly they are valid under the same conditions (5.2), (5.3), and (5.8) as the Tamm formula (5.42), but include, in addition, the charge finite acceleration (or deceleration). Another reason is that experimentalists want to know what, in fact, they measure. For this they need quite transparent analytic formulae to distinguish contributions from the uniform and accelerated (decelerated) charge motions. The formulae presented in the next section satisfy these requirements and may be used for the rough estimation of the acceleration effects. After this stage the explicit formulae presented in this section may be applied (as was done in [22]) to take into account the effect of finite distances. Our experience [23] tells us that exact numerical calculations without preliminary analytical consideration are not very productive.

In what follows we intend to investigate the deviation from the Tamm formula arising from the charge deceleration. Let us consider particular cases.

5.3.3. PARTICULAR CASES

Decelerated and accelerated motion on a finite interval
Let a charge move in the interval (z_1, z_2) according to the law shown in Fig. 5.15(a):

$$z = z_1 + v_1(t - t_1) + \frac{1}{2}a(t - t_1)^2. \tag{5.43}$$

The motion begins at the instant t_1 and terminates at the instant t_2. The charge velocity varies linearly with time from the value $v = v_1$ at $t = t_1$ down to value $v = v_2$ at $t = t_2$: $v = v_1 + a(t - t_1)$. It is convenient to express the acceleration a and the motion interval through z_1, z_2, v_1, v_2:

$$a = \frac{v_1^2 - v_2^2}{2(z_1 - z_2)}, \quad t_2 - t_1 = \frac{2(z_2 - z_1)}{v_2 + v_1}.$$

For the case treated the function $\tau(z)$ entering (5.41) is given by

$$\tau(z) = t_1 - 2v_1 \frac{z_2 - z_1}{v_2^2 - v_1^2} \left[1 - \left(1 + \frac{z - z_1}{z_2 - z_1} \frac{v_2^2 - v_1^2}{v_1^2} \right)^{1/2} \right]. \tag{5.44}$$

When the conditions (5.2),(5.3) and (5.8) are fulfilled (i.e., ψ_1 is of the form (5.41)), the radiation intensity can be taken in a closed form. For this we should evaluate integrals

$$I_c(z_1, v_1; z_2, v_2) = \int_{z_1}^{z_2} \cos \psi_1 dz \quad \text{and} \quad I_s(z_1, v_1; z_2, v_2) = \int_{z_1}^{z_2} \sin \psi_1 dz \tag{5.45}$$

entering into (5.40), where ψ_1 is the same as in (5.41). We write them in a manifest form for the motion beginning at the point z_1, at the instant t_1 with the velocity v_1 and ending at the point $z_2 > z_1$ with the velocity v_2. There are four possibilities depending on the signs of $\cos \theta$ and $(v_1 - v_2)$. Obviously $v_2 > v_1$ and $v_1 > v_2$ correspond to accelerated and decelerated motions, respectively; $\cos \theta > 0$ and $\cos \theta < 0$ correspond to the observational angles lying in front and back semispheres, respectively.

1) $v_2 > v_1$, $\cos \theta > 0$

$$I_c = \frac{1}{kn \cos \theta} \{\sin(u_2^2 - \gamma) - \sin(u_1^2 - \gamma) + a\sqrt{2\pi}[\cos \gamma(C_2 - C_1) + \sin \gamma(S_2 - S_1)]\},$$

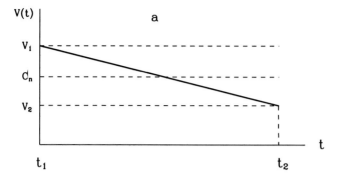

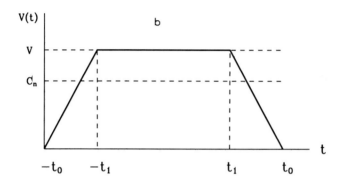

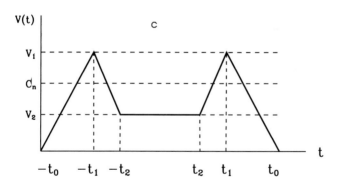

Figure 5.15.   Time dependences of charge velocities treated in the text.
(a): Charge deceleration in a finite interval. v_1, v_2 and c_n are the charge initial and final velocities and velocity of light in medium, respectively.
(b): Charge acceleration followed by the uniform motion and deceleration. This case allows one to estimate contributions to the radiation intensity from the accelerated, uniform, and decelerated parts of a charge trajectory.
(c): This motion permits one to estimate how the radiation intensity changes when the transition from a velocity greater to a velocity smaller than the velocity of light in medium takes place.

$$I_s = \frac{1}{kn\cos\theta}\{\cos(u_2^2-\gamma)-\cos(u_1^2-\gamma)-\alpha\sqrt{2\pi}[\cos\gamma(S_2-S_1)-\sin\gamma(C_2-C_1)]\},$$

2) $v_2 > v_1$, $\cos\theta < 0$

$$I_c = -\frac{1}{kn\cos\theta}\{\sin(u_2^2+\gamma)-\sin(u_1^2+\gamma)-\alpha\sqrt{2\pi}[\cos\gamma(C_2-C_1)-\sin\gamma(S_2-S_1)]\},$$

$$I_s = \frac{1}{kn\cos\theta}\{\cos(u_2^2+\gamma)-\cos(u_1^2+\gamma)+\alpha\sqrt{2\pi}[\cos\gamma(S_2-S_1)+\sin\gamma(C_2-C_1)]\},$$

3) $v_1 > v_2$, $\cos\theta > 0$

$$I_c = -\frac{1}{kn\cos\theta}\{\sin(u_2^2+\gamma)-\sin(u_1^2+\gamma)+\alpha\sqrt{2\pi}[\cos\gamma(C_2-C_1)-\sin\gamma(S_2-S_1)]\},$$

$$I_s = \frac{1}{kn\cos\theta}\{\cos(u_2^2+\gamma)-\cos(u_1^2+\gamma)-\alpha\sqrt{2\pi}[\cos\gamma(S_2-S_1)+\sin\gamma(C_2-C_1)]\},$$

4) $v_1 > v_2$, $\cos\theta < 0$

$$I_c = \frac{1}{kn\cos\theta}\{\sin(u_2^2-\gamma)-\sin(u_1^2-\gamma)-\alpha\sqrt{2\pi}[\cos\gamma(C_2-C_1)+\sin\gamma(S_2-S_1)]\},$$

$$I_s = \frac{1}{kn\cos\theta}\{\cos(u_2^2-\gamma)-\cos(u_1^2-\gamma)+\alpha\sqrt{2\pi}[\cos\gamma(S_2-S_1)-\sin\gamma(C_2-C_1)]\}.$$

Here we put

$$C_1 = C(u_1), \quad C_2 = C(u_2), \quad S_1 = S(u_1), \quad S_2 = S(u_2),$$

$$\alpha = \left[\frac{k(z_2-z_1)}{n|\cos\theta(\beta_2^2-\beta_1^2)|}\right]^{1/2},$$

$$u_1 = \sqrt{\frac{k(z_2-z_1)n|\cos\theta|}{|\beta_2^2-\beta_1^2|}}\left(\beta_1 - \frac{1}{n\cos\theta}\right),$$

$$u_2 = \sqrt{\frac{k(z_2-z_1)n|\cos\theta|}{|\beta_2^2-\beta_1^2|}}\left(\beta_2 - \frac{1}{n\cos\theta}\right),$$

$$\gamma = \omega t_1 + \frac{k(z_2-z_1)}{(\beta_2^2-\beta_1^2)n\cos\theta} - kn\cos\theta\frac{\beta_2^2 z_1 - \beta_1^2 z_2}{\beta_2^2-\beta_1^2} - 2\beta_1\frac{k(z_2-z_1)}{(\beta_2^2-\beta_1^2)}.$$

C and S are Fresnel integrals defined as

$$S(x) = \sqrt{\frac{2}{\pi}}\int_0^x dt\,\sin t^2 \quad\text{and}\quad C(x) = \sqrt{\frac{2}{\pi}}\int_0^x dt\,\cos t^2.$$

Obviously I_c and I_s are the elements from which the total radiation intensity for the charge motion consisting of any superposition of accelerated, decelerated, and uniform parts can be constructed.

Using them we evaluate the intensity of radiation:

$$\sigma_r(\theta) = \frac{e^2 k^2 n \sin^2 \theta}{4\pi^2 c} [(\int_{z_1}^{z_2} dz' \cos \psi_1)^2 + (\int_{z_1}^{z_2} dz' \sin \psi_1)^2]$$

$$= \frac{e^2 \sin^2 \theta}{2\pi^2 cn \cos^2 \theta} \{1 - \cos(u_2^2 - u_1^2) + \pi \alpha^2 [(C_2 - C_1)^2 + (S_2 - S_1)^2]$$

$$\pm \sqrt{2\pi} \alpha [(C_2 - C_1)(\sin u_2^2 - \sin u_1^2) - (S_2 - S_1)(\cos u_2^2 - \cos u_1^2)]\}. \quad (5.46)$$

The plus and minus signs in (5.46) refer to $\cos \theta > 0$ and $\cos \theta < 0$, respectively. Furthermore $\beta_1 = v_1/c$ and $\beta_2 = v_2/c$. When $v_1 \to v_2 = v$ the intensity (5.46) goes into the Tamm formula (5.42) in which one should put $t_0 = (z_2 - z_1)/2v$.

Figure 5.16 shows angular radial distributions for the fixed initial velocity $\beta_1 = 1$ and various final velocities β_2. The length of the sample was chosen to be $L = 0.5$ cm, the wavelength $\lambda = 4 \times 10^{-5}$ cm, the refractive index of the sample $n = 1.392$. For β_2 close to β_1 ($\beta_2 = 0.99$) the angular distribution strongly resembles the Tamm one. When β_2 diminishes ($\beta_2 = 0.9$ and $\beta_2 = 0.8$) a kind of a plateau appears. Its edges are at the Cherenkov angles corresponding to β_1 and β_2 ($\cos \theta_1 = 1/\beta_1 n$, $\cos \theta_2 = 1/\beta_2 n$). On the Cherenkov threshold ($\beta_2 = 1/n$), σ_r has a peculiar form with fast oscillations at large angles. This form remains the same for the velocities below the Cherenkov threshold, but the oscillations disappear for $\beta_2 = 0$.

An important case is the decelerated motion with a final zero velocity. Experimentally it is realized in heavy water reactors where electrons arising in β decay are decelerated down to a complete stop, in neutrino experiments, in the original Cherenkov experiments, etc.. Radiation intensities for various initial velocities are shown in Fig. 5.17. It is easy to check that their maxima, despite the highly non-uniform character of this motion, are always at the Cherenkov angle θ_1 defined by $\cos \theta_1 = 1/\beta_1 n$ and corresponding to the initial velocity v_1. The angular dependences of the radial intensity are always smooth for $\beta_2 = 0$. Analytically these radiation intensities are described by Eq.(5.46) in the whole angular interval. For completeness, we have collected in Fig. 5.18 the radiation intensities corresponding to a number of initial velocities and zero final velocity.

An important quantity is the total energy radiated per unit frequency. It is obtained by integration of the angular-frequency distribution over the

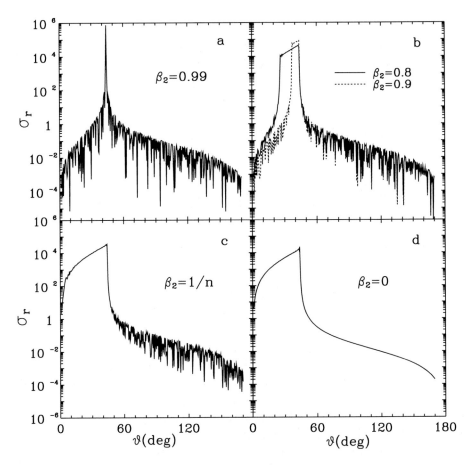

Figure 5.16. Radiation intensities (in units of e^2/c) corresponding to Fig. 5.15(a) for $\beta_1 = 1$ fixed and various β_2. (a): For $\beta_2 = 0.99$ the radiation spectrum is close to that described by the Tamm formula (5.42). (b): For smaller β_2 a kind of plateau appears in the radiation intensity. Its edges are at the Cherenkov angles corresponding to β_1 and β_2. (c): For $\beta_2 = 1/n$, the distribution of radiation has a specific form without oscillations to the left of the maximum. (d): This form remains essentially the same for smaller β_2, but the tail oscillations disappear. In all these cases the main radiation maximum is at $\cos\theta = 1/\beta_1 n$. All these results are confirmed analytically in section 5.3.4. These intensities were evaluated for the following parameters: the wavelength $\lambda = 4 \times 10^{-5}$ cm, the motion length $L = 0.5$ cm, the refractive index $n = 1.392$.

solid angle:

$$\sigma_r(\omega) = \frac{d\mathcal{E}}{d\omega} = \int \sigma_r(\omega, \theta) d\Omega. \qquad (5.47)$$

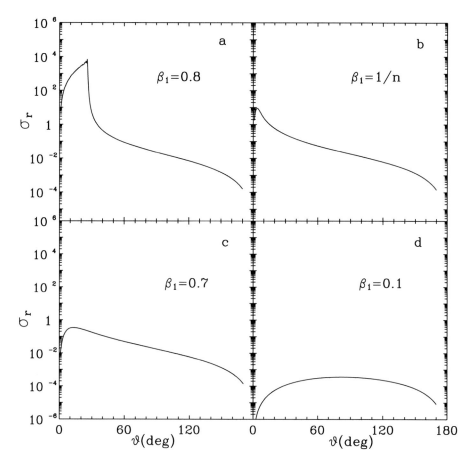

Figure 5.17. Radiation intensities corresponding to Fig. 5.15(a) for $\beta_2 = 0$ fixed and various β_1. For $\beta_1 = 1$ the radiation spectrum is shown in Fig. 5.16(d). For smaller β_1 the maximum of intensity shifts to smaller angles (a) reaching zero angle at the Cherenkov threshold $\beta_1 = 1/n$ (b). The maximum is at the Cherenkov angle corresponding to β_1. Below the Cherenkov threshold the form of the radiation spectrum remains practically the same, but its amplitude decreases (c,d). Other parameters are the same as in Fig. 5.16.

The integration of the Tamm intensity (5.42) over the solid angle gives the frequency distribution of the radiated energy $\sigma(\omega)$. It was written out explicitly in [8] (see also Chapter 2, Eq. (2.109)). In the limit $\omega t_0 \to \infty$, it is transformed into the following expression given by Tamm [1]:

$$\sigma_T(\omega) = \frac{e^2 kL}{c}(1 - \frac{1}{\beta_n^2})\Theta(\beta n - 1) + \frac{4e^2}{\pi cn}(\frac{1}{2\beta_n} \ln \frac{1+\beta_n}{|\beta_n - 1|} - 1).$$

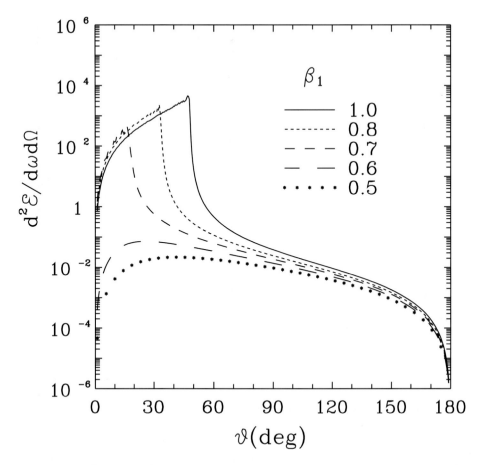

Figure 5.18.   Angular radiation intensities corresponding to the charge motion with a complete stop for a number of initial velocities β_1. It is seen that these intensities do not oscillate. The angle where they are maximal increases with increase of β_1. The motion interval $L = 0.1$ cm, the wavelength $\lambda = 4 \times 10^{-5}$ cm, the refractive index $n = 1.5$.

Here $k = \omega/c$, $\quad \beta_n = \beta n$, and $L = 2z_0$ is the motion interval. This equation has a singularity at $\beta = 1/n$, whilst $\sigma(\omega)$ given by (2.109) is not singular there.

We integrate now angular distributions corresponding to the decelerated motion with a final zero velocity and shown in Fig. 5.18, and relate them to the Tamm integral intensivity. Fig. 5.19 demonstrates that, despite their quite different angular distributions, the ratio R of these integral intensities does not depend on the frequency except for the neighbourhood of $\beta = 1/n$ where $\sigma_T(\omega)$ is not valid. For the charge velocity v above the light velocity

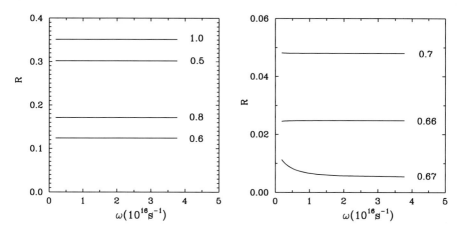

Figure 5.19.   The ratio R of the integral intensity for a motion with a zero final velocity to the Tamm integral intensity for a number of initial velocities v_1. Although R does not depend on the frequency (except for the velocity $\beta_1 = 0.67$ close to the Cherenkov threshold $1/n$), it strongly depends on β_1 being minimal at the threshold. The analytical formula (5.63) given below shows that $R \to 0.5$ for small β_1. To this frequency interval there corresponds the wavelength interval $(5 \times 10^{-6} \mathrm{cm} < \lambda < 10^{-4}\ \mathrm{cm})$ which encompasses the visible light interval $(4 \times 10^{-5}\mathrm{cm} < \lambda < 8 \times 10^{-5}\mathrm{cm})$. Numbers on curves are β_1.

c_n in medium (where the Tamm intensity is approximately proportional to ω), this ratio decreases as v approaches c_n. For $v < c_n$ (where the ω dependence given by the Tamm formula is logarithmic) R begins to rise. The analytical considerations (see Eq. (5.63) given below) show that the radiation intensity (5.47) is one half of $\sigma_T(\omega)$ for $\beta_1 n < 1$. Therefore R tends to $1/2$ for small β_1. We see that integral intensities for the decelerated motion, up to a factor independent of ω, coincide with the Tamm intensity. Therefore the total energy, for the decelerated motion,

$$\mathcal{E} = \int_{\omega_1}^{\omega_2} d\omega \frac{d\mathcal{E}}{d\omega}$$

radiated in the frequency interval (ω_1, ω_2) up to the same factor coincides with the Tamm integral intensity.

Tamm [1] obtained the following condition

$$\frac{t_0^2}{2}\left|\frac{dv}{dt}\right| \ll \lambda \tag{5.48}$$

for the frequency spectrum $\sigma(\omega)$ to be the linear function of frequency. For the decelerated motion treated, this condition takes the form

$$\frac{v_1 - v_2}{v_1 + v_2} \ll \frac{\lambda}{L}, \tag{5.49}$$

where $L = z_2 - z_1$ is the motion interval. When the final velocity is zero (5.49) is reduced to $L \ll \lambda$, which for $L = 0.1$ cm and $\lambda = 4 \times 10^{-5}$ cm takes the form $1 \ll 4 \times 10^{-4}$. Figure 5.19 demonstrates that the frequency independence of the above ratio R takes place despite the strong violation of the Tamm condition.

The radiation intensity (5.46) disappears for the fixed wavelength if the acceleration length $L = z_2 - z_1$ tends to zero. At first glance, this disagrees with results of Chapter 2, in which it was mentioned many times about the BS shock waves arising at the beginning and end of motion. The following simple consideration underlines this controversy. It is known that the energy radiated by a non-uniformly moving charge for the whole its motion is given by

$$W = \frac{2e^2}{3c^3} \int\limits_{-\infty}^{\infty} |\vec{a}(t)|^2 dt$$

($\vec{a}$ is the charge acceleration). In the case treated, the acceleration has a constant value

$$a = \frac{v_2^2 - v_1^2}{2(z_2 - z_1)}$$

in the time interval

$$t_2 - t_1 = 2\frac{z_2 - z_1}{v_1 + v_2}.$$

Substituting all this into W, one finds

$$W = \frac{e^2}{3L}(\beta_1 - \beta_2)^2(\beta_1 + \beta_2).$$

It is seen that $W \to \infty$ for $L \to 0$. To see the reason for this, we fix the acceleration length L and let the radiated frequency tend to ∞. The radiation intensity then tends to the analytical angular intensity $\sigma_r(\theta, \omega)$ given by (5.58) and (5.59). It is infinite at the angles θ_1 and θ_2 defined by $\cos\theta_1 = 1/\beta_1 n$ and $\cos\theta_2 = 1/\beta_2 n$ (it is, therefore, suggested that both β_1 and β_2 are larger than $1/n$). To obtain the energy radiated for the whole charge motion, one should integrate $\sigma_r(\theta, \omega)$ over angles and frequency. The $\sigma_r(\omega)$ (5.47) tends to ∞ for $\omega \to \infty$, and therefore, the total radiated energy

$$\sigma = \int\limits_0^{\infty} \sigma_r(\omega) d\omega$$

is also infinite. Therefore, the infinite value of W, in the limit of a small length L of acceleration, is owed to the contribution of high frequencies. If L is so small that for visible light (where the VC is usually observed) $kL \ll 1$, then the disappearance of (5.46) tells us that for this frequency there is no contribution to the radiation intensity. This contribution reappears for high frequencies.

It was shown explicitly [23] in the time representation that for the accelerated charge motion, the Cherenkov shock wave and the shock wave closing the Cherenkov cone arise at the instant when the charge velocity coincides with the velocity of light in medium. The content of this section then may be viewed as the translation of [23] into the frequency language (which is more frequently used by experimentalists).

The calculations of this section were performed with analytical formula (5.46) which is valid both for the decelerated ($v_1 > v_2$) and accelerated ($v_2 > v_1$) charge motion in medium. The results of this section may be useful for the study of the VC radiation arising from the decelerated motion of heavy ions in medium (for them the energy losses are large owing to their large atomic number) [15].

Simplest superposition of accelerated, decelerated, and uniform motions.
We also consider another problem corresponding to the motion shown in Fig. 5.15(b). A charge is at rest at the spatial point $z = -z_0$ up to an instant $t = -t_0$. In the time interval $-t_0 < t < -t_1$ it moves with acceleration a up to reaching the velocity v at the spatial point $z = -z_1$:

$$z = -z_0 + \frac{1}{2}a(t + t_0)^2, \quad v(t) = a(t + t_0).$$

In the time interval $-t_1 < t < t_1$ a charge moves with the constant velocity v: $z = vt$. Finally, in the time interval $t_1 < t < t_0$ a charge moves with deceleration a down to reaching the state of rest at the instant t_0 at the spatial point $z = z_0$:

$$z = z_0 - \frac{1}{2}a(t - t_0)^2, \quad v(t) = -a(t - t_0).$$

It is convenient to express t_0, t_1, and a through z_0, z_2 and v:

$$a = \frac{v^2}{2(z_0 - z_1)}, \quad t_0 = \frac{2z_0 - z_1}{v}, \quad t_1 = \frac{z_1}{v}.$$

After the instant $t = t_0$ the charge is at rest at the point $z = z_0$.
The radiation intensity is

$$\sigma_r(\omega, \theta) = \frac{e^2 k^2 n \sin^2 \theta}{4\pi^2 c}[(I_c)^2 + (I_s)^2]. \tag{5.50}$$

Here

$$I_c = \sum_i I_c^{(i)}, \quad I_s = \sum_i I_s^{(i)},$$

$$I_c^{(i)} = \int dz' \cos \psi_i, \quad I_s^i = \int dz' \sin \psi_i \quad (i = 1, 2, 3)$$

and $\psi_i = -knz' \cos \theta + \omega \tau_i$. The superscripts $1, 2$ and 3 refer to the accelerated $(-z_0 < z' < -z_1)$, uniform $(-z_1 < z' < z_1)$, and decelerated $(z_1 < z' < z_0)$ parts of a charge trajectory. The functions $\tau_i(z)$ entering into ψ_i are equal to

$$\tau_1 = -\frac{2z_0 - z_1}{v} + \frac{2}{v}\sqrt{(z + z_0)(z_0 - z_1)} \quad \text{for} \quad -z_0 < z < -z_1,$$

$$\tau_2 = \frac{z}{v} \quad \text{for} \quad -z_1 < z < z_1,$$

$$\tau_3 = \frac{2z_0 - z_1}{v} - \frac{2}{v}\sqrt{(z_0 - z)(z_0 - z_1)} \quad \text{for} \quad z_1 < z < z_0. \tag{5.51}$$

We rewrite I_c and I_s in a manifest form

$$I_c = I_c(-z_0, 0; -z_1, v) + I_c(-z_1, v; z_1, v) + I_c(z_1, v; z_0, 0),$$

$$I_s = I_s(-z_0, 0; -z_1, v) + I_s(-z_1, v; z_1, v) + I_s(z_1, v; z_0, 0), \tag{5.52}$$

where the functions $I_c(z_1, v_1; z_2, v_2)$ and $I_s(z_1, v_1; z_2, v_2)$ are the same as in (5.45). Owing to the symmetry of the problem,

$$I_s(-z_0, 0; -z_1, v) = -I_s(z_1, v; z_0, 0), \quad I_c(-z_0, 0; -z_1, v) = I_c(z_1, v; z_0, 0),$$

$$I_s(-z_1, v; z_1, v) = 0,$$

$$I_c(-z_1, v; z_1, v) = \frac{2\beta}{(1 - \beta n \cos \theta)} \sin \left[\frac{\omega z_1}{v} (1 - \beta n \cos \theta) \right]. \tag{5.53}$$

Using (5.50) we evaluated a number of angular dependences for $\beta = 1$ and various values of the non-uniform motion lengths z_1 (Figs. 5.20 and 5.21). Each of these figures contains three curves depicting the total intensity σ_t given by (5.50), its bremsstrahlung part σ_{BS} obtained by dropping in (5.52) the term $I_c(-z_1, v; z_1, v)$ corresponding to the uniform motion on the interval $(-z_1, z_1)$, and the Tamm intensity σ_T obtained by dropping in (5.52) the terms $I_c(-z_0, 0; -z_1, v)$ and $I_c(z_1, v; z_0, 0)$ corresponding to the non-uniform motion. For the motion shown in Fig. 5.15 (b) u_1 and u_2 are given by

$$u_1 = -\sqrt{k(z_0 - z_1)n|\cos \theta|}\frac{1}{\beta n \cos \theta},$$

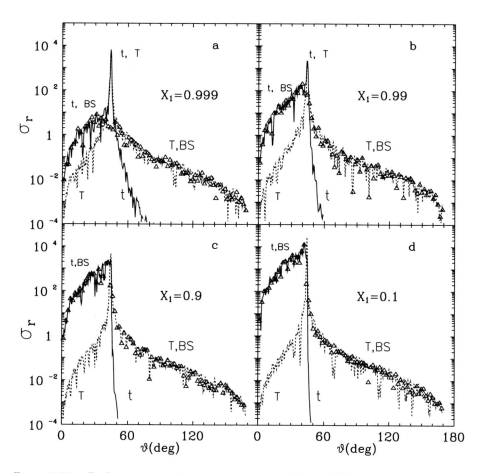

Figure 5.20. Radiation intensities corresponding to Fig. 5.15(b) for $\beta = 1$ and various x_1. Here $x_1 = z_1/z_0$ is the part of a charge trajectory on which it moves uniformly. Other parameters are the same as in Fig. 5.16. Solid and dotted lines refer to the total intensity and the intensity associated with the charge uniform motion in the interval $(-z_1, z_1)$, respectively. Triangles refer to the intensity associated with a charge non-uniform motion on the intervals $(-z_0, -z_1)$ and (z_1, z_0). Since these lines overlap, we have supplied them with letters t (total), T (Tamm) and BS (bremsstrahlung). To make radiation intensities more visible, we have averaged them over three neighbouring points, thus considerably smoothing the oscillations. The same is true for Figs. 5.21 and 5.22. The main maximum of the total radiation intensity is at the Cherenkov angle defined by $\cos\theta = 1/\beta n$. Its sudden drop above this angle is owed to the interference of the VC radiation and BS (see section 5.3.4).

$$u_2 = \sqrt{k(z_0 - z_1)n|\cos\theta|}\left(1 - \frac{1}{\beta n \cos\theta}\right).$$

It follows from this that for $z_1 \to z_0$ (this corresponds to the vanishing interval for the non-uniform motion), $u_1 \to 0$, $u_2 \to 0$ and $I_c(-z_0, 0; -z_1, v)$ and $I_c(z_1, v; z_0, 0)$ also tend to zero (despite that acceleration and deceleration become infinite in this limit), and the whole intensity is reduced to the contribution arising from a charge uniform motion in the interval $(-z_0, z_0)$. The parameter x_1 in Figs. 5.20-5.22 means z_1/z_0. It shows on which part of the total path a charge moves uniformly. For example, $x_1 = 0.999$ means that uniform and non-uniform motions take place on the 0.999 and 0.001 parts of the total motion length, respectively.

We turn to Fig. 5.20(a) corresponding to $x_1 = 0.999$. We see that the total intensity σ_t coincides with the Tamm intensity σ_T only in the immediate neighbourhood of the main maximum (which, in turn, consists of many peaks). To the right of this maximum, the intensity of the BS radiation practically coincides with the Tamm intensity, whilst the total intensity is much smaller. To the left of the main maximum, σ_t practically coincides with σ_{BS}, whilst σ_T is an order smaller. This looks more pronounced for $x_1 = 0.99$, at which the total and BS intensities increase to the left of the main maximum. Let $x_1 = 0.9$ (Fig. 5.20(c)). We observe that σ_{BS} coincides with σ_T to the right of the main maximum and with σ_t to the left of it. At the main maximum σ_t, σ_{BS} and σ_T are of the same order. This picture remains the same for smaller x_1, up to $x_1 = 0.1$ (Fig. 5.20(d)). Beginning from $x_1 = 0.01$, the maximum of the Tamm intensity begins to decrease (Fig. 5.21(a)). This is more pronounced for smaller x_1 (Fig. 5.21(b)) where it is shown that for $x_1 = 0.001$ both σ_T and σ_{BS} begin to oscillate to the right of the main maximum. For very small x_1, σ_T degenerates into

$$\sigma_T(\theta) = \frac{4e^2 n z_1^2}{\lambda^2 c} \sin^2 \theta$$

whilst σ_{BS} coincides with σ_t everywhere except for large angles, where σ_{BS} is very small (Fig. 5.21(c)). Finally, for $x_1 = 0$, $\sigma_T = 0$ and $\sigma_{BS} = \sigma_t$ everywhere (Fig. 5.21 d).

What can we learn from these figures?

1. The total intensity coincides with BS to the left of the main maximum.

2. The Tamm formula satisfactorily describes BS to the right of the main radiation maximum.

3. The Tamm formula coincides with the total intensity only in the immediate vicinity of the main maximum. It disagrees sharply with BS and with the total intensity to the left of the main maximum.

4. The BS maximum is at the angle $\cos\theta = 1/\beta n$ coinciding with the VC radiation angle. This takes place even for Fig. 5.21(d) which describes the accelerated and decelerated charge motions and does not include the uniform motion.

5. The radiation from accelerated and decelerated paths of the charge trajectory tends to zero when the lengths of these paths tend to zero (de-

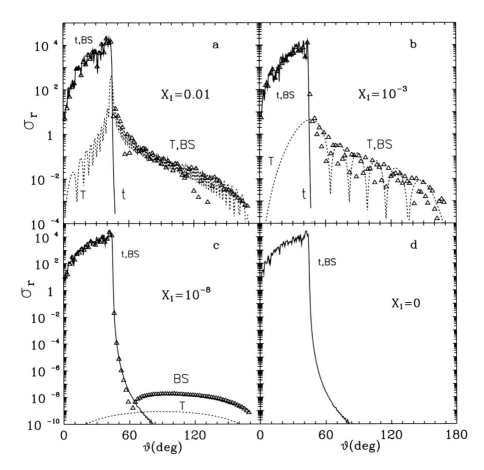

Figure 5.21.   The same as in Fig. 5.20, but for smaller x_1. It is seen that with the diminishing of the uniform motion interval, the Tamm radiation intensity tends to zero, whilst the total intensity approaches the BS intensity. Again, the main maximum of the total radiation intensity is at the Cherenkov angle defined by $\cos\theta = 1/\beta n$.

spite the infinite acceleration and deceleration). There are no jumps of the charge velocity for arbitrarily small (yet, finite) acceleration and deceleration paths. Therefore in this limit the Tamm formula describes the radiation of a charge moving uniformly in the finite interval without recourse to the velocity jumps at the ends of the motion interval. However, some reservation is needed. Although there are no jumps in velocity and the acceleration is everywhere finite for the smoothed Tamm problem, there are jumps in acceleration at the instants corresponding to the beginning and end of motion and at the instants when the uniform and non-uniform charge motions meet with each other. At these instants the third order time derivatives of the charge trajectory are infinite and they, in principle, can give a contribution to the Tamm formula. To exclude this possibility the everywhere continuous charge trajectory should be considered (this will be done below in this chapter).

The problem treated in this section describes the same physical situation as the original Tamm problem. Since the acceleration and deceleration exhibited by a charge are always finite in reality, the problem treated in this section is more physical.

We consider in some detail the relation of the smoothed Tamm problem to the original Tamm problem [1]. If the acceleration and deceleration lengths L of the charge trajectory tend to zero, the total radiation intensity reduces to the integral over the uniform motion interval

$$\sigma_r(\omega, \theta) = \frac{e^2 k^2 n \sin^2 \theta}{4\pi^2 c} (I_c)^2,$$

where

$$I_c = \int\limits_{-z_0}^{z_0} dz' \cos \psi$$

with $\psi = kz'(1/\beta - n \cos \theta)$. Integrating over z', one gets the Tamm angular intensity (5.5). We have seen in Chapter 2, that:

i) in the exactly soluble Tamm problem the BS and Cherenkov shock waves certainly exist in the time and spectral representations;

ii) the approximate Tamm radiation intensity (5.5) contains the BS shock waves and does not describe properly the Cherenkov shock wave originated from the charge motion in the interval $(-z_0, z_0)$.

Let kL be arbitrary small, but finite (L is the length through which a charge moves non-uniformly). The contribution of the accelerated (decelerated) part of the charge trajectory to the radiation intensity then also tends to zero and the total radiation intensity coincides with the approximate Tamm intensity (5.5). Since there are no velocity jumps now, a question arises what kind of radiation contributes to the total intensity. This intriguing situation can be resolved in the following way. Although there are no velocity jumps, there are acceleration jumps at the start and end of the motion, and at the instants when the accelerated part of the charge trajectory meets with the uniform part. We associate the non-vanishing total radiation intensity for $kL \ll 1$ with these acceleration jumps. This is valid only under the approximations (5.2), (5.3), (5.6), and (5.8) which lead to ψ_1 given by (5.41), and which result in the disappearance of the Cherenkov shock wave. As we have seen in Chapters 2 and 3, the Cherenkov shock wave certainly exists in the exactly solvable original and smoothed Tamm problem.

Let the observed wavelength λ lie in the optical region. Then, for $kL \ll 1$, the optical and lower frequencies do not contribute to the integral over the accelerated and decelerated parts of the charge trajectory. However, as we have learned from Chapter 2, the BS shock waves exist even for the instantaneous jumps of the charge velocity. This means that for small acceleration lengths L, the BS shock wave is formed mainly from high frequencies. To see this explicitly, we now fix L and change λ. For $L \ll \lambda$, the total radiation intensity reduces to the Tamm one. On the other hand, for the very short wavelengths satisfying $\lambda \ll L$, both uniform and non-uniform parts of the charge trajectory contribute to the radiation intensity. Analytic estimates made in subsection 5.3.4 confirm this. In fact, the radiation intensity (for $\lambda \ll L$) equals (5.68) for $\theta < \theta_c$ and zero for $\theta > \theta_c$. Here $\cos \theta_c = 1/\beta n$. This radiation intensity disagrees sharply with the Tamm formula (5.5).

It should be noted that in the time representation the space-time evolution of the shock waves arising in the problem treated was studied in the past in [23]. It was shown there that a complex consisting of the Cherenkov shock wave and the shock wave (not BS shock wave) closing the Cherenkov cone is created at the instant when the charge velocity coincides with the velocity of light in medium. On the part of the trajectory, corresponding to the uniform charge motion (Fig. 5.15(b)) the dimensions of this complex grow, but its form remains the same. On the decelerated part of the charge trajectory it leaves the charge at the instant when the charge velocity again coincides with the velocity of light in medium. After this instant, it propagates with the velocity of light in medium. In this section, meeting the experimentalists demands, we have translated results of [23] into the frequency language. In fact, experimentalists ask questions like these: how many photons with frequency ω should be observed, what is their angular distribution? Analytic formulae of this section answer these questions.

More complicated superposition of accelerated, decelerated, and uniform motions

We also consider another problem corresponding to the motion shown in Fig. 5.15(c). This is needed to investigate how the radiation intensity looks when the velocity v_2 changes from the value above c_n to the value below it. A charge is at rest at the spatial point $z = -z_0$ up to an instant $t = -t_0$. In the time interval $-t_0 < t < -t_1$ it moves with an acceleration a up to reaching the velocity v_1 at the spatial point $z = -z_1$:

$$z = -z_0 + \frac{1}{2}a(t + t_0)^2, \quad v = a(t + t_0).$$

It is convenient to express t_1 and a through z_1 and v_1:

$$a = \frac{v_1^2}{2(z_0 - z_1)}, \qquad t_0 - t_1 = \frac{2(z_0 - z_1)}{v_1}.$$

In the time interval $-t_1 < t < -t_2$ a charge moves with deceleration a down to reaching the velocity v_2 at the spatial point $z = -z_2$:

$$z = -z_1 + v_1(t + t_1) - \frac{1}{2}a(t + t_1)^2, \quad v = a(t + t_1).$$

It is convenient to express t_2 and z_2 through v_2:

$$z_2 = z_0 - (z_0 - z_1)(2 - \frac{\beta_2^2}{\beta_1^2}), \quad t_2 = t_0 - 2\frac{2v_1 - v_2}{v_1^2}(z_0 - z_1). \qquad (5.54)$$

In the time interval $-t_2 < t < t_2$ a charge moves uniformly with the velocity v_2 up to reaching the spatial point $z = z_2$:

$$z = -z_2 + v_2(t + t_2), \quad v = v_2.$$

Therefore $z_2 = v_2 t_2$. Substituting z_2 and t_2 from (3.10) we find t_0

$$t_0 = \frac{1}{v_2}[z_0 - (z_0 - z_1)(2 - 4\frac{v_2}{v_1} + \frac{v_2^2}{v_1^2})].$$

In the time interval $t_2 < t < t_1$ a charge moves with acceleration a up to reaching the velocity v_1 at the spatial point $z = z_1$:

$$z = z_2 + v_2(t - t_2) + \frac{1}{2}a(t - t_2)^2, \quad v = v_2 + a(t - t_2).$$

Finally, in the time interval $t_1 < t < t_0$ a charge moves with deceleration a down to reaching the state of rest at the instant t_0 at the spatial point $z = z_0$:

$$z = z_1 + v_1(t - t_1) - \frac{1}{2}a(t - t_1)^2, \quad v = v_1 - a(t - t_1).$$

After the instant t_0, the charge is at rest at the point $z = z_0$. For that motion the Fourier transform of the current density reduces to the following sum

$$j_\omega = \frac{e}{2\pi}\delta(x)\delta(y)[\Theta(z + z_0)\Theta(-z - z_1)\exp(-i\omega\tau_1)$$

$$+\Theta(z + z_1)\Theta(-z - z_2)\exp(-i\omega\tau_2) + \Theta(z + z_2)\Theta(z_2 - z)\exp(-i\omega\tau_3)$$

$$+\Theta(z - z_2)\Theta(z_1 - z)\exp(-i\omega\tau_4) + \Theta(z - z_1)\Theta(z_0 - z)\exp(-i\omega\tau_5)],$$

where

$$\tau_1 = -t_0 + \frac{2}{v_1}\sqrt{(z+z_0)(z_0-z_1)},$$

$$\tau_2 = -t_0 + \frac{2}{v_1}[2(z_0-z_1) - \sqrt{(z_0-z_1)(z_0-z-2z_1)}],$$

$$\tau_3 = \frac{z}{v_2}, \quad \tau_4 = t_0 - \frac{2}{v_1}[2(z_0-z_1) - \sqrt{(z_0-z_1)(z_0+z-2z_1)}],$$

$$\tau_5 = t_0 - \frac{2}{v_1}\sqrt{(z_0-z)(z_0-z_1)}. \tag{5.55}$$

If the conditions (5.2),(5.3),(5.6) and (5.7) are satisfied then the radiation intensity can be evaluated analytically:

$$\sigma_r(\omega,\theta) = \frac{e^2 \sin^2\theta}{n\pi^2 c}[(I_c)^2 + (I_s)^2], \tag{5.56}$$

where:

$$I_c = I_c(-z_0,0;-z_1,v_1) + I_c(-z_1,v_1;-z_2,v_2) + I_c(-z_2,v_2;z_2,v_2)$$

$$+I_c(z_2,v_2;z_1,v_1) + I_c(z_1,v_1;z_0,0),$$

$$I_s = I_s(-z_0,0;-z_1,v_1) + I_s(-z_1,v_1;-z_2,v_2) + I_s(-z_2,v_2;z_2,v_2)$$

$$+I_s(z_2,v_2;z_1,v_1) + I_s(z_1,v_1;z_0,0). \tag{5.57}$$

Again, owing to the symmetry of the problem

$$I_c(-z_0,0;-z_1,v_1) = I_c(z_1,v_1;z_0,0),$$

$$I_c(-z_1,v_1;-z_2,v_2) = I_c(z_2,v_2;z_1,v_1),$$

$$I_c(-z_2,v_2;z_2,v_2) = \frac{2\beta_2}{(1-\beta_2 n\cos\theta)}\sin\left[\frac{\omega z_2}{v_2}(1-\beta_2 n\cos\theta)\right],$$

$$I_s(-z_0,0;-z_1,v_1) = -I_s(z_1,v_1;z_0,0),$$

$$I_s(-z_1,v_1;-z_2,v_2) = -I_s(z_2,v_2;z_1,v_1),$$

$$I_s(-z_2,v_2;z_2,v_2) = 0, \quad I_s = 0.$$

Now we choose $\beta_1 = 1$, $x_1 = 0.99$, and change β_2. The case $\beta_2 = 1$ is shown in Fig. 5.20(b). Smaller values of β_2 are shown in Fig. 5.22. Consider Fig. 5.22(a), corresponding to $\beta_2 = 0.8$. We see two Cherenkov maxima at the angles $\theta_1 = \arccos(1/\beta_1 n)$ and $\theta_2 = \arccos(1/\beta_2 n)$. As in Figs. 5.20 and 5.21, we observe that the Tamm formula satisfactorily describes BS in the back part of the angular spectrum (for $\beta = 0.8$ this agreement begins from

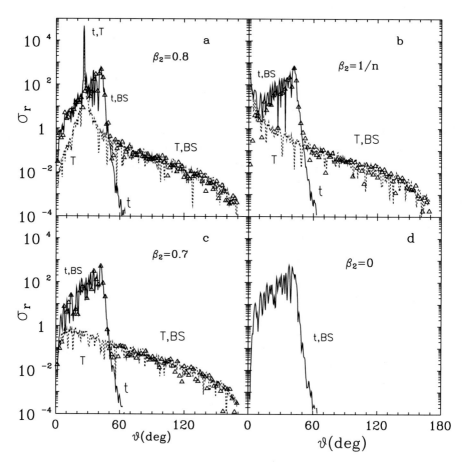

Figure 5.22. Total, Tamm, and BS radiation intensities corresponding to Fig. 5.15(c) for $\beta = 1, x_1 = 0.99$ and various β_2. The case $\beta_2 = 1$ is considered in Fig. 5.20(b). Other parameters are the same as in Fig. 5.16. For β_1 and β_2 greater than $1/n$ the total intensity has two maxima at the Cherenkov angles defined by $\cos\theta = 1/\beta_1 n$ and $\cos\theta = 1/\beta_2 n$ (a,b). At the Cherenkov threshold these maxima have the same height. For $\beta_2 < 1/n$ only one maximum corresponding to $\cos\theta = 1/\beta_1 n$ survives (c,d). For $\beta_2 = 0$ the Tamm intensity is zero, and $\sigma_t = \sigma_{BS}$.

$\theta \approx 50^0$). The total intensity is satisfactorily reproduced by the BS intensity everywhere in the front angular region $(0 < \theta < 50^0)$ except for the immediate neighbourhood of the Cherenkov angle. In this angular region the Tamm formula disagrees both with the total and BS intensities everywhere except for angles close to the Cherenkov angle. An important case is $\beta_2 = 1/n$ corresponding to the Cherenkov threshold (Fig. 5.22 b). The total intensity has two maxima of the same magnitude: one corresponding to the Cherenkov maximum (at $\theta = 0^0$) and other corresponding to the BS maximum. For β_2 below the Cherenkov threshold one Cherenkov maxima disappears (Fig. 5.22(c)), whilst the Tamm intensity decreases coinciding at large angles with the intensity of BS. In the forward direction the total intensity does not differ from the BS intensity. Finally, for $\beta_2 = 0$ the Tamm intensity disappears, whilst the total intensity coincides with the BS intensity (Fig. 5.22(d)).

What can we learn from this section? There are two characteristic velocities β_1 and β_2 in Fig. 5.22. Correspondingly, there are two Cherenkov maxima defined by $\cos\theta = 1/\beta_1 n$ and $\cos\theta = 1/\beta_1 n$ when both β_1 and β_1 are greater than $1/n$ (Fig. 5.22 (a,b)). When β_2 becomes smaller than $1/n$, only one Cherenkov maximum corresponding to $\cos\theta = 1/\beta_1 n$ survives (Fig. 5.22(c,d)).

5.3.4. ANALYTIC ESTIMATES

In this section the radiation intensities written out in a previous section in terms of Fresnel integrals will be expressed through elementary functions. This is possible when the arguments of Fresnel integrals are large. Physically this means that the product kl_a is large (k is the wave number and l_a is the spatial interval in which a charge moves non-uniformly). For large arguments, $C(x)$ and $S(x)$ behave as

$$C(x) \rightarrow \frac{1}{2} + \frac{1}{\sqrt{2\pi}}\frac{\sin x^2}{x}, \quad S(x) \rightarrow \frac{1}{2} - \frac{1}{\sqrt{2\pi}}\frac{\cos x^2}{x}$$

for $x \rightarrow +\infty$ and

$$C(x) \rightarrow -\frac{1}{2} + \frac{1}{\sqrt{2\pi}}\frac{\sin x^2}{x}, \quad S(x) \rightarrow -\frac{1}{2} - \frac{1}{\sqrt{2\pi}}\frac{\cos x^2}{x}$$

for $x \rightarrow -\infty$.

Pure decelerated motion
For the decelerated motion shown in Fig. 5.15 (a) and corresponding to $\beta_1 n > 1$ and $\beta_2 n > 1$, one finds that for $k(z_2 - z_1) \gg 1$ the radiation intensity is given by:

$$\sigma_r = \frac{e^2 n \sin^2\theta}{\pi^2 c} \left\{ \frac{1}{4} \left[\frac{\beta_2 - \beta_1}{(1 - \beta_1 n \cos\theta)(1 - \beta_2 n \cos\theta)} \right]^2 \right.$$

$$\left. + \frac{\beta_1 \beta_2}{(1 - \beta_1 n \cos\theta)(1 - \beta_2 n \cos\theta)} \sin^2\psi \right\} \tag{5.58}$$

for $0 < \theta < \theta_2$ and $\theta > \theta_1$. Here we put

$$\cos\theta_1 = 1/\beta_1 n, \quad \cos\theta_2 = 1/\beta_2 n, \quad \psi = \frac{k(z_2 - z_1)}{\beta_1 + \beta_2}(\frac{\beta_1 + \beta_2}{2}n\cos\theta - 1).$$

On the other hand, for $\theta_2 < \theta < \theta_1$ one has

$$\sigma_r = \sigma_r(5.58) + \frac{e^2 \sin^2\theta}{\pi c n \cos^2\theta}$$

$$\times \left[\alpha^2 + \frac{\alpha n \cos\theta}{\sqrt{2\pi}}\left(\beta_2 \frac{\cos u_2^2 - \sin u_2^2}{\beta_2 n \cos\theta - 1} - \beta_1 \frac{\cos u_1^2 - \sin u_1^2}{\beta_1 n \cos\theta - 1}\right)\right], \qquad (5.59)$$

where α, u_1 and u_2 are the same as in (5.46). The term proportional to α^2 is much larger than others everywhere except for the angles close to θ_1 and θ_2. For these angles the above expansion of Fresnel integrals fails (since u_1 and u_2 vanish at these angles). These formulae mean that radiation intensity oscillates with decreasing amplitude for $0 < \theta < \theta_2$ and $\theta > \theta_1$ (oscillations are due to $\sin^2\psi$), and has a plateau

$$\frac{e^2\alpha^2 \sin^2\theta}{\pi c n \cos^2\theta} \qquad (5.60)$$

for $\theta_2 < \theta < \theta_1$. The oscillating terms (the first term in (5.59) and the term proportional to α) are much smaller than the non-oscillating term (5.60). Exactly such behaviour of σ_r with maxima at θ_1 and θ_2 and a rather flat region between them demonstrates Fig. 5.16 (b).

For $\beta_2 = 1/n$ these formulae predict intensity oscillations for $\theta > \theta_1$ and their absence for $\theta < \theta_1$ (see Fig. 5.16 (c)).

A particular interesting case having numerous practical applications corresponds to the complete termination of motion ($\beta_2 = 0$). In this case

$$\sigma_r = \frac{e^2 n \beta_1^2}{4\pi^2 c} \frac{\sin^2\theta}{(1 - \beta_1 n \cos\theta)^2} \qquad (5.61)$$

for $\theta > \theta_1$ and

$$\sigma_r = \sigma_r(5.61) + \frac{e^2 \sin^2\theta}{\pi c n \cos^2\theta}\left(\alpha^2 - \frac{\beta_1 \alpha n \cos\theta}{\sqrt{2\pi}}\frac{\cos u_1^2 - \sin u_1^2}{\beta_1 n \cos\theta - 1}\right) \qquad (5.62)$$

for $\theta < \theta_1$. Here α and u_1 are the same as in (5.46) if one puts $\beta_2 = 0$ in them:

$$\alpha = \frac{1}{\beta_1}\sqrt{\frac{k(z_2 - z_1)}{n\cos\theta}}, \quad u_1 = \sqrt{k(z_2 - z_1)n\cos\theta}\left(1 - \frac{1}{\beta_1 n \cos\theta}\right).$$

Since $\alpha \gg 1$, the radiation intensity for $\theta > \theta_1$ is much smaller than for $\theta < \theta_1$. There are no intensity oscillations for $\theta > \theta_1$ and very small oscillations for $\theta < \theta_1$ (they are owed to the last term in (5.62) proportional α which is much smaller than the term proportional α^2). Figures 5.16(d) and 5.17 agree with this prediction. When $\beta_1 n < 1$ the same Eq. (5.61) is valid for all angles. In this case, the integration over the solid angle can be performed analytically:

$$\sigma_r(\omega) = \int \sigma_r(\theta, \omega) d\Omega = \frac{2e^2}{\pi c n}(\frac{1}{2\beta_1 n} \ln \frac{1 + \beta_1 n}{1 - \beta_1 n} - 1), \qquad (5.63)$$

that is two times smaller then the Tamm frequency intensity (5.33) This expression is not valid for β_1 close to $1/n$.

The singularities occurring in (5.58), (5.59), (5.61), and (5.62) are owed to the condition $k(z_2 - z_1) \gg 1$ used. The initial radiation intensity (5.46) is finite both for $\cos \theta = 1/\beta_1 n$ and $\cos \theta = 1/\beta_2 n$.

Smoothed Tamm problem
We now evaluate asymptotic radiation intensities for the motion shown in Fig. 5.15(b) (the smooth Tamm problem). For this aim we should evaluate the integrals $I_c = \int v d\tau \cos \psi$ and $I_s = \int v d\tau \sin \psi$ entering (5.50). In terms of Fresnel integrals, they are given by (5.52). Owing to the symmetry of the problem treated, $I_s = 0$ while I_c is reduced to

$$I_c = I_c^a + I_c^d + I_c^u = 2I_c^a + I_c^u. \qquad (5.64)$$

Here I_c^a, I_c^d, and I_c^u are the integrals over the accelerated $(-z_0 < z < -z_1)$, decelerated $(z_1 < z < z_0)$ and uniform $(-z_1 < z < z_1)$ parts of a charge trajectory, respectively. Again, it was taken into account that $I_c^a = I_c^d$ owing to the symmetry of the problem. The integral I_c^u corresponding to the uniform motion on the interval $(-z_1 < z < z_1)$ is

$$I_c^u = \frac{2\beta}{k(1 - \beta n \cos \theta)} \sin[\frac{kz_1}{\beta}(1 - \beta n \cos \theta)]. \qquad (5.65)$$

Then for $\theta < \pi/2$ one has

$$I_c^a = \int_{-z_0}^{-z_1} dz \cos \psi = \frac{1}{kn \cos \theta}\{\sin(u_2^2 - \gamma) - \sin(u_1^2 - \gamma)$$

$$+ \alpha\sqrt{2\pi}[\cos \gamma(C_2 - C_1) + \sin \gamma(S_2 - S_1)]\}. \qquad (5.66)$$

For the motion shown in Fig. 5.15 (b), u_1, u_2, α and γ are given by

$$u_1 = -\sqrt{k(z_0 - z_1)n \cos \theta} \frac{1}{\beta n \cos \theta},$$

$$u_2 = \sqrt{k(z_0 - z_1)n \cos \theta} \left(1 - \frac{1}{\beta n \cos \theta}\right),$$

$$\alpha = \frac{1}{\beta} \left[\frac{k(z_0 - z_1)}{n \cos \theta}\right]^{1/2}, \quad \gamma = kz_0 n \cos \theta + \frac{k(z_0 - z_1)}{\beta^2 n \cos \theta} - \frac{k(2z_0 - z_1)}{\beta}.$$

Replacing Fresnel integrals by their asymptotic values, we obtain for $k(z_0 - z_1) \gg 1$ and $\theta < \theta_c$ ($\cos \theta_c = 1/\beta n$):

$$I_c^a = -\alpha\sqrt{2\pi}\frac{\cos \gamma + \sin \gamma}{kn \cos \theta} + \frac{\beta n}{k(\beta n \cos \theta - 1)} \sin[kz_1(1 - \beta n \cos \theta)]. \quad (5.67)$$

To obtain I_c one should double I_c^a (since $I_c^a = I_c^d$) and add I_c^u given by (5.65). This gives

$$I_c = 2I_c^a + I_c^u = -\alpha\sqrt{2\pi}\frac{\cos \gamma + \sin \gamma}{kn \cos \theta}$$

and

$$\sigma_r = \frac{e^2}{2\pi cn^2\beta^2} k(z_0 - z_1)\frac{\sin^2 \theta}{\cos^3 \theta}(1 + \sin 2\gamma). \quad (5.68)$$

We see that for $\theta < \theta_c$ the part of I_c^a is compensated by the Tamm amplitude I_c^u. In this angular region the oscillations are owed to the $(1+\sin 2\gamma)$ factor. For $\theta > \theta_c$ one finds

$$I_c^a = \frac{\beta n}{k(\beta n \cos \theta - 1)} \sin[kz_1(1 - \beta n \cos \theta)]. \quad (5.69)$$

Inserting (5.65) and (5.69) into (5.64) we find

$$I_c = 2I_c^a + I_c^u = 0 \quad \text{and} \quad \sigma_r = 0.$$

We see that for $\theta > \theta_c$ the total contribution of the accelerated and decelerated parts of the charge trajectory is compensated by the contribution of its uniform part. The next terms arising from the expansion of Fresnel integrals are of the order $1/k(z_0 - z_1)$, and therefore are negligible for $k(z_0 - z_1) \gg 1$. This behaviour of radiation intensities is confirmed by Figs. 5.20 and 5.21 which demonstrate that radiation intensities suddenly drop for $\theta > \theta_c$.

For $\theta > \theta_c$ the radiation intensity disappears for arbitrary z_1 satisfying the condition $k(z_0 - z_1) \gg 1$ and, in particular, for $z_1 = 0$. In this case, there is no uniform motion, and the accelerated motion in the interval $-z_0 < z < 0$ is followed by the decelerated motion in the interval $0 < z < z_0$. The radiation intensity is obtained from (5.68) by setting $z_1 = 0$ in it. For $kz_0 \gg 1$ it reduces to

$$\sigma_r = \frac{e^2 k z_0 \sin^2 \theta}{2\pi n^2 c \cos^3 \theta}(1 + \sin 2\gamma) \tag{5.70}$$

for $\theta < \theta_c$. Here $\gamma = (1 - 1/\beta n \cos \theta)^2 k z_0 n \cos \theta$. For $\theta > \theta_c$, σ_r is small (it is of the order $1/kz_0$). Owing to the factor $(1 + \sin 2\gamma)$, σ_r is a fast oscillating function of θ for $\theta < \theta_c$ (see Fig. 5.21(d)) with a large amplitude (since $kz_0 \gg 1$). Fig 5.21 (d) confirms this.

For $\beta n < 1$ the condition $\theta < \theta_c$ cannot be satisfied and radiation intensities are of the order $1/k(z_0 - z_1) \ll 1$ for all angles.

In the opposite case ($kz_0 \to 0$), the radiation intensity tends to zero:

$$\sigma_r = \frac{e^2 \mu n k^2 z_0^2 \sin^2 \theta}{\pi^2 c}. \tag{5.71}$$

This particular case indicates that the disappearance of radiation intensities at high frequencies above some critical angle has a more general reason. It will be shown in the next two subsections that radiation intensities describing the absolutely continuous charge motion in medium are exponentially small outside some angular region. It should be stressed again that formulae obtained in this section are not valid near the angle θ_c where the arguments of the Fresnel integrals vanish.

5.3.5. THE ABSOLUTELY CONTINUOUS CHARGE MOTION.

When the conditions (5.2), (5.3), and (5.8) are satisfied, the vector potential (3.1) is reduced to

$$A_\omega = \frac{\mu e}{2\pi cr} \exp(iknr)I, \tag{5.72}$$

where

$$I = \int v(t') \exp[i(\omega t' - kn \cos \theta z(t'))]dt'.$$

Electromagnetic field strengths contributing to the radial energy flux are

$$H_\phi = -\frac{ienk \sin \theta}{2\pi cr} \exp(iknr)I, \quad E_\theta = -\frac{ie\mu k \sin \theta}{2\pi cr} \exp(iknr)I.$$

The radiation intensity is given by

$$\sigma_r(\theta, \omega) = \frac{d^2 \mathcal{E}}{d\omega d\Omega} = \frac{e^2 k^2 n\mu}{4\pi^2 c} \sin^2 \theta |I|^2. \tag{5.73}$$

This means that all information on the radiation intensity is contained in I. In the quasi-classical approximation,

$$I = v(t_c)\sqrt{\frac{2\pi}{|\dot{v}(t_c)kn\cos\theta|}}\exp(\pm i\pi/4)\exp(i\psi_c), \qquad (5.74)$$

where $\psi_c = \omega t_c - kn z_c \cos\theta$, $z_c = z(t_c)$ and t_c is found from the equation

$$1 - n\beta(t_c)\cos\theta = 0. \qquad (5.75)$$

The $\pm$ signs in (5.74) coincide with the sign of $\dot{v}(t_c)kn\cos\theta$. Under the conditions (5.2), (5.3), and (5.8), the charge uniformly moving in the interval $(-z_0, z_0)$ radiates with the intensity given by the famous Tamm formula (5.5).

Simplest absolutely continuous charge motion.
A charge moves according to the law (Fig. 5.23)

$$v(t) = \frac{v_0}{\cosh^2(t/t_0)}. \qquad (5.76)$$

Obviously $v(t) = v_0$ for $t = 0$ and $v(t) \to 0$ for $t \to \pm\infty$. The charge position at the instant t is given by $z(t) = v_0 t_0 \tanh(t/t_0)$. Therefore the charge motion is confined to $-L/2 < z < L/2$, where $L = 2v_0 t_0$ is the motion interval. The velocity, being expressed through the current charge position, is

$$v(z) = v_0(1 - 4z^2/L^2) \qquad (5.77)$$

The drawback of this motion is that one can not change t_0 without changing the motion interval L.

For the motion law shown in Fig. 5.23, the amplitude I entering in (5.72) is given by

$$I = \frac{\pi v_0 \omega t_0^2}{\sinh(\pi\omega t_0/2)}\exp(i\omega t_0\beta_0 n\cos\theta)$$

$$\times \Phi(1 + i\omega t_0/2; 2; -2i\omega t_0\beta_0 n\cos\theta), \qquad (5.78)$$

where $\Phi(\alpha; \beta; z)$ is the confluent hypergeometric function. Correspondingly the radiation intensity is

$$\sigma_r(\theta, \omega) = \frac{e^2 n\mu\beta_0^2\omega^4 t_0^4}{4c\sinh^2(\pi\omega t_0/2)}|\Phi|^2. \qquad (5.79)$$

When $\omega t_0 \ll 1$,

$$I = 2v_0 t_0 \quad \text{and} \quad \sigma_r(\theta, \omega) = \frac{n\mu}{\pi^2 c}e^2\beta_0^2\omega^2 t_0^2 \sin^2\theta. \qquad (5.80)$$

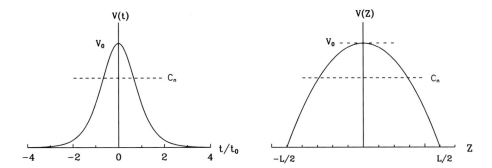

Figure 5.23.   The motion corresponding to (5.76). Left and right parts correspond to $v(t)$ and $v(z)$, where z is the charge position at the time t. It is seen that the charge position is confined to a finite spatial interval $(-L/2, L/2)$.

This coincides with (5.71).

In the opposite case ($\omega t_0 \gg 1$), by applying the quasi-classical approximation one finds that I is exponentially small for all angles if $\beta_0 < 1/n$. If $\beta_0 > 1/n$, I is exponentially small for $\theta > \theta_c$ ($\cos \theta_c = 1/\beta_0 n$) and

$$|I|^2 = \frac{\pi c t_0 \sqrt{\beta_0}}{(n \cos \theta)^{3/2} k (\beta_0 n \cos \theta - 1)^{1/2}} \cos^2 \psi_c \tag{5.81}$$

for $\theta < \theta_c$. Here

$$\psi_c = \omega \left(t_c - \frac{n z_c}{c} \cos \theta \right) + \frac{\pi}{4}, \quad \cosh \frac{t_c}{t_0} = \sqrt{\beta_0 n \cos \theta},$$

$$z_c = v_0 t_0 \left(1 - \frac{1}{n \beta_0 \cos \theta} \right)^{1/2}.$$

When evaluating $|I|^2$ it was taken into account that Eq. (5.75) has two real roots for $\beta_0 n > 1$:

$$t_c = \pm t_0 \left(\sqrt{\beta_0 n \cos \theta} + \sqrt{\beta_0 n \cos \theta - 1} \right).$$

The radiation intensity (5.73), with $|I|^2$ given by (5.81), is the analogue of the Tamm formula (5.5) for the motion law (5.76).

Radiation intensities $\sigma_r(\theta)$ corresponding to the charge motion shown in Fig. 5.23 are presented in Fig. 5.24 for a number of $\beta_0 = v_0/c$ together with the Tamm intensities σ_T corresponding to the same $L = 0.1 cm$, $\lambda = 4 \times 10^{-5}$ cm, $n = 1, 5$ and β_0. It is seen that the positions of main maxima of σ_r and σ_T coincide for $v_0 > c_n$ and are at the Cherenkov angle defined by $\cos \theta_c = 1/\beta_0 n$. For $v_0 < c_n$, σ_r is much smaller than σ_T (d). For $v_0 > c_n$

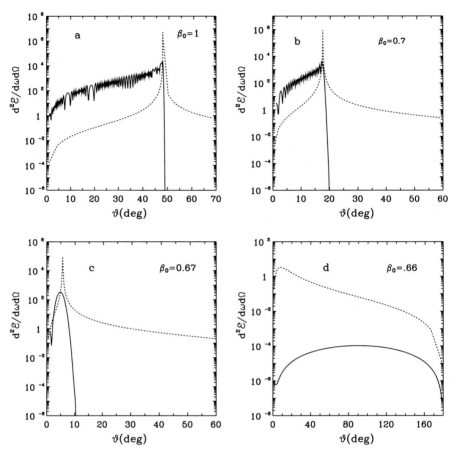

Figure 5.24. Angular radiation intensities corresponding to the charge motion shown in Fig. 5.23 (solid curves) and the Tamm intensities (dotted lines) for a number of v_0. For $v_0 > c_n$ the maximum of intensity is at the Cherenkov angle θ_c defined by $\cos\theta_c = 1/\beta_0 n$. The angle θ_c decreases with decreasing v_0. For $\beta_0 > 1/n$ the radiation intensity falls almost instantly for $\theta > \theta_c$. For $\beta_0 < 1/n$ the radiation intensity is exponentially small for all angles. The original angular intensities are highly oscillating functions. To make them more visible, we draw the Tamm angular intensity through its maxima. Other intensities, for which the maxima positions are not explicitly known, are obtained by averaging over three neighbouring points, thus, considerably smoothing the oscillations. This is valid also for Fig. 5.26.

and $\theta > \theta_c$, σ_r falls very rapidly and σ_T dominates in this angular region (a,b,c). For $\theta < \theta_c$, σ_r is much larger than σ_T (a,b) (except for $\theta = \theta_c$). This is in complete agreement with quasi-classical formula (5.81) which predicts the exponential decrease of σ_r for $\theta > \theta_c$ and its oscillations described by (5.81) for $\theta < \theta_c$.

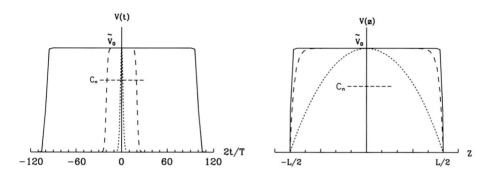

Figure 5.25.   The motion corresponding to (5.82). Dotted, broken and dotted lines correspond to $\tau_0 = T_0/T = 0.5$, 10 and 25, respectively. For large τ_0 the interval where a charge moves with almost constant velocity increases. The charge position is confined to a finite spatial interval $(-L/2, L/2)$. This motion is much richer than the one shown in Fig. 5.23.

In the past, analytical radiation intensities for the charge motion in vacuum shown in Fig. 5.24, were obtained in [24]. In this case (5.75) has no real roots, and at high frequencies the quasi-classical radiation intensity is exponentially small for all angles.

More complicated absolutely continuous charge motion.
A charge moves according to the law (Fig. 5.25)

$$v = \frac{1}{2}v_0\left(\tanh\frac{t+T_0}{T} - \tanh\frac{t-T_0}{T}\right),\tag{5.82}$$

The maximal velocity (at $t = 0$) is $\tilde{v}_0 = v_0\tanh(T_0/T)$. Equation (5.82) is slightly inconvenient. When we change either T or T_0, the maximal velocity, the interval to which the motion is confined, and the behaviour of the velocity inside this interval are also changed. We rewrite this expression in a slightly different form, more suitable for applications

$$v(t) = \tilde{v}_0\frac{1 + \cosh(2T_0/T)}{\cosh(2t/T) + \cosh(2T_0/T)},\tag{5.83}$$

The charge position at the instant t is given by

$$z(t) = \frac{LT}{4T_0}\ln\frac{\cosh(t+T_0)/T}{\cosh(t-T_0)/T},\tag{5.84}$$

where

$$L = 2v_0T_0 = 2\tilde{v}_0T_0\coth(T_0/T)\tag{5.85}$$

is the motion interval. We reverse this expression, thus obtaining

$$\frac{T_0}{T} = \frac{1}{2}\ln\frac{1 + 2T_0\tilde{v}_0/L}{1 - 2T_0\tilde{v}_0/L}. \tag{5.86}$$

It is seen that the fixing of $\tilde{v}_0$ and L leaves only one free parameter. If we identify it with T_0 then (5.86) defines T as a function of T_0 (for the fixed L and $\tilde{v}_0$). For $T_0 \ll T$ the r.h.s. of (5.86) should also be small. This is possible if $2T_0\tilde{v}_0/L \ll 1$. The r.h.s. of (5.86) then tends to $2T_0\tilde{v}_0/L$. Equating both sides of (5.86) we find that $T = L/2\tilde{v}_0$ in this limit. For $T_0 \to L/2\tilde{v}_0$ the r.h.s. of (5.86) tends to ∞. Therefore $T/T_0 \to 0$. It follows from this that the available interval for T and T_0 is $(0, L/2\tilde{v}_0)$ (for fixed L and $\tilde{v}_0$). We express the charge velocity through its current position z. For this we first express $\cosh(2t/T)$ through z:

$$\cosh\frac{2t}{T} = \frac{\sinh[T_0(1 + 2z/L)/T]}{2\sinh[T_0(1 - 2z/L)/T]} + \frac{\sinh[T_0(1 - 2z/L)/T]}{2\sinh[T_0(1 + 2z/L)/T]}. \tag{5.87}$$

Substituting this into (5.83) we obtain $v(z)$. For $T_0 \ll T$, $v(z)$ reduces to

$$v(z) = \tilde{v}_0(1 - \frac{4z^2}{L^2}), \tag{5.88}$$

which coincides with (5.77) if we identify $\tilde{v}_0$ with v_0. In the opposite case $(T \ll T_0)$

$$v(z) = \frac{\tilde{v}_0}{1 + \exp(-2T_0/T)\cosh(2t/T)}. \tag{5.89}$$

If z is so close to $(L/2)$ that

$$1 - \frac{2z}{L} \ll \frac{T}{T_0},$$

then (5.87) gives

$$\cosh\frac{2t}{T} = \frac{T}{4T_0(1 - 2z/L)}\exp\left[\frac{T_0}{T}(1 + 2z/L)\right]$$

and

$$v(z) = \frac{2\tilde{v}_0 T_0}{T}(1 - 2z/L). \tag{5.90}$$

On the other hand, if

$$1 - \frac{2z}{L} \gg \frac{T}{T_0}$$

then

$$\cosh(2t/T) = 1 \quad \text{and} \quad v = \tilde{v}_0. \tag{5.91}$$

Since, according to our assumption, $T/T_0 \ll 1$, the transition from (5.90) to (5.91) is realized in a very narrow z interval. For example, for $T/T_0 = 10^{-6}$, it takes place in the interval $(1 - 10^{-5}) < 2z/L < (1 - 10^{-7})$. The same considerations are valid in the neighbourhood of another boundary point $z = -L/2$. We conclude: the horizontal part (where $v \approx \tilde{v}_0$) of the charge trajectory exists if $T \ll T_0$ (see (5.91)) and does not exist if $T_0 \ll T$ (see (5.88)). However, in both cases ($T \ll T_0$ and $T \gg T_0$) $v(z)$ decreases linearly when z approaches boundary points.

The law of motion (5.83) is much richer than (5.76). It is extensively used in nuclear physics to parametrize the nuclear densities [25,26].

For the law of motion shown in Fig. 5.25 the amplitude I entering into (5.72) equals

$$I = \frac{1}{2} v_0 T \exp[-i\omega T_0(1 - \beta_0 n \cos \theta)] \frac{\omega \pi T/2}{\sinh(\omega \pi T/2)} [1 - \exp(-4T_0/T)]$$

$$\times {}_2F_1[1 - \frac{i\omega T}{2} \beta_0 n \cos \theta, 1 + \frac{i\omega T}{2}; 2; 1 - \exp(-4T_0/T)]. \tag{5.92}$$

Here ${}_2F_1(\alpha, \beta; \gamma; z)$ is the usual hypergeometric function. The radiation intensity is

$$\sigma_r(\theta, \omega) = \frac{e^2 \mu n \beta_0^2 \omega^4 T^4 \sin^2 \theta}{64c \sinh^2(\pi \omega T/2)} [1 - \exp(-4T_0/T)]^2 |F|^2. \tag{5.93}$$

Consider particular cases.

Let T be much smaller than T_0 (ωT is arbitrary). Then,

$$\sigma_r = \frac{e^2 \mu \beta_0 \omega T \sin^2 \theta}{8\pi^3 c \cos \theta (n\beta_0 \cos \theta - 1)}$$

$$\times \frac{\sinh[(\pi \omega T n \beta_0 \cos \theta)/2]}{\sinh(\pi \omega T/2) \sinh[\pi \omega T (\beta_0 n \cos \theta - 1)/2]}. \tag{5.94}$$

If, in addition, the frequency is so large that $\omega T \gg 1$, then (5.94), for $\beta_0 n < 1$, is exponentially small for all angles:

$$\sigma_r = \frac{e^2 \mu \beta_0 \omega T}{4\pi^3 c \cos \theta (1 - n\beta_0 \cos \theta)} \sin^2 \theta \exp[-\pi \omega T (1 - n\beta_0 \cos \theta)]. \tag{5.95}$$

For $\beta_0 n > 1$ and $\theta > \theta_c$ ($\cos \theta_c = 1/\beta_0 n$), the radiation intensity (5.94) coincides with (5.95). On the other hand, for $\theta < \theta_c$

$$\sigma_r = \frac{e^2 \mu \beta_0 \omega T}{4\pi^3 c \cos \theta (n\beta_0 \cos \theta - 1)} \sin^2 \theta. \tag{5.96}$$

In this angular region there is no exponential damping.

Let T_0 be much smaller than T. We should first express v_0 through $\tilde{v}_0$ and then take the limit $T_0/T \to 0$. The hypergeometric function $_2F_1$ is then transformed into the confluent hypergeometric function Φ, and (5.93) is transformed into (5.79) if we identify T and $\tilde{v}_0$ entering (5.93) (after expressing v_0 through $\tilde{v}_0$) with t_0 and v_0 entering (5.79).

In the limit $\omega T \to 0$, (5.93) goes into

$$\sigma_r = \frac{e^2 \mu n \beta_0^2 \omega^2 T_0^2}{\pi^2 c} \sin^2 \theta,$$

which coincides with (5.80) and (5.71). The quasi-classical approximation being applied to I gives

$$\sigma_r(\theta, \omega) = \frac{e^2 \tilde{\beta}_0 \omega T \mu \sin^2 \theta}{4\pi^2 c s_c \cos \theta} \cos^2 \psi_c \tag{5.97}$$

for $\theta < \theta_c$ and σ_r is exponentially small outside this angular region. Here

$$s_c = (n\tilde{\beta}_0 \cos\theta - 1)^{1/2} \left(n\tilde{\beta}_0 \cos\theta - \tanh^2 \frac{T_0}{T} \right)^{1/2}, \quad \psi_c = \omega t_c - knz_c \cos\theta + \frac{\pi}{4};$$

t_c is found from the equation

$$\cosh \frac{2t_c}{T} = \tilde{\beta}_0 n \cos\theta \left(1 + \cosh \frac{2T_0}{T} \right) - \cosh \frac{2T_0}{T},$$

where $z_c = z(t_c)$, and z is given by (5.84).

Unfortunately, we have not succeeded in obtaining the Tamm formula (5.5) from the radiation intensity (5.93). It should appear in the limit $T/T_0 \to 0$ (when the horizontal part of the charge trajectory (where $v \approx \tilde{v}_0$) is large). Equations (5.94) and (5.96) are infinite at the Cherenkov angle, but do not oscillate, contrary to the Tamm intensity (5.5). The quasi-classical expression (5.97) oscillates, but it is also infinite at the Cherenkov angle (again, contrary to the Tamm intensity). Probably, the inability to obtain the Tamm formula (5.5) from (5.93) in the limit $T/T_0 \to 0$ (when the dependence v(z), given by (5.83), is visually indistinguishable from that of the Tamm (see Fig. 5.25)) points to the importance of the velocity discontinuities. In fact, there are two velocity jumps in the Tamm problem and no velocity jumps for the absolutely continuous motion shown in Fig. 5.25.

Radiation intensities $\sigma_r(\theta)$ corresponding to Fig. 5.25, for fixed $\beta_0 = 1$, $L = 0.1$ cm, $\lambda = 4 \times 10^{-5}$ cm and a number of diffuseness parameters $\tau_0 = T_0/T$, are shown in Fig. 5.26. The positions of the main maxima are

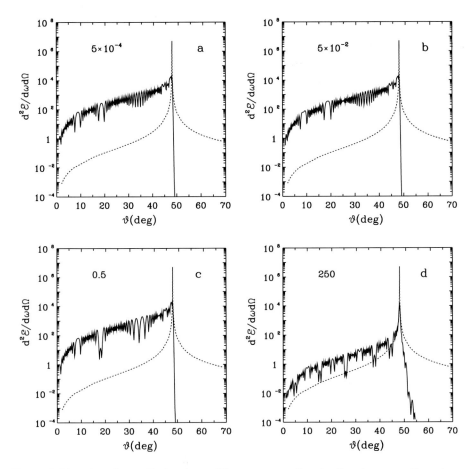

Figure 5.26.   Angular radiation intensities corresponding to the charge motion shown in Fig. 5.25 (solid lines) for $\tilde{\beta}_0 = 1$ and a number of diffuseness parameters $\tau_0 = T_0/T$. Angular intensities approach the Tamm one (dotted line) rather slowly even for large values of τ_0. This is due to their different asymptotic behaviour.

at the Cherenkov angle θ_c. The fast angular oscillations in the region $\theta < \theta_c$ are described by the quasi-classical formula (5.97). Again we observe that σ_r falls almost instantaneously for $\theta > \theta_c$. The reason for this is due to different asymptotical behaviour of radiation intensities which fall exponentially for the absolutely continuous motion presented in Fig. 5.25 and do not decrease with frequency (except for $\cos\theta = 1/\beta n$) for the original Tamm problem involving two velocity jumps.

In the past, analytical radiation intensities for the charge motion in vacuum shown in Fig. 5.25 were obtained in [27] and discussed in [28]. In

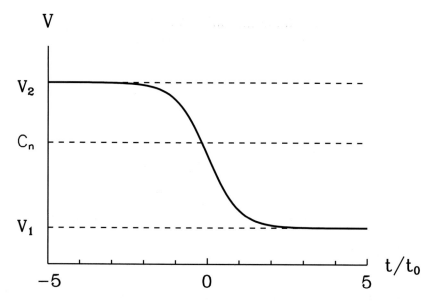

Figure 5.27. The unbounded charge motion corresponding to (5.98) and describing the smooth transition from the velocity v_2 at $t = -\infty$ to the velocity v_1 at $t = \infty$.

this case the radiation intensity at high frequencies is exponentially small for all angles.

Smooth infinite charge motion
Let a charge moves according to the law (Fig. 5.27)

$$v = v_+ + v_- \tanh \frac{t}{t_0}, \quad v_\pm = \frac{v_1 \pm v_2}{2}, \quad -\infty < t < \infty. \qquad (5.98)$$

The current charge position is $z = v_+ t + v_- t_0 \ln \cosh(t/t_0)$. For $t \to \pm\infty$, $v \to v_{1,2}$ and $z \to v_{1,2} t$.

For the motion shown in Fig. 5.27 one obtains

$$I = \frac{c t_0}{2n \cos \theta} \exp\left[\frac{1}{2} i\omega t_0 n(\beta_1 - \beta_2) \cos \theta\right] \frac{\Gamma(\alpha_2)\Gamma(-\alpha_1)}{\Gamma[i\omega t_0 n \cos \theta(\beta_2 - \beta_1)/2]},$$

where $\Gamma(z)$ is the gamma function and

$$\alpha_1 = i\omega t_0(1 - n\beta_1 \cos \theta)/2, \quad \alpha_2 = i\omega t_0(1 - n\beta_2 \cos \theta)/2.$$

Correspondingly,

$$|I|^2 = \frac{c^2 t_0 \pi(\beta_2 - \beta_1)}{2\omega n \cos \theta(1 - n\beta_1 \cos \theta)(1 - n\beta_2 \cos \theta)}$$

$$\times \frac{\sinh[\pi n \cos\theta \omega t_0 (\beta_2 - \beta_1)/2]}{\sinh[\pi(1 - \beta_1 n \cos\theta)\omega t_0/2]\sinh[\pi(1 - \beta_2 n \cos\theta)\omega t_0/2]}$$

and

$$\sigma_r(\theta, \omega) = \frac{e^2 \mu \omega t_0 \sin^2\theta}{8\pi c \cos\theta} F, \tag{5.99}$$

where

$$F = \frac{(\beta_2 - \beta_1)}{(1 - n\beta_1 \cos\theta)(1 - n\beta_2 \cos\theta)}$$

$$\times \frac{\sinh[\pi n \cos\theta \omega t_0 (\beta_2 - \beta_1)/2]}{\sinh[\pi(1 - \beta_1 n \cos\theta)\omega t_0/2]\sinh[\pi(1 - \beta_2 n \cos\theta)\omega t_0/2]}.$$

Here we put $\cos\theta_1 = 1/\beta_1 n$ and $\cos\theta_2 = 1/\beta_2 n$.

High-frequency limit of the radiation intensity. Consider the behaviour of the radiation intensity for $\omega t_0 \gg 1$. Obviously $\theta_1 < \theta_2$ for the decelerated motion $(\beta_2 > \beta_1)$.

Let $\beta_1 n > 1$ and $\beta_2 n > 1$. Then for $\theta < \theta_1$

$$F = \frac{2(\beta_2 - \beta_1)}{(\beta_1 n \cos\theta - 1)(\beta_2 n \cos\theta - 1)} \exp[-\pi\omega t_0(\beta_1 n \cos\theta - 1)]. \tag{5.100}$$

For $\theta > \theta_2$

$$F = \frac{2(\beta_2 - \beta_1)}{(\beta_1 n \cos\theta - 1)(\beta_2 n \cos\theta - 1)} \exp[-\pi\omega t_0(1 - \beta_2 n \cos\theta)]. \tag{5.101}$$

Finally, for $\theta_1 < \theta < \theta_2$

$$F = \frac{2(\beta_2 - \beta_1)}{(1 - \beta_1 n \cos\theta)(\beta_2 n \cos\theta - 1)}. \tag{5.102}$$

We see that two maxima should be observed at the Cherenkov angles θ_1 and θ_2. Between these maxima the radiation intensity is a smooth function of θ.

For $\theta < \theta_1$ and $\theta > \theta_2$, the radiation intensity is exponentially small.

For $\beta_1 n < 1$ and $\beta_2 n > 1$, F is equal to (5.102) for $0 < \theta < \theta_2$, and to (5.101) for $\theta > \theta_2$.

For $\beta_1 n < 1$ and $\beta_2 n < 1$, F has the form (5.101) and the radiation intensity is exponentially small for all angles.

Sharp transition from v_2 to v_1. For $\omega t_0 \ll 1$ (this corresponds either to the sharp change of the charge velocity near $t = 0$ or to large observed wavelengths) one finds

$$\sigma_r(\theta, \omega) = \frac{e^2 n \mu \sin^2 \theta}{4\pi c} \frac{(\beta_2 - \beta_1)^2}{(1 - \beta_1 n \cos \theta)^2 (1 - \beta_2 n \cos \theta)^2}. \qquad (5.103)$$

In this case $\sigma_r(\theta, \omega)$ has two maxima at the Cherenkov angles θ_1 and θ_2 if both $\beta_1 n > 1$ and $\beta_2 n > 1$ and one maximum at θ_2 if $\beta_1 n < 1$ and $\beta_2 n > 1$.

Strictly speaking, the validity of Eqs. (5.99)-(5.103) is slightly in doubt. When obtaining them we used Eq. (5.72) the validity of which implies that a charge motion takes place in an interval much smaller than the radius of the observational sphere S. However, Eq. (5.98) describes the unbounded charge motion. For a sufficiently large time, when a charge will be outside S, the validity of (5.72) will break down.

In the past, analytical radiation intensities for the charge motion in vacuum, shown in Fig. 5.27, were obtained in [29] and discussed in [28]. In this case the radiation intensity at high frequencies is exponentially small for all angles.

5.3.6. SUPERPOSITION OF UNIFORM AND ACCELERATED MOTIONS

To avoid the trouble occurring in a previous subsection, we consider the following problem. A charge is at rest at the point $-z_0$ up to an instant $-t_0$. In the time interval $-t_0 < t < -t_1$, a charge moves uniformly with the velocity v_1 until it reaches the spatial point $-z_1$. In the time interval $-t_1 < t < t_1$ a charge moves with deceleration a until it reaches the spatial point z_1. In the time interval $t_1 < t < t_0'$, a charge moves uniformly with the velocity v_2 until it reaches the spatial point z_0 where it is at rest for $t > t_0'$. It is easy to express t_1, t_0, t_0' and a through z_1, z_0, v_1 and v_2:

$$t_1 = \frac{2z_1}{v_1 + v_2}, \quad t_0 = \frac{z_0}{v_1} + \frac{v_1 - v_2}{v_1 + v_2} \frac{z_1}{v_1}, \quad t_0' = \frac{z_0}{v_2} - \frac{v_1 - v_2}{v_1 + v_2} \frac{z_1}{v_2}, \quad a = \frac{v_1^2 - v_2^2}{4z_1}.$$

The radiation intensity is

$$\sigma_r(\omega, \theta) = \frac{e^2 k^2 n \sin^2 \theta}{4\pi^2 c} [(I_c)^2 + (I_s)^2].$$

Here

$$I_c = \sum_i I_c^{(i)}, \quad I_s = \sum_i I_s^{(i)}$$

and

$$I_c^{(i)} = \int dz' \cos \psi_i, \quad I_s^i = \int dz' \sin \psi_i, \quad i = 1, 2, 3,$$

where $\psi_i = -knz'\cos\theta + \omega\tau_i$. The superscripts $1, 2$ and 3 refer to the uniform motion with the velocity v_1 $(-z_0 < z' < -z_1)$, the decelerated motion $(-z_1 < z' < z_1)$, and to the uniform motion with the velocity v_2 $(z_1 < z' < z_0)$, respectively. The functions $\tau_i(z)$ entering ψ_i are equal to

$$\tau_1 = \frac{z}{v_1} - \frac{z_1}{v_1}\frac{v_1 - v_2}{v_1 + v_2}, \quad \tau_2 = \frac{2z_1}{v_1 - v_2} - \sqrt{\frac{8z_1}{v_1^2 - v_2^2}}\sqrt{z_1\frac{v_1^2 + v_2^2}{v_1^2 - v_2^2}} - z,$$

$$\tau_3 = \frac{z}{v_2} - \frac{z_1}{v_2}\frac{v_1 - v_2}{v_1 + v_2}.$$

The integrals I_c and I_s are given by

$$I_c = \frac{2\beta_1}{k(1 - \beta_1 n\cos\theta)}\sin\left[\frac{k(z_0 - z_1)}{2}\left(\frac{1}{\beta_1} - n\cos\theta\right)\right]$$

$$\times\cos\left[\frac{k(z_0 + z_1)}{2}\left(\frac{1}{\beta_1} - n\cos\theta\right) + \alpha_1\right]$$

$$+ \frac{2\beta_2}{k(1 - \beta_2 n\cos\theta)}\sin\left[\frac{k(z_0 - z_1)}{2}\left(\frac{1}{\beta_2} - n\cos\theta\right)\right]$$

$$\times\cos\left[\frac{k(z_0 + z_1)}{2}\left(\frac{1}{\beta_2} - n\cos\theta\right) - \alpha_2\right]$$

$$- \frac{2}{kn\cos\theta}\sin\left[kz_1\left(\frac{2}{\beta_1 + \beta_2} - n\cos\theta\right)\right]$$

$$- \frac{2}{k(n\cos\theta)^{3/2}}\sqrt{\frac{\pi kz_1}{\beta_1^2 - \beta_2^2}}[\cos\gamma(C_2 - C_1) - \sin\gamma(S_2 - S_1)],$$

$$I_s = \frac{2\beta_2}{k(1 - \beta_2 n\cos\theta)}\sin\left[\frac{k(z_0 - z_1)}{2}\left(\frac{1}{\beta_2} - n\cos\theta\right)\right]$$

$$\times\sin\left[\frac{k(z_0 + z_1)}{2}\left(\frac{1}{\beta_2} - n\cos\theta\right) - \alpha_2\right]$$

$$- \frac{2\beta_1}{k(1 - \beta_1 n\cos\theta)}\sin\left[\frac{k(z_0 - z_1)}{2}\left(\frac{1}{\beta_1} - n\cos\theta\right)\right]$$

$$\times\sin\left[\frac{k(z_0 + z_1)}{2}\left(\frac{1}{\beta_1} - n\cos\theta\right) + \alpha_1\right]$$

$$- \frac{2}{k(n\cos\theta)^{3/2}}\sqrt{\frac{\pi kz_1}{\beta_1^2 - \beta_2^2}}[\cos\gamma(S_2 - S_1) + \sin\gamma(C_2 - C_1)]. \qquad (5.104)$$

Here

$$C_1 = C(u_1), \quad C_2 = C(u_2), \quad \alpha_1 = \frac{v_1 - v_2}{v_1 + v_2}\frac{kz_1}{\beta_1}, \quad \alpha_2 = \frac{v_1 - v_2}{v_1 + v_2}\frac{kz_1}{\beta_2},$$

$$\gamma = -\frac{kz_1 n \cos\theta}{\beta_1^2 - \beta_2^2}\left[\left(\beta_1 - \frac{1}{n\cos\theta}\right)^2 + \left(\beta_2 - \frac{1}{n\cos\theta}\right)^2\right],$$

$$u_1 = \sqrt{\frac{2kz_1 n \cos\theta}{\beta_1^2 - \beta_2^2}}\left(\beta_1 - \frac{1}{n\cos\theta}\right), \quad u_2 = \sqrt{\frac{2kz_1 n \cos\theta}{\beta_1^2 - \beta_2^2}}\left(\beta_2 - \frac{1}{n\cos\theta}\right).$$

$C_1 = C(u_1)$, $C_2 = C(u_2)$, $S_1 = S(u_1)$, $S_2 = S(u_2)$ are the Fresnel integrals.

Particular cases
Sharp transition between velocities. When the transition from v_1 to v_2 is very sharp ($kz_1 \ll 1$), one gets

$$I_c^2 + I_s^2 = \frac{4}{k^2}\{\frac{1}{(1/\beta_1 - n\cos\theta)^2}\sin^2\left[\frac{kz_0}{2}\left(\frac{1}{\beta_1} - n\cos\theta\right)\right]$$

$$+\frac{1}{(1/\beta_2 - n\cos\theta)^2}\sin^2\left[\frac{kz_0}{2}\left(\frac{1}{\beta_2} - n\cos\theta\right)\right]$$

$$+\frac{2}{(1/\beta_1 - n\cos\theta)(1/\beta_2 - n\cos\theta)}\sin\left[\frac{kz_0}{2}\left(\frac{1}{\beta_1} - n\cos\theta\right)\right]$$

$$\times\sin\left[\frac{kz_0}{2}\left(\frac{1}{\beta_2} - n\cos\theta\right)\right]\cos\left[\frac{kz_0}{2}\left(\frac{1}{\beta_1} + \frac{1}{\beta_2} - 2n\cos\theta\right)\right]\}. \quad (5.105)$$

That is, the radiation intensity reduces to the sum of the Tamm intensities for v_1 and v_2 and to their interference.

High frequency limit. In the high-frequency limit ($kz_1 \gg 1$), one gets

$$I_c^2 + I_s^2 = \frac{1}{k^2}\times[\frac{1}{(1/\beta_1 - n\cos\theta)^2} + \frac{1}{(1/\beta_2 - n\cos\theta)^2}$$

$$-\frac{2\cos\psi}{(1/\beta_1 - n\cos\theta)(1/\beta_2 - n\cos\theta)}] \quad (5.106)$$

for $\theta > \theta_1$ and $\theta < \theta_2$ and

$$I_c^2 + I_s^2 = (5.106) + \frac{1}{k^2}$$

$$\times\left[\frac{4\pi\alpha^2}{n^2\cos^2\theta} + \frac{2\sqrt{2\pi}\alpha}{n\cos\theta}\left(\frac{\sin\gamma_1 + \cos\gamma_1}{1/\beta_1 - n\cos\theta} + \frac{\sin\gamma_2 - \cos\gamma_2}{1/\beta_2 - n\cos\theta}\right)\right] \quad (5.107)$$

for $\theta_2 < \theta < \theta_1$. Here we put

$$\alpha = \sqrt{\frac{2kz_1}{n(\beta_1^2 - \beta_2^2)\cos\theta}},$$

$$\psi = kz_0\left(\frac{1}{\beta_1} + \frac{1}{\beta_2} - 2n\cos\theta\right) - kz_1\frac{(\beta_1 - \beta_2)^2}{\beta_1\beta_2(\beta_1 + \beta_2)},$$

$$\gamma_1 = kz_0\left(\frac{1}{\beta_1} - n\cos\theta\right) + \alpha_1 + \gamma, \quad \gamma_2 = kz_0\left(\frac{1}{\beta_2} - n\cos\theta\right) - \alpha_2 - \gamma.$$

Furthermore, θ_1 and θ_2 are defined by $\cos\theta_1 = 1/\beta_1 n$ and $\cos\theta_2 = 1/\beta_2 n$. Since $\alpha \gg 1$, the radiation intensity for $\theta_2 < \theta < \theta_1$ is much larger than for $\theta < \theta_2$ and $\theta > \theta_1$. Thus, the radiation intensity has a plateau for $\theta_2 < \theta < \theta_1$, where it changes quite slowly (since the non-oscillating term proportional to α^2 is much larger than the oscillating terms proportional to α and (5.106)). For $\theta < \theta_2$ and $\theta > \theta_1$ the radiation intensity is kz_1 times smaller than for $\theta_2 < \theta < \theta_1$. The singularities of the radiation intensity at $\theta = \theta_1$ and $\theta = \theta_2$ are owed to the approximations involved. More accurately, they are owed to the replacement of the Fresnel integrals by their asymptotic values. In fact, the integrals I_c and I_s defined by (5.104) are finite at $\theta = \theta_1$ and $\theta = \theta_2$.

Comparison with smooth infinite charge motion (5.98)

We observe that the qualitative behaviour of the angular intensity for the motion treated is very similar to that given by (5.98). For example, in the high-frequency limit both of them are maximal at the Cherenkov angles θ_1 and θ_2 corresponding to the velocities β_1 and β_2, respectively, have a plateau between θ_1 and θ_2 and sharply decrease outside this plateau. The difference is in their asymptotic behaviour: the radiation intensities are exponentially small for the absolutely continuous motion (5.98) and are quite smooth angular functions for the finite charge motion discussed in this subsection. The other difference is that the radiation intensity (5.99) corresponding to the motion law (5.98) is infinite at the Cherenkov angles θ_1 and θ_2, whilst the radiation intensity (5.104) corresponding to the finite motion discussed in this subsection is everywhere finite (its infinities in the high-frequency limit is a result of the approximations involved).

5.3.7. SHORT DISCUSSION OF THE SMOOTHED TAMM PROBLEM

We have considered a number of versions of the smoothed Tamm problem allowing analytical solutions. They have the common property that for the charge velocity greater than the velocity of light in medium, an angular region exists where the radiation intensity is proportional to the frequency and the region where the radiation intensity is small for high frequencies.

This investigation is partly inspired by the influential paper [14] in which the charge motion with a velocity decreasing linearly with time was investigated numerically. The behaviour of radiation intensities obtained there strongly resembles the behaviour of analytical intensities (5.58)-(5.62). In addition the authors of [14] correctly guessed that the Tamm radiation intensity (5.5) is somehow related to the velocity jumps at the start and end of the motion.

Our understanding of this problem coincides with that given in [24,27, 28,29] for the charge motion in vacuum where it was shown that radiation intensities for the absolutely continuous motion are exponentially decreasing functions of ω. The modification for a charge moving in medium looks as follows. The asymptotic behaviour of the radiation intensity depends on how much the charge motion is discontinuous. For example, for the absolutely continuous charge motions shown in Figs. 5.23, 5.25, and 5.27, the radiation intensities decrease exponentially with ω for θ above some critical angle θ_c, and are proportional to ω for $\theta < \theta_c$. For the motion without velocity jumps (but with the acceleration jumps) shown in Fig. 5.15(b), the radiation intensity falls as $1/\omega$ for $\theta > \theta_c$ and is proportional to ω for $\theta < \theta_c$. For the charge motion with velocity and acceleration jumps shown in Fig. 5.15(a), the radiation intensity does not depend on the frequency for $\theta > \theta_c$, although it is much smaller than for $\theta < \theta_c$ (again, in this angular region, σ_r is proportional to ω).

A question arises what kind of the radiation fills the angular region $\theta < \theta_c$ (see Figs. 5.18, 5.20, 5.21, 5.24(a,b), 5.26(a-c)). For this, we again turn to [23] where the exact radiation fields were obtained for the charge accelerated and decelerated motions. At the start of motion ($t = 0$), the spherically symmetric Bremsstrahlung shock wave (BSW) arises which propagates with the velocity of light in medium. At the instant t_0 when the charge velocity coincides with the charge velocity in medium, a complex arises consisting of the finite Cherenkov shock wave SW1 and the shock wave SW2 closing the Cherenkov cone. The singularities carried by these two shock waves are the same and are much stronger than the singularity carried by BSW (for details see again [23]). The SW1 attached to a moving charge intersects the motion axis at the angle $\pi/2 - \theta_{Ch}$, where θ_{Ch} is the Cherenkov angle corresponding to the current charge velocity ($\cos\theta_{Ch} = 1/\beta n$). Obviously $\theta_{Ch} = 0$ at $t = t_0$ and $\theta_{Ch} = \theta_c$ at the end of acceleration. Here θ_c is the Cherenkov angle corresponding to the maximal charge velocity. The SW2 detached from a charge and intersecting the motion axis behind the charge at a right angle, differs from zero in the angular sector $0 < \theta < \theta_{Ch}$. The angular distribution in the spectral representation (since transition to it involves integration over all times) fills the angular region $0 < \theta < \theta_c$.

We conclude: The radiation intensity in the $0 < \theta < \theta_c$ angular region consists of the Cherenkov shock wave, the shock wave closing the Cherenkov cone and the Bremsstrahlung shock wave.

5.3.8. HISTORICAL REMARKS ON THE VC RADIATION AND BREMSSTRAHLUNG

Cherenkov at first followed the Vavilov explanation of the nature of radiation observed in his experiments. We quote him [30]:

> All the facts stated above unambiguously testify that the nature of the γ luminescence is the electromagnetic deceleration of electrons moving in a fluid. The facts that γ luminescence is partially polarized, and that its brightness has a highly pronounced asymmetry, strongly resemble the similar picture for the bremsstrahlung of fast electrons in the Roentgen region. However, in the case of the γ luminescence the complete theoretical interpretation encounters with a number of difficulties.

(our translation from Russian).

Collins and Reiling [31] shared this viewpoint:

> It is to be understood that the electron in its passage through the medium gradually loses nearly all its energy through ionization and excitation processes, and the resulting acceleration is responsible for the VC radiation.

Later, Cherenkov changed his opinion in favour of the Tamm-Frank theory. What were the reasons for this?

At first we clarify conditions under which the Cherenkov experiments are performed. According to him ([32], p.24),

> ...the absorption of electrons in fluids was complete.

This means that we should apply the numerical and analytic results of Chapter 5 relating to the charge motion with a zero final velocity.

There are three main reasons why Cherenkov abandoned the original viewpoint. We consider them step by step. In page 33 of [32] he writes

> For the radiation produced by electrons in fluids, the angle θ (measured away from the direction of the electron motion) for which the maximum of radiation is observed increases with increasing electron velocity. This dependence of θ is just the opposite of that expected if one suggests that radiation in fluids is owed to deceleration. For the bremsstrahlung it is characteristic that the position of the intensity maximum shifts towards the initial beam with rising electron energy

However, numerical and analytic results obtained and Fig. 5.18 demonstrate that the maximum of the radiation intensity for the decelerated motion in medium behaves exactly in the same way as in the Tamm-Frank theory.

Concerning decreasing of the radiation intensity at large angles. Again, we quote P.A. Cherenkov ([32], p.34):

> To the aforesaid about the azimuthal distribution of the intensity should be added that the asymmetry of radiation relative to the plane perpendicular to the electron beam is more pronounced for the observed radiation of fluids than for the bremsstrahlung

Turning to the motion law presented in Fig. 5.15(a), it was shown numerically and analytically (see e.g., Fig. 5.16) that the radiation intensity falls more rapidly than that described by the Tamm formula (which is almost symmetrical relative to the Cherenkov angle). For the decelerated motion with a zero final velocity, the decrease of radiation is determined either by the exact equation (5.46) (where one should set $\beta_2 = 0$) or by the analytic Eqs. (5.61) and (5.62). The latter is infinite at $\cos\theta = 1/\beta n$, whilst (5.46) gives there $\sigma_r(\cos\theta = 1/\beta n) = e^2 Ln(1 - 1/\beta_n^2)/2c\lambda$ (L and λ are the motion interval and wavelength). The Tamm intensity at the same angle is much larger for $L/\lambda \gg 1$: $\sigma_T(\cos\theta = 1/\beta n) = e^2 L^2 n(1 - 1/\beta_n^2)/c\lambda^2$ Comparing (5.5) and (5.61) we see that for $\theta > \theta_c$, σ_r and σ_T decrease in the same way, with the exception that σ_T oscillates, whilst σ_r does not (Figs. 5.17 and 5.18). It should be mentioned that no oscillations in the angular intensity were observed in the original Cherenkov experiments.

The last Cherenkov objection concerns the frequency dependence of the integral intensity. According to him ([32, p.33)

> In both of the cases the same qualitative result is obtained: the energy of the bremsstrahlung spectrum decreases at large frequencies. For our purposes it is enough to say that it does not rise with energy. On the other hand, the experiment shows that for the radiation induced by fast electrons the energy rises in proportion to the frequency, which, obviously, disagrees with results following from the bremsstrahlung theory

Turning to Fig. 5.19, we observe that the ratio of the BS integral intensity to that of Tamm does not depend on the frequency. Since the Tamm integral intensity rises in proportion to the frequency, the same is valid for the BS integral intensity.

Let us summarize the discussion: Since the Tamm condition (5.48) is strongly violated, the radiation observed in the original Cherenkov experiments cannot be attributed uniquely to the uniform motion of the charge. This fact was intuitively guessed by and Collins and Reiling [31]:

> In conclusion it may be stated that the experimental results reported here are in complete agreement with the classical explanation as developed by Frank and Tamm. It would be expected, however, that at very short wavelengths a determination of the intensity would result in a deviation from the classical theory in much the same way that the classical theory of Rayleigh-Jeans fails at short wave-lengths.

Indeed for high frequencies the formulae (5.68), (5.69) and numerical results (Figs. (5.20) and (5.21)) corresponding to the smooth Tamm problem disagree drastically with the Tamm radiation intensity.

Thus the Vavilov explanation of these experiments supported initially by Cherenkov, was at least partly, correct. A sharp distinction of angular intensities shown in Fig. 5.18 from the Tamm intensity given by (5.5) supports this claim. Probably the beauty of the Tamm-Frank theory, concretely predicting the position of the radiation maximum, its dependence on the electron energy and the medium properties, the frequency proportionality of the total radiated energy, the absence of concrete calculations on the radiation of decelerated electron in medium (Cherenkov used references treating BS in vacuum), and the similarity of the predictions of the Tamm-Frank theory and the BS theory in medium, enabled him to change his opinion.

The aforesaid is related to the original Cherenkov experiments in which the Compton electrons knocked out by photons are completely absorbed in medium. In modern experiments high-energy charged particles move through a medium almost without energy loss. In this case the Tamm condition (5.48) is valid and one can use either the original Tamm formula (5.5) or its modifications (5.18) and (5.26) valid for finite observational distances and small decelerations.

5.4. Short résumé of Chapter 5

We briefly summarize the main results obtained:

1) The analytic formula (5.18) has been found describing the intensity of the VC radiation at finite distances from a moving charge. It is shown that under the conditions close to the experimental ones the Cherenkov angular spectrum broadens enormously. The analytic formula obtained is in reasonable agreement with the exact formula (5.14) and sharply disagrees with the Tamm formula (which does not depend on the distance). When the observational distance tends to infinity, the above formula passes into the Tamm formula.

2) Also, another closed formula (5.26) has been obtained which takes into account both the possible deceleration of a charge owed to the energy losses and the finite distance of the observational point from a moving charge. For very large observational distances this formula is transformed into that found in [9]. Previously, the broadening of the Cherenkov angular spectrum experimentally observed in the heavy ions experiments was attributed to the deceleration of heavy ions in a dielectric slab [15]. Our consideration shows that finite distances of the point of observation con-

tribute to the above broadening as well. In particular, it should be observed in high-energy electron experiments (for which the energy losses are negligible) if the measurements are performed at finite distances from a dielectric slab.

3) The above formulae are applied to the description of the VC radiation observed in the recent Darmstadt experiments with heavy ions.

4) The analytic solution (5.46) describing the charge motion in medium with arbitrary acceleration (deceleration) (Fig. 5.15 (a)) is found. The total radiation intensity has one maximum at the Cherenkov angle corresponding to β_1 (see Fig. 5.16 (a,c,d)) or two maxima at the Cherenkov angles corresponding to β_1 and β_2 (Fig. 5.16 (b)). This solution may be applied to study the radiation produced by electrons moving uniformly in heavy-water reactors (the electron arising from the β decay of some nucleus, moves with deceleration, and then is absorbed by another nucleus). Another possible application is to experiments with heavy ions moving in medium [15] (due to large atomic numbers, the energy losses for heavy ions are also large).

5) Analytic expressions are found for the electromagnetic field and the energy flux radiated by a charge moving along the trajectory which consists of accelerated, decelerated, and uniform motion parts (Fig. 5.15 (b)). It is shown that when the lengths of accelerated and decelerated parts tend to zero their contribution to the radiated energy flux also tends to zero despite the infinite value of acceleration along them. The total radiation intensity has a maximum at the Cherenkov angle defined by $\cos\theta = 1/\beta n$ (Figs. 5.20 and 5.21). The possible applications of this model are the same as those of the original Tamm problem.

6) Analytic expressions are obtained for the electromagnetic field and the energy flux radiated by a charge moving along the trajectory shown in Fig. 5.15(c). The total radiation intensity has two maxima at the Cherenkov angles defined by $\cos\theta = 1/\beta_1 n$ and $\cos\theta = 1/\beta_2 n$ if both β_1 and β_2 are greater than $1/n$ (Fig. 5.22 (a,b)). Only one maximum corresponding to $\cos\theta = 1/\beta_1 n$ survives if $\beta_2 < 1/n$ (Fig. 5.22 (c,d)).

It follows from Figs. 5.20 and 5.21 that angular distributions corresponding to finite accelerations are highly non-symmetrical relative to the Cherenkov angle, whilst distributions described by the Tamm formula are almost symmetrical. The angular distributions observed by Cherenkov were also highly non-symmetrical (see, e.g., [32]). They strongly resemble the radiation intensities shown in Fig. 5.18 and corresponding to the zero final energy.

7) We have evaluated the radiation intensity for the Tamm problem with absolute continuous time dependence of a charge velocity. It is shown that the radiation intensity cannot be reduced to the intensity corresponding to the Tamm problem when the length of acceleration region tends to zero.

8) The fact that the maximum of the radiation intensity lies at the Cherenkov angle does not necessarily testify to the charge uniform motion with a velocity greater than the velocity of light in medium. In fact, we have shown numerically and analytically that the maximum of the radiation intensity lies at the Cherenkov angle even if the motion is highly non-uniform.

9) It is shown for the motion beginning with a velocity v_1 and terminating with a velocity v_2 that there are two Cherenkov maxima if both $\beta_1 n$ and $\beta_2 n$ are greater than 1. Only one Cherenkov maximum survives if one of these quantities is smaller than 1.

10) The radiation intensity for a charge coming to a complete stop in a medium does oscillate. Its maximum is at the Cherenkov angle θ_c defined by $\cos \theta_c = 1/\beta n$, where β is the initial velocity. The integral intensity is a linear function of frequency.

References

1. Tamm I.E. (1939) Radiation Induced by Uniformly Moving Electrons, *J. Phys. USSR*, **1, No 5-6**, pp. 439-461.
2. Lawson J.D. (1954) On the Relation between Cherenkov Radiation and Bremsstrahlung *Phil. Mag.*, **45**, pp.748-750.
3. Lawson J.D. (1965) Cherenkov Radiation, "Physical" and "Unphysical", and its Relation to Radiation from an Accelerated Electron *Amer. J. Phys.*, **33**, pp. 1002-1005.
4. Zrelov V.P. and Ruzicka J. (1989) Analysis of Tamm's Problem on Charge Radiation at its Uniform Motion over a Finite Trajectory *Czech. J. Phys.*, **B 39**, pp. 368-383.
5. Zrelov V.P. and Ruzicka J. (1992) Optical Bremsstrahlung of Relativistic Particles in a Transparent Medium and its Relation to the Vavilov-Cherenkov Radiation *Czech. J. Phys.*, **42**, pp. 45-57.
6. Afanasiev G.N., Beshtoev Kh. and Stepanovsky Yu.P. (1996) Vavilov-Cherenkov Radiation in a Finite Region of Space *Helv. Phys. Acta*, **69**, pp. 111-129.
7. Afanasiev G.N., Kartavenko V.G. and Stepanovsky Yu.P. (1999) On Tamm's Problem in the Vavilov-Cherenkov Radiation Theory *J.Phys. D: Applied Physics*, **32**, pp. 2029-2043.
8. Afanasiev G.N., Kartavenko V.G. and Ruzicka J, (2000) Tamm's Problem in the Schwinger and Exact Approaches *J. Phys. A: Mathematical and General*, **33**, pp. 7585-7606.
9. Kuzmin E.S. and Tarasov A.V. (1993) Diffraction-like Effects in Angular Distribution of Cherenkov Radiation from Heavy Ions *Rapid Communications JINR*, **4/61/-93**, pp. 64-69.
10. Frank I.M. (1988) *Vavilov-Cherenkov Radiation*, Nauka, Moscow.
11. Zrelov V.P. (1970) *Vavilov-Cherenkov Radiation in High-Energy Physics*, vols. 1 and 2, Israel Program for Scientific Translations.
12. Aitken D.K. et al. (1963) Transition Radiation in Cherenkov Detectors *Proc. Phys. Soc.*, **83**, pp. 710-722.
13. Zrelov V.P., Klimanova M., Lupiltsev V.P. and Ruzicka J. (1983) Calculations of Threshold Characteristics of Vavilov-Cherenkov Radiation Emitted by Ultrarelativistic Particles in Gaseous Cherenkov Detector *Nucl. Instr. and Meth.*, **215**, pp. 141-146;
    Zrelov V.P., Lupiltsev V.P. and Ruzicka J. (1988) *Nucl. Instr. and Meth.*, **A270**,

pp. 62-68.

14. Krupa L., Ruzicka J. and Zrelov V.P. (1995) Is the Criterion of Constant Particle Velocity Necessary for the Vavilov-Cherenkov Effect? *JINR Preprint* **P2-95-381**.

15. Ruzicka J. et al. (1999) The Vavilov-Cherenkov Radiation Arising at Deceleration of Heavy Ions in a Transparent Medium *Nucl. Instr. and Meth.*, **A431**, pp. 148-153.

16. Dedrick K.G. (1952) The Influence of Multiplr Scattering on the Angular Width of Cherenkov Radiation *Phys.Rev.*, **87**, pp. 891-896.

17. Bowler M.G. (1996) Effects of Electron Scattering on Cherenkov Light Output *Instr. and Meth.*, **A378**, pp.463-467.

18. Bowler M.G. and Lay M.D. (1996) Angular Distribution of Cherenkov Light from Electrons both Produced and Stopping in Water *Instr. and Meth.*, **A378**, pp. 468-471.

19. Schwinger J. (1949) On the Classical Radiation of Accelerated Electrons *Phys.Rev.*,**A 75**, pp. 1912-1925.

20. Smith G.S. (1993) Cherenkov Radiation from a Charge of Finite Size or a Bunch of Charges *Amer. J. Phys.*, **61**, pp. 147-155.

21. Vavilov S.I. (1934) On Possible Reasons for the Blue γ Radiation in Fluids, *Dokl. Akad, Nauk*, **2, 8**, pp. 457-459.

22. Afanasiev G.N. and Shilov V.M. (2000) New Formulae for the Radiation Intensity in the Tamm Problem *J. Phys.D: Applied Physics*, **33**, pp. 2931-2940.

23. Afanasiev G.N. and Shilov V.M. (2000) On the Smoothed Tamm Problem *Physica Scripta*, **62**, pp. 326-330.
    Afanasiev G.N., Eliseev S.M. and Stepanovsky Yu.P. (1998) Transition of the Light Velocity in the Vavilov-Cherenkov Effect *Proc. Roy. Soc. London*, **A 454**, pp. 1049-1072.
    Afanasiev G.N. and Kartavenko V.G. (1999) *Cherenkov-like shock waves associated with surpassing the light velocity barrier Canadian J. Phys.*, **77**, pp. 561-569.

24. Abbasov I.I. (1982) Radiation Emitted by a Charged Particle Moving for a Finite Interval of Time under Continuous Acceleration and Deceleration *Kratkije soobchenija po fizike FIAN*, **No 1**, pp. 31-33; English translation: (1982) *Soviet Physics-Lebedev Institute Reports* **No1**, pp.25-27.

25. Lukyanov V.K., Eldyshev Yu.N. and Poll Yu.S. (1972), Analysis of Elastic Electron Scattering in Light Nuclei on the Basis of Symmetrized Fermi-Density Distribution *Yadernaya Fiz.*, **16**, pp. 506-514.

26. Grypeos M.E., Koutroulos C.G., Lukyanov V.K. and Shebeko A.V. (2001) Properties of Fermi and Symmetrized Fermi Functions and Applications in Nuclear Physics *Phys. Elementary Particles and Atomic Nuclei*, **32**, pp. 1494-1562.

27. Abbasov I.I. (1985) Radiation of a Charged Particle Moving Uniformly in a Given Bounded Segment with Allowance for Smooth Acceleration at the Beginning of the Path, and Smooth Deceleration at the End *Kratkije soobchenija po fizike FIAN*, **No 8**, pp. 33-36. English translation: (1985) *Soviet Physics-Lebedev Institute Reports*, **No 8**, pp. 36-39.

28. Abbasov I.I., Bolotovskii B.M. and Davydov V.A. (1986) High-Frequency Asymptotics of Radiation Spectrum of the Moving Charged Particles in Classical Electrodynamics *Usp. Fiz. Nauk*, **149**, pp. 709-722. English translation: Sov. Phys. Usp., 29 (1986), 788.

29. Bolotovskii B.M. and Davydov V.A. (1981) Radiation of a Charged Particle with Acceleration at a Finite Path Length *Izv. Vuzov, Radiofizika*, **24** , pp. 231-234.

30. Cherenkov P.A. (1936) Influence of Magnetic Field on the Observed Luminescence of Fluids Induced by Gamma Rays, *Dokl. Akad, Nauk*, **3, 9**, pp. 413-416.

31. Collins G.B. and Reiling V.G. (1938) Cherenkov Radiation *Phys. Rev*, **54**, pp. 499-503.

32. Cherenkov P.A. (1944) Radiation of Electrons Moving in Medium with Superluminal Velocity, *Trudy FIAN*, **2, No 4**, pp. 3-62.

CHAPTER 6

RADIATION OF ELECTRIC, MAGNETIC AND TOROIDAL DIPOLES MOVING IN A MEDIUM

6.1. Introduction.

The radiation of Compton electrons moving in water was observed by Cherenkov in 1934 (see his Doctor of Science dissertation published in [1]). During 1934-1937 Tamm and Frank associated it with the radiation of electrons moving with a velocity v greater than the velocity of light in medium c_n (see, e.g., the Frank monograph [2]).

The radiation of electric and magnetic dipoles moving uniformly in medium with $v > c_n$ was first considered by Frank in [3,4]. The procedure used by him is as follows. The Maxwell equations are rewritten in terms of electric and magnetic Hertz vector potentials. The electric and magnetic field strengths are expressed through them uniquely. In the right hand sides of these equations there enter electric and magnetic polarizabilities which are expressed through the laboratory frame (LF) electric (π) and magnetic (μ) moments of a moving particle. These moments are related to the electric (π') and magnetic (μ') moments in the dipole rest frame (RF) via the relations [5]

$$\vec{\pi} = \vec{\pi}' - (1 - \gamma^{-1})(\vec{\pi}'\vec{n}_v)\vec{n}_v + \beta(\vec{n}_v \times \vec{\mu}'),$$

$$\vec{\mu} = \vec{\mu}' - (1 - \gamma^{-1})(\vec{\mu}'\vec{n}_v)\vec{n}_v - \beta(\vec{n}_v \times \vec{\pi}'). \tag{6.1}$$

Here $\beta = v/c$, $\gamma = 1/\sqrt{1-\beta^2}$, $\vec{n}_v = \vec{v}/v$, v is the velocity of a dipole relative to the LF.

Let there be only the electric dipole ($\mu' = 0$) in the RF. Then

$$\vec{\pi} = \vec{\pi}' - (1 - \gamma^{-1})(\vec{\pi}'\vec{n}_v)\vec{n}_v, \quad \vec{\mu} = -\vec{\beta}(\vec{n}_v \times \vec{\pi}'). \tag{6.2}$$

Excluding π' one finds in the LF

$$\vec{\mu} = -\beta(\vec{n}_v \times \vec{\pi}). \tag{6.3}$$

Similarly, if only the magnetic moment differs from zero in the RF, then in the LF

$$\vec{\mu} = \vec{\mu}' - (1 - \gamma^{-1})(\vec{\mu}'\vec{n}_v)\vec{n}_v, \quad \vec{\pi} = \beta(\vec{n}_v \times \vec{\mu}). \tag{6.4}$$

Using these relations Frank evaluated the electromagnetic field (EMF) strengths and the energy flux per unit frequency and per unit length of the cylinder surface coaxial with the motion axis. These quantities depended on the dipole spatial orientation. For the electric dipole and for the magnetic dipole parallel to the velocity Frank obtained expressions which satisfied him. For a magnetic dipole perpendicular to the velocity, the radiated energy did not disappear for $v = c_n$. Its vanishing is intuitively expected and is satisfied, e.g., for the electric charge and dipole and for the magnetic dipole parallel to the velocity. On these grounds Frank declared [6] the formula for the radiation intensity of the magnetic dipole perpendicular to the velocity as to be incorrect. He also admitted that the correct expression for the above intensity is obtained if the second of Eqs.(6.4) is changed to

$$\vec{\pi} = n^2 \beta (\vec{n}_v \times \vec{\mu}), \qquad (6.5)$$

whilst (6.3) remains the same. Here n is the medium refractive index.

Equation (6.5) was supported by Ginzburg [7] who pointed out that the internal structure of a moving magnetic dipole and the polarization induced inside it are essential. This idea was further elaborated in [8].

In [9] the radiation of toroidal dipoles (i.e., the elementary (infinitesimally small) toroidal solenoids (TS)) moving uniformly in a medium was considered. It was shown that the EMF of a TS moving in medium a extends beyond its boundaries. This seemed to be surprising since the EMF of a TS either at rest in medium (or vacuum) or moving in vacuum is confined to its interior.

After many years Frank returned [10,11] to the original transformation law (6.2)-(6.4). In particular, in [11] a rectangular current frame moving uniformly in medium was considered. The evaluated electric moment of the moving current distribution was in agreement with (6.4).

Another transformation law for the magnetic moment, grounding on the proportionality between the magnetic and mechanical moments was suggested in [12]. This proportionality taking place, e.g., for an electron, was confirmed experimentally to a great accuracy in $g - 2$ experiments. In them the electron spin precession is described by the Bargmann-Michel-Telegdi equation. In this theory the spin is a three-vector $\vec{s}$ in its rest frame. In another inertial frame (and, in particular, in the laboratory frame relative to which a particle with spin moves with the velocity $\vec{v}$), the spin has four components $(\vec{S}, S_0)$ defined by

$$\vec{S} = \vec{s} + \frac{\gamma^2}{\gamma + 1}(\vec{\beta} \cdot \vec{s})\vec{\beta}, \quad S_0 = \gamma(\vec{\beta} \cdot \vec{s}).$$

A nice exposition of these questions may be found in [13].

The goal of this consideration is to obtain EMF potentials and strengths for point-like electric and magnetic dipoles and an elementary toroidal dipole moving in a medium with an arbitrary velocity v greater or smaller than the velocity of light c_n in medium. In the reference frame attached to a moving source we have a finite static distribution of charge and current densities. We postulate that charge and current densities in the laboratory frame, relative to which the source moves with a constant velocity, can be obtained from the rest frame densities via the Lorentz transformations, the same as in vacuum. The further procedure is in decreasing the dimensions of the LF charge-current sources to zero, in a straightforward solution of the Maxwell equations for the EMF potentials with the LF point-like charge-current densities in their r.h.s., and in a subsequent evaluation of the EMF strengths. In the time and spectral representations, this was done in [14,15]. The reason for using the spectral representation which is extensively used by experimentalists is to compare our results with those of [1-10] written in the frequency representation.

The plan of this exposition is as follows. In section 6.3 the electromagnetic field strengths are evaluated in the time representation for electric, magnetic and toroidal dipoles moving uniformly in an unbounded non-dispersive medium. In section 6.4 the same radiation intensities are evaluated in the spectral representation. A lot of misprints in previous publications is recovered. It is not our aim to recover these misprints, but we need reliable working formulae which can be applied to concrete physical problems. In the same section the electromagnetic fields of electric, magnetic and toroidal dipoles moving uniformly in a finite medium interval are obtained. In section 6.5 the EMF of a precessing magnetic dipole is obtained. This can be applied to astrophysical problems. A brief discussion of the results obtained and their summary is given in section 6.6.

6.2. Mathematical preliminaries: equivalent sources of the electromagnetic field

This section is essentially an extract of [16]. It is needed for the understanding of subsequent exposition.

6.2.1. A PEDAGOGICAL EXAMPLE: CIRCULAR CURRENT.

According to the Ampére hypothesis, the distribution of magnetic dipoles $\vec{M}(\vec{r})$ is equivalent to the current distribution $\vec{J}(\vec{r}) = \mathrm{curl}\vec{M}(\vec{r})$. For example, a circular current flowing in the $z = 0$ plane

$$\vec{J} = I\vec{n}_\phi \delta(\rho - d)\delta(z) \tag{6.6}$$

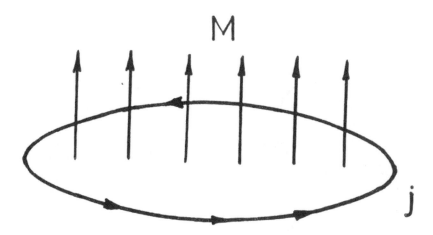

Figure 6.1.   The circular current $\vec{j}$ is equivalent to the magnetization perpendicular to the current plane.

is equivalent to the magnetization (see Fig. 6.1)

$$\vec{M} = I\vec{n}\Theta(d - \rho)\delta(z) \qquad (6.7)$$

different from zero in the same plane and directed along its normal $\vec{n}$ ($\Theta(x)$ is a step function). In what follows, by magnetic and toroidal dipoles we understand infinitesimal circular loop and toroidal winding with a constant current flowing in them. When the radius d of the circumference along which the current flows tends to zero, the current $\vec{J}$ becomes ill-defined (it is not clear what the vector $\vec{n}_\phi$ means at the origin). On the other hand, the vector $\vec{M}$ is still well-defined. In this limit the elementary current (6.6) turns out to be equivalent to the magnetic dipole oriented normally to the plane of this current:

$$\vec{M} = I\pi d^2 \vec{n}\delta^3(\vec{r}), \quad (\delta^3(\vec{r}) = \delta(\rho)\delta(z)/2\pi\rho) \qquad (6.8)$$

and

$$\vec{J} = I\pi d^2 \mathrm{curl}(\vec{n}\delta^3(\vec{r})) \qquad (6.9)$$

Equations (6.8) and (6.9) define the magnetization and current density corresponding to the elementary magnetic dipole.

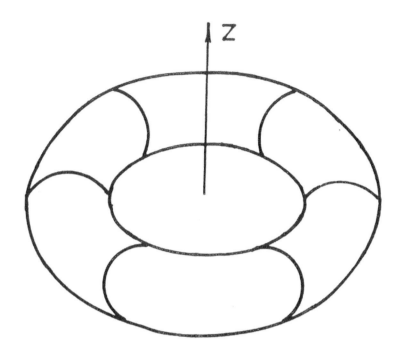

Figure 6.2. The poloidal current flowing on the torus surface.

6.2.2. THE ELEMENTARY TOROIDAL SOLENOID.

The case next in complexity is the poloidal current flowing in the winding of TS (Fig. 6.2):

$$\vec{j} = -\frac{gc}{4\pi}\vec{n}_\psi\frac{\delta(R - \tilde{R})}{d + \tilde{R}\cos\psi}. \tag{6.10}$$

The coordinates $\tilde{R}, \psi$ and ϕ are related to the Cartesian ones as follows:

$$x = (d + \tilde{R}\cos\psi)\cos\phi, \quad y = (d + \tilde{R}\cos\psi)\sin\phi, \quad z = \tilde{R}\sin\psi. \tag{6.11}$$

The condition $\tilde{R} = R$ defines the surface of a particular torus (Fig. 6.3). For $\tilde{R}$ fixed and ψ, ϕ varying, the points x, y, z given by (6.11) fill the surface of the torus $(\rho - d)^2 + z^2 = R^2$. The choice $\vec{j}$ in the form (6.10) is convenient, because in the static case a magnetic field H is equal to g/ρ inside the torus and vanishes outside it. In this case g may be also expressed either through the magnetic flux Φ penetrating the torus or through the total number N

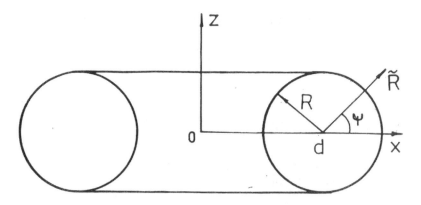

Figure 6.3. The coordinates $\widetilde{R}, \psi$ parametrizing the torus.

of turns in the toroidal winding and the current I in a particular turn

$$g = \frac{\Phi}{2\pi(d - \sqrt{d^2 - R^2}} = \frac{2NI}{c}.$$

We write out the differential operators div and curl in $\widetilde{R}, \psi,$ and ϕ coordinates:

$$\mathrm{div}\vec{A} = \frac{1}{\widetilde{R}(d + \widetilde{R}\cos\psi)}$$

$$\times \left[\frac{\partial}{\partial\widetilde{R}}\widetilde{R}(d + \widetilde{R}\cos\psi)A_{\widetilde{R}} + \frac{\partial}{\partial\psi}(d + \widetilde{R}\cos\psi)A_\psi + \frac{\partial}{\partial\phi}\widetilde{R}A_\phi \right],$$

$$(\mathrm{curl}\vec{A})_{\widetilde{R}} = \frac{1}{\widetilde{R}(d + \widetilde{R}\cos\psi)}\left[\frac{\partial}{\partial\phi}(\widetilde{R}A_\psi) - \frac{\partial}{\partial\psi}(d + \widetilde{R}\cos\psi)A_\phi \right],$$

$$(\mathrm{curl}\vec{A})_\phi = \frac{1}{\widetilde{R}}\left[\frac{\partial\widetilde{R}}{\partial\psi} - \frac{\partial}{\partial\widetilde{R}}(\widetilde{R}A_\psi) \right],$$

$$(\mathrm{curl}\vec{A})_\psi = \frac{1}{d + \widetilde{R}\cos\psi}\left[\frac{\partial}{\partial\widetilde{R}}(d + \widetilde{R}\cos\psi)A_\phi - \frac{\partial A_{\widetilde{R}}}{\partial\phi} \right]. \tag{6.12}$$

As $\mathrm{div}\vec{j} = 0$, the current $\vec{j}$ can be presented as the curl of a certain vector $\vec{M}$:

$$\vec{j} = \mathrm{curl}\vec{M}. \qquad (6.13)$$

Or, in a manifest form:

$$-\frac{gc}{4\pi}\frac{\delta(R-\tilde{R})}{d+\tilde{R}\cos\psi} = \frac{1}{d+\tilde{R}\cos\psi}\left[\frac{\partial}{\partial\tilde{R}}(d+\tilde{R}\cos\psi)M_\phi - \frac{\partial M_{\tilde{R}}}{\partial\phi}\right].$$

Due to the axial symmetry of the problem, the term involving ϕ differentiation drops out, and one obtains

$$-\frac{gc}{4\pi}\frac{\delta(R-\tilde{R})}{d+\tilde{R}\cos\psi} = \frac{1}{d+\tilde{R}\cos\psi}\frac{\partial}{\partial\tilde{R}}(d+\tilde{R}\cos\psi)M_\phi.$$

Contracting by the factor $d + \tilde{R}\cos\psi$ one has

$$-\frac{gc}{4\pi}\delta(R-\tilde{R}) = \frac{\partial}{\partial\tilde{R}}(d+\tilde{R}\cos\psi)M_\phi.$$

It follows from this that

$$M_\phi = \frac{gc}{4\pi}\frac{\Theta(R-\tilde{R})}{d+\tilde{R}\cos\psi}, \qquad (6.14)$$

i.e., M_ϕ is confined to the interior of the torus (Fig. 6.4).

We rewrite M_ϕ in cylindrical coordinates:

$$M_\phi = \frac{gc}{4\pi\rho}\Theta[R - \sqrt{(\rho-d)^2 + z^2}]. \qquad (6.15)$$

Since $\mathrm{div}\vec{M} = 0$ the magnetization vector $\vec{M}$ can, in its turn, be presented as a curl of a certain vector $\vec{T}$. It turns out that only the z component of $\vec{T}$ differs from zero:

$$T_z = -\frac{gc}{4\pi}[\Theta(d - \sqrt{R^2 - z^2} - \rho)\ln\frac{d - \sqrt{R^2 - z^2}}{d + \sqrt{R^2 - z^2}}$$

$$+\Theta(d + \sqrt{R^2 - z^2} - \rho)\Theta(\rho - d + \sqrt{R^2 - z^2})\ln\frac{\rho}{d + \sqrt{R^2 - z^2}}]. \qquad (6.16)$$

Thus T_z differs from zero in two spatial regions:

a) Inside the torus hole defined as $0 \le \rho \le d - \sqrt{R^2 - z^2}$, where T_z does not depend on ρ:

$$T_z = -\frac{gc}{4\pi}\ln\frac{d - \sqrt{R^2 - z^2}}{d + \sqrt{R^2 - z^2}}. \qquad (6.17)$$

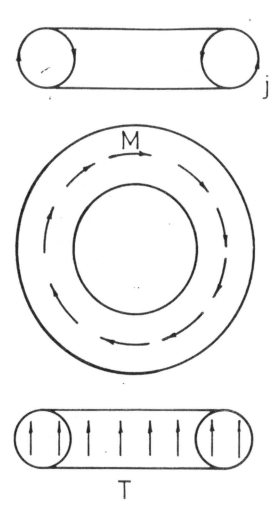

Figure 6.4. The poloidal current $\vec{j}$ flowing on the torus surface is equivalent to the magnetization $\vec{M}$ confined to the interior of the torus and to the toroidization $\vec{T}$ directed along the torus symmetry axis.

b) Inside the torus itself $\left(d - \sqrt{R^2 - z^2} \le \rho \le d + \sqrt{R^2 - z^2}\right)$ where

$$T_z = -\frac{gc}{4\pi} \ln \frac{\rho}{d + \sqrt{R^2 - z^2}}. \qquad (6.18)$$

In other spatial regions $T_z = 0$.

Now let the minor radius R of a torus tend to zero (this corresponds to an infinitely thin torus). The second term in (6.16) then drops out, whilst the first reduces to

$$T_z \to \frac{gc}{2\pi d}\Theta(d-\rho)\sqrt{R^2 - z^2}. \tag{6.19}$$

For infinitesimal R

$$\sqrt{R^2 - z^2} \to \frac{1}{2}\pi R^2 \delta(z).$$

Therefore in this limit

$$\vec{j} = \mathrm{curlcurl}\vec{T}, \quad \vec{T} = \vec{n}_z \frac{gcR^2}{4d}\delta(z)\Theta(d-\rho), \tag{6.20}$$

i.e., the vector $\vec{T}$ is confined to the equatorial plane of a torus and is perpendicular to it.

Let now $d \to 0$ (in addition to $R \to 0$). In this limit

$$\frac{1}{d}\Theta(d-\rho) \to \frac{d}{2\rho}\delta(\rho)$$

and the current of an elementary (i.e., infinitely small) TS is

$$\vec{j} = \mathrm{curlcurl}\vec{T}, \quad \vec{T} = \frac{1}{4}\pi cgdR^2 \delta^3(\vec{r})\vec{n}_z. \tag{6.21}$$

The elementary current flowing in the winding of the elementary TS is then given by

$$\vec{j} = f\mathrm{curl}^{(2)}(\vec{n}\delta^3(\vec{r})) \tag{6.22}$$

where $\mathrm{curl}^{(2)} = \mathrm{curlcurl}$, $\vec{n}$ means the unit vector normal to the equatorial plane of TS and $f = \pi cgdR^2/4$.

Physically, Eqs. (6.10), (6.13) and (6.20)-(6.22) mean that the poloidal current $\vec{j}$ given by Eq.(6.10) is equivalent (i.e., produces the same magnetic field) to the toroidal tube with the magnetization $\vec{M}$ defined by (6.14) and to the toroidization $\vec{T}$ given by (6.16). This illustrates Fig. 6.4.

Another remarkable property of these configurations is that they interact in the same way with the time-dependent magnetic or electric field ([16]). For example, the usual current loop interacts with an external magnetic field in the same way as the magnetic dipole orthogonal to it. The poloidal current shown in the upper part of Fig. 6.4, the magnetized ring corresponding to the magnetization M in its middle part and the toroidal distribution T in its lower part, all of them interact in the same way with the external electromagnetic field. Obviously, the equivalence between current distributions and magnetizations (toroidizations) is a straightforward generalization of the original Ampére hypothesis.

In what follows we need the Lorentz transformation formulae for the charge and current densities and for electromagnetic strengths. They may be found in any textbook on electrodynamics (see, e.g., [13,17]). Let ρ' and $\vec{j}'$ be charge and current densities in the rest frame S' which moves with a constant velocity $\vec{v}$ relative to the laboratory frame (LF) S. Then

$$\rho = \gamma(\rho' + \vec{\beta}\vec{j}'/c), \quad \vec{j} = \vec{j}' + \frac{\gamma - 1}{\beta^2}\vec{\beta}(\vec{\beta}\vec{j}') + \gamma\vec{v}\rho'. \qquad (6.23)$$

Here $\gamma = (1 - \beta^2)^{-1/2}$, $\vec{\beta} = \vec{v}/c$. If there is no charge density in S' then

$$\rho = \gamma\vec{\beta}\vec{j}'/c, \quad \vec{j}_{\parallel} = \gamma\vec{j}'_{\parallel}, \quad j_{\perp} = j'_{\perp}, \qquad (6.24)$$

where $\vec{j}_{\parallel}$ and $j_{\perp}$ are the components of $\vec{j}$ parallel and perpendicular to $\vec{v}$. If there is no current density in S' then

$$\rho = \gamma\rho', \quad \vec{j} = \gamma\vec{v}\rho'. \qquad (6.25)$$

Let $\vec{E}, \vec{D}, \vec{H}, \vec{B}$ and $\vec{E}', \vec{D}', \vec{H}', \vec{B}'$ be electromagnetic strengths and inductions in the LF and in S', resp. Then,

$$\vec{E} = \gamma(\vec{E}' - \vec{\beta} \times \vec{B}') - \frac{\gamma^2}{\gamma + 1}\vec{\beta}(\vec{\beta}\vec{E}'),$$

$$\vec{B} = \gamma(\vec{B}' + \vec{\beta} \times \vec{E}') - \frac{\gamma^2}{\gamma + 1}\vec{\beta}(\vec{\beta}\vec{B}'),$$

$$\vec{D} = \gamma(\vec{D}' - \vec{\beta} \times \vec{H}') - \frac{\gamma^2}{\gamma + 1}\vec{\beta}(\vec{\beta}\vec{D}'),$$

$$\vec{H} = \gamma(\vec{H}' + \vec{\beta} \times \vec{D}') - \frac{\gamma^2}{\gamma + 1}\vec{\beta}(\vec{\beta}\vec{H}'). \qquad (6.26)$$

We also need constitutive relations [18] in the reference frame which moves with the velocity $\vec{v}$ relative to the laboratory frame (in the latter the surrounding matter is at rest)

$$\vec{D}' = \frac{1}{1 - \beta_n^2}\{\epsilon[\vec{E}'(1 - \beta^2) + \vec{\beta}(\vec{\beta}\vec{E}')(1 - n^2)] + \vec{\beta} \times \vec{H}'(1 - n^2)\},$$

$$\vec{B}' = \frac{1}{1 - \beta_n^2}\{\mu[\vec{H}'(1 - \beta^2) + \vec{\beta}(\vec{\beta}\vec{H}')(1 - n^2)] - \vec{\beta} \times \vec{E}'(1 - n^2)\}, \quad (6.27)$$

where $\beta_n = v/c_n$, $c_n = c/n$ is the velocity of light in medium, $n = \sqrt{\epsilon\mu}$ is its refractive index, ϵ and μ are electric permittivity and magnetic permeability, respectively.

For the sake of completeness, we write out Maxwell equations and wave equations for the electromagnetic potentials corresponding to charge $\rho(\vec{r}, t)$ and current $\vec{j}(\vec{r}, t)$ densities imbedded into a non-dispersive medium with constant ϵ and μ:

$$\operatorname{div}\vec{D} = 4\pi\rho, \quad \vec{D} = \epsilon\vec{E}, \quad \operatorname{div}\vec{B} = 0, \quad \vec{B} = \mu\vec{H},$$

$$\operatorname{curl}\vec{E} = -\frac{1}{c}\frac{\partial\vec{B}}{\partial t}, \quad \operatorname{curl}\vec{H} = \frac{1}{c}\frac{\partial\vec{D}}{\partial t} + \frac{4\pi}{c}\vec{j},$$

$$\vec{B} = \operatorname{curl}\vec{A}, \quad \vec{E} = -\operatorname{grad}\Phi - \frac{1}{c}\frac{\partial\vec{A}}{\partial t}, \quad \operatorname{div}\vec{A} + \frac{\epsilon\mu}{c}\frac{\partial\Phi}{\partial t} = 0,$$

$$\left(\Delta - \frac{1}{c_n^2}\frac{\partial^2}{\partial t^2}\right)\Phi = -\frac{4\pi}{\epsilon}\rho_{Ch}, \quad \left(\Delta - \frac{1}{c_n^2}\frac{\partial^2}{\partial t^2}\right)\vec{A} = -\frac{4\pi\mu}{c}\vec{j}.$$

In what follows, by the term 'magnetic dipole' we mean the magnetic moment carried by an infinitesimal circular loop. The alternative to it is the magnetic moment composed of two magnetic poles. These two different realizations of magnetic dipoles interact with magnetic media in a different way (see, e.g., [8]) .

We also use the fields of electric $\vec{p}$ and magnetic $\vec{m}$ dipoles which rest at the origin

$$\vec{E} = -\frac{\vec{p}}{r^3} + 3\vec{r}\frac{\vec{r}\vec{p}}{r^5}, \quad \vec{B} = -\frac{\vec{m}}{r^3} + 3\vec{r}\frac{\vec{r}\vec{m}}{r^5}. \tag{6.28}$$

6.3. Electromagnetic field of electric, magnetic, and toroidal dipoles in time representation.

6.3.1. ELECTROMAGNETIC FIELD OF A MOVING POINT-LIKE CURRENT LOOP

The velocity is along the loop axis
Consider a conducting loop $\mathcal{L}$ moving uniformly in a non-dispersive medium with the velocity v directed along the loop axis (Fig. 6.5 a).

Let in this loop a constant current I flows. In the reference frame attached to the moving loop, the current density is equal to

$$\vec{j} = I\vec{n}_\phi\delta(\rho' - d)\delta(z'), \quad \rho' = \sqrt{x'^2 + y'^2}. \tag{6.29}$$

In accordance with (6.24) one obtains in the LF

$$j = I\vec{n}_\phi\delta(\rho - d)\delta(\gamma(z - vt)) = \frac{I}{\gamma}\vec{n}_\phi\delta(\rho - d)\delta(z - vt). \tag{6.30}$$

Here $\vec{n}_\phi = \vec{n}_y\cos\phi - \vec{n}_x\sin\phi$, $\gamma = 1/\sqrt{1 - \beta^2}$. Since the current direction is perpendicular to the velocity, no charge density arises in the LF.

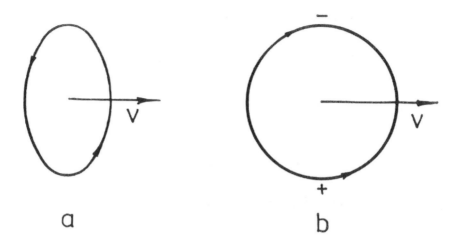

Figure 6.5. a) There is no induced charge density when the symmetry axis of the current loop is along the velocity; b) The induced charge density arises when the symmetry axis of the current loop is perpendicular to the velocity.

The solution of Eq. (6.28) for electromagnetic potentials is given by

$$\Phi = \frac{1}{\epsilon} \int \frac{1}{R} \rho_{Ch}(\vec{r}\,', t')\delta(t' - t + R/c_n)dV'dt',$$

$$\vec{A} = \frac{\mu}{c} \int \frac{1}{R} \vec{j}(\vec{r}\,', t')\delta(t' - t + R/c_n)dV'dt', \quad R = |\vec{r} - \vec{r}\,'|.$$

Like for a charge at rest the current $\vec{j}$ may be expressed through the magnetization

$$\vec{j} = \text{curl}\vec{M}. \tag{6.31}$$

The magnetization $\vec{M}$ is perpendicular to the plane of a current loop:

$$M_z = \frac{I_0}{\gamma}\Theta(d - \rho)\delta(z - vt). \tag{6.32}$$

Substituting this into the vector potential and integrating by parts one finds

$$\vec{A} = \frac{\mu}{c}\text{curl} \int \frac{1}{R}\delta(t' - t + R/c_n)\vec{M}dV'dt. \tag{6.33}$$

The electric scalar potential is zero.

Now let the loop radius d tend to zero. Then,

$$\Theta(d - \rho) \to \pi d^2\delta(x)\delta(y) \quad \text{and} \quad M_z \to \frac{I_0\pi d^2}{\gamma}\delta(x)\delta(y)\delta(z - vt).$$

Substituting this into (6.33) and integrating over the spatial variables one obtains

$$A_\phi = -\frac{\mu I_0\pi d^2}{c\gamma}\frac{\partial\alpha}{\partial\rho}, \tag{6.34}$$

where

$$\alpha = \int \frac{1}{R}\delta(t' - t + R/c_n)dt', \quad R = \sqrt{\rho^2 + (z - vt')^2}. \tag{6.35}$$

This integral can be taken in a closed form (see, e.g., [19]):

$$\alpha = \frac{1}{r_m} \quad \text{for} \quad v < c_n$$

and

$$\alpha = \frac{2}{r_m}\Theta(vt - z - \rho/\gamma_n) \quad \text{for} \quad v > c_n. \tag{6.36}$$

Here $r_m = [(z - vt)^2 + \rho^2(1 - \beta_n^2)]^{1/2}$, $\gamma_n = |1 - \beta_n^2|^{-1/2}$, $\beta_n = v/c_n$. The equality $vt - z - \rho/\gamma_n = 0$ defines the surface of the Cherenkov cone attached to the moving magnetic dipole. Therefore for $\beta_n < 1$, α differs from zero everywhere, whilst for $\beta_n > 1$ it differs from zero only inside the Cherenkov cone where $vt - z - \rho/\gamma_n > 0$. Performing differentiation in (6.34) one finds

$$A_\phi = \frac{\mu m(1 - \beta_n^2)\rho}{\gamma r_m^3}$$

for $\beta < \beta_n$ and

$$A_\phi = \frac{2\mu m(1 - \beta_n^2)\rho}{\gamma r_m^3}\Theta(vt - z - \rho/\gamma_n) + \frac{2\mu m}{\gamma\gamma_n r_m}\delta(vt - z - \rho/\gamma_n) \tag{6.37}$$

for $\beta_n > 1$. Here $m = I_0\pi d^2/c$.

Therefore for $\beta_n < 1$, A_ϕ differs from zero everywhere except for the motion axis. It is infinite at the position of a moving charge and decreases as r^{-2} at large distances.

For $\beta_n > 1$, A_ϕ vanishes outside the Cherenkov cone, being infinite on its surface and falling as r^{-2} inside it.

Electromagnetic field strengths are obtained by differentiating A_ϕ:

$$E_x = \frac{\mu\beta m}{\gamma}\frac{\partial^2\alpha}{\partial z\partial y}, \quad E_y = -\frac{\mu\beta m}{\gamma}\frac{\partial^2\alpha}{\partial z\partial x}, \quad E_z = 0,$$

$$B_x = \frac{\mu m}{\gamma}\frac{\partial^2\alpha}{\partial z\partial x}, \quad B_y = \frac{\mu m}{\gamma}\frac{\partial^2\alpha}{\partial z\partial y},$$

$$B_z = -\frac{\mu m}{\gamma}\left[\Delta - (1-\beta_n^2)\frac{\partial^2\alpha}{\partial z^2}\right], \quad \Delta = \frac{\partial^2}{\partial x^2} + \frac{\partial^2}{\partial y^2} + (1-\beta_n^2)\frac{\partial^2}{\partial z^2}. \quad (6.38)$$

The action of Δ and $\partial^2/\partial z^2$ on α gives for $\beta_n < 1$:

$$\Delta\alpha = -4\pi\delta(x)\delta(y)\delta(z - vt),$$

$$(1-\beta_n^2)\frac{\partial^2\alpha}{\partial z^2} = -\frac{1-\beta_n^2}{r_m^3}\left[1 - 3\frac{(z-vt)^2}{r_m^2}\right] - \frac{4\pi}{3}\delta^3(\vec{r}).$$

Here $\delta^3(\vec{r}) = \delta(x)\delta(y)\delta(z - vt)$. These relations result from the identity (see, e.g., [20])

$$\frac{\partial^2}{\partial x_i\partial x_j}\frac{1}{r} = -\frac{1}{r^3}\left(\delta_{ij} - 3\frac{x_ix_j}{r^2}\right) - \frac{4\pi}{3}\delta_{ij}\delta^3(\vec{r}). \quad (6.39)$$

Higher derivatives of $1/r$ are obtained by differentiating (6.39).

For $\beta_n < 1$ the EMF strengths of a moving point-like current loop are given by

$$E_x = 3m\mu\beta\frac{\gamma_n^3}{\gamma}\frac{y(z-vt)}{r^5}, \quad E_y = -3m\beta\mu\frac{\gamma_n^3}{\gamma}\frac{y(z-vt)}{r^5},$$

$$B_x = 3m\mu\frac{\gamma_n^3}{\gamma}\frac{x(z-vt)}{r^5}, \quad B_y = 3m\mu\frac{\gamma_n^3}{\gamma}\frac{y(z-vt)}{r^5},$$

$$B_z = \frac{m\mu}{\gamma}\left\{\frac{8\pi}{3}\delta^3(\vec{r}) - \frac{\gamma_n}{r^3}\left[1 - 3\gamma_n^2\frac{(z-vt)^2}{r^2}\right]\right\}, \quad (6.40)$$

where

$$m = I_0\pi d^2/c, \quad r^2 = x^2 + y^2 + (z-vt)^2\gamma_n^2, \quad \delta^3(\vec{r}) = \delta(x)\delta(y)\delta(z-vt).$$

In what follows, in order not to overload the exposition we drop the δ-function terms corresponding to the current position of a moving dipole. They are easily restored from Eq.(6.39).

It is seen that $\vec{B}$ in (6.40) strongly resembles the field of magnetic dipole. On the other hand, the electric field $\vec{E}$ having only two Cartesian components, cannot be reduced to the field of an electric dipole.

We conclude: for $\beta_n < 1$ the EMF strengths differ from zero everywhere, falling like r^{-3} at large distances. For $\beta_n > 1$ they are equal to zero outside the Cherenkov cone ($vt - z - \rho/\gamma_n < 0$), infinite on its surface, and fall as r^{-3} inside the Cherenkov cone ($vt - z - \rho/\gamma_n > 0$). As a result, only the moving EMF singularity coinciding with the Cherenkov cone will be observed in the wave zone.

In the rest frame of the magnetic dipole the EMF is given by

$$\vec{E}' = 0, \quad B'_x = 3\frac{m\mu\gamma_n^3}{\gamma^3}\frac{x'z'}{r'^5}, \quad B'_y = 3\frac{m\mu\gamma_n^3}{\gamma^3}\frac{y'z'}{r'^5},$$

$$B'_z = -m\mu\frac{\gamma_n}{\gamma r'^3}\left(1 - 3\frac{\gamma_n^2}{\gamma^2}\frac{z'^2}{r'^2}\right),$$

$$H'_x = 3m\frac{\gamma_n}{\gamma}\frac{x'z'}{r'^5}, \quad H'_y = 3m\frac{\gamma_n}{\gamma}\frac{y'z'}{r'^5}, \quad H'_z = -m\frac{\gamma_n}{\gamma r'^3}\left(1 - 3\frac{\gamma_n^2}{\gamma^2}\frac{z'^2}{r'^2}\right),$$

$$D'_x = 3m(n^2 - 1)\frac{\gamma_n^3\beta}{\gamma}\frac{y'z'}{r'^5}, \quad D'_y = -3m(n^2 - 1)\frac{\gamma_n^3\beta}{\gamma}\frac{x'z'}{r'^5}, \quad (6.41)$$

where $r'^2 = (x'^2 + y'^2) + \gamma_n^2 z'^2/\gamma^2$ and $x' = x$, $y' = y$, $z' = \gamma(z - vt)$. Since in this reference frame the medium has the velocity $-\vec{v}$, the familiar constitutive relations $\vec{B}' = \mu\vec{H}'$, $\vec{D}' = \epsilon\vec{E}'$ are not longer valid. Instead, Eqs. (6.27) should be used.

In vacuum, Eqs. (6.40) and (6.41) reduce to

$$E_x = 3m\gamma^2\frac{y(z - vt)}{r_0^5}, \quad E_y = -3m\gamma^2\frac{x(z - vt)}{r_0^5},$$

$$H_x = 3m\gamma^2\frac{x(z - vt)}{r_0^5}, \quad H_y = 3m\gamma^2\frac{y(z - vt)}{r_0^5},$$

$$H_z = -\frac{m}{r_0^3}\left[1 - 3\frac{\gamma^2(z - vt)^2}{r_0^2}\right], \quad (6.42)$$

$$\vec{E}' = 0, \quad H'_x = 3m\frac{x'z'}{r'^5},$$

$$H'_y = 3m\frac{y'z'}{r'^5}, \quad H'_z = -\frac{m}{r'^3}\left(1 - 3\frac{z'^2}{r'^2}\right), \quad (6.43)$$

where $r_0^2 = \gamma^2(z - vt)^2 + x^2 + y^2$ and $r'^2 = x'^2 + y'^2 + x'^2$. Equations (6.42) and (6.43) are connected by the Lorentz transformation.

The velocity is in the plane of loop

Let a circular loop move in the direction perpendicular to the symmetry axis (say, along the x axis, see Fig.5 (b)). Then in the LF one has

$$j_x = -I_0\delta(z)\frac{y\gamma}{d}\delta(\rho_1 - d), \quad j_y = I_0\delta(z)\frac{(x - vt)\gamma}{d}\delta(\rho_1 - d),$$

$$\rho_{Ch} = -I_0\delta(z)\frac{yv\gamma}{c^2d}\delta(\rho_1 - d).$$

Here $\rho_1 = [(x - vt)^2\gamma^2 + y^2]^{1/2}$. The charge density arises because on a part of the loop, the current has a non-zero projection on the direction of motion. It is easy to check that

$$j_x = I_0\gamma\delta(z)\frac{\partial}{\partial y}M_z, \quad j_y = -I_0\frac{1}{\gamma}\delta(z)\frac{\partial}{\partial x}M_z,$$

$$\rho_{Ch} = I_0\frac{v\gamma}{c^2}\delta(z)\frac{\partial}{\partial y}M_z, \tag{6.44}$$

where $M_z = \Theta(d - \rho_1)$. In the limit of an infinitesimal loop,

$$M_z = \Theta(d - \rho_1) \to \delta(x - vt)\delta(y)\pi d^2/\gamma. \tag{6.45}$$

For the electromagnetic potentials one easily finds

$$\Phi = \frac{m\beta}{\epsilon}\frac{\partial\alpha_1}{\partial y}, \quad A_x = m\mu\frac{\partial\alpha_1}{\partial y}, \quad A_y = -\frac{m\mu}{\gamma^2}\frac{\partial\alpha_1}{\partial x}.$$

Here

$$\alpha_1 = \int dt'\frac{1}{R_1}\delta(t' - t + R_1/c_n), \quad R_1 = [(x - vt')^2 + y^2 + z^2]^{1/2}.$$

Again, this integral can be taken in a closed form:

$$\alpha_1 = \frac{1}{r_m^{(1)}} \quad \text{for} \quad \beta_n < 1$$

and

$$\alpha_1 = \frac{2}{r_m^{(1)}}\Theta\left(vt - x - \frac{1}{\gamma_n}\sqrt{y^2 + z^2}\right) \tag{6.46}$$

for $\beta_n > 1$. Here $r_m^{(1)} = [(x - vt)^2 + (y^2 + z^2)(1 - \beta_n^2)]^{1/2}$. Therefore

$$\Phi = -\frac{m\beta}{\epsilon}\frac{y}{(r_m^{(1)})^3}(1 - \beta_n^2), \quad A_x = -m\mu\frac{y}{(r_m^{(1)})^3}(1 - \beta_n^2),$$

$$A_y = \frac{m\mu}{\gamma^2} \frac{x - vt}{(r_m^{(1)})^3}$$

for $\beta_n < 1$ and

$$\Phi = -2\frac{m\beta}{\epsilon} \frac{y}{(r_m^{(1)})^3}(1 - \beta_n^2)\Theta(vt - x - \rho'/\gamma_n)$$

$$- \frac{2m\beta}{\epsilon\gamma_n} \frac{y}{r_m^{(1)}\rho'}\delta(vt - x - \rho'/\gamma_n),$$

$$A_x = -2m\mu\frac{y}{(r_m^{(1)})^3}(1 - \beta_n^2)\Theta(vt - x - \rho'/\gamma_n)$$

$$- \frac{2m\mu}{\gamma_n} \frac{y}{r_m^{(1)}\rho'}\delta(vt - x - \rho'/\gamma_n),$$

$$A_y = \frac{2m\mu}{\gamma^2} \frac{x - vt}{(r_m^{(1)})^3}\Theta(vt - x - \rho'/\gamma_n) + \frac{2m\mu}{\gamma^2} \frac{1}{r_m^{(1)}}\delta(vt - x - \rho'/\gamma_n). \quad (6.47)$$

for $\beta_n > 1$. Here $\rho' = \sqrt{y^2 + z^2}$. Electromagnetic field strengths are given by

$$E_x = -\frac{m\beta}{\epsilon}(1 - n^2)\frac{\partial^2\alpha_1}{\partial x \partial y}, \quad E_y = -\frac{m\beta}{\epsilon}\left(\frac{\partial^2\alpha_1}{\partial y^2} + \frac{n^2}{\gamma^2}\frac{\partial^2\alpha_1}{\partial x^2}\right),$$

$$E_z = -\frac{m\beta}{\epsilon}\frac{\partial^2\alpha_1}{\partial z \partial y}, \quad B_x = \frac{m\mu}{\gamma^2}\frac{\partial^2\alpha_1}{\partial z \partial x}, \quad B_y = m\mu\frac{\partial^2\alpha_1}{\partial z \partial y},$$

$$B_z = -m\mu\left(\frac{1}{\gamma^2}\frac{\partial^2\alpha_1}{\partial x^2} + \frac{\partial^2\alpha_1}{\partial y^2}\right). \quad (6.48)$$

For $\beta_n < 1$ the EMF falls as r^{-3} at large distances. For $\beta_n > 1$ the EMF strengths vanish outside the Cherenkov cone ($vt - x - \rho'/\gamma_n < 0$), they decrease like r^{-3} at large distances inside the Cherenkov cone ($vt - x - \rho'/\gamma_n > 0$), and they are infinite on the Cherenkov cone. Thus in the wave zone the electromagnetic field is confined to the Cherenkov cone ($vt - x - \rho'/\gamma_n = 0$) where it is infinite.

We write out EMF in the manifest form for $\beta_n < 1$:

$$E_x = -3m\frac{\beta}{\epsilon}(1 - n^2)\gamma_n^3\frac{(x - vt)y}{r'^5}, \quad E_z = -3m\frac{\beta}{\epsilon}\gamma_n\frac{yz}{r'^5},$$

$$E_y = \frac{m\beta\gamma_n}{\epsilon r'^3}\left\{\left(1 - 3\frac{y^2}{r'^2}\right) + n^2\frac{\gamma_n^2}{\gamma^2}\left[1 - 3\gamma_n^2\frac{(x - vt)^2}{r'^2}\right]\right\},$$

$$B_x = 3m\mu\frac{\gamma_n^3}{\gamma^2}\frac{(x - vt)z}{r'^5}, \quad B_y = 3m\mu\gamma_n\frac{yz}{r'^5},$$

$$B_z = \frac{m\mu\gamma_n}{r'^3} \left\{ \left(1 - 3\frac{y^2}{r'^2}\right) + \frac{\gamma_n^2}{\gamma^2}\left[1 - 3\gamma_n^2\frac{(x-vt)^2}{r'^2}\right]\right\}, \qquad (6.49)$$

where $r'^2 = y^2 + z^2 + (x - vt)^2\gamma_n^2$.

For the motion in the vacuum this reduces to

$$E_x = 0, \quad E_z = -3\frac{m\beta\gamma}{c}\frac{yz}{r_1^5}, \quad E_y = -\frac{m\beta\gamma}{cr_1^3}(1 - 3z^2/r_1^2),$$

$$H_x = 3m\gamma\frac{z(x-vt)}{r_1^5}, \quad H_y = 3m\gamma\frac{yz}{r_1^5}, \quad H_z = -m\gamma\frac{1}{r_1^3}(1 - 3z^2/r_1^2). \quad (6.50)$$

Here $r_1^2 = y^2 + z^2 + (x - vt)^2\gamma^2$. Again, Eqs. (6.50) can be obtained by applying a suitable Lorentz transformation to the EMF strengths in the dipole rest frame.

6.3.2. ELECTROMAGNETIC FIELD OF A MOVING POINT-LIKE TOROIDAL SOLENOID

Consider the poloidal current (Fig. 6.2) flowing on the surface of a torus

$$(\rho - d)^2 + z^2 = R_0^2$$

(R_0 and d are the minor and large radii of torus). It is convenient to introduce coordinates $\rho = d + R\cos\psi$, $z = R\sin\psi$ (Fig. 6.3). In these coordinates the poloidal current flowing on the torus surface is given by

$$\vec{j} = j_0\frac{\delta(R_0 - R)}{d + R_0\cos\psi}\vec{n}_\psi.$$

Here $\vec{n}_\psi = \vec{n}_z\cos\psi - \vec{n}_\rho\sin\psi$ is the vector lying on the torus surface and defining the current direction, $R = \sqrt{(\rho - d)^2 + z^2}$. The cylindrical components of $\vec{j}$ are

$$j_z = j_0\frac{\delta(R_0 - R)}{d + R_0\cos\psi}\cos\psi, \quad j_\rho = -\frac{\delta(R_0 - R)}{d + R_0\cos\psi}\sin\psi.$$

The velocity is along the torus axis

Let this current distribution move uniformly along the z axis (directed along the torus symmetry axis) with the velocity v (Fig. 6.6(a)). In the LF the non-vanishing charge and current components are

$$\rho_{Ch} = j_0\gamma\beta\frac{\rho - d}{c\rho R_0}\delta(R_0 - R_2), \quad j_\rho = -j_0\gamma\frac{z - vt}{\rho R_0}\delta(R_0 - R_2),$$

$$j_z = j_0\gamma\frac{\rho - d}{\rho R_0}\delta(R_0 - R_2). \qquad (6.51)$$

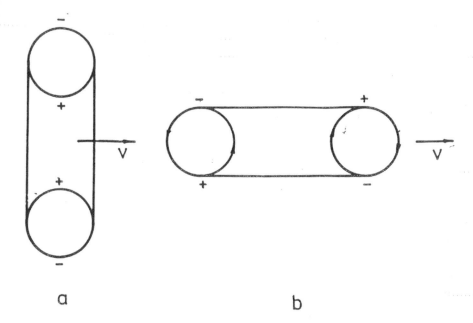

a
b

Figure 6.6. The induced charge densities for the cases in which the symmetry axes of a moving toroidal solenoid are along the velocity (a) or perpendicular to it (b).

Here $R_2 = \sqrt{(\rho - d)^2 + (z - vt)^2 \gamma^2}$. These components may be represented in the form

$$j_z = \frac{1}{\rho}\frac{\partial}{\partial\rho}(\rho M_\phi), \quad j_\rho = -\frac{1}{\gamma^2}\frac{\partial M_\phi}{\partial z}, \quad \rho_{Ch} = \frac{\beta}{c\rho}\frac{\partial}{\partial\rho}(\rho M_\phi). \tag{6.52}$$

Here

$$M_\phi = -j_0\gamma\frac{1}{\rho}\Theta(R_0 - R_2).$$

The Cartesian components of $\vec{M}$ are

$$M_x = j_0\gamma\frac{y}{\rho^2}\Theta(R_0 - R_2), \quad M_y = -j_0\gamma\frac{x}{\rho^2}\Theta(R_0 - R_2). \tag{6.53}$$

Let the minor radius R_0 tend to zero. Then

$$\Theta\left[R_0 - \sqrt{(\rho - d)^2 + (z - vt)^2\gamma^2}\right] \to \frac{\pi R_0^2}{\gamma}\delta(\rho - d)\delta(z - vt)$$

and

$$M_x = -\frac{j_0}{d}\pi R_0^2 \frac{\partial}{\partial y}\Theta(d - \rho)\delta(z - vt),$$

$$M_y = \frac{j_0}{d}\pi R_0^2 \frac{\partial}{\partial x}\Theta(d - \rho)\delta(z - vt). \tag{6.54}$$

Therefore

$$j_x = -\frac{1}{\gamma^2}\frac{\partial M_y}{\partial z} = -\frac{j_0\pi R_0^2}{\gamma^2 d}\frac{\partial^2}{\partial z\partial x}\Theta(d - \rho)\delta(z - vt),$$

$$j_y = \frac{1}{\gamma^2}\frac{\partial M_x}{\partial z} = -\frac{j_0\pi R_0^2}{\gamma^2 d}\frac{\partial^2}{\partial z\partial y}\Theta(d - \rho)\delta(z - vt),$$

$$j_z = \frac{\partial M_y}{\partial x} - \frac{\partial M_x}{\partial y} = \frac{j_0\pi R_0^2}{d}\left(\frac{\partial^2}{\partial x^2} + \frac{\partial^2}{\partial y^2}\right)\Theta(d - \rho)\delta(z - vt),$$

$$\rho_{Ch} = \frac{\beta}{c}\left(\frac{\partial M_y}{\partial x} - \frac{\partial M_x}{\partial y}\right)$$

$$= \frac{\beta j_0\pi R_0^2}{cd}\left(\frac{\partial^2}{\partial x^2} + \frac{\partial^2}{\partial y^2}\right)\Theta(d - \rho)\delta(z - vt). \tag{6.55}$$

Let the major torus radius also tend to zero. Then

$$\Theta(d - \rho) = \pi d^2\delta(x)\delta(y)$$

and

$$j_x = -\frac{j_0\pi^2 R_0^2 d}{\gamma^2}\frac{\partial^2}{\partial z\partial x}\delta(x)\delta(y)\delta(z - vt),$$

$$j_y = -\frac{j_0\pi R_0^2 d}{\gamma^2 d}\frac{\partial^2}{\partial z\partial y}\delta(x)\delta(y)\delta(z - vt),$$

$$j_z = j_0\pi^2 R_0^2 d\left(\frac{\partial^2}{\partial x^2} + \frac{\partial^2}{\partial y^2}\right)\delta(x)\delta(y)\delta(z - vt),$$

$$\rho_{Ch} = \frac{\beta j_0\pi^2 R_0^2 d}{c}\left(\frac{\partial^2}{\partial x^2} + \frac{\partial^2}{\partial y^2}\right)\delta(x)\delta(y)\delta(z - vt). \tag{6.56}$$

From this one easily obtains the electromagnetic potentials

$$\Phi = \frac{\beta m_t}{\epsilon}\left[\Delta - (1 - \beta_n^2)\frac{\partial^2}{\partial z^2}\right]\alpha, \quad A_x = -\frac{m_t\mu}{\gamma^2}\frac{\partial^2}{\partial z\partial x}\alpha,$$

$$A_y = -\frac{m_t\mu}{\gamma^2}\frac{\partial^2}{\partial z\partial y}\alpha, \quad A_z = m_t\mu\left[\Delta - (1 - \beta_n^2)\frac{\partial^2}{\partial z^2}\right]\alpha, \tag{6.57}$$

where α is the same as in Eqs. (6.35) and (6.36) and $m_t = \pi^2 j_0 d R_0^2/c$. Being written in a manifest form, the electromagnetic potentials are

$$\Phi = \frac{\beta m_t}{\epsilon}(1 - \beta_n^2)\frac{1}{r_m^3}\left[1 - 3\frac{(z - vt)^2}{r_m^2}\right], \quad A_x = -3\frac{m_t\mu}{\gamma^2}(1 - \beta_n^2)\frac{x(z - vt)}{r_m^5},$$

$$A_y = -3\frac{m_t\mu}{c\gamma^2}(1 - \beta_n^2)\frac{y(z - vt)}{r_m^5}, \quad A_z = \mu m_t(1 - \beta_n^2)\frac{1}{r_m^3}\left[1 - 3\frac{(z - vt)^2}{r_m^2}\right]$$

for $\beta_n < 1$ and

$$\Phi = \frac{2\beta m_t}{\epsilon}\{\frac{1 - \beta_n^2}{r_m^3}\left[1 - 3(1 - \beta_n^2)\frac{(z - vt)^2}{r_m^2}\right]\Theta(vt - z - \rho/\gamma_n)$$

$$+2(1 - \beta_n^2)\frac{\rho}{\gamma_n r_m^3}\delta(vt - z - \rho/\gamma_n)$$

$$+\frac{1}{r_m}\left[\frac{1}{\gamma_n^2}\dot{\delta}(vt - z - \rho/\gamma_n) - \frac{1}{\gamma_n\rho}\delta(vt - z - \rho/\gamma_n)\right]\},$$

$$A_z = 2\mu m_t\{\frac{1 - \beta_n^2}{r_m^3}\left[1 - 3(1 - \beta_n^2)\frac{(z - vt)^2}{r_m^2}\right]\Theta(vt - z - \rho/\gamma_n)$$

$$+2(1 - \beta_n^2)\frac{\rho}{\gamma_n r_m^3}\delta(vt - z - \rho/\gamma_n)$$

$$+\frac{1}{r_m}\left[\frac{1}{\gamma_n^2}\dot{\delta}(vt - z - \rho/\gamma_n) - \frac{1}{\gamma_n\rho}\delta(vt - z - \rho/\gamma_n)\right]\},$$

$$A_\rho = -\frac{2m_t\mu\rho}{\gamma^2}[3(1 - \beta_n^2)\frac{z - vt}{r_m^5}\Theta(vt - z - \rho/\gamma_n)$$

$$+\frac{1}{r_m^3}\left(\frac{z - vt}{\rho} + 1 - \beta_n^2\right)\delta(vt - z - \rho/\gamma_n) + \frac{1}{r_m\rho\gamma_n}\dot{\delta}(vt - z - \rho/\gamma_n)]. \quad (6.58)$$

for $\beta_n > 1$ (the dot above delta function means a derivative over its argument). In the past, the scalar electric potential Φ for $\beta_n < 1$ was found in [9]. The electromagnetic field strengths are equal to

$$E_x = -\frac{\beta m_t}{\epsilon}\left[\Delta + (n^2 - 1)\frac{\partial^2}{\partial z^2}\right]\frac{\partial\alpha}{\partial x},$$

$$E_y = -\frac{\beta m_t}{\epsilon}\left[\Delta + (n^2 - 1)\frac{\partial^2}{\partial z^2}\right]\frac{\partial\alpha}{\partial y},$$

$$E_z = \frac{\beta m_t}{\epsilon}(n^2 - 1)\left[\Delta + (\beta_n^2 - 1)\frac{\partial^2}{\partial z^2}\right]\frac{\partial\alpha}{\partial z},$$

$$B_x = m_t\mu\left[\Delta + \beta^2(n^2 - 1)\frac{\partial^2}{\partial z^2}\right]\frac{\partial\alpha}{\partial y},$$

$$B_y = -m_t \mu \left[\Delta + \beta^2 (n^2 - 1) \frac{\partial^2}{\partial z^2} \right] \frac{\partial \alpha}{\partial x},$$

$$B_z = 0, \quad \Delta = \frac{\partial^2}{\partial x^2} + \frac{\partial^2}{\partial y^2} + (1 - \beta_n^2) \frac{\partial^2}{\partial z^2}. \tag{6.59}$$

For $\beta_n < 1$ the EMF falls as r^{-4} at large distances. For $\beta_n > 1$ the EMF field strengths are equal to zero outside the Cherenkov cone; inside this cone, they fall like r^{-4} for $r \to \infty$ and they are infinite on the Cherenkov cone.

We write out the EMF in a manifest form for $\beta_n < 1$:

$$E_x = -\frac{\beta m_t}{\epsilon} \frac{3x}{r^5}(n^2 - 1)\gamma_n^3 F, \quad E_y = -\frac{\beta m_t}{\epsilon} \frac{3y}{r^5}(n^2 - 1)\gamma_n^3 F,$$

$$E_z = -\frac{\beta m_t}{\epsilon}(n^2 - 1)\frac{3(z - vt)}{r^5}\gamma_n^3 F, \quad B_x = m_t \mu \frac{3y}{r^5}\gamma_n^3 \beta^2(n^2 - 1)F,$$

$$B_y = -m_t \mu \frac{3x}{r^5}\beta^2 \gamma_n^3(n^2 - 1)F, \quad B_z = 0, \quad F = 1 - 5\frac{\gamma_n^2(z - vt)^2}{r^2}. \tag{6.60}$$

It is seen that the electric field of an elementary toroidal solenoid moving in the non-dispersive medium strongly resembles the field of an electric quadrupole. As the magnetic field in (6.60) has only the ϕ component, it cannot be reduced to the field of a magnetic quadrupole. Provisionally. it may be called the field of the moving toroidal moment.

The electromagnetic strengths and inductions in the reference frame in which the toroidal dipole is at rest and the medium moves with the velocity $-\vec{v}$, are equal to

$$B'_x = \frac{m_t \gamma}{\epsilon} \beta^2 \gamma_n^3(n^2 - 1)^2 \frac{3y'}{r'^5}F', \quad F' = 1 - 5\frac{\gamma_n^2 z'^2}{\gamma^2 r'^2},$$

$$B'_y = -\frac{m_t \gamma}{\epsilon} \beta^2 \gamma_n^3(n^2 - 1)^2 \frac{3x'}{r'^5}F', \quad B'_z = 0, \quad \vec{H}' = 0,$$

$$E'_x = \frac{m_t \gamma \beta}{\epsilon}(1 - n^2)\gamma_n \frac{3x'}{r'^5}F', \quad E'_y = \frac{m_t \gamma \beta}{\epsilon}(1 - n^2)\gamma_n \frac{3y'}{r'^5}F',$$

$$E'_z = \frac{m_t \beta}{\epsilon}(1 - n^2)\gamma_n^3 \frac{3z'}{\gamma r'^5}F', \quad D'_x = -\beta m_t \gamma(n^2 - 1)\gamma_n^3 \frac{3x'}{r'^5}F',$$

$$D'_y = -\beta m_t \gamma(n^2 - 1)\gamma_n^3 \frac{3y'}{r'^5}F', \quad D'_z = -\beta m_t(n^2 - 1)\frac{\gamma_n^3}{\gamma} \frac{3z'}{r'^5}F'. \tag{6.61}$$

Here $r'^2 = (x'^2 + y'^2) + z'^2 \gamma_n^2/\gamma^2$. It is seen that $\vec{H}'$ differs from zero only at the toroidal dipole position (the term with δ function is omitted), whilst $\vec{B}'$, $\vec{D}'$, and $\vec{E}'$ differ from zero everywhere. In this reference frame there are no relations $\vec{B}' = \mu\vec{H}'$, $\vec{D}' = \epsilon\vec{E}'$ which are valid only in the reference frame where the medium is at rest. Instead Eqs. (6.27) should be used.

From the inspection of Eqs. (6.59)-(6.61) we conclude:

i) For a TS being at rest either in the vacuum or in the medium the EMF differs from zero only inside the TS.

ii) For a TS moving in vacuum with a constant velocity the EMF differs from zero only inside the TS. Without any calculations this can be proved by applying the Lorentz transformation to the EMF strengths of a TS at rest. Since this transformation is linear and since the EMF strengths vanish for a TS at rest, they vanish for a moving TS as well.

iii) The EMF of a TS moving in the medium differs from zero both inside and outside the TS. At first glance this seems to be incorrect. In fact, let TS initially be at rest in the medium. Let us pass to the Lorentz reference frame 1 in which the TS velocity is v. In this frame the EMF strengths vanish outside the TS. Both the TS and medium move with the velocity V relative this frame. However, Eqs. (6.59),(6.60) are valid in the frame 2 relative to which the medium is at rest whilst a TS moves with the velocity v. Therefore, these reference frames are not equivalent. There is no Lorentz transformation relating them. In the spectral representation, these important facts were established previously in [9].

The velocity is perpendicular to the torus axis

Let a toroidal solenoid move in medium with the velocity perpendicular to the torus symmetry axis (Fig. 6(b)). For definiteness, let the TS move along the x axis. Then in the LF

$$\rho_{Ch} = -\frac{j_0 v \gamma^2}{c^2 R_0}\frac{z(x-vt)}{\rho_1^2}\delta(R_1 - R_0), \quad j_x = -j_0\frac{\gamma^2}{R_0}\frac{z(x-vt)}{\rho_1^2}\delta(R_1 - R_0),$$

$$j_y = -j_0\frac{zy}{\rho_1^2}\frac{\delta(R_1 - R_0)}{R_0}, \quad j_z = j_0\frac{\rho_1 - d}{\rho_1}\frac{\delta(R_1 - R_0)}{R_0}.$$

Here

$$\rho_1 = \sqrt{(x-vt)^2\gamma^2 + y^2}, \quad R_1 = \sqrt{(\rho_1 - d)^2 + z^2}.$$

It is easy to check that

$$j_x = -\frac{\partial M_y}{\partial z}, \quad j_y = \frac{\partial M_x}{\partial z}, \quad j_z = \frac{1}{\gamma^2}\frac{\partial M_y}{\partial x} - \frac{\partial M_x}{\partial y},$$

$$\rho_{Ch} = -\frac{\beta}{c}\frac{\partial M_y}{\partial z}, \tag{6.62}$$

where

$$M_y = -j_0\gamma^2\frac{x-vt}{\rho_1^2}\Theta(R_0 - R_1), \quad M_x = j_0\frac{y}{\rho_1^2}\Theta(R_0 - R_1), \quad M_z = 0.$$

Let the minor radius R_0 of a torus tend to zero. Then

$$\Theta(R_0 - R_1) = \pi R_0^2 \delta(\rho_1 - d)\delta(z)$$

and

$$M_x = -j_0 \frac{\pi R_0^2}{d} \frac{\partial}{\partial y} \Theta(d - \rho_1)\delta(z), \quad M_y = j_0 \frac{\pi R_0^2}{d} \frac{\partial}{\partial x} \Theta(d - \rho_1)\delta(z).$$

Therefore

$$\rho_{Ch} = -\frac{\beta j_0 \pi R_0^2}{cd} \frac{\partial^2}{\partial x \partial z} \Theta(d - \rho_1)\delta(z), \quad j_x = -\frac{j_0 \pi R_0^2}{d} \frac{\partial^2}{\partial x \partial z} \Theta(d - \rho_1)\delta(z),$$

$$j_y = -\frac{j_0 \pi R_0^2}{d} \frac{\partial^2}{\partial y \partial z} \Theta(d - \rho_1)\delta(z),$$

$$j_z = \frac{j_0 \pi R_0^2}{d\gamma^2} \frac{\partial^2}{\partial x^2} \Theta(d - \rho_1)\delta(z) + \frac{j_0 \pi R_0^2}{d} \frac{\partial^2}{\partial y^2} \Theta(d - \rho_1)\delta(z).$$

Now we let the major radius d go to zero. Then

$$\Theta(d - \rho_1) = \frac{\pi d^2}{\gamma} \delta(x - vt)\delta(y),$$

$$\rho_{Ch} = -\frac{\beta j_0 \pi^2 dR_0^2}{c\gamma} \frac{\partial^2}{\partial x \partial z} \delta(x - vt)\delta(y)\delta(z),$$

$$j_x = -\frac{j_0 \pi^2 dR_0^2}{\gamma} \frac{\partial^2}{\partial x \partial z} \delta(x - vt)\delta(y)\delta(z),$$

$$j_y = -\frac{j_0 \pi^2 dR_0^2}{\gamma} \frac{\partial^2}{\partial y \partial z} \delta(x - vt)\delta(y)\delta(z),$$

$$j_z = \frac{j_0 \pi^2 dR_0^2}{\gamma^3} \frac{\partial^2}{\partial x^2} \delta(x-vt)\delta(y)\delta(z) + \frac{j_0 \pi^2 dR_0^2}{\gamma} \frac{\partial^2}{\partial y^2} \delta(x-vt)\delta(y)\delta(z). \quad (6.63)$$

As a result we arrive at the following electromagnetic potentials:

$$\Phi = -\frac{\beta m_t}{\gamma \epsilon} \frac{\partial^2}{\partial x \partial z} \alpha_1, \quad A_x = -\frac{m_t \mu}{\gamma} \frac{\partial^2}{\partial x \partial z} \alpha_1,$$

$$A_y = -\frac{m_t \mu}{\gamma} \frac{\partial^2}{\partial y \partial z} \alpha_1, \quad A_z = \frac{m_t \mu}{\gamma^3} \frac{\partial^2}{\partial x^2} \alpha_1 + \frac{m_t \mu}{\gamma} \frac{\partial^2}{\partial y^2} \alpha_1,$$

where α_1 is given by (6.46). In the manifest form the EMF potentials are given by

$$\Phi = -\frac{3\beta m_t}{\epsilon \gamma \gamma_n^2} \frac{(x - vt)z}{r_m^5}, \quad A_x = -\frac{3\mu m_t}{\gamma \gamma_n^2} \frac{(x - vt)z}{r_m^5},$$

$$A_y = -\frac{3\mu m_t}{\gamma\gamma_n^4}\frac{yz}{r_m^5}, \quad A_z = \frac{m_t\mu}{\gamma}\left[\frac{1}{\gamma^2 r_m^3}\left(1 - 3\frac{x^2}{r_m^2}\right) + \frac{1}{\gamma_n^2 r_m^3}\left(1 - 3\frac{y^2}{r_m^2\gamma_n^2}\right)\right].$$

Electromagnetic field strengths are

$$E_x = \frac{\beta m_t}{\gamma\epsilon}(1 - n^2)\frac{\partial^3 \alpha_1}{\partial x^2 \partial z}, \quad E_y = \frac{\beta m_t}{\gamma\epsilon}(1 - n^2)\frac{\partial^3 \alpha_1}{\partial x \partial y \partial z},$$

$$E_z = \frac{\beta m_t}{\gamma\epsilon}\left[(n^2 - 1)\left(\frac{\partial^2}{\partial x^2} + \frac{\partial^2}{\partial y^2}\right) + \tilde{\Delta}\right]\frac{\partial \alpha_1}{\partial x},$$

$$B_x = \frac{\mu m_t}{\gamma}\left[\tilde{\Delta} + \beta^2(n^2 - 1)\frac{\partial^2}{\partial x^2}\right]\frac{\partial \alpha_1}{\partial y},$$

$$B_y = -\frac{\mu m_t}{\gamma}\left[\tilde{\Delta} + \beta^2(n^2 - 1)\frac{\partial^2}{\partial x^2}\right]\frac{\partial \alpha_1}{\partial x},$$

$$B_z = 0, \quad \tilde{\Delta} = (1 - \beta_n^2)\frac{\partial^2}{\partial x^2} + \frac{\partial^2}{\partial y^2} + \frac{\partial^2}{\partial z^2}. \tag{6.64}$$

It is seen that electromagnetic field strengths are equal to zero outside the Cherenkov cone, fall like r^{-4} at large distances inside this cone, and are infinite on the Cherenkov cone. Since for $\beta_n < 1$, $\tilde{\Delta}\alpha_1 = -4\pi\delta(x - vt)\delta(y)\delta(z)$ one may drop the $\tilde{\Delta}$ operators in (6.64). This confirms the previous result that EMF goes beyond a TS moving in medium.

6.3.3. ELECTROMAGNETIC FIELD OF A MOVING POINT-LIKE ELECTRIC DIPOLE

Consider an electric dipole consisting of point electric charges:

$$\rho_d = e[\delta^3(\vec{r} + a\vec{n}) - \delta^3(\vec{r} - a\vec{n})].$$

Here $\vec{r}$ defines the dipole center of mass, $2a$ is the distance between charges and vector $\vec{n}$ defines the dipole orientation. Let the dipole move uniformly along the z axis (Fig. 6.7). Then,

$$\rho_d = e\gamma\{\delta(x + an_x)\delta(y + an_y)\delta[(z - vt)\gamma + an_z]$$

$$-\delta(x - an_x)\delta(y - an_y)\delta[(z - vt)\gamma - an_z]\}, \quad j_z = v\rho_d.$$

Let the distance between charges tend to zero. Then

$$\rho_d = 2ea(\vec{n}\vec{\nabla})\delta(x)\delta(y)\delta(z - vt), \quad j_z = v\rho_d. \tag{6.65}$$

Here

$$(\vec{n}\vec{\nabla}) = \vec{n}_x\nabla_x + \vec{n}_y\nabla_y + \frac{1}{\gamma}\vec{n}_z\nabla_z, \quad \nabla_i = \frac{\partial}{\partial x_i}.$$

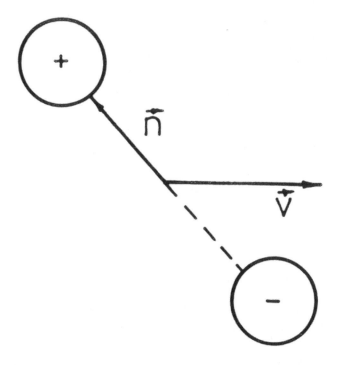

Figure 6.7.   A moving electric dipole with arbitrary orientation relative to its velocity.

The electromagnetic potentials are equal to

$$\Phi = \frac{2ea}{\epsilon}(\vec{n}\vec{\nabla})\alpha, \quad A_z = 2ea\mu\beta(\vec{n}\vec{\nabla})\alpha,$$

where α is the same as in (6.36). In a manifest form the electromagnetic potentials are

$$\Phi = -\frac{2ea}{\sqrt{1-\beta_n^2}\,\epsilon r^3}(\vec{n}\vec{r}), \quad A_z = -\frac{2ea\mu\beta}{\sqrt{1-\beta_n^2}\,r^3}(\vec{n}\vec{r}),$$

$$(\vec{n}\vec{r}) = xn_x + yn_y + n_z(z - vt)\frac{\sqrt{1-\beta^2}}{1-\beta_n^2}, \quad r^2 = x^2 + y^2 + \frac{(z-vt)^2}{1-\beta_n^2} \quad (6.66)$$

for $\beta_n < 1$ and

$$\Phi = \frac{4ea(\vec{n}\vec{r})}{\epsilon r_1^3}R, \quad A_z = \frac{4ea\mu\beta(\vec{n}\vec{r})}{r_1^3}R \quad (6.67)$$

for $\beta_n > 1$. Here

$$R = \left[\frac{1}{\sqrt{\beta_n^2 - 1}} \Theta \left(vt - z - \rho\sqrt{\beta_n^2 - 1} \right) - \frac{r_1^2}{\rho} \delta \left(vt - z - \rho\sqrt{\beta_n^2 - 1} \right) \right]$$

and

$$r_1^2 = \frac{(z - vt)^2}{\beta_n^2 - 1} - \rho^2.$$

The non-vanishing electromagnetic field strengths are

$$E_x = -\frac{2ea}{\epsilon} \frac{\partial}{\partial x} (\vec{n}\vec{\nabla})\alpha, \quad E_y = -\frac{2ea}{\epsilon} \frac{\partial}{\partial y} (\vec{n}\vec{\nabla})\alpha,$$

$$E_z = -\frac{2ea}{\epsilon} (1 - \beta_n^2) \frac{\partial}{\partial z} (\vec{n}\vec{\nabla})\alpha,$$

$$B_x = 2ea\mu\beta \frac{\partial}{\partial y} (\vec{n}\vec{\nabla})\alpha, \quad B_y = -2ea\mu\beta \frac{\partial}{\partial x} (\vec{n}\vec{\nabla})\alpha. \tag{6.68}$$

It is seen that electromagnetic field strengths vanish outside the Cherenkov cone, inside this cone they fall as r^{-3} at large distances, and they are infinite on the Cherenkov cone.

We limit ourselves to the $\beta_n < 1$ case. The EMF is equal to

$$E_x = \frac{2ea}{\epsilon} \frac{\gamma_n}{r^3} \left[n_x - 3\frac{x}{r^2} (\vec{n}\vec{r}) \right], \quad E_y = \frac{2ea}{\epsilon} \frac{\gamma_n}{r^3} \left[n_y - 3\frac{y}{r^2} (\vec{n}\vec{r}) \right],$$

$$E_z = \frac{2ea}{\epsilon\gamma} \frac{\gamma_n}{r^3} \left[n_z - 3\gamma\frac{z - vt}{r^2} (\vec{n}\vec{r}) \right], \quad B_x = -2ea\mu\beta\gamma_n \frac{1}{r^3} \left[n_y - 3\frac{y}{r^2} (\vec{n}\vec{r}) \right],$$

$$B_y = 2ea\mu\beta\gamma_n \frac{1}{r^3} \left[n_x - 3\frac{x}{r^2} (\vec{n}\vec{r}) \right], \quad B_z = 0. \tag{6.69}$$

We see that $\vec{E}$ resembles the field of an electric dipole, whilst $\vec{H}$, having only two Cartesian components, cannot be interpreted as the field of a magnetic dipole.

In the reference frame in which the electric dipole is at rest

$$B'_x = \frac{2ea(1 - n^2)\beta\gamma\gamma_n}{\epsilon} \frac{1}{r'^3} \left[n_y - 3\frac{y'}{r'^2} (\vec{n}\vec{r}\,') \right],$$

$$B'_y = -\frac{2ea(1 - n^2)\beta\gamma\gamma_n}{\epsilon} \frac{1}{r'^3} \left[n_x - 3\frac{x'}{r'^2} (\vec{n}\vec{r}\,') \right], \quad \vec{H}' = 0,$$

$$E'_x = 2ea\frac{\gamma}{\gamma_n\epsilon} \frac{1}{r'^3} \left[n_x - 3\frac{x'}{r'^2} (\vec{n}\vec{r}\,') \right], \quad E'_y = 2ea\frac{\gamma}{\gamma_n\epsilon} \frac{1}{r'^3} \left[n_y - 3\frac{y'}{r'^2} (\vec{n}\vec{r}\,') \right],$$

$$E'_z = 2ea\frac{\gamma_n}{\gamma\epsilon}\frac{1}{r'^3}\left[n_z - 3\frac{z'}{r'^2}(\vec{n}\vec{r}\,')\right], \quad D'_x = \frac{2ea\gamma_n}{\gamma\epsilon}\frac{1}{r'^3}\left[n_x - 3\frac{x'}{r'^2}(\vec{n}\vec{r}\,')\right],$$

$$D'_y = \frac{2ea\gamma_n}{\gamma\epsilon}\frac{1}{r'^3}\left[n_y - 3\frac{y'}{r'^2}(\vec{n}\vec{r}\,')\right],$$

$$D'_z = 2ea\frac{\gamma_n}{\gamma r'^3}\left[n_z - 3\frac{z'}{r'^3}(\vec{n}\vec{r}\,')\right]. \tag{6.70}$$

In this reference frame the constitutive relations (6.27) should be used. For the vector $\vec{n}$ oriented along the motion axis, one gets

$$E_x = -6ea\gamma_n^3\frac{x(z-vt)}{\gamma\epsilon r^5}, \quad E_y = -6ea\gamma_n^3\frac{y(z-vt)}{\gamma\epsilon r^5},$$

$$E_z = \frac{2ea}{\epsilon}\left[\frac{\gamma_n}{r^3}(1 - 3\frac{\gamma_n^2(z-vt)^2}{r^2})\right],$$

$$B_x = 6\mu\beta ea\gamma_n^3\frac{y(z-vt)}{\gamma r^5}, \quad B_y = -6\mu\beta ea\gamma_n^3\frac{x(z-vt)}{\gamma r^5}, \tag{6.71}$$

where $r^2 = x^2 + y^2 + (z-vt)^2\gamma_n^2$ is the same as in (6.36).

For the vector $\vec{n}$ perpendicular to the motion axis (say, $\vec{n}$ is in the x direction) the field strengths are

$$E_x = \frac{2ea}{\epsilon}\frac{\gamma_n}{r^3}\left[1 - 3\frac{x^2}{r^2}\right], \quad E_y = -6ea\gamma_n\frac{xy}{\epsilon r^5}, \quad E_z = -6ea\gamma_n\frac{x(z-vt)}{\epsilon r^5},$$

$$B_x = 6ea\mu\gamma_n\beta\frac{xy}{r^5}, \quad B_y = 2ea\mu\beta\frac{\gamma_n}{r^3}\left[1 - 3\frac{x^2}{r^2}\right]. \tag{6.72}$$

6.3.4. ELECTROMAGNETIC FIELD OF INDUCED DIPOLE MOMENTS

Now we apply the formalism developed by Frank to evaluate the EMF of moving magnetic and electric dipoles.

Electromagnetic field of a moving magnetic dipole
In our translation from Russian, the Frank prescription for the evaluation of EMF of the moving dipole, may be formulated as follows ([6], p. 190):

> It is suggested that a moving electric dipole p'_1 is equivalent to some dipoles at rest, namely, to the electric p_1 and magnetic m_1 placed at the point coinciding with the instantaneous position of a moving dipole. The same is suggested for a magnetic dipole.

According to this prescription the moving magnetic dipole $\vec{m}'$ creates the following magnetic $\vec{m}$ and electric $\vec{p}$ dipole moments in the LF:

$$\vec{m} = \vec{m}' - (1 - \sqrt{1 - \beta^2})\vec{v}(\vec{v}\vec{m}')/v^2, \quad \vec{p} = (\vec{\beta} \times \vec{m}'), \quad \vec{\beta} = \vec{v}/c. \quad (6.73)$$

For the $\vec{m}'$ directed along the motion axis, (6.73) passes into

$$m_x = m_y = 0, \quad m_z \equiv m = m'/\gamma, \quad \vec{p} = 0. \quad (6.74)$$

The EMF of induced dipoles (6.74) which are at rest in the instantaneous position of the moving magnetic dipole (this is essentially the Frank prescription) is given by

$$\vec{E}_d = 0, \quad B_x^d = 3m\frac{\gamma x(z - vt)}{r^5}, \quad B_y^d = 3m\frac{\gamma y(z - vt)}{r^5},$$

$$B_z^d = -m\left(\frac{1}{r^3} - 3\frac{\gamma^2(z - vt)^2}{r^5}\right). \quad (6.75)$$

Here $r = [x^2 + y^2 + \gamma^2(z - vt)^2]^{1/2}$. By comparing (6.75) with (6.40) we conclude that the magnetic field of a moving point-like current loop resembles (but not coincides with) that of a magnetic dipole. The non-trivial dependence on γ_n in (6.40) tells us that the magnetic field of a moving magnetic dipole cannot be obtained by the simple Frank prescription (6.73).

Furthermore, the Frank prescription (6.74) gives a zero electric field, while the exact electric field (6.40) differs from zero. Another way to see this is to write out the electric field created by the induced electric dipole $\vec{p}$ which is at rest in the instantaneous position of a moving magnetic dipole:

$$(E_d)_x = -\frac{1}{r^3}p_x + 3x\frac{xp_x + yp_y + \gamma(z - vt)p_z}{r^5},$$

$$(E_d)_y = -\frac{1}{r^3}p_y + 3y\frac{xp_x + yp_y + \gamma(z - vt)p_z}{r^5},$$

$$(E_d)_z = -\frac{1}{r^3}p_z + 3\gamma z\frac{xp_x + yp_y + \gamma(z - vt)p_z}{r^5}, \quad (6.76)$$

where $r^2 = x^2 + y^2 + (z - vt)^2\gamma^2$.

The exact electric field of a moving point-like current loop has only the ϕ component (see (6.40)). It is easy to check that it is impossible to vanish simultaneously E_ρ and E_z for any choice of p_x, p_y, p_z. This means that the electric field (6.40) produced by a moving magnetic dipole cannot be associated with the field of the induced electric dipole.

For the $\vec{m}'$ perpendicular to the motion axis (for definiteness, let the motion and symmetry axes be along the x and z axes, respectively.) Eq.(6.73) gives

$$m_x = m_y = 0, \quad m_z = m = m', \quad p_y = -\beta m.$$

The EMF generated by this dipole moment is

$$(E_d)_x = -3\beta m\gamma\frac{y(x - vt)}{r^5}, \quad (E_d)_y = \frac{\beta m}{r^3} - 3\beta m\frac{y^2}{r^5}, \quad (E_d)_z = -3\beta m\frac{yz}{r^5},$$

$$(B_d)_x = 3\gamma m\frac{z(x - vt)}{r^5}, \quad (B_d)_y = 3\gamma m\frac{yz}{r^5}, \quad (B_d)_z = -\frac{m}{r^3} + 3m\frac{z^2}{r^5}.$$

$$(6.77)$$

These expressions slightly resemble the exact ones (6.49), but not reduce to them (again, owing to the nontrivial γ_n dependence in (6.49)).

The situation remains essentially the same if instead of $\vec{p}$ given by (6.1), the modified Frank formula ([6])

$$\vec{p} = n^2(\beta \times m'), \quad n^2 = \epsilon\mu \tag{6.78}$$

is used.

Electromagnetic field of a moving electric dipole
According to Frank a moving electric dipole $\vec{p}'$ creates the following magnetic $\vec{m}$ and electric $\vec{p}$ dipole moments in the LF:

$$\vec{p} = \vec{p}' - (1 - \sqrt{1 - \beta^2})\vec{v}(\vec{v}\vec{p}')/v^2, \quad \vec{m} = -\vec{\beta} \times \vec{p}'. \tag{6.79}$$

For $\vec{p}'$ aligned along the motion axis z this reduces to

$$p_x = p_y = 0, \quad p_z = p'/\gamma, \quad \vec{m} = 0. \tag{6.80}$$

The EMF of induced dipoles (6.80) at rest in the instant position of the moving electric dipole is given by

$$E_x = p\frac{x\gamma(z - vt)}{r^5}, \quad E_y = p\frac{y\gamma(z - vt)}{r^5},$$

$$E_z = -\frac{p}{r^3} + 3p\frac{\gamma^2(z - vt)^2}{r^5}, \quad \vec{B} = 0. \tag{6.81}$$

By comparing this with (6.71) we conclude that the electric field (6.81) of an induced electric dipole resembles (but does not reduce to) the exact electric field (6.71) of a moving electric dipole. On the other hand, the magnetic field vanishes for the induced magnetic moment (6.80) which disagrees with the behaviour of the exact magnetic field (6.71) of the moving electric dipole. The latter cannot be attributed to the magnetic dipole.

For an electric dipole oriented perpendicularly (say, in the x direction) to the motion direction z, one obtains from (6.79) for the non-vanishing components of induced dipole moments

$$p_x \equiv p = p', \quad m_y = -\beta p. \tag{6.82}$$

The corresponding EMF is

$$E_x = -\frac{p}{r^3} + 3p\frac{x^2}{r^2}, \quad E_y = 3p\frac{xy}{r^5}, \quad E_z = 3p\frac{x\gamma(z - vt)}{r^5},$$

$$B_x = -3\beta p\frac{xy}{r^5}, \quad B_y = \frac{\beta p}{r^3} - 3\beta p\frac{y^2}{r^5}, \quad B_z = -3\beta p\frac{y\gamma(z - vt)}{r^5}. \quad (6.83)$$

By comparing this with (6.72) we conclude that the electric field of an induced dipole moment resembles the exact electric field (6.72) of a moving electric dipole. On the other hand, there are three components of the magnetic field of the induced moment (6.82) and only two exact non-vanishing components in (6.72). Therefore the exact magnetic field (6.72) of a moving electric dipole cannot be attributed to the induced magnetic dipole (6.82).

6.4. Electromagnetic field of electric, magnetic, and toroidal dipoles in the spectral representation

We consider the radiation of electric, magnetic, and toroidal dipoles moving uniformly in an unbounded medium (this corresponds to the Tamm-Frank problem). They are obtained from the corresponding charge-current densities in an infinitesimal limit. The behaviour of radiation intensities in the neighbourhood of the Cherenkov threshold $\beta = 1/n$ is investigated. The frequency and velocity regions are defined where radiation intensities are maximal. The comparison with previous attempts is given. We consider also the radiation of electric, magnetic, and toroidal dipoles moving uniformly in medium, in a finite spatial interval (this corresponds to the Tamm problem). The properties of radiation arising from the precession of a magnetic dipole are also studied.

6.4.1. UNBOUNDED MOTION OF MAGNETIC, TOROIDAL, AND ELECTRIC DIPOLES IN MEDIUM

Pedagogical example: Uniform unbounded charge motion in medium
Consider first the uniform unbounded charge motion in medium along the z axis. Charge and current densities are given by

$$\rho_{Ch} = e\delta(z - vt)\delta(x)\delta(y), \quad j_z = e\rho_{Ch}.$$

Their Fourier components are given by

$$\rho_\omega = \frac{1}{2\pi}\int \rho_{Ch}\exp(i\omega t)dt = \frac{e}{2\pi v}\delta(x)\delta(y)\exp\left(\frac{ikz}{\beta}\right),$$

$$j_\omega = v\rho_\omega, \quad k = \frac{\omega}{c}.$$

The electromagnetic potentials corresponding to these densities are

$$\Phi = \frac{1}{2\pi v\epsilon} \int \exp\left[ik\left(\frac{z'}{\beta} + nR\right)\right]\frac{dz'}{R}, \quad A_z = \mu\epsilon\beta\Phi. \tag{6.84}$$

Here $R = [x^2 + y^2 + (z - z')^2]^{1/2}$, ϵ and μ are the electric and magnetic constants of the medium, $n = \sqrt{\epsilon\mu}$ is its refractive index. Making the change of the integration variable $z' = z + \rho\sinh\chi$, we rewrite (6.84) in the form

$$\Phi = \frac{1}{2\pi v\epsilon}\alpha, \quad A_z = \mu\epsilon\beta\Phi,$$

where

$$\alpha = \exp(i\psi)I,$$

$$I = \int_{-\infty}^{\infty} \exp\left[ik\rho\left(\frac{\sinh\chi}{\beta} + n\cosh\chi\right)\right]d\chi, \quad \psi = \frac{kz}{\beta}. \tag{6.85}$$

The integral I can be evaluated in a closed form [21] (see also Chapter 2). It is given by

$$I = 2K_0 \quad \text{for} \quad v < c_n$$

and

$$I = i\pi H_0^{(1)} \quad \text{for} \quad v > c_n, \tag{6.86}$$

where the arguments of all Bessel functions are $k\rho/\beta\gamma_n$, $\gamma_n = |1 - \beta_n^2|^{-1/2}$, $\beta_n = \beta n$ and $c_n = c/n$ is the velocity of light in medium. The scalar electric potential is given by

$$\Phi = \frac{e}{\pi v\epsilon}\exp(i\psi)K_0$$

for $v < c_n$ and

$$\Phi = \frac{ie}{2v\epsilon}\exp(i\psi)H_0^{(1)}$$

for $v > c_n$. The magnetic potential is $A_z = \beta\epsilon\mu\Phi$. Correspondingly, the electromagnetic field strengths are equal to

$$E_\rho = \frac{ek}{\pi v\epsilon\beta\gamma_n}\exp(i\psi)K_1, \quad E_z = -\frac{iek}{\pi v\epsilon\beta}(1 - \beta^2 n^2)\exp(i\psi)K_0,$$

$$H_\phi = \frac{ek}{\pi v\gamma_n}\exp(i\psi)K_1$$

for $\beta_n < 1$ and

$$E_\rho = i\frac{ek}{2v\epsilon\beta\gamma_n}\exp(i\psi)H_1^{(1)}, \quad E_z = \frac{ek}{2v\epsilon\beta}(1 - \beta^2 n^2)\exp(i\psi)H_0^{(1)},$$

$$H_\phi = i\frac{ek}{2v\gamma_n}\exp(i\psi)H_1^{(1)}$$

for $\beta_n > 1$.

The radial energy flux per unit length and per unit frequency through the surface of the cylinder of radius ρ coaxial with the motion axis is given by

$$\sigma_\rho = \frac{d^2\mathcal{E}}{d\omega dz} = -\pi\rho c(E_z H_\phi^* + E_z^* H_\phi).$$

It is equal to zero for $\beta_n < 1$ and

$$\sigma_\rho = \frac{e^2\omega\mu}{c^2}\left(1 - \frac{1}{\beta^2 n^2}\right) \tag{6.87}$$

for $\beta_n > 1$, which coincides with the frequency distribution of radiation given by Tamm and Frank.

Radiation of magnetic dipole uniformly moving in medium
The magnetic dipole is parallel to the velocity. Let a constant current I flow in a current loop. In the time representation the current density in the LF is given by Eqs. (6.30)-(6.32). The Fourier components of this current density are

$$j_x(\omega) = \partial M_z(\omega)/\partial y, \quad j_y(\omega) = -\partial M_z(\omega)/\partial x, \quad j_z(\omega) = 0,$$

where

$$M_z(\omega) = \frac{Id^2}{2\gamma v}\delta(x)\delta(y)\exp(i\psi)$$

and ψ is the same as in (6.85). The vector magnetic potential satisfies the equation

$$\triangle \vec{A}_\omega + k_n^2 \vec{A}_\omega = -\frac{4\pi\mu}{c}\vec{j}_\omega, \quad k_n = kn.$$

Its non-vanishing components are given by

$$A_x = \frac{\mu m_d}{2\pi\gamma v}\frac{\partial\alpha}{\partial y}, \quad A_y = -\frac{\mu m_d}{2\pi\gamma v}\frac{\partial\alpha}{\partial x},$$

where α is the same as in (6.85) and $m_d = I\pi d^2/c$ is the magnetic moment of the current loop in its rest frame. It is seen that only the ϕ component of $\vec{A}_\omega$ differs from zero:

$$A_\omega = -\frac{m_d\mu}{2\pi\gamma v}\frac{\partial\alpha}{\partial\rho}.$$

The electromagnetic field strengths are

$$E_\phi = -\frac{ikm_d\mu}{2\pi\gamma v}\frac{\partial\alpha}{\partial\rho}, \quad H_\rho = \frac{ikm_d}{2\pi\gamma\beta v}\frac{\partial\alpha}{\partial\rho}, \quad H_z = \frac{m_d}{2\pi\gamma v\beta^2}k^2(\beta_n^2 - 1)\alpha.$$

In a manifest form, they are equal to

$$E_\phi = \frac{ik^2 m_d \mu}{\pi \beta \gamma_n \gamma v} \exp(i\psi) K_1, \quad H_\rho = -\frac{ik^2 m_d}{\pi \gamma_n \gamma \beta^2 v} \exp(i\psi) K_1,$$

$$H_z = -\frac{m_d k^2}{\pi \gamma v \beta^2 \gamma_n^2} \exp(i\psi) K_0$$

for $\beta_n < 1$ and

$$E_\phi = -\frac{k^2 m_d \mu}{2\beta \gamma_n \gamma v} \exp(i\psi) H_1^{(1)}, \quad H_\rho = -\frac{k^2 m_d}{2\gamma_n \gamma \beta^2 v} \exp(i\psi) H_1^{(1)},$$

$$H_z = i\frac{m_d k^2}{2\gamma v \beta^2 \gamma_n^2} \exp(i\psi) H_0^{(1)}$$

for $\beta_n > 1$. The energy emitted in the radial direction per unit length and per unit frequency

$$\sigma_\rho = \frac{d^2 \mathcal{E}}{d\omega dz} = -\pi \rho c (E_\phi H_z^* + H_z E_\phi^*)$$

is equal to zero for $v < c_n$ and

$$\sigma_\rho = \frac{\omega^3 m_d^2 \mu}{v^4 \gamma^2 \gamma_n^2} \tag{6.88}$$

for $v > c_n$. In the past, this equation was obtained by Frank in [6,9], but without the factor γ^2 in the denominator. It is owed to the factor γ in the denominator of (6.30). On the other hand, this factor is presented in [3, 4, 22]. When obtaining (6.88) it was suggested that the current density is equal to (6.29) in the reference frame attached to a moving current loop. The current density in the LF is obtained from (6.29) by the Lorentz transformation. It follows from (6.88) that the intensity of radiation produced by a magnetic dipole parallel to the velocity differs from zero in the velocity window $c_n < v < c$. Therefore, v should not be too close to either c_n or c. For this, n should differ appreciably from unity. Probably, the best candidate for observing this radiation is a neutron moving in a medium with large n. By comparing (6.90) with the radiation intensity of a moving charge ($\sigma_e = e^2 \omega \mu / c^2 \gamma_n^2$) we see that there is a chance of observing the radiation from a neutron moving in medium only for very high frequencies.

The magnetic dipole is perpendicular to the velocity. Let the current loop lie in the $z = 0$ plane with its velocity along the x axis (magnetic dipole is along the z axis). In the time representation the current density in the

LF is given by Eqs. (6.44) and (6.45). The Fourier components of these densities are

$$j_x(\omega) = \frac{m_d}{2\pi\beta}\exp(i\psi_1)\delta(z)\frac{\partial}{\partial y}\delta(y), \quad j_y(\omega) = -\frac{m_d}{2\pi\beta\gamma^2}\frac{\partial}{\partial x}\delta(z)\delta(y)\exp(i\psi_1),$$

$$\rho_{Ch}(\omega) = \frac{m_d}{2\pi c}\delta(z)\exp(i\psi_1)\frac{\partial}{\partial y}\delta(y), \tag{6.89}$$

where $\psi_1 = kx/\beta$. The electromagnetic potentials are equal to

$$\Phi = \frac{m_d}{2\pi c\epsilon}\frac{\partial\alpha_1}{\partial y}, \quad A_x = \frac{m_d\mu}{2\pi v}\frac{\partial\alpha_1}{\partial y}, \quad A_y = -\frac{m_d\mu}{2\pi\gamma^2 v}\frac{\partial}{\partial x}\alpha_1. \tag{6.90}$$

Here

$$\alpha_1 = \exp(i\psi_1)\int_{-\infty}^{\infty}\exp\left[ik\rho_1\left(\frac{\sinh\chi}{\beta} + n\cosh\chi\right)\right]d\chi,$$

$$\rho_1 = \sqrt{y^2 + z^2}. \tag{6.91}$$

This integral is evaluated along the same lines as α in (6.85). It is equal to $2K_0$ for $v < c_n$ and $i\pi H_0^{(1)}$ for $v > c_n$. The arguments of these Bessel functions are $k\rho_1/\beta\gamma_n$. The electromagnetic field strengths are

$$E_x = \frac{im_dk\cos\phi}{2\pi v\epsilon}(n^2 - 1)\frac{\partial\alpha_1}{\partial\rho_1},$$

$$E_y = \frac{m_d}{2\pi c\epsilon}\left\{\frac{\cos 2\phi}{\rho_1}\frac{\partial\alpha_1}{\partial\rho_1} + \left[\cos^2\phi\frac{k^2(\beta_n^2 - 1)}{\beta^2} + \frac{k^2n^2}{\gamma^2\beta^2}\right]\alpha_1\right\},$$

$$E_z = \frac{m_d}{2\pi c\epsilon}\sin\phi\cos\phi\left[\frac{k^2(\beta_n^2 - 1)}{\beta^2}\alpha_1 + \frac{2}{\rho_1}\frac{\partial\alpha_1}{\partial\rho_1}\right],$$

$$H_x = \frac{ikm_d\sin\phi}{2\gamma^2 v\beta}\frac{\partial\alpha_1}{\partial\rho_1}, \quad H_y = -\frac{m_d\sin\phi\cos\phi}{2\pi v}\left[\frac{k^2(\beta_n^2 - 1)}{\beta^2}\alpha_1 + \frac{2}{\rho_1}\frac{\partial\alpha_1}{\partial\rho_1}\right],$$

$$H_z = \frac{m_d}{2\pi v}\left[\frac{k^2\alpha_1}{\gamma^2\beta^2} + \frac{\cos 2\phi}{\rho_1}\frac{\partial\alpha_1}{\partial\rho_1} + \cos^2\phi\frac{k^2(\beta_n^2 - 1)}{\beta^2}\alpha_1\right].$$

The angle ϕ ($\cos\phi = y/\rho_1$, $\sin\phi = z/\rho_1$) defines the azimuthal position of the observational point in the yz plane. It is counted from the y axis. In a manifest form the field strengths are equal to

$$E_x = -\frac{im_dk^2\cos\phi}{\pi v\beta\gamma_n\epsilon}(n^2 - 1)\exp(i\psi_1)K_1,$$

$$E_y = -\frac{km_dc}{\pi\epsilon}\left[\frac{\cos 2\phi}{\rho_1\beta\gamma_n}K_1 - \frac{k}{\beta^2}\left(\frac{n^2}{\gamma^2} - \frac{\cos^2\phi)}{\gamma_n^2}\right)K_0\right]\exp(i\psi_1),$$

$$E_z = -\frac{m_d k \sin\phi \cos\phi}{\pi\beta\gamma_n c\epsilon}\left(\frac{2}{\rho_1}K_1 + \frac{k}{\beta\gamma_n}K_0\right)\exp(i\psi_1),$$

$$H_x = -\frac{im_d k^2 \sin\phi}{\gamma^2\gamma_n v\beta^2}K_1\exp(i\psi_1),$$

$$H_y = \frac{m_d k \sin\phi \cos\phi}{\pi v\beta\gamma_n}\left(\frac{k}{\beta\gamma_n}K_0 + \frac{2}{\rho_1}K_1\right)\exp(i\psi_1),$$

$$H_z = \frac{m_d k}{\pi v\beta}\left[k\left(\frac{1}{\beta\gamma^2} - \frac{\cos^2\phi}{\gamma_n^2}\right)K_0 - \frac{\cos2\phi}{\gamma_n\rho_1}K_1\right]\exp(i\psi_1) \qquad (6.92)$$

for $v < c_n$ and

$$E_x = \frac{m_d k^2 \cos\phi}{2v\beta\gamma_n}(n^2-1)H_1^{(1)}\exp(i\psi_1), \quad H_x = \frac{m_d k^2 \sin\phi}{2\gamma^2\gamma_n\beta^2 v}H_1^{(1)}\exp(i\psi_1),$$

$$E_y = -\frac{im_d k}{2\epsilon v}\left[\frac{\cos2\phi}{\rho_1\gamma_n}H_1^{(1)} - \frac{k}{\beta}\left(\frac{\cos^2\phi}{\gamma_n^2} + \frac{n^2}{\gamma^2}\right)H_0^{(1)}\right]\exp(i\psi_1),$$

$$E_z = \frac{im_d k \sin\phi \cos\phi}{2v\epsilon\gamma_n}\left(\frac{k}{\beta\gamma_n}H_0^{(1)} - \frac{2}{\rho_1}H_1^{(1)}\right)\exp(i\psi_1),$$

$$H_y = -\frac{im_d k \sin\phi \cos\phi}{2v\beta\gamma_n}\left[\frac{k}{\beta\gamma_n}H_0^{(1)} - \frac{2}{\rho_1}H_1^{(1)}\right]\exp(i\psi_1),$$

$$H_z = \frac{im_d k}{2v\beta}\left[\frac{k}{\beta}\left(\frac{\cos^2\phi}{\gamma_n^2} + \frac{1}{\gamma^2}\right)H_0^{(1)} - \frac{\cos2\phi}{\rho_1\gamma_n}H_1^{(1)}\right]\exp(i\psi_1) \qquad (6.93)$$

for $v > c_n$. To evaluate the energy flux in the radial direction (that is, perpendicular to the motion axis), one should find the components of field strengths tangential to the surface of a cylinder coaxial with the motion axis and perpendicular it. They are given by

$$E_\phi = E_z\cos\phi - E_y\sin\phi, \quad H_\phi = H_z\cos\phi - H_y\sin\phi.$$

We rewrite them in a manifest form. It is easy to check that

$$E_\phi = -\frac{m_d k \sin\phi}{\pi v\epsilon}\left(\frac{1}{\rho_1\gamma_n}K_1 + \frac{kn^2}{\beta\gamma^2}K_0\right)\exp(i\psi_1),$$

$$H_\phi = \frac{km_d \cos\phi}{\pi v\beta}\left[k\beta(n^2-1)K_0 - \frac{1}{\rho_1\gamma_n}K_1\right]\exp(i\psi_1) \qquad (6.94)$$

for $v < c_n$ and

$$E_\phi = -\frac{im_d k \sin\phi}{2v\epsilon}\left[\frac{kn^2}{\beta\gamma^2}H_0^{(1)} + \frac{1}{\rho_1\gamma_n}H_1^{(1)}\right]\exp(i\psi_1),$$

$$H_\phi = +\frac{im_d k \cos\phi}{2v\beta}\left[k\beta(n^2-1)H_0^{(1)} - \frac{1}{\rho_1\gamma_n}H_1^{(1)}\right]\exp(i\psi_1) \qquad (6.95)$$

for $v > c_n$. The energy flux per unit length and per unit frequency through the cylindrical surface of radius ρ_1 is equal to

$$\frac{d^2\mathcal{E}}{dxd\omega} = \int_0^{2\pi} \sigma(\omega,\phi)d\phi,$$

where

$$\sigma(\omega,\phi) = \frac{d^3\mathcal{E}}{dxd\omega d\phi} = \frac{c}{2}\rho_1(E_\phi^* H_x + E_\phi H_x^* - H_\phi^* E_x - H_\phi E_x^*). \qquad (6.96)$$

Substituting field strengths here, one obtains that the differential intensity is zero for $v < c_n$ and

$$\sigma(\omega,\phi) = \frac{m_d^2 k^3}{2\pi\epsilon\beta v}\left[\frac{n^2}{\gamma^4\beta^2}\sin^2\phi + (n^2-1)^2\cos^2\phi\right] \qquad (6.97)$$

for $v > c_n$. The integration over ϕ gives

$$\sigma(\omega) = \frac{m_d^2 k^3}{2\beta\epsilon v}\left[\frac{n^2}{\gamma^4\beta^2} + (n^2-1)^2\right]. \qquad (6.98)$$

Equations (6.97) and (6.98) coincide with those obtained by Frank [3,4,22], who noted that in the limit $\beta \to 1/n$ these intensities do no vanish as it is intuitively expected. On these grounds Frank declared them as to be incorrect [6]. 30 years later Frank returned to the same problem [10]. He attributed the non-vanishing of the intensities (6.97) and (6.98) to the specific polarization of the medium.

We analyse this question in some detail. The intensity (6.96) is non-zero for $\beta = 1/n + \epsilon$ and zero for $\beta = 1/n - \epsilon$, where $\epsilon \ll 1$. Since it consists of EMF strengths (see (6.96)), the latter should exhibit a jump at $\beta = 1/n$ too. Turning to Eqs. (6.92) and (6.93) defining the EMF strengths we observe that E_x and H_x are continuous at $\beta = 1/n$, while E_ϕ and H_ϕ entering into (6.96) exhibit jump. Further examination shows that this jump is due to the fact that first terms in the definition of E_ϕ and H_ϕ in (6.94) and (6.95) are not transformed into each other when β changes from $1/n - \epsilon$ to $1/n + \epsilon$. Further reflection shows that this is owed to Eqs. (6.86). Separating in them real and imaginary parts, one has

$$I_1 = \int_0^\infty \cos(\frac{k\rho}{\beta}\sinh\chi)\cos(k\rho n\cosh\chi) = K_0$$

for $\beta < 1/n$,

$$I_1 = -\frac{\pi}{2} N_0 \tag{6.99}$$

for $\beta > 1/n$;

$$I_2 = \int\limits_0^\infty \cos(\frac{k\rho}{\beta} \sinh \chi) \sin(k\rho n \cosh \chi) = 0$$

for $\beta < 1/n$,

$$I_2 = \frac{\pi}{2} J_0 \tag{6.100}$$

for $\beta > 1/n$, where the arguments of all Bessel functions are $k\rho/\beta\gamma_n$. Now, I_1 is continuous at $\beta = 1/n$, whilst I_2 is zero for $\beta < 1/n$ and tends to $\pi/2$ as $\beta \to 1/n$.

Furthermore, for $\beta = 1/n$, I_2 looks like ($y = k\rho n$):

$$I_2 = \int\limits_0^\infty \cos(y \sinh \chi) \sin(y \cosh \chi) d\chi = \frac{1}{2} \int\limits_{-\infty}^\infty \cos(y \sinh \chi) \sin(y \cosh \chi) d\chi$$

$$= \frac{1}{2} \mathrm{Im} \int\limits_{-\infty}^\infty \exp[iy(\sinh \chi + \cosh \chi)] d\chi = \frac{1}{2} \mathrm{Im} \int\limits_{-\infty}^\infty \exp[iy \exp \chi] d\chi.$$

Putting $t = \exp(\chi)$ one obtains

$$\int\limits_{-\infty}^\infty \exp[iy \exp(\chi)] d\chi = \int\limits_0^\infty \exp(iyt) \frac{dt}{t}.$$

and

$$\mathrm{Im} \int\limits_0^\infty \exp(iyt) \frac{dt}{t} = \int\limits_0^\infty \sin(yt) \frac{dt}{t} = \frac{\pi}{2}$$

Therefore I_2 is equal to

$$I_2 = \frac{\pi}{2}$$

for $\beta = 1/n + \epsilon$,

$$I_2 = \frac{\pi}{4}$$

for $\beta = 1/n$ and

$$I_2 = 0$$

for $\beta = 1/n - \epsilon$, $\epsilon \ll 1$. As a result, the radiation intensities are equal one half of (6.97) or (6.98) for $\beta = 1/n$.

Again, a neutron moving in a dielectric medium with n appreciably different from unity is the best candidate for observing this radiation. The absence of the overall $1/\gamma$ factor in (6.97) and (6.98) makes it easier to observe radiation from a neutron with the spin perpendicular to the velocity than from a neutron with the spin directed along it.

Electromagnetic field of a point-like toroidal solenoid uniformly moving in unbounded medium
The velocity is along the torus symmetry axis. Let this current distribution move uniformly along the z axis (directed along the torus symmetry axis) with the velocity v. In the time representation, in the laboratory frame, the non-vanishing charge and current components are given by (6.56). The spectral representations of these densities are

$$\rho_{Ch}(\omega) = \frac{m_t}{2\pi c}\left(\frac{\partial^2}{\partial x^2} + \frac{\partial^2}{\partial y^2}\right)D, \quad j_z(\omega) = \frac{m_t}{2\pi\beta}\left(\frac{\partial^2}{\partial x^2} + \frac{\partial^2}{\partial y^2}\right)D,$$

$$j_x = -\frac{m_t}{2\pi\beta\gamma^2}\frac{\partial^2}{\partial z\partial x}D, \quad j_y = -\frac{m_t}{2\pi\beta\gamma^2}\frac{\partial^2}{\partial z\partial y}D,$$

where $D = \delta(x)\delta(y)\exp(i\psi)$, $\psi = kz/\beta$ and $m_t = \pi^2 j_0 dR_0^2/c$ is the toroidal moment. Electromagnetic potentials are given by

$$\Phi = \frac{m_t}{2\pi\epsilon c}\left(\frac{\partial^2}{\partial x^2} + \frac{\partial^2}{\partial y^2}\right)\alpha, \quad A_z = \frac{\mu m_t}{2\pi v}\exp(i\psi)\left(\frac{\partial^2}{\partial x^2} + \frac{\partial^2}{\partial y^2}\right)\alpha,$$

$$A_x = -\frac{\mu m_t}{2\pi\gamma^2 v}\frac{\partial^2\alpha}{\partial z\partial x}, \quad A_y = -\frac{\mu m_t}{2\pi\gamma^2 v}\frac{\partial^2\alpha}{\partial z\partial y},$$

where α is the same as in (6.85). Electromagnetic field strengths are

$$E_x = \frac{m_t k^2}{2\pi\epsilon v\beta}(n^2 - 1)\frac{\partial\alpha}{\partial x}, \quad E_y = \frac{m_t k^2}{2\pi\epsilon v\beta}(n^2 - 1)\frac{\partial\alpha}{\partial y},$$

$$E_z = \frac{ik^3 m_t}{2\pi\epsilon v\beta^2}(n^2 - 1)(1 - \beta_n^2)\alpha, \quad H_x = -\frac{m_t k^2}{2\pi v}(n^2 - 1)\frac{\partial\alpha}{\partial y},$$

$$H_y = \frac{m_t k^2}{2\pi v}(n^2 - 1)\frac{\partial\alpha}{\partial x}, \quad H_z = 0.$$

Or, explicitly

$$E_\rho = -\frac{m_t k^3}{\pi\epsilon v\beta^2\gamma_n}(n^2 - 1)\exp(i\psi)K_1,$$

$$E_z = \frac{ik^3 m_t}{\pi\epsilon v\beta^2}(n^2 - 1)\exp(i\psi)(1 - \beta_n^2)K_0,$$

$$H_\phi = -\frac{m_t k^3}{\pi v \beta \gamma_n}(n^2 - 1)\exp(i\psi)K_1$$

for $\beta_n < 1$ and

$$E_\rho = -i\frac{m_t k^3}{2\epsilon v \beta^2 \gamma_n}(n^2 - 1)\exp(i\psi)H_1^{(1)},$$

$$E_z = \frac{k^3 m_t}{2\epsilon v \beta^2}(n^2 - 1)\exp(i\psi)(\beta_n^2 - 1)H_0^{(1)},$$

$$H_\phi = -i\frac{m_t k^3}{2 v \beta \gamma_n}(n^2 - 1)\exp(i\psi)H_1^{(1)}$$

for $\beta_n > 1$. The energy loss through the cylinder surface of the radius ρ coaxial with the motion axis per unit frequency and per unit length is

$$\sigma_\rho(\omega) = \frac{d^2\mathcal{E}}{dz d\omega} = -\pi c\rho(E_z H_\phi^* + E_z^* H_\phi).$$

It is equal to zero for $v < c_n$ and

$$\sigma_\rho(\omega) = \frac{k^5 m_t^2}{\epsilon v \beta^3}(\beta_n^2 - 1)(n^2 - 1)^2 \qquad (6.101)$$

for $v > c_n$. In the past, this equation was obtained in [9]. The absence of the overall $1/\gamma$ factor in (6.101) and its proportionality to ω^5 show that the radiation intensity for the toroidal dipole directed along the velocity is maximal for large frequencies and $v \sim c$.

The velocity is perpendicular to the torus axis. Let a toroidal solenoid move in a medium with a velocity perpendicular to the torus symmetry axis (coinciding with the z axis). For definiteness let the TS move along the x axis. Then, in the LF, in the time representation, the charge and current densities are given by (6.63). The Fourier transforms of these densities are

$$\rho_{Ch} = -\frac{m_t}{2\pi c\gamma}\frac{\partial^2}{\partial x \partial z}\exp(i\psi_1)\delta(y)\delta(z),$$

$$j_x = -\frac{m_t}{2 v\pi\gamma}\frac{\partial^2}{\partial x \partial z}\exp(i\psi_1)\delta(y)\delta(z),$$

$$j_y = -\frac{m_t}{2 v\pi\gamma}\frac{\partial^2}{\partial y \partial z}\exp(i\psi_1)\delta(y)\delta(z),$$

$$j_z = \frac{m_t}{2 v\pi\gamma}[\frac{1}{\gamma^2}\frac{\partial^2}{\partial x^2} + \frac{\partial^2}{\partial y^2}]\exp(i\psi_1)\delta(y)\delta(z).$$

Here $\psi_1 = kx/\beta$. As a result we arrive at the following electromagnetic potentials:

$$\Phi = -\frac{\beta m_t}{2c\pi\gamma\epsilon}\frac{\partial^2}{\partial x\partial z}\alpha_1, \quad A_x = -\frac{m_t\mu}{2v\pi\gamma}\frac{\partial^2}{\partial x\partial z}\alpha_1,$$

$$A_y = -\frac{m_t\mu}{2v\pi\gamma}\frac{\partial^2}{\partial y\partial z}\alpha_1, \quad A_z = \frac{m_t\mu}{2v\pi\gamma}\left[\frac{1}{\gamma^2}\frac{\partial^2}{\partial x^2} + \frac{\partial^2}{\partial y^2}\right]\alpha_1,$$

where α_1 is the same as in (6.91). We give without derivation the EMF strengths

$$E_x = \frac{k^2 m_t}{2\pi v\beta\gamma\epsilon}(n^2 - 1)\sin\phi\frac{\partial\alpha_1}{\partial\rho_1},$$

$$E_y = \frac{ikm_t}{2\pi\gamma\epsilon v}(n^2 - 1)\sin\phi\cos\phi\left[\frac{k^2(\beta_n^2 - 1)}{\beta^2}\alpha_1 + \frac{2}{\rho_1}\frac{\partial\alpha_1}{\partial\rho_1}\right],$$

$$E_z = -\frac{ikm_t}{2\pi\gamma\epsilon v}(n^2 - 1)\left[\frac{k^2}{\beta^2}(1 + (\beta_n^2 - 1)\cos^2\phi)\alpha_1 + \frac{\cos 2\phi}{\rho_1}\frac{\partial\alpha_1}{\partial\rho_1}\right],$$

$$E_\phi = -\frac{ikm_t}{2\pi\gamma\epsilon v}(n^2 - 1)\cos\phi\left[k^2 n^2\alpha_1 + \frac{1}{\rho_1}\frac{\partial\alpha_1}{\partial\rho_1}\right],$$

$$H_x = -\frac{m_t}{2v\pi\gamma}k^2(n^2 - 1)\cos\phi\frac{\partial\alpha_1}{\partial\rho_1}, \quad H_y = \frac{im_t k^3}{2\pi v\gamma\beta}(n^2 - 1)\alpha_1, \quad H_z = 0,$$

$$H_\phi = -\frac{im_t k^3\alpha_1}{2\pi v\gamma\beta}(n^2 - 1)\sin\phi,$$

where ϕ is the angle defining the observational point in the yz plane. It is counted from the y axis and is defined by (6.92) and (6.93). In a manifest form the EMF strengths are given by

$$E_x = -\frac{k^3 m_t}{\pi v\beta^2\gamma\gamma_n\epsilon}(n^2 - 1)\sin\phi K_1\exp(i\psi_1),$$

$$E_y = \frac{ikm_t}{\pi v\gamma\epsilon}(n^2 - 1)\sin\phi\cos\phi\left[\frac{k^2}{\beta^2}(\beta_n^2 - 1)K_0 - \frac{2k}{\rho\beta\gamma_n}K_1\right]\exp(i\psi_1),$$

$$E_z = -\frac{ikm_t}{\pi v\gamma\epsilon}(n^2-1)\left[\frac{k^2}{\beta^2}(1 + \cos^2\phi(\beta_n^2 - 1))K_0 - \frac{k}{\rho\beta\gamma_n}\cos 2\phi K_1\right]\exp(i\psi_1),$$

$$E_\phi = -\frac{ikm_t}{\pi v\gamma\epsilon}(n^2 - 1)\cos\phi\left[k^2 n^2 K_0 - \frac{k}{\rho\beta\gamma_n}K_1\right]\exp(i\psi_1),$$

$$H_x = \frac{m_t k^3}{\pi v\beta\gamma\gamma_n}(n^2 - 1)\cos\phi K_1\exp(i\psi_1), \quad H_y = \frac{im_t k^3}{\pi v\gamma\beta}(n^2 - 1)K_0\exp(i\psi_1),$$

$$H_z = 0, \quad H_\phi = -\frac{im_t k^3}{\pi v \gamma \beta}(n^2 - 1)\sin\phi K_0 \exp(i\psi_1)$$

for $v < c_n$ and

$$E_x = -\frac{im_t k^3}{2v\epsilon\beta^2\gamma\gamma_n}\sin\phi(n^2 - 1)H_1^{(1)}\exp(i\psi_1),$$

$$E_y = -\frac{km_t}{2v\gamma\epsilon}(n^2 - 1)\sin\phi\cos\phi\left[\frac{k^2}{\beta^2}(\beta_n^2 - 1)H_0^{(1)} - \frac{2k}{\rho\beta\gamma_n}H_1^{(1)}\right]\exp(i\psi_1),$$

$$E_z = \frac{km_t}{2v\gamma\epsilon}(n^2 - 1)$$

$$\times\left[\frac{k^2}{\beta^2}(1 + \cos^2\phi(\beta_n^2 - 1))H_0^{(1)} - \frac{k}{\rho\beta\gamma_n}\cos 2\phi H_1^{(1)}\right]\exp(i\psi_1),$$

$$E_\phi = \frac{km_t}{2v\gamma\epsilon}(n^2 - 1)\cos\phi\left[k^2 n^2 H_0^{(1)} - \frac{k}{\rho\beta\gamma_n}H_1^{(1)}\right]\exp(i\psi_1),$$

$$H_x = \frac{im_t k^3}{2v\beta\gamma\gamma_n}(n^2 - 1)\cos\phi H_1^{(1)}\exp(i\psi_1),$$

$$H_y = -\frac{m_t k^3}{2v\gamma\beta}(n^2 - 1)H_0^{(1)}\exp(i\psi_1),$$

$$H_z = 0, \quad H_\phi = \frac{m_t k^3}{2v\gamma\beta}(n^2 - 1)\sin\phi H_0^{(1)}\exp(i\psi_1)$$

for $v > c_n$. Again, E_ϕ and H_ϕ are tangential to the torus surface and perpendicular to the torus velocity directed along the x axis.

The energy flux through the cylindrical surface of the radius ρ_1 per unit length and per unit frequency is equal to

$$\frac{d^2\mathcal{E}}{dxd\omega} = \int_0^{2\pi}\sigma(\omega, \phi)d\phi,$$

where

$$\sigma(\omega, \phi) = \frac{d^3\mathcal{E}}{dxd\omega d\phi} = \frac{c}{2}\rho_1(E_\phi^* H_x + E_\phi H_x^* - H_\phi^* E_x - H_\phi E_x^*).$$

Substituting here field strengths, one obtains that the differential intensity is zero for $v < c_n$ and

$$\sigma(\omega, \phi) = \frac{k^5 m_t^2}{2\epsilon v \beta \pi \gamma^2}(n^2 - 1)^2\left(n^2\cos^2\phi + \frac{1}{\beta^2}\sin^2\phi\right) \tag{6.102}$$

for $v > c_n$. The integration over ϕ gives

$$\sigma(\omega) = \frac{k^5 m_t^2}{2\epsilon v \beta \gamma^2}(n^2 - 1)^2 \left(n^2 + \frac{1}{\beta^2}\right). \qquad (6.103)$$

As far as we know, the radiation intensities (6.102) and (6.103) are obtained here for the first time. They are discontinuous: in fact, they decrease from (6.102) or (6.103) for $\beta_n > 1$ to their one-half for $\beta = 1/n$ and to zero for $\beta < 1/n$. Also, we observe the appearance of the velocity window $c_n < v < c$ in which the radiation differs from zero.

Unbounded motion of a point-like electric dipole
Fourier components of the charge and current densities (6.65) are

$$\rho_d(\omega) = \frac{ea}{\pi v}(\vec{n}\vec{\nabla})\delta(x)\delta(y)\exp\left(\frac{ikz}{\beta}\right), \quad j_z(\omega) = v\rho_d(\omega). \qquad (6.104)$$

The electromagnetic potentials are equal to

$$\Phi = \frac{ea}{\pi v \epsilon}(\vec{n}\vec{\nabla})\alpha, \quad A_z = \frac{ea\mu}{\pi c}(\vec{n}\vec{\nabla})\alpha,$$

where α is the same as in (6.85). The non-vanishing components of EMF strengths are

$$E_x = -\frac{ea}{\pi v \epsilon}\frac{\partial}{\partial x}(\vec{n}\vec{\nabla})\alpha, \quad E_y = -\frac{ea}{\pi v \epsilon}\frac{\partial}{\partial y}(\vec{n}\vec{\nabla})\alpha,$$

$$E_z = -\frac{ea}{\pi v \epsilon}(1 - \beta_n^2)\frac{\partial}{\partial z}(\vec{n}\vec{\nabla})\alpha,$$

$$H_x = \frac{ea}{\pi c}\frac{\partial}{\partial y}(\vec{n}\vec{\nabla})\alpha, \quad H_y = -\frac{ea}{\pi c}\frac{\partial}{\partial x}(\vec{n}\vec{\nabla})\alpha.$$

In a manifest form we write out only those components of field strengths which are needed for the evaluation of the radial cylindric energy flux. They are equal to

$$E_z = \frac{2ek^2 a}{\pi \epsilon \beta^2 v}(1 - \beta_n^2)\left(\frac{n_z}{\gamma}K_0 + i\frac{\tilde{n}_\rho}{\gamma_n}K_1\right)\exp(i\psi),$$

$$H_\phi = \frac{2eak}{\pi v}\left\{\tilde{n}_\rho\left[\frac{k}{\beta}(\beta_n^2 - 1)K_0 - \frac{1}{\gamma_n \rho}K_1\right] + \frac{ikn_z}{\beta \gamma \gamma_n}K_1\right\}\exp(i\psi)$$

for $v < c_n$ and

$$E_z = \frac{ek^2 a}{\epsilon \beta^2 v}(\beta_n^2 - 1)\left(\frac{\tilde{n}_\rho}{\gamma_n}H_1^{(1)} - i\frac{n_z}{\gamma}H_0^{(1)}\right)\exp(i\psi),$$

$$H_\phi = \frac{eak}{v}\left\{i\tilde{n}_\rho\left[\frac{k}{\beta}(\beta_n^2 - 1)H_0^{(1)} - \frac{1}{\rho\gamma_n}H_1^{(1)}\right] - n_z\frac{k}{\beta\gamma\gamma_n}H_1^{(1)}\right\}\exp(i\psi)$$

for $v > c_n$. Here $\psi = kz/\beta$, $\tilde{n}_\rho = \sin\theta_0\cos(\phi - \phi_0)$; θ_0 is the angle between the symmetry axis of the electric dipole and its velocity; ϕ is the azimuthal position of the observational point on the cylinder surface and ϕ_0 defines the orientation of the electric dipole in the plane perpendicular to the motion axis.

The radiation intensity per unit length of the cylindrical surface coaxial with the motion axis, per unit azimuthal angle and per unit frequency is

$$\sigma(\phi, \omega) = \frac{d^3\mathcal{E}}{dz d\phi\, d\omega} = -\frac{c\rho}{2}(E_z H_\phi^* + E_z^* H_\phi).$$

It is equal to

$$\sigma_\rho(\phi, \omega) = \frac{4e^2 a^2 k^3 n_z \tilde{n}_\rho}{\pi^2 \epsilon \beta^3 v\gamma}(1 - \beta_n^2)$$

$$\times\left[\frac{k\rho}{\beta}(1 - \beta_n^2)(K_0^2 + K_1^2) + \frac{1}{\gamma_n}K_0 K_1\right] \qquad (6.105)$$

for $v < c_n$ and

$$\sigma_\rho(\phi, \omega) = \frac{2e^2 a^2 k^3}{\pi\epsilon\beta^3 v}(\beta_n^2 - 1)\{\tilde{n}_\rho^2(\beta_n^2 - 1) + n_z^2(1 - \beta^2) + \tilde{n}_\rho n_z\frac{\pi}{2\gamma}$$

$$\times\left[\frac{k\rho}{\beta}(\beta_n^2 - 1)(J_0^2 + N_0^2 + J_1^2 + N_1^2) - \frac{1}{\gamma_n}(N_0 N_1 + J_0 J_1)\right]\} \qquad (6.106)$$

for $v > c_n$. Integrating over the azimuthal angle ϕ one finds that $\sigma_\rho(\omega) = 0$ for $v < c_n$ and

$$\sigma_\rho(\omega) = \frac{2e^2 a^2 k^3}{\pi\epsilon\beta^3 v}(\beta_n^2 - 1)[(\beta_n^2 - 1)\sin^2\theta_0 + 2(1 - \beta^2)\cos^2\theta_0] \qquad (6.107)$$

for $v > c_n$. For the symmetry axis along the velocity ($\theta_0 = 0$) and perpendicular to it ($\theta_0 = \pi/2$) one finds

$$\sigma_\rho(\omega, \theta_0 = 0) = \frac{4e^2 a^2 k^3}{\epsilon\beta^3 v}(\beta_n^2 - 1)(1 - \beta^2) \qquad (6.108)$$

and

$$\sigma_\rho(\omega, \theta_0 = \theta/2) = \frac{2e^2 a^2 k^3}{\epsilon\beta^3 v}(\beta_n^2 - 1)^2, \qquad (6.109)$$

respectively.

Again, the same confusion with (6.108) and (6.109) takes place in the physical literature. In [6, 10, 23], the factor $(1 - \beta^2)$ in (6.108) is absent.

Yet, it presents in [3, 4, 22]. In [22], $(\beta_n^2 - 1)$, instead of $(\beta_n^2 - 1)^2$, enters (6.109). The expression given in [23] is two times larger than (6.109). The correct expression for (6.109) is given in [3, 4, 6, 10].

It is rather surprising that for $\beta_n < 1$ the non-averaged radiation intensities are equal to zero when the symmetry axis is either parallel or perpendicular to the velocity, but differs from zero for the intermediate inclination of the symmetry axis (see (6.105)). Integration over the azimuthal angle gives $\sigma_\rho(\omega, \theta) = 0$ for $\beta_n < 1$.

Again, it should be mentioned that we did not intend to demonstrate misprints in the papers of other authors. What we need are the reliable formulae suitable for practical applications.

6.4.2. THE TAMM PROBLEM FOR ELECTRIC CHARGE, MAGNETIC, ELECTRIC, AND TOROIDAL DIPOLES

Pedagogical example: the Tamm problem for the electric charge
Tamm considered the following problem [24]. A point charge is at rest at the point $z = -z_0$ of the z axis up to an instant $t = -t_0$ and at the point $z = z_0$ after the instant $t = t_0$. In the time interval $-t_0 < t < t_0$, it moves uniformly along the z axis with the velocity v greater or smaller than the velocity $c_n = c/n$ of light in medium. The non-vanishing z Fourier component of the vector potential (VP) is given by

$$A_z(x, y, z) = \frac{e\mu}{2\pi c}\alpha_T, \qquad (6.110)$$

where

$$\alpha_T = \int_{-z_0}^{z_0} \frac{dz'}{R}\exp\left[ik\left(\frac{z'}{\beta} + nR\right)\right], \quad R = [\rho^2 + (z-z')^2]^{1/2}, \quad \rho^2 = x^2 + y^2.$$

Tamm presents R in the form $R = r - z'\cos\theta$, thus disregarding the second order terms relative to z'.

Imposing the conditions:

i) $r \gg z_0$ (this means that the observational distance is much larger than the motion interval);

ii) $k_n r \gg 1$, $k_n = \omega/c_n$ (this means that the observations are made in the wave zone);

iii) $nz_0^2/2r\lambda \ll 1$, $\lambda = 2\pi c/\omega$ (this means that the second-order terms in the expansion of R should be small compared with π since they enter as a phase in α_T; λ is the observed wavelength), Tamm obtained the following expression for α_T

$$\alpha_T = \frac{2}{kr}\exp(ik_n r)q$$

and for the vector magnetic potential

$$A_z = \frac{e\mu}{\pi\omega r}\exp(iknr)q. \tag{6.111}$$

Here

$$q = \frac{1}{1/\beta - n\cos\theta}\sin[kz_0(1/\beta - n\cos\theta)].$$

In the limit $kz_0 \to \infty$,

$$q \to \pi\delta(1/\beta - n\cos\theta) \quad \text{and} \quad A_z \to \frac{e\mu}{\omega n r}\exp(iknr)\delta(\cos\theta - 1/\beta n).$$

Using (6.111) Tamm evaluated the EMF strengths and the energy flux through the sphere of the radius r for the whole time of observation

$$\mathcal{E} = r^2\int S_r d\Omega dt = \int \frac{d^2\mathcal{E}}{d\Omega d\omega}d\Omega d\omega, \quad d\Omega = \sin\theta d\theta d\phi, \quad S_r = \frac{c}{4\pi}E_\theta H_\phi,$$

where

$$\frac{d^2\mathcal{E}}{d\Omega d\omega} = \frac{e^2\mu n}{\pi^2 c}[\sin\theta\frac{\sin kz_0(1/\beta - n\cos\theta)}{n\cos\theta - 1/\beta}]^2, \quad \beta_n = \beta n. \tag{6.112}$$

is the energy emitted into the solid angle $d\Omega$, in the frequency interval $d\omega$. This famous formula obtained by Tamm is frequently used by experimentalists for the identification of the charge velocity. When kz_0 is large,

$$\frac{d^2\mathcal{E}}{d\Omega d\omega} = \frac{e^2\mu kz_0}{\pi c}(1 - 1/\beta_n^2)\delta(\cos\theta - 1/\beta n). \tag{6.113}$$

Integrating this equation over the solid angle one finds

$$\frac{d\mathcal{E}}{d\omega} = \frac{2e^2\mu kz_0}{c}(1 - 1/\beta_n^2). \tag{6.114}$$

Correspondingly, the energy radiated per unit frequency and per unit length (obtained by dividing (6.114) by the motion interval $L = 2z_0$) is

$$\frac{d\mathcal{E}}{d\omega dL} = \frac{e^2\mu}{c}(1 - 1/\beta_n^2). \tag{6.115}$$

The typical experimental situations described by the Tamm formula are:

i) β decay of a nucleus at one spatial point accompanied by a subsequent absorption of the emitted electron at another point;

ii) A high energy electron consequently moves in the vacuum, enters the dielectric slab, leaves the slab and propagates again in vacuum. Since the electron moving uniformly in vacuum does not radiate (apart from the transition radiation arising at the boundaries of the dielectric slab), the experimentalists describe this situation via the Tamm formula, assuming that the electron is created at one side of the slab and is absorbed at the other.

The Tamm problem for the magnetic dipole
The magnetic dipole is parallel to the velocity. In this case the Fourier components of the current density differ from zero only in the motion interval $(-z_0, z_0)$. Correspondingly, the magnetic potential and the field strengths are given by

$$A_\phi = -\frac{\mu m_d}{2\pi v \gamma} \frac{\partial \alpha_T}{\partial \rho}, \quad \mu H_\theta = -\frac{\partial A_\phi}{\partial r} - \frac{\cot \theta}{r} A_\phi,$$

where α_T is the same as in (6.110). Using approximations i)-iii), one gets

$$H_\theta = -\frac{m_d k^2 n^2 \sin \theta}{2\pi \gamma v} \alpha_T.$$

The electric field strengths are obtained from the relation

$$\text{curl}\vec{H} = -ik\epsilon \vec{E}$$

valid outside the motion interval. This gives

$$E_\phi = \frac{k^2 n \mu m_d}{2\pi \gamma v} \alpha_T \sin \theta.$$

When evaluating field strengths we have dropped the terms which decrease at infinity faster than $1/r$ and which do not contribute to the radiation flux. The distribution of the radial energy flux on the sphere of the radius r is given by

$$\sigma_r(\theta, \phi) = \frac{d^2\mathcal{E}}{d\Omega d\omega} = -\frac{c}{2} r^2 (E_\phi H_\theta^* + E_\phi^* H_\theta) = \frac{m_d^2 k^2 n^3 \mu \sin^2 \theta}{\pi^2 \gamma^2 \beta v} q^2. \quad (6.116)$$

In the limit $kz_0 \to \infty$ one has

$$\frac{d^2\mathcal{E}}{d\Omega d\omega} = \frac{m_d^2 k^2 n^2 \mu k z_0}{\pi \gamma^2 \beta v} (1 - 1/\beta_n^2) \delta(\cos \theta - 1/\beta_n). \quad (6.117)$$

Integration over the solid angle gives the frequency distribution of the emitted radiation per unit frequency and per unit length

$$\frac{d\mathcal{E}}{dLd\omega} = \frac{m_d^2 \omega^3 \mu}{v^4 \gamma^2 \gamma_n^2}. \quad (6.118)$$

This coincides with (6.88).

The magnetic dipole is perpendicular to the velocity. Let the magnetic dipole directed along the z axis move on the interval $(-x_0, x_0)$ of the x axis with a constant velocity v. We write out without derivation the electromagnetic field strengths contributing to the radial energy flux

$$E_\theta = \frac{m_d k^2 \mu n}{2\pi v}\alpha'_T(1 - \beta^2 \cos^2\theta)\cos\phi, \quad E_\phi = -\frac{m_d k^2 \mu n}{2\pi v \gamma^2}\alpha'_T \cos\theta \sin\phi,$$

$$H_\theta = \frac{m_d k^2 n^2}{2\pi v \gamma^2}\alpha'_T \cos\theta \sin\phi, \quad H_\phi = \frac{m_d k^2 n^2}{2\pi v}\alpha'_T(1 - \beta^2 \cos^2\theta)\cos\phi.$$

where

$$\alpha'_T = (2/kr)q\exp(ik_n r), \quad q = (1/\beta - n\cos\theta)^{-1}\sin[kx_0(1/\beta - n\cos\theta)].$$

The θ is the angle between the radius vector of the observational point and the motion axis (which is the x axis). The ϕ is the observational azimuthal angle in the yz plane. The value $\phi = 0$ corresponds to the y axis, the magnetic moment is along the z axis.

The distribution of the radial energy flux on the sphere of the radius r is given by

$$\sigma_r(\theta, \phi, \omega) = \frac{d^2\mathcal{E}}{d\Omega d\omega} = \frac{c}{2}r^2(E_\theta H_\phi^* + E_\theta^* H_\phi - E_\phi H_\theta^* - E_\phi^* H_\theta)$$

$$= \frac{m_d^2 k^2 n^3 \mu}{\pi^2 \beta v}\left[\cos^2\phi(1 - \beta^2\cos^2\theta)^2 + \gamma^{-4}\sin^2\phi\cos^2\theta\right]q^2. \qquad (6.119)$$

In the limit $kz_0 \to \infty$ this gives

$$\frac{d^2\mathcal{E}}{d\Omega d\omega} = \frac{m_d^2 k^3 z_0 n^2 \mu}{\pi \beta v}$$

$$\times \left[\cos^2\phi(1 - 1/n^2)^2 + \frac{1}{\gamma^4 \beta_n^2}\sin^2\phi\right]\delta(\cos\theta - 1/\beta_n). \qquad (6.120)$$

Integration over the solid angle gives

$$\frac{d^2\mathcal{E}}{dL d\omega} = \frac{m_d^2 k^3 n^2 \mu}{2\beta v}\left[(1 - 1/n^2)^2 + \frac{1}{\gamma^4 \beta_n^2}\right]. \qquad (6.121)$$

This coincides with (6.98).

The Tamm problem for the toroidal dipole
The toroidal dipole is parallel to the velocity. The direction of the toroidal dipole coincides with the direction of its symmetry axis. The electromagnetic vector potential and field strengths contributing to the radial energy flux are given by

$$E_\theta = \frac{im_t k^3 n^2 \mu}{2\pi v} \sin\theta (1 - \beta^2 \cos^2\theta)\alpha_T,$$

$$H_\phi = \frac{im_t k^3 n^3}{2\pi v} \sin\theta (1 - \beta^2 \cos^2\theta)\alpha_T,$$

where α_T is the same as above. The distribution of the radial energy flux on the sphere of the radius r is given by

$$\sigma_r = \frac{d^2\mathcal{E}}{d\Omega d\omega} = \frac{c}{2}r^2(E_\theta H_\phi^* + E_\theta^* H_\phi)$$

$$= \frac{m_t^2 k^4 n^5 \mu}{\pi^2 \beta v} \sin^2\theta (1 - \beta^2 \cos^2\theta)^2 q^2. \tag{6.122}$$

Here θ is the polar angle of the observational point. In the limit $kz_0 \to \infty$, (6.122) goes into

$$\frac{d^2\mathcal{E}}{d\Omega d\omega} = \frac{m_t^2 k^5 z_0 n^4 \mu}{\pi \beta v}(1 - 1/\beta_n^2)(1 - 1/n^2)^2 \delta(\cos\theta - 1/\beta_n). \tag{6.123}$$

Integration over the solid angle gives

$$\frac{d^2\mathcal{E}}{dL d\omega} = \frac{m_t^2 k^5 n^4 \mu}{\beta v}(1 - 1/\beta_n^2)(1 - 1/n^2)^2. \tag{6.124}$$

This coincides with (6.101).

The symmetry axis is perpendicular to the velocity. In this case the electromagnetic field strengths contributing to the radial energy flux are given by

$$E_\theta = -\frac{i\mu m_t k^3 n^2 \alpha_T'}{2v\pi\epsilon\gamma}(1 - \beta^2 \cos^2\theta)\cos\theta \sin\phi,$$

$$E_\phi = -\frac{i\mu m_t k^3 n^2 \alpha_T'}{2v\pi\epsilon\gamma}(1 - \beta^2 \cos^2\theta)\cos\phi,$$

$$H_\theta = \frac{im_t k^3 n^3 \alpha_T'}{2v\pi\gamma}(1 - \beta^2 \cos^2\theta)\cos\phi,$$

$$H_\phi = -\frac{im_t k^3 n^3 \alpha'_T}{2v\pi\gamma}(1 - \beta^2 \cos^2\theta)\cos\theta\sin\phi.$$

Correspondingly, the radial energy flux is

$$\sigma_r(\theta, \phi, \omega) = \frac{d^2\mathcal{E}}{d\omega d\Omega} = \frac{1}{2}cr^2(E_\theta H_\phi^* + E_\theta^* H_\phi - E_\phi H_\theta^* - E_\phi^* H_\theta)$$

$$= \frac{m_t^2 k^4 n^5 \mu}{\gamma^2 \pi^2 v\beta}(1 - \beta^2 \cos^2\theta)^2(\cos^2\theta\sin^2\phi + \cos^2\phi)q^2. \tag{6.125}$$

Again, θ is the polar angle of the observational point; the toroidal dipole is along the z axis, the angle ϕ defining the position of the observational point in the yz plane perpendicular to the velocity, is counted from the y axis.

In the limit $kz_0 \to \infty$, (6.125) goes into

$$\frac{d^2\mathcal{E}}{d\omega d\Omega} = \frac{m_t^2 k^5 z_0 n^4 \mu}{\gamma^2 \pi v\beta}(1 - 1/n^2)^2(\frac{1}{\beta_n^2}\sin^2\phi + \cos^2\phi)\delta(\cos\theta - 1/\beta_n). \tag{6.126}$$

The integration over the solid angle ϕ gives

$$\frac{d^2\mathcal{E}}{d\omega dL} = \frac{m_t^2 k^5 n^4 \mu}{2\gamma^2 v\beta}(1 - 1/n^2)^2\left(\frac{1}{\beta_n^2} + 1\right). \tag{6.127}$$

This coincides with (6.103).

Tamm's problem for the electric dipole with arbitrary orientation of the symmetry axis

Let the electric dipole move along the z axis and let it be directed along the vector $\vec{n} = (n_x, n_y, n_z)$ defining the direction of its symmetry axis in the laboratory reference frame. In this case the vector potential and electromagnetic field strengths contributing to the radial energy flux are given by

$$A_z = \frac{iea\mu}{\pi c}(\vec{n}\vec{\nabla})\alpha_T, \quad E_\theta = \frac{eak^2 n\mu}{\pi c}\sin\theta\left(\tilde{n}_\rho\sin\theta + \frac{1}{\gamma}n_z\cos\theta\right)\alpha_T,$$

$$H_\phi = \frac{eak^2 n^2}{\pi c}\sin\theta\left(\tilde{n}_\rho\sin\theta + \frac{1}{\gamma}n_z\cos\theta\right)\alpha_T,$$

where $\tilde{n}_\rho = \sin\theta_0\cos(\phi - \phi_0)$ and $n_z = \cos\theta_0$; θ and ϕ define the position of the observational point; θ_0 and ϕ_0 define the orientation of the electric dipole. Correspondingly the radial energy flux is

$$\sigma_r(\theta, \phi, \omega) = \frac{d^2\mathcal{E}}{d\omega d\Omega} = \frac{1}{2}cr^2(E_\theta H_\phi^* + E_\theta^* H_\phi)$$

$$= \frac{4e^2 a^2 k^2 n^3 \mu}{\pi^2 c}\left(\tilde{n}_\rho\sin\theta + \frac{1}{\gamma}n_z\cos\theta\right)^2\sin^2\theta q^2. \tag{6.128}$$

For the electric dipole oriented along the velocity ($\tilde{n}_\rho = 0, n_z = 1$) (6.128) is reduced to

$$\sigma_r^\|(\theta, \phi, \omega) = \frac{4e^2 a^2 k^2 n^3 \mu}{\gamma^2 \pi^2 c} \cos^2 \theta \sin^2 \theta q^2. \tag{6.129}$$

Correspondingly for the electric dipole orientation perpendicular to the motion axis ($\tilde{n}_\rho = \cos(\phi - \phi_0), n_z = 0$), one has

$$\sigma_r^\perp(\theta, \phi, \omega) = \frac{4e^2 a^2 k^2 n^3 \mu}{\pi^2 c} q^2 \sin^4 \theta \cos^2(\phi - \phi_0). \tag{6.130}$$

In the limit $kz_0 \to \infty$ one finds

$$\frac{d^2 \mathcal{E}}{d\omega d\Omega} = \frac{4e^2 a^2 k^3 z_0 n^2 \mu}{\pi c}$$

$$\times \left(\tilde{n}_\rho \sqrt{1 - 1/\beta_n^2} + \frac{1}{\gamma \beta_n} n_z \right)^2 (1 - 1/\beta_n^2) \delta(\cos \theta - 1/\beta_n). \tag{6.131}$$

$$\sigma_r^\|(\theta, \phi, \omega) = \frac{4e^2 a^2 k^3 z_0 n^2 \mu}{\gamma^2 \pi c} \frac{1}{\beta_n^4 \gamma_n^2} \delta(\cos \theta - 1/\beta_n), \tag{6.132}$$

$$\sigma_r^\perp(\theta, \phi, \omega) = \frac{4e^2 a^2 k^3 z_0 n^2 \mu}{\pi c \beta_n^4 \gamma_n^4} \cos^2(\phi - \phi_0) \delta(\cos \theta - 1/\beta_n). \tag{6.133}$$

The integration over the solid angle gives

$$\frac{d^2 \mathcal{E}}{d\omega dL} = \frac{2e^2 a^2 k^3 n^2 \mu}{c}$$

$$\times \left[\sin^2 \theta_0 (1 - 1/\beta_n^2) + \frac{2}{\gamma^2 \beta_n^2} \cos^2 \theta_0 \right] (1 - 1/\beta_n^2). \tag{6.134}$$

$$\left(\frac{d^2 \mathcal{E}}{d\omega dL} \right)_\| = \frac{4e^2 a^2 k^3 n^2 \mu}{\gamma^2 c} \frac{1}{\beta_n^2} \left(1 - \frac{1}{\beta_n^2} \right), \tag{6.135}$$

$$\left(\frac{d^2 \mathcal{E}}{d\omega dL} \right)_\perp = \frac{2e^2 a^2 k^3 n^2 \mu}{c} \left(1 - \frac{1}{\beta_n^2} \right)^2. \tag{6.136}$$

These equations coincide with (6.107)-(6.109).

Concluding remarks on the dipoles moving in medium. As expected, the integral Tamm intensities (that is, integrated over the solid angle) in the limit $kz_0 \to \infty$ (large motion interval) coincide with the radiation intensities corresponding to the unbounded motion treated in section 2. The radiation intensities for the Tamm problem differ considerably from those given by Frank in [3,4]. There is an essential difference between our derivation and that of [3,4].

The method used by Frank is quite complicated. He writes the Maxwell equations in terms of electric and magnetic vector Hertz potentials which are related to the electromagnetic field strengths. In the right hand sides of the Maxwell equations there are electric and magnetic polarizations proportional to the LF electric and magnetic moment, respectively. Electric and magnetic moments in the LF are connected with those in the dipole RF through the well-known linear relations (see, e.g. [5]). When in the dipole RF there is only electric or magnetic dipole one may exclude from these relations the non-zero magnetic moment of the RF, thus obtaining the relation between the electric and magnetic moments of the LF.

On the other hand, we define the charge and current densities in the RF. Using the Lorentz transformation, the same as in vacuum, we recalculate them into the LF. We then let the dimensions of these distributions tend to zero, thus obtaining infinitesimal the charge and current distributions corresponding to the electric, magnetic, or toroidal dipoles. With these infinitesimal charge and current distributions we solve the Maxwell equations finding the electromagnetic potentials and field strengths. Using them we evaluate the radiated energy flux.

6.5. Electromagnetic field of a precessing magnetic dipole

Consider an infinitely thin circular turn with a constant current flowing in it. Let the center of this current loop coincide with the origin, whilst its symmetry axis precesses around the z axis with a constant angular velocity ω_0. We choose the rest frame (RF) of this loop as follows. Let $\vec{n}_x, \vec{n}_y$, and $\vec{n}_z$ be the orthogonal basis vectors of the laboratory frame (LF). The $\vec{e}_z$ vector of RF we align along the loop symmetry axis $\vec{n}$. Being expressed in terms of the LF basis vectors it is given by

$$\vec{n} = \vec{e}_z = \cos\theta_0 \vec{n}_z + \sin\theta_0 \vec{n}_\rho = \vec{n}_r,$$

where $\vec{n}_\rho = \cos\omega_0 t\,\vec{n}_x + \sin\omega_0 t\,\vec{n}_y$ and θ_0 is the inclination angle of the loop symmetry axis towards the laboratory z axis. Other two basis vectors of RF lying in the plane of loop, we choose in the following way

$$\vec{e}_x = \frac{1}{\sin\theta_0}(\vec{n} \times \vec{n}_z) = \cos\omega_0 t\,\vec{n}_y - \sin\omega_0 t\,\vec{n}_x = \vec{n}_\phi,$$

$$\vec{e}_y = \frac{1}{\sin\theta_0}(\vec{n} \times (\vec{n} \times \vec{n}_z)) = \cos\omega_0 t\,\vec{n}_\rho - \sin\omega_0 t\,\vec{n}_z = \vec{n}_\theta, \qquad (6.137)$$

that is, $\vec{e}_x, \vec{e}_y$ and $\vec{e}_z$ coincide with the spherical basis vectors.

Let x, y, z and x', y', z' be the coordinates of the same point in the laboratory and proper reference frames, respectively. They are related as follows

$$x' = x \sin \omega_0 t - y \cos \omega_0 t, \quad y' = \rho \cos \theta_0 - z \sin \theta_0,$$
$$z' = \rho \sin \theta_0 + z \cos \theta_0, \tag{6.138}$$

where $\rho = x \cos \omega_0 t + y \sin \omega_0 t$. The current density in the RF is given by

$$\vec{j}' = \vec{e}_\psi I_0 \delta(z') \delta(\rho' - d),$$

where $\rho' = \sqrt{x'^2 + y'^2}$; $e_\psi = \vec{e}_x \cos \psi - \vec{e}_y \sin \psi$ is the vector lying in the plane of the loop and defining the direction of current and ψ is the azimuthal angle in the plane of the loop defined by $\cos \psi = x'/d$, $\sin \psi = y'/d$. In the LF, the components of the current density are given by

$$j_x = \left(\cos \theta_0 \frac{\partial}{\partial y} - \sin \omega_0 t \sin \theta_0 \frac{\partial}{\partial z} \right) M,$$

$$j_y = \left(-\cos \theta_0 \frac{\partial}{\partial x} + \cos \omega_0 t \sin \theta_0 \frac{\partial}{\partial z} \right) M,$$

$$j_z = \sin \theta_0 \left(\sin \omega_0 t \frac{\partial}{\partial x} - \cos \omega_0 t \frac{\partial}{\partial y} \right) M, \tag{6.139}$$

where

$$M = I_0 \delta(z') \Theta(d - \sqrt{x'^2 + y'^2}).$$

x', y' and z' should be expressed through the coordinates (x, y, z, t) of the LF via the relations (6.138). We are interested in studying the point-like $(d \to 0)$ current loop, which is equivalent to the magnetic dipole. In this limit

$$M = \pi d^2 I_0 \delta(x) \delta(y) \delta(z).$$

The vector magnetic potential is given by

$$\vec{A} = \frac{1}{c} \int \frac{1}{R} \vec{j}(\vec{r}', t') \delta(t' - t + R/c) dV' dt'.$$

After integration one finds for the spherical components of $\vec{A}$:

$$A_r = 0, \quad A_\theta = -\frac{\pi d^2 I_0}{c} \sin \theta_0 \frac{\partial}{\partial r} \frac{\sin \psi}{r},$$

$$A_\phi = \frac{\pi d^2 I_0}{c} \left(\frac{1}{r^2} \cos \theta_0 \sin \theta + \sin \theta_0 \cos \theta \frac{\partial}{\partial r} \frac{\sin \psi}{r} \right). \tag{6.140}$$

Here $\psi = \omega_0 t - k_0 r - \phi$. The non-vanishing components of the field strengths are

$$E_r = 0, \quad E_\phi = \frac{\pi d^2 I_0 k_0}{c} \sin \theta_0 \cos \theta \frac{\partial}{\partial r} \frac{\sin \psi}{r}, \quad E_\theta = \frac{\pi d^2 I_0 k_0}{c} \sin \theta_0 \frac{\partial}{\partial r} \frac{\cos \psi}{r},$$

$$H_r = \frac{2\pi d^2 I_0}{cr}\left(\frac{1}{r^2}\cos\theta_0\cos\theta - \sin\theta_0\sin\theta\frac{\partial}{\partial r}\frac{\cos\psi}{r}\right),$$

$$H_\phi = -\frac{\pi d^2 I_0}{c}\sin\theta_0\frac{1}{r}\frac{\partial}{\partial r}r\frac{\partial}{\partial r}\frac{\sin\psi}{r},$$

$$H_\theta = -\frac{\pi d^2 I_0}{cr}\frac{\partial}{\partial r}\left(\frac{1}{r}\cos\theta_0\sin\theta + r\sin\theta_0\cos\theta\frac{\partial}{\partial r}\frac{\cos\psi}{r}\right). \qquad (6.141)$$

To evaluate the radiation field one should leave in (6.141) the terms which decrease no faster than $1/r$ for $r \to \infty$:

$$E_r = 0, \quad E_\theta = -H_\phi \approx \frac{\pi d^2 k_0^2 I_0}{cr}\sin\theta_0\sin\psi,$$

$$H_r \approx 0, \quad E_\phi = H_\theta \approx \frac{\pi d^2 k_0^2 I_0}{cr}\sin\theta_0\cos\theta\cos\psi. \qquad (6.142)$$

The radial energy flux per unit time through a surface element $r^2 d\Omega$ is

$$S_r = \frac{d\mathcal{E}}{dtd\Omega} = \frac{cr^2}{4\pi}(E_\theta H_\phi - H_\theta E_\phi)$$

$$= \frac{\pi}{4c}(d^2 k_0^2 I_0\sin\theta_0)^2(\sin^2\psi + \cos^2\theta\cos^2\psi). \qquad (6.143)$$

However, experimentalists usually measure not the time distribution of the energy flux flowing through the observational sphere, but photons with definite frequency. For this we evaluate the Fourier transforms of the field strengths

$$\vec{E}(\omega) = \frac{1}{2\pi}\int\limits_{-\infty}^{\infty}\exp(i\omega t)\vec{E}(t)dt, \quad \vec{H}(\omega) = \frac{1}{2\pi}\int\limits_{-\infty}^{\infty}\exp(i\omega t)\vec{H}(t)dt.$$

In the wave zone where $kr \gg 1$ one finds

$$E_\theta(\omega) = H_\phi(\omega)$$

$$= -\frac{i\pi k_0^2 I_0 d^2}{2cr}\sin\theta_0[\exp(-i\Phi_0)\delta(\omega + \omega_0) - \exp(i\Phi_0)\delta(\omega - \omega_0)],$$

$$E_\phi(\omega) = -H_\theta(\omega) = -\frac{\pi k_0^2 I_0 d^2}{2cr}\sin\theta_0\cos\theta$$

$$\times[\exp(-i\Phi_0)\delta(\omega + \omega_0) + \exp(i\Phi_0)\delta(\omega - \omega_0)], \qquad (6.144)$$

where $\Phi_0 = k_0 r + \phi$. The energy radiated into the unit solid angle, per unit frequency is

$$\frac{d^2\mathcal{E}}{d\omega d\Omega} = \frac{cr^2}{4\pi}(E_\theta H_\phi^* - H_\theta^* E_\phi + \text{c.c.})$$

$$= \frac{\pi k_0^4 I_0^2 d^4}{8c} \sin^2 \theta_0 (1 + \cos^2 \theta)[\delta(\omega - \omega_0)]^2). \tag{6.145}$$

This means that only the photons with an energy ω_0 should be observed.

A question arises of why we did not use the instantaneous Lorentz transformation when transforming the charge and current densities from the dipole non-inertial RF to the inertial LF.

The reason for this may be illustrated using the circular loop with the current density $j = \vec{j_0}\delta(\rho - a)\delta(z)/2\pi a$ in its RF as an example. Let this loop rotate with a constant angular velocity ω around its symmetry axis. Then in the LF the charge density $\sigma = a\omega j\gamma/c^2$ and the charge

$$q = \int \sigma dV = a\omega j_0 \gamma/c^2$$

arise. Here a is the loop radius, $\gamma = 1/\sqrt{1 - \beta^2}$, $\beta = a\omega/c$.

This absurd result is because that it is not always possible to apply the instantaneous Lorentz transformation for the transformation between the inertial and non-inertial reference frames. The correct approach is as follows. In the inertial reference frame (that is, in the laboratory frame) there is only the static current density. In the non-inertial reference frame (attached to a rotating current loop) both charge and current densities differ from zero. There is no charge in this reference frame since a charge is no longer a spatial integral over the charge density, but includes integration over other hypersurfaces [25].

The content of this section may be applied to the explanation of radiation observed from neutron stars (magnetars) with super-strong magnetic fields (see e.g., [26].

6.6. Discussion and Conclusion

In this Chapter we have evaluated the electromagnetic fields of electric, magnetic, and toroidal dipoles moving im medium. We use the following procedure. First, in the dipole reference rest frame we consider finite charge and current densities which in the infinitesimal limit reduce to electric, magnetic, and toroidal dipoles. Then, we transform these finite charge-current densities to the laboratory frame using the Lorentz transformation, the same as in vacuum. Then, we let the dimensions of these densities tend to zero, thus obtaining densities describing moving electric, magnetic and toroidal dipoles. With these densities we solve the Maxwell equations, find electromagnetic potentials, field strengths, and the radiated energy flux. This procedure is straightforward, without any ambiguities. On the other hand, complications arise when one formulates the same problem in terms of electric and magnetic polarizations (see Introduction). The ambiguity is owed to the transformation laws between electric and magnetic moments in two inertial reference frames. Since these two approaches should be equivalent, the question arises of whether the same ambiguity takes place for the charge and current densities. Or, more exactly: Is it true that charge and current densities in two inertial reference frames placed in a medium are related via the vacuum Lorentz transformation? It should be noted that a standard electrodynamics of moving bodies (see, e.g., [27]-[29]) definitely supports the same transformation law for the charge and current densities both in medium and vacuum.

Another ambiguity is that there is another formulation of relativistic spin theory. We mean the so-called Bargmann-Michel-Telegdi theory. In it there are three spin components in the spin rest frame, four components in any other reference frame, and there is no electric moment in this reference frame.

We briefly summarize the main results obtained:

1. The exact electromagnetic fields of point-like electric and magnetic dipoles moving in a non-dispersive medium are obtained in the time representation. The formalism of induced electric and magnetic moments suggested by Frank does not describe properly the exact electromagnetic fields.

2. The exact electromagnetic field of a point-like toroidal solenoid moving in a non-dispersive medium is obtained. For the velocity of an elementary toroidal solenoid smaller than the velocity of light in medium, the electric field of moving TS is similar to the field of an electric quadrupole.

3. In the spectral representation, treating electric, magnetic and toroidal dipoles as an infinitesimal limit of corresponding charge and current densities, we study how they radiate when moving uniformly in an unbounded medium. The frequency and velocity domains where radiation intensities are maximal are defined. The behaviour of radiation intensities near the Cherenkov threshold is investigated in some detail.

4. Radiation intensities are obtained for electric, magnetic, and toroidal dipoles moving uniformly in a medium, in a finite spatial interval (Tamm problem).

5. The electromagnetic field arising from the precession of the point-like magnetic dipole around a fixed spatial axis is found. It turns out that the precessing magnetic dipole radiates the sole frequency coinciding with that of the precession.

References

1.   Cherenkov P.A. (1944) Radiation of Electrons Moving in Medium with Superluminal Velocity, *Trudy FIAN*, **2, No 4**, pp. 3-62.

2. Frank I.M. (1988) *Vavilov-Cherenkov Radiation*, Nauka, Moscow.
3. Frank I.M. (1942) Doppler Effect in Refractive Medium *Izv. Acad. Nauk SSSR, ser.fiz.*, **6**, pp.3-31.
4. Frank I.M. (1943) Doppler Effect in Refractive Medium *Journal of Physics USSR*, **7, No 2**, pp.49-67.
5. Frenkel J. (1956) *Electrodynamics*, Izdat. AN SSSR, Moscow-Leningrad, in Russian.
6. Frank I.M. (1953) Cherenkov Radiation for Multipoles, In the book: *To the memory of S.I. Vavilov*, pp.172-192, Izdat. AN SSSR, Moscow, 1953, in Russian.
7. Ginzburg V.L. (1953) On the Cherenkov Radiation of the Magnetic Dipole, In the book: *To the memory of S.I. Vavilov*, pp.172-192, Izdat. AN SSSR, Moscow, 1953, in Russian.
8. Ginzburg V.L. (1984) On Fields and Radiation of 'true' and current magnetic Dipoles in a Medium, *Izv. Vuz., ser. Radiofizika*, **27**, pp. 852-872.
9. Ginzburg V.L. and Tsytovich V.N. (1985) Fields and Radiation of Toroidal Dipole Moments Moving Uniformly in a Medium *Zh. Eksp. Theor. Phys.*, bf 88, pp. 84-95.
10. Frank I.M. (1984) Vavilov-Cherenkov Radiation for Electric and Magnetic Multipoles *Usp.Fiz.Nauk*, **144**, pp. 251-275.
11. Frank I.M. (1989) On Moments of magnetic Dipole Moving in Medium *Usp.Fiz.Nauk* **158**, pp. 135-138.
12. Streltzov V.N. (1990) Relativistic Dipole Moment *JINR Communication* **P2-90-101**, Dubna.
13. Jackson J.D. (1975) *Classical Electrodynamics*, J.Wiley, New York.
14. Afanasiev G.N. and Stepanovsky Yu.P. (2000) Electromagnetic Fields of Electric, Magnetic and Toroidal Dipoles Moving in Medium *Physica Scripta*, **61**, pp.704-716.
15. Afanasiev G.N. and Stepanovsky Yu.P. (2002) On the Radiation of Electric, Magnetic and Toroidal Dipoles *JINR Preprint*,**E2-2002-142**, pp. 1-30.
16. Afanasiev G.N. and Stepanovsky Yu.P. (1995) The Electromagnetic Field of Elementary Time-Dependent Toroidal Sources *J.Phys.A*, **28**, pp.4565-4580;
    Afanasiev G.N., Nelhiebel M. and Stepanovsky Yu.P. (1996) The Interaction of Magnetization with an External Electromagnetic field and a Time-Dependent Magnetic Aharonov-Bohm Effect *Physica Scripta*,**54**, pp. 417-427;
    Afanasiev G.N. and Dubovik V.M. (1998) Some Remarkable Charge-Current Configurations *Physics of Particles and Nuclei*, **29**, pp. 366-391; Afanasiev G.N., (1999) *Topological Effects in Quantum Mechanics*, Kluwer, Dordrecht. 1999).
17. Landau L.D. and Lifshitz E.M. (1962) *The Classical Theory of Fields* Pergamon, New York.
18. Landau L.D. and Lifshitz E.M (1960) *Electrodynamics of Continuous Media*, Pergamon, Oxford.
19. Afanasiev G.N., Beshtoev Kh. and Stepanovsky Yu.P. (1996) Vavilov-Cherenkov Radiation in a Finite Region of Space *Helv. Phys. Acta*, **69**, pp. 111-129; Afanasiev G.N. and Kartavenko V.G. (1998) Radiation of a Point Charge Uniformly Moving in a Dielectric Medium *J. Phys. D: Applied Physics*, **31**, pp.2760-2776.
20. Frahm C.P. (1982) Some Novel Delta-Function Identities *Am. J. Phys.*, **51**, pp. 826-829).
21. Afanasiev G.N., Kartavenko V.G. and Stepanovsky Yu.P. (1999) On Tamm's Problem in the Vavilov-Cherenkov Radiation Theory *J.Phys. D: Applied Physics*, **32**, pp. 2029-2043.
22. Frank I.M. (1946) Radiation of Electrons Moving in Medium with Superlight Velocity *Usp. Fiz. Nauk*, **30, No 3-4**, pp. 149-183.
23. Villavicencio M., Jimenez J.L. and Roa-Neri J.A.E. (1998) Cherenkov Effect for an Electric Dipole *Foundations of Physics*, **5**, pp. 445-459.
24. Tamm I.E. (1939) Radiation Induced by Uniformly Moving Electrons, *J. Phys. USSR*, **1, No 5-6**, pp. 439-461.
25. Rohrlich F. 1965) *Classical Charged Particles* Addison, Massachusetts.
26. Ziolkovski J. (2000) Magnetars, in *Proc. Int. Workshop "Hot Points in Astro-*

physics", pp.176-192, Dubna.
27. Pauli W. (1958) *Theory of Relativity*, Pergamon, New York.
28. Moller C. (1972) *The Theory of Relativity*, Clarendon, Oxford.
29. Sommerfeld A. (1949) *Electrodynamik*, Geest@Portig, Leipzig.

CHAPTER 7

QUESTIONS CONCERNING OBSERVATION OF THE VAVILOV-CHERENKOV RADIATION

7.1. Introduction

It is known that the frequency spectrum of a point-like charge moving uniformly with a velocity v greater than the velocity of light in medium extends to infinity. The total radiated energy and the photon number are infinite. This is because of the point-like structure of a moving charge whose infinite self-energy is a reservoir allowing charge to move uniformly despite the energy losses from the radiation, ionization, and the polarization of the surrounding medium. The easiest way of obtaining the finite frequency spectrum is to consider a charge of finite dimensions. This was done in a nice paper [1] in which the charge density having a zero dimension in the transverse direction and a Gaussian distribution along the motion axis was considered. The frequency spectrum obtained there, extended up to v/a, where a is the parameter of the Gaussian distribution. Obviously, this charge distribution is quite unphysical. The next attempt was made in [2] in which the charge distributions were chosen in the form of a spherical shell, a Yukawa distribution, and that of [1]. It should be noted that the authors of [1] and [2] related their charge densities to the laboratory frame. It seems to us that it is more natural to relate charge densities to the rest frame of the moving charge. There are two reasons for this. First, the charge form factor of a moving charge is the Fourier transform of a charge density related to the rest frame of a moving charge. Second, in another laboratory frame moving relative the initial frame with a constant velocity, the charge density is no longer spherically symmetric. So we prefer to define the charge density in its rest frame. Charge and current densities in the laboratory frame are then obtained by the Lorentz transformation. Solving the Maxwell equations with these densities, we find electromagnetic field strengths and the radiated energy flux. This is essentially the procedure adopted by us. In addition to the current densities studied in [1,2] we considered the charge density uniformly distributed inside the sphere and the spherical Gaussian distribution. A charge moving uniformly in medium radiates if its velocity exceeds the velocity of light in medium. If there is no external force supporting this motion, the charge should be decelerated. In the absence of dispersion the total energy (obtained by the integration over the frequency spectrum) is infinite for a point-like charge. For a charge of finite dimensions this quantity is finite. Equating it to the kinetic energy loss, one can find how a charge moves when it loses the energy as a result of the Cherenkov radiation. This is done in subsection (7.2).

Another way of obtaining a finite radiated energy is to take into account the medium dispersion. The crucial step was made by Frank and Tamm in 1937 [3] who presented general formulae for the electromagnetic field strengths and radiation intensity in the spectral representation without specifying the concrete form of the medium dispersion law. Their formulae, predicting a concrete angular position of the maximum of the radiation intensity for a particular charge velocity and wavelength, are extensively used by experimentalists. The taking into account of the medium dispersion is important since the refractive index and absorption depend on the frequency. The disregarding of the medium dispersion is possible only in a restricted frequency region. For example, for the pure water the refractive index is almost constant in the visible light region $(4 \times 10^{-5}\text{cm} < \lambda < 7 \times 10^{-5}\text{cm})$ and for $\lambda > 1$ km [4]. The visible light region is surrounded by two absorption peaks (at $\lambda = 3 \times 10^{-4}$ cm and at $\lambda = 5 \times 10^{-7}$ cm).

The next important case was made by Fermi [5], who considered an uniform motion of a charge in medium with a complex dielectric permittivity chosen in a standard one-pole form (4.1) extensively used in optics. From general formulae presented by him it follows that a charge moving in such a medium should radiate at each velocity. Physically this can be understood as follows. The current density of the uniformly moving charge contains all frequencies. A particular spectral component of the electromagnetic wave propagates in medium without damping if the Tamm-Frank radiation condition $\beta^2 n^2(\omega) > 1$ is satisfied. For the ϵ parametrization (4.1) there always exists a frequency interval for which this condition is satisfied. Since the transition from the time representation to the spectral representation involves integration over all frequencies, a charge uniformly moving in medium described by (4.1) radiates at each velocity. The question arises of the space-time distribution of this radiation. Analytically and numerically this was investigated in [6-8], where it was shown that for the charge velocity v above some critical v_c, the switching of the medium dispersion leads to the appearance of several finite height maxima in the neighbourhood of the singular Cherenkov cone corresponding to the non-dispersive medium. For the charge velocity below v_c the bunch of the radiation intensity appears behind the moving charge at a sufficiently large distance from it. Recent measurements [9] of the space-time distribution of the radiation intensity for $v < v_c$ are in satisfactory agreement with results of [6-8] in which the finite expressions were obtained for the total (that is, integrated over frequency) energy and the number of photons radiated per unit length of the charge path for the dielectric permittivity (4.1) with $p = 0$.

In this treatment (subsection (7.3)), equating this energy to the kinetic energy loss, we find how the Vavilov-Cherenkov (VC) radiation affects the velocity of a point-like charge moving in a dispersive medium.

So far we have implicitly assumed that the measuring device is in the same medium in which the charge moves. However, the charge usually moves in one medium while observations are performed in another. For example, in the initial Cherenkov experiments the electrons moved in water, whilst the observations were made in air. Complications and ambiguities arising from such experimental procedure are also discussed.

The other problem which will be considered in this chapter is the transition and VC radiation on dielectric and metallic spheres. The notion of transition radiation was introduced by Frank and Ginzburg [10] who studied radiation arising from a uniformly moving charge passing from one medium to another. They considered the plane boundary between media 1 and 2. A thorough exposition of transition radiation may be found in [11]. In this chapter we consider the charge motion which begins and terminates in medium 2 and which passes through the dielectric sphere filled with medium 1. The energy flux is evaluated in medium 2. As far as we know, the transition radiation for the spherical boundary is considered here for the first time. In the past, transition radiation was considered in the physical literature only for plane interfaces. For the problem treated we have evaluated angular and frequency radiation intensities for a number of charge velocities and of medium properties. These expressions contain transition and Cherenkov radiation as well as charge radiation from the charge instantaneous beginning and termination of motion.

We also analyse attempts to explain transition radiation in terms of the charge instantaneous stop in one medium and its instantaneous beginning of motion in another medium [12-14]. We prove that their contribution to the radiation intensity disappears if the motion with instantaneous velocity jumps can be considered as a limiting case of the charge smooth motion. We also consider the interpretation of transition radiation in terms of semi-infinite charge motions, with an instantaneous stop of a charge in one medium and with its instantaneous start in another medium [10,11]. We show that for the charge velocity greater than the velocity of light in medium, the terms corresponding to the Cherenkov radiation should be taken into account.

7.2. Cherenkov radiation from a charge of finite dimensions

Consider a charge of finite dimensions moving uniformly in a medium with a velocity v directed along the z axis. Let the charge density in the reference frame in which it is at rest be spherically symmetric: $e\rho_{Ch}(r')$, where $r' = \sqrt{x'^2 + y'^2 + z'^2}$. In the laboratory frame (relative to which a charge moves with a velocity v), the charge and current densities are given by

$$\rho_L = e\gamma\rho_{Ch}(r), \quad j_z = v\rho_L,$$

where $r = [\rho^2 + \gamma^2(z - vt)^2]^{1/2}$, $\rho = \sqrt{x^2 + y^2}$, $\gamma = (1 - \beta^2)^{-1/2}$ and $\beta = v/c$. The Fourier transform of ρ_L is defined as

$$\rho_\omega = \frac{1}{2\pi} \int\limits_{-\infty}^{\infty} dt \exp(i\omega t)\rho_L(t).$$

Making the change of variables $(t = z/v + \rho x/\gamma v)$ we transform ρ_ω to the form

$$\rho_\omega = \frac{e}{\pi v} \exp(i\psi)f(\rho), \quad \psi = \omega z/v$$

where

$$f(\rho) = \rho \int\limits_{0}^{\infty} \cos(\omega\rho x/v\gamma)\rho_{Ch}(\rho\sqrt{1 + x^2})dx.$$

The electric scalar and magnetic vector (only its z component differs from zero) potentials are

$$\Phi_\omega(x', y', z') = \frac{1}{\epsilon} \int \frac{1}{R} \exp(ik_n R)\rho_\omega(x', y', z')dV', \quad A_\omega = \beta\mu\epsilon\Phi_\omega.$$

Here $R = [(x - x')^2 + (y - y')^2 + (z - z')^2]^{1/2}$, $k_n = kn$, $k = \omega/c$, and $n = \sqrt{\epsilon\mu}$ is the refractive index of the medium with parameters ϵ and μ.

Now we take into account the expansion

$$\frac{1}{R} \exp(ik_n R) = \sum_{m=0}^{\infty} \epsilon_m \cos m(\phi - \phi') \tag{7.1}$$

$$\times \left\{ i \int\limits_{-k_n}^{k_n} dk_z \exp[ik_z(z - z')]G_m^{(1)} + \frac{2}{\pi} \left(\int\limits_{-\infty}^{-k_n} + \int\limits_{k_n}^{\infty} \right) dk_z \exp[ik_z(z - z')]G_m^{(2)} \right\},$$

where

$$G_m^{(1)} = J_m(\sqrt{k_n^2 - k_z^2}\rho_<)H_m^{(1)}(\sqrt{k_n^2 - k_z^2}\rho_>),$$

$$G_m^{(2)} = I_m(\sqrt{k_z^2 - k_n^2}\rho_<)K_m(\sqrt{k_z^2 - k_n^2}\rho_>), \quad \epsilon_m = \frac{1}{1 + \delta m, 0}.$$

Furthermore, J_m, $H_m^{(1))}$, I_m, and K_m are the Bessel, Hankel, modified Bessel and Macdonald functions, respectively. Substituting this expansion into Φ, and integrating over z' and ϕ', one gets

$$\Phi(x, y, z) = \frac{2\pi e}{\epsilon v}\exp(i\omega z/v)\left[i\Theta(\beta_n - 1)H_0^{(1)}\Phi_1 + \frac{2}{\pi}\Theta(1 - \beta_n)K_0\Phi_2\right],$$

$$A_z = \beta\epsilon\mu\Phi,$$

where

$$\Phi_1 = \int\int \rho^2 d\rho dt \cos(\omega\rho t/\gamma v)J_0\rho_{Ch}(\rho\sqrt{1 + t^2}),$$

and

$$\Phi_2 = \int\int \rho^2 d\rho dt \cos(\omega\rho t/\gamma v)I_0\rho_{Ch}(\rho\sqrt{1 + t^2}).$$

Here and later we drop the arguments of the usual and modified Bessel functions if they are $k\rho\sqrt{n^2 - 1/\beta^2}$ and $k\rho\sqrt{1/\beta^2 - n^2}$, respectively. The integration over ρ and t runs over the $(0, \infty)$ interval.

We intend to find the energy flux in the radial direction through the surface of a cylinder of the radius ρ coaxial with the motion axis. It coincides with the energy radiated per unit cylinder length and per unit frequency, and is given by

$$S_\rho = \frac{d^2\mathcal{E}}{dzd\omega} = -\pi\rho c(E_z H_\phi^* + E_z^* H_\phi).$$

Thus we need E_z and H_ϕ. They are equal to

$$E_z = \frac{2\pi e i\mu\omega}{c^2}(1 - 1/\beta^2 n^2)\exp(i\omega z/v)$$

$$\times\left[i\Theta(\beta_n - 1)H_0^{(1)}\Phi_1 + \frac{2}{\pi}\Theta(1 - \beta_n)K_0\Phi_2\right],$$

$$H_\phi = \frac{2\pi e}{c}\exp(i\omega z/v)|k|\sqrt{|n^2 - 1/\beta^2|}$$

$$\times\left[i\Theta(\beta_n - 1)H_1^{(1)}\Phi_1 + \frac{2}{\pi}\Theta(1 - \beta_n)K_1\Phi_2\right].$$

Substituting them into S_ρ one finds

$$S_\rho(\omega) = F \cdot S_{TF}, \tag{7.2}$$

where

$$S_{TF} = \frac{e^2 \mu \omega}{c^2} (1 - 1/\beta^2 n^2) \tag{7.3}$$

is the Tamm-Frank frequency distribution of the energy radiated by the uniformly moving point-like charge per unit length and per unit frequency [15], and

$$F = 16\pi^2 \Phi_1^2 \tag{7.4}$$

is the factor taking into account the finite dimension of a charge (form factor, for short).

The number of photons radiated by a moving charge per unit length of the cylindrical surface and per unit frequency is given by

$$N_\rho(\omega) = \frac{d^2 N}{dz d\omega} = F \cdot N_{TF}, \tag{7.5}$$

where N_{TF} is the corresponding Tamm-Frank frequency distribution of the photon number

$$N_{TF} = \frac{\alpha \mu}{c} (1 - 1/\beta^2 n^2), \tag{7.6}$$

and $\alpha = e^2/\hbar c$ is the fine structure constant.

The total energy and number of photons radiated per unit length of the cylindrical surface are obtained by integrating $S_\rho(\omega)$ and $N_\rho(\omega)$ over ω

$$S_\rho = \frac{d\mathcal{E}}{dz} = \int_0^\infty S_\rho(\omega) d\omega, \quad N_\rho = \frac{dN}{dz} = \int_0^\infty N_\rho(\omega) d\omega. \tag{7.7}$$

In what follows, when integrating (7.7) we assume the medium to be dispersion-free, that is, n does not depend on frequency. Consider particular cases.

1. Let the charge be uniformly distributed inside the sphere of the radius a:

$$\rho_{Ch}(r) = \rho_0 \Theta(a - r), \quad \rho_0 = \frac{1}{(4\pi a^3/3)}. \tag{7.8}$$

Then,

$$\rho_L = e\gamma \rho_0 \Theta\{a - [\rho^2 + \gamma^2(z - vt)^2]^{1/2}\},$$

$$\rho_\omega = \frac{e\gamma \rho_0}{\pi \omega} \exp(i\omega z/v) \sin\left(\frac{\omega}{\gamma v}\sqrt{a^2 - \rho^2}\right).$$

The form factor F entering into (7.2) is given by

$$F = \frac{9}{2}\pi \frac{J_{3/2}^2(y)}{y^3}, \tag{7.9}$$

where $y = ka\sqrt{n^2 - 1}$. The total radiated energy and the number of photons defined by (7.7) are given by

$$S_\rho = \frac{9e^2\mu}{4a^2}\frac{1 - 1/\beta^2 n^2}{n^2 - 1}, \quad N_\rho = \frac{3\alpha\pi\mu}{5a}\frac{1 - 1/\beta^2 n^2}{\sqrt{n^2 - 1}} \tag{7.10}$$

for $\beta > 1/n$ and zero otherwise.

2. The charge is distributed over the surface of the sphere

$$\rho(r) = \rho_0\delta(a - r), \quad \rho_0 = 1/(4\pi a^2). \tag{7.11}$$

The form factor F is

$$F = \left(\frac{\sin y}{y}\right)^2, \quad y = ka\sqrt{n^2 - 1}. \tag{7.12}$$

The total energy

$$S_\rho = \int_0^\infty S_\rho(\omega)d\omega = \frac{e^2\mu(1 - \beta^2 n^2)}{a^2(n^2 - 1)}\int_0^\infty \frac{dy}{y}\sin^2 y \tag{7.13}$$

diverges whilst the total number of photons is finite:

$$N_\rho = \int_0^\infty N_\rho(\omega)d\omega = \frac{\alpha\mu\pi}{2a}\frac{1 - 1/\beta^2 n^2}{\sqrt{n^2 - 1}}. \tag{7.14}$$

The divergence of S_ρ is owed to the contribution of high frequencies. In the past, frequency distribution $S_\rho(\omega)$ was obtained in [2] but with the form factor given by

$$F' = \left(\frac{\sin y'}{y'}\right)^2, \quad \text{where} \quad y' = ka\sqrt{\epsilon}.$$

This leads to different physical predictions: for n slightly greater than 1, the form factor F also tends to 1 and the frequency distribution $S_\rho(\omega)$ tends to the Tamm-Frank distribution whilst the form factor F' and the frequency distribution $S_\rho(\omega)$, found in [2], are rapidly oscillating functions of ω when $\epsilon \to 1$.

3. The charge is distributed according to the Gauss law

$$\rho_{Ch}(r) = \rho_0\exp(-r^2/a^2), \quad \rho_0 = 12/(\pi^{3/2}a^3). \tag{7.15}$$

Then,

$$\rho_\omega = \frac{e\gamma}{2\pi^2 a^2 v}\exp(i\psi)\exp(-\rho^2/a^2)\exp(-k^2 a^2/4\beta^2).$$

The form factor F is

$$F = \exp(-k^2 a^2 n^2/2). \qquad (7.16)$$

The total radiated energy and the number of photons are finite now

$$S_\rho = \frac{e^2 \mu}{a^2 n^2}(1 - 1/\beta^2 n^2), \quad N_\rho = \frac{\alpha\mu\pi^{3/2}}{an}(1 - 1/\beta^2 n^2). \qquad (7.17)$$

4. For the Yukawa charge distribution

$$\rho_{Ch}(r) = \rho_0 \frac{\exp(-r/a)}{r}, \quad \rho_0 = \frac{1}{4\pi a^2}, \qquad (7.18)$$

one has

$$\Phi_1 = \frac{1}{4\pi}\frac{1}{1 + k^2 a^2(n^2 - 1)}, \quad F = \frac{1}{[1 + k^2 a^2(n^2 - 1)]^2},$$

$$N_\rho(\omega) = N_{TF}F, \quad S_\rho(\omega) = S_{TF}F, \qquad (7.19)$$

The integral number of emitted photons and the integral radiated energy are given by

$$N_\rho = \int d\omega N_\rho(\omega) = \frac{\pi\mu\alpha}{4a\sqrt{n^2 - 1}}(1 - 1/\beta^2 n^2),$$

$$S_\rho = \int d\omega S_\rho(\omega) = \frac{e^2 \mu}{2a^2(n^2 - 1)}(1 - 1/\beta^2 n^2). \qquad (7.20)$$

The following $S_\rho(\omega)$ was found in [2] for the Yukawa distribution

$$S_\rho = S_{TF}F, \quad \text{where} \quad F = \frac{1}{16\pi^2 c^2(1 + k^2 a^2 \epsilon)}.$$

Obviously, this F is not reduced to 1 in the limit $a \to 0$ (as it should be). This is due to the extra factor $1/16\pi^2 c^2$.

There are two reasons why we cannot compare our results step by step with those obtained in [1,2]. The first reason is purely technical: the authors of [1,2] carried out the double Fourier transform over space and time variables, and then returned to the frequency distribution using integration in k space. The advantage of our approach is that we always operate in a space-frequency representation, no intermediate steps are needed. The second reason is a result of different definitions of charge densities. For example, we define the spherical charge density ρ_{Ch} in a moving system attached to a moving charge and then recalculate it into the laboratory frame using the Lorentz transformations, thus obtaining ρ_L. On the other hand, the authors of [2] postulate the spherical charge density ρ_{Ch} in the laboratory frame. It should be noted that in the laboratory frame the charge density owed to the γ factors cannot be spherically symmetrical (this fact is observed experimentally).

7.2.1. CHERENKOV RADIATION AS THE ORIGIN OF THE CHARGE DECELERATION

The following ambiguity arises. The Cherenkov radiation is usually associated with the radiation of a charge moving uniformly in a medium. Since the moving charge radiates, its kinetic energy should decrease. The energy radiated per unit length is equal to

$$\frac{d\mathcal{E}}{dz} = C(1 - 1/\beta^2 n^2) \tag{7.21}$$

for $\beta > 1/n$ and zero otherwise. The constant C, independent of β, is defined by one of Eqs. (7.10), (7.17) or (7.20). Obviously, (7.21) should be equal to the loss of kinetic energy:

$$\frac{dT}{dz} = m_0 c^2 \frac{d}{dz} \frac{1}{\sqrt{1 - \beta^2}} = -C(1 - 1/\beta^2 n^2). \tag{7.22}$$

Or, introducing the dimensionless variable $\tilde{z} = z/L$, $L = m_0 c^2/C$, one obtains

$$\frac{d}{d\tilde{z}} \frac{1}{\sqrt{1 - \beta^2}} = -(1 - 1/\beta^2 n^2). \tag{7.23}$$

Integrating this equation we find

$$(n^2 - 1)(\tilde{z} - \tilde{z}_0) = \frac{1}{2\alpha} \ln \left[\left(\frac{\alpha + \gamma^{-1}}{\alpha + \gamma_0^{-1}} \right)^2 \cdot \frac{n^2 \beta_0^2 - 1}{n^2 \beta^2 - 1} \right] - n^2(\gamma - \gamma_0). \tag{7.24}$$

Here $\gamma = 1/\sqrt{1 - \beta^2}$, $\gamma_0 = 1/\sqrt{1 - \beta_0^2}$, $\alpha = \sqrt{1 - 1/n^2}$, and β_0 is the charge velocity at the spatial point z_0. This equation, being resolved relative to β, defines the charge velocity $\beta(z)$ at the particular point z of the motion axis. It follows from (7.24) that

$$\beta \to 1 - \frac{1}{2(1 - 1/n^2)^2 \tilde{z}^2}$$

for $\tilde{z} \to -\infty$ and

$$\beta \to \frac{1}{n} \left\{ 1 + \frac{1}{2} \exp[-2(n^2 - 1)\tilde{z}] \right\}$$

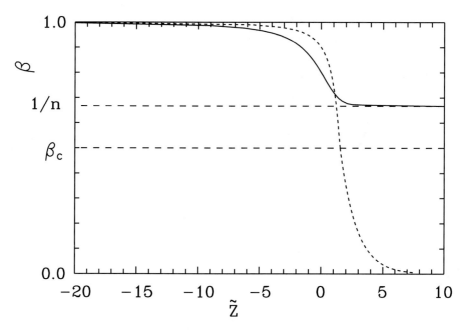

Figure 7.1.   This figure shows how a moving charge is decelerated when all its energy losses are owed to the Cherenkov radiation. The solid curve corresponds to a charge of finite dimensions moving in a dispersion-free medium. The charge velocity approaches $1/n$ for $\tilde{z} \to \infty$. The pointed curve corresponds to a point-like charge moving in a dispersive medium. Its velocity is equal to β_c at $\tilde{z} = \tilde{z}_c$. Below β_c the asymptotic form of β given by $\beta \sim \beta_c \exp[-(\tilde{z} - \tilde{z}_c)/4\beta_c^2\gamma_c^2]$ was used.

for $\tilde{z} \to \infty$. The dependence $\beta(\tilde{z})$ for typical parameters $n = 1.5, \beta_0 = 0.8$ and $z_0 = 0$ is shown in Fig. 7.1.

7.3. Cherenkov radiation in dispersive medium

Another way of obtaining the finite value of the radiated energy and the number of photons is to take into account the medium dispersion. The energy flux in the radial direction through the cylinder surface of the radius ρ is given by

$$\frac{d^3\mathcal{E}}{\rho d\phi dz dt} = -\frac{c}{4\pi}E_z(t)H_\phi(t).$$

Integrating this expression over the whole time of a charge motion and over the azimuthal angle ϕ, and multiplying it by ρ, one obtains the energy radiated for the whole charge motion per unit length of the cylinder surface

$$\frac{d\mathcal{E}}{dz} = -\frac{c\rho}{2} \int E_z H_\phi dt.$$

Substituting here instead of E_z and H_ϕ their Fourier transforms and performing the time integration, one finds

$$\frac{d\mathcal{E}}{dz} = \int\limits_0^\infty d\omega \sigma_\rho(\omega), \tag{7.25}$$

where

$$\sigma_\rho(\omega) = \frac{d^2\mathcal{E}}{dzd\omega} = -\pi\rho c E_z(\omega) H_\phi^*(\omega) + c.c. \tag{7.26}$$

is the energy radiated in the radial direction per unit frequency and per unit length of the observational cylinder. The identification of the energy flux with σ_ρ is typical in the Tamm-Frank theory [15] describing the unbounded charge motion in medium.

If the dielectric permittivity is chosen in the form

$$\epsilon(\omega) = 1 + \frac{\omega_L^2}{\omega_0^2 - \omega^2},$$

then $\sigma_\rho(\omega)$ is given by (7.3). The integration in (7.25) runs over the frequency region corresponding $1 - \beta^2\epsilon < 0$, which corresponds to the Tamm-Frank condition $\beta n > 1$. It is easy to check that for $\beta > \beta_c = 1/\sqrt{1 + \omega_L^2/\omega_0^2}$ this condition is satisfied for $0 < \omega < \omega_0$. For $\beta < \beta_c$ this condition is satisfied for $\omega_c < \omega < \omega_0$, where $\omega_c = \omega_0\sqrt{1 - \beta^2\gamma^2/\beta_c^2\gamma_c^2}$ and $\gamma_c = 1/\sqrt{1 - \beta_c^2}$. This frequency window narrows as β diminishes. For $\beta \to 0$ the frequency spectrum is concentrated near the frequency ω_0. The total energy radiated per unit length of the observational cylinder is equal to [6,7] (see also Chapter 4)

$$\frac{d\mathcal{E}}{dz} = \int\limits_0^\infty S_\rho(\omega)d\omega = \frac{e^2\omega_0^2}{2c^2}\left[1 - 1/\beta^2 - \frac{1}{\beta^2\beta_c^2\gamma_c^2}\ln(1 - \beta_c^2)\right] \tag{7.27}$$

for $\beta > \beta_c$ and

$$\frac{d\mathcal{E}}{dz} = -\frac{e^2\omega_L^2}{2c^2}\left[1 + \frac{1}{\beta^2}\ln(1 - \beta^2)\right] \tag{7.28}$$

for $\beta < \beta_c$.

Energy balance as a result of the medium dispersion.
According to Section 2 the influence of the charge finite dimension becomes essential for $ka \sim 1$. If for a we take 1 fm, then $\omega_f \sim 10^{23} \mathrm{s}^{-1}$. On the other hand, in the presence of dispersion, the frequency spectrum of the radiation intensity extends up to ω_0. If we identify ω_0 with the ultraviolet frequency $\sim 10^{16} \mathrm{s}^{-1}$, then $\omega_0 \ll \omega_f$. This means that the influence of the dispersion begins at a much smaller frequency than that owed due to the finite charge dimensions.

Since in the presence of dispersion $d\mathcal{E}/dz$ is finite (see (7.27) and (7.28)), one can extract $v(z)$ from the energy balance condition $dT/z = -d\mathcal{E}/dz$, similarly as was done for a charge of finite dimensions. The following equations are valid now

$$\frac{d}{d\tilde{z}} \frac{1}{\sqrt{1 - \beta^2}} = -\left(1 - \frac{1}{\beta^2 \tilde{n}^2}\right) \tag{7.29}$$

for $\beta > \beta_c$ and

$$\frac{d}{d\tilde{z}} \frac{1}{\sqrt{1 - \beta^2}} = \left(\frac{1}{\beta_c^2} - 1\right)\left[1 + \frac{1}{\beta^2}\ln(1 - \beta^2)\right] \tag{7.30}$$

for $\beta < \beta_c$. Here we put

$$\tilde{n}^2 = \left[1 + \left(\frac{1}{\beta_c^2} - 1\right)\ln(1 - \beta_c^2)\right]^{-1}, \quad \tilde{z} = \frac{z}{L}, \quad L = \frac{2m_0 c^4}{e^2 \omega_L^2}\left(\frac{1}{\beta_c^2} - 1\right).$$

For $\beta > \beta_c$ one then finds the following equation

$$(\tilde{n}^2 - 1)(\tilde{z} - \tilde{z}_0) = \frac{1}{2\alpha}\ln\left[\left(\frac{\alpha + \gamma^{-1}}{\alpha + \gamma_0^{-1}}\right)^2 \cdot \frac{\tilde{n}^2 \beta_0^2 - 1}{\tilde{n}^2 \beta^2 - 1}\right] - \tilde{n}^2(\gamma - \gamma_0). \tag{7.31}$$

Here $\alpha = \sqrt{1 - 1/\tilde{n}^2}$; γ, γ_0, and z_0 are the same as in (7.24). It follows from (7.31) that

$$\beta \to 1 - \frac{1}{2(1 - 1/\tilde{n}^2)^2 \tilde{z}^2}$$

for $\tilde{z} \to -\infty$. The velocity β_c is reached at

$$\tilde{z}_c = \tilde{z}_0 + \frac{1}{2\alpha(\tilde{n}^2 - 1)}\ln\left[\left(\frac{\alpha + \gamma_c^{-1}}{\alpha + \gamma_0^{-1}}\right)^2 \cdot \frac{\tilde{n}^2 \beta_0^2 - 1}{\tilde{n}^2 \beta_c^2 - 1}\right] - \frac{\tilde{n}^2}{\tilde{n}^2 - 1}(\gamma_c - \gamma_0). \tag{7.32}$$

For $\beta > \beta_c$ the dependence $\beta(\tilde{z})$ extracted from (7.31) is shown in Fig.1 for typical parameters $\beta_c = 0.5$, $\beta_0 = 0.9$ and $\tilde{z}_0 = 0$. Below β_c, the asymptotic form of $\tilde{\beta}$ given by $\tilde{\beta} \sim \exp[-(\tilde{z} - \tilde{z}_c)/4\beta_c^2 \gamma_c^2]$ and obtained from (7.30) is presented.

Energy balance as a result of the ionization losses.
Although the energy balance is important from the theoretical viewpoint, it is slightly academic. The reason is that the energy losses owed to the ionization of medium atoms are much larger than the Cherenkov radiation losses. To a good accuracy they are described by

$$dT/dz = -\frac{C}{\beta^2}F, \qquad (7.33)$$

where C is a constant dependent on the charge of a moving particle and on the medium properties, and F is a function weakly dependent on β. For the electrons propagating in water $C \approx 1.65$ Mev/cm. On the other hand, the constant $e^2\omega_0^2/2c^2$ entering (7.27) is $\sim 10^{-2}$ Mev/cm for $\omega_0 \approx 10^{16}\mathrm{s}^{-1}$. Since $e^2\omega_0^2/2c^2 \ll C$ the ionization energy losses are much larger than those owed to the Cherenkov radiation. This means that usually one can disregard the Cherenkov energy losses in (7.33). The notable exceptions are: i) gases, in which ionization energy losses are small; ii) substances with the large boundary frequency ω_0 (lying, e.g., in the Roentgen part of the frequency spectrum); iii) substances with a refractive index different from unity for $\omega \to \infty$ (the typical example is ZnSE discussed in Chapter 4).

Eq. (7.33) can be solved analytically if one sets $F = 1$. Then

$$\beta(z) = \frac{\sqrt{2}[x(x+4)]^{1/4}}{[\sqrt{x(x+4)}+x+2]^{1/2}}. \qquad (7.34)$$

Here $x = (z_f - z)/L$ and $L = m_0c^2/C$; z_f is the spatial point at which $\beta = 0$. For $x \to 0$, $\beta \sim \sqrt{2}x^{1/4}$ whilst $\beta \sim 1 - 1/x^2$ for $x \to \infty$. The function $\beta(x)$ is shown in Fig. 7.2. The velocity β, as a function of z, drops almost instantly for small L. This justifies the validity of the Tamm problem [16] which involves a sudden transition from the charge uniform motion to the state of rest. On the other hand, for large L (e.g., for heavy particles with not very large charges Z (for Z large , $C \sim Z^2$ is also large and, therefore, L is small)) the transition to the state of rest will be smooth and the deviation from the Tamm picture is to be expected. The radiation intensity corresponding to the energy losses (7.33) and to the velocity dependence (7.34) is given by

$$\sigma_r = \frac{e^2\mu nk^2 \sin^2\theta}{4\pi^2 c}(I_c^2 + I_s^2), \qquad (7.35)$$

where

$$I_c = \int_{z_1}^{z_2} \cos\psi dz, \quad I_s = \int_{z_1}^{z_2} \sin\psi dz, \quad \psi = \omega t(z) - knz\cos\theta.$$

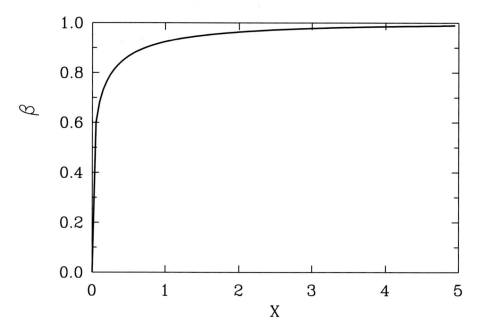

Figure 7.2. This figure shows how a moving charge is decelerated as a result of the ionization energy losses described by (7.29). Here $x = (z_f - z)/L$, z_f is the spatial point at which $\beta = 0$ and L is the same as in (7.31).

$$\tau(z) = c(t_f - t)/L = \sqrt{y^2 - 1} + \frac{\pi}{2} - 2\arctan(y + \sqrt{y^2 - 1}),$$

$$y = \frac{1}{2}x + 1 + \frac{1}{2}\sqrt{x(x+4)}. \qquad (7.36)$$

To what can the results of this section be applied?

First, we mention the VC radiation from the electron bunches produced in linear electronic accelerators. According to [17] the typical bunch dimension is about 1 cm which corresponds to a cut off frequency $\omega \sim 3 \times 10^{10}\text{s}^{-1}$ lying in the far infrared region.

The other application is owed to [9], in which predictions of the position of the radiation maximum made in [6-8] were checked experimentally. In this reference, the dimension of an electric dipole layer propagating in ZnSe crystal, was about 10^{-3} cm. The corresponding cut off frequency $\omega \sim 3 \times 10^{13}\text{s}^{-1}$ lies, again, in the infrared region.

When discussing the VC radiation from extended charges we implicitly implied that they are structureless. However, the electronic bunches and electric dipole layers mentioned above consist of electrons and electric dipoles, respectively. For frequencies larger than c/d (d is the distance between particular charges or dipoles) their internal structure becomes essential. Consequently the VC radiation from particular charges and dipoles takes place and the above consideration is no longer valid.

7.4. Radiation of a charge moving in a cylindrical dielectric sample

Up to now we have implicitly suggested that the radiation intensity is observed in the same medium where a charge moves. However, a charge usually moves in one medium (water, glass) while the observations are made in another medium (air, vacuum) (see, e.g., the nice Cherenkov review [18]). We intend now to consider arising complications. Consider a cylindrical sample C of radius a filled with a medium with the parameters ϵ_1 and μ_1. This sample is surrounded by another medium with parameters ϵ_2 and μ_2 such that $n_2 < n_1$. Let a charge move with a constant velocity v along the axis of C with a constant velocity v satisfying the inequality $1/n_1 < \beta < 1/n_2$ (that is, the medium inside C is optically more dense than outside it). In the past, this problem was considered by Frank and Ginzburg [19] who, having written the general solution for arbitrary n_1 and n_2, applied it to the concrete case when the medium inside C was vacuum, while outside C was a medium with the refractive index n_2. They obtained the remarkable result that despite the absence of the energy flux inside C it reappears outside C if $\beta n_2 > 1$.

As to other possibilities, they remark that

> Similarly, as it was done above, one may easily consider other particular cases ($\beta n_1 > 1, \beta n_2 < 1; \beta n_1 > 1, \beta n_2 > 1$), which will not be considered here. We note only, that for $\beta n_2 < 1$, there are no radiation energy losses for both $\beta n_1 < 1$ and $\beta n_1 > 1$.

We consider in some detail the case corresponding to $n_2 < n_1, \beta n_1 > 1, \beta n_2 < 1$. One easily finds that the electromagnetic field arising from an unbounded charge motion along the axis of C is equal to

$$A_z = C_2\mu_2 \exp(i\psi)K_0(2), \quad H_\phi = C_2 k \exp(i\psi)\sqrt{1/\beta_2^2 - 1}K_1(2),$$

$$E_z = -ikC_2\mu_1(1/\beta_2^2 - 1)\exp(i\psi)K_0(2), \quad E_\rho = H_\phi/\beta\epsilon_2 \qquad (7.37)$$

outside C, and

$$A_z = \mu_1 \exp(i\psi)[\frac{ie}{2c}H_0^{(1)}(1) + C_1 J_0(1)],$$

$$H_\phi = \exp(i\psi)kn_1\sqrt{1 - 1/\beta_1^2}[\frac{ie}{2c}H_1^{(1)}(1) + C_1 J_1(1)], \quad E_\rho = H_\phi/\beta\epsilon_1,$$

$$E_z = ik\mu_1 \exp(i\psi)(1 - 1/\beta_1^2)[\frac{ie}{2c}H_0^{(1)}(1) + C_1 J_0(1)] \tag{7.38}$$

inside it. Here $\psi = kz/\beta$, $\beta_1 = \beta n_1$, $\beta_2 = \beta n_2$. The arguments of the Bessel functions are $2 = k\rho\sqrt{1/\beta^2 - n_2^2}$ for $\rho > a$ and $1 = k\rho\sqrt{n_1^2 - 1/\beta^2}$ for $\rho < a$. The coefficients C_1 and C_2 are found from the continuity of H_ϕ and E_z at $\rho = a$:

$$C_1 = \frac{e}{2c}\left[\frac{1}{\triangle}(n_1\mu_2\sqrt{1/\beta_2^2 - 1}K_0 N_1 + n_2\mu_1\sqrt{1 - 1/\beta_1^2}K_1 N_0) - i\right], \tag{7.39}$$

$$C_2 = \frac{e\mu_1}{\pi cka\triangle}\sqrt{\frac{1 - 1/\beta_1^2}{1/\beta_2^2 - 1}},$$

$$\triangle = n_1\mu_2\sqrt{1/\beta_2^2 - 1}K_0 J_1 + \mu_1 n_2\sqrt{1 - 1/\beta_1^2}K_1 J_0.$$

The arguments of the usual and modified Bessel functions entering into (7.39) are $kan_1\sqrt{1 - 1/\beta_1^2}$ and $kan_2\sqrt{1/\beta_2^2 - 1}$, respectively. We evaluate now the energy fluxes.

7.4.1. RADIAL ENERGY FLUX

The radial energy flux is

$$\sigma_\rho = \frac{d^2\mathcal{E}}{dz d\omega} = -\pi\rho c(E_z H_\phi^* + \text{c.c.}). \tag{7.40}$$

Obviously, it is equal to zero outside C and

$$\sigma_\rho = -\pi\rho ck^2\mu_1 n_1(1 - 1/\beta_1^2)^{3/2}$$

$$\times\{\left[-\frac{e}{2c}H_0^{(1)}(1) + iC_1 J_0(1)\right]\left[-\frac{ie}{2c}H_1^{(2)}(1) + C_1^* J_1(1)\right]$$

$$+ \left[-\frac{e}{2c}H_0^{(2)}(1) - iC_1^* J_0(1)\right]\left[\frac{ie}{2c}H_1^{(1)}(1) + C_1 J_1(1)\right]\}$$

$$= -\pi\rho ck^2\mu_1 n_1(1 - 1/\beta_1^2)^{3/2}$$

$$\times\left\{-\frac{e^2}{\pi c^2 kn_1\rho\sqrt{1 - 1/\beta_1^2}} + \frac{ie}{2c}[J_1(1)N_0(1) - J_0(1)N_1(1)](C_1 - C_1^*)\right\} = 0$$

inside C (it was taken into account that $\text{Im}C_1 = -e/2c$). Thus the radial energy flux is equal to zero inside C too. This is because the contribution of the terms with a product of Hankel functions in the energy flux is compensated by the terms with a product of Bessel and Hankel functions.

The following complication arises. Let the detector be placed outside C, that is, in the medium where $\beta n_2 < 1$. In fact, this is a typical situation in

Cherenkov experiments. For example, in classical Cherenkov experiments [18] the electrons moved in a vessel filled with water, whilst the observations of the Cherenkov light were made in air, in a dark room, by a human eye. There is no radial energy flux outside C. Then, how can the Cherenkov radiation be observed there? One may argue that since the human eye is filled with a substance having the refractive index approximately equal to that of water, the Cherenkov radiation reappears in it (similarly to the appearance of the radiation in the medium surrounding a vacuum channel with a charge moving along its axis [19]), and therefore it could be detected. However, there are now known substances with large refractive indices. Does this mean that the radial energy flux cannot be detected outside C (for this it is enough to use a collimator selecting only the photons emitted in the radial direction) if the measuring device is fabricated from the substance with a refractive index n_2 smaller than $1/\beta$ and n_1? A possible answer is given in the following section in which the energy flux in the direction parallel to the motion axis will be evaluated.

7.4.2. ENERGY FLUX ALONG THE MOTION AXIS

The energy flux parallel to the motion axis is

$$\sigma_z = \frac{d^2\mathcal{E}}{d\rho d\omega} = \pi\rho c(E_\rho H_\phi^* + \text{c.c.}). \tag{7.41}$$

It is equal to

$$\sigma_z = \frac{\rho e^2 \mu_1^2 \mu_2 (1 - 1/\beta_1^2)}{\pi v a^2 \triangle^2}[K_1(2)]^2 \tag{7.42}$$

outside C and

$$\sigma_z = \pi\rho c \mu_1 k^2 \left(1 - \frac{1}{\beta_1^2}\right) \times$$

$$\times \left\{\frac{e^2}{4c^2}[J_1^2(1) + N_1^2(1)] + J_1^2(1)|C_1|^2 - \frac{e}{c}[N_1(1)C_{1r} - J_1(1)C_{1i})]\right\}, \tag{7.43}$$

inside it. Here C_{1r} and C_{1i} are the real and imaginary parts of C_1:

$$C_{1r} = \frac{e}{2c}\frac{1}{\triangle}\left(n_1\mu_2\sqrt{1/\beta_2^2 - 1}K_0 N_1 + n_2\mu_1\sqrt{1 - 1/\beta_1^2}K_1 N_0\right), \quad C_{1i} = -\frac{e}{2c}.$$

In general, σ_z is exponentially small outside C, except for ω satisfying $\triangle = 0$. For these ω, σ_z is infinite. For large ka the equation $\triangle = 0$ reduces to

$$kan_1\sqrt{\beta_1^2 - 1} = \frac{\pi}{4} - \arctan\frac{\epsilon_2\sqrt{\beta_1^2 - 1}}{\epsilon_1\sqrt{1 - \beta_2^2}} + m\pi, \tag{7.44}$$

where m is integer. The distance between the neighbouring maxima of σ_z is $\Delta\omega = \pi c/(an_1\sqrt{\beta_1^2 - 1})$. For a cylinder radius $a \sim 10$ cm, the $\Delta\omega$ is about 10^{10}s^{-1}. The typical optical frequency is about 5×10^{15}s^{-1}. Since a real Cherenkov detector has the finite frequency resolution width (several 10^{15}s^{-1} units) it inevitably covers many maxima of σ_z, and therefore a measuring device oriented parallel to the direction of motion will detect the almost continuous radiation. Inside C, σ_z given by (7.43) is also singular at the frequencies defined by (7.43). For other frequencies σ_z is not exponentially small. As a function of ρ it is infinite on the axis of motion ($\rho = 0$) (along which a charge moves) and oscillates with increasing of ρ. There is no radiation maximum in the $z =$const plane at the Cherenkov angle defined by $\cos\theta_c = 1/\beta n$. The physical interpretation is as follows. A moving charge emits the Cherenkov gamma ray at the Cherenkov angle. This Cherenkov gamma ray intersects the particular $z =$const at some radius ρ. Depending on the charge position, ρ changes from 0, when the charge intersects the above the $z =$const plane, up to $\rho = a$, when the charge is at a distance $a\tan\theta_c$ in front this plane. A photographic plate placed at the $z =$const plane perpendicularly to the motion axis will be darkened, with a main maximum at $\rho = 0$ (the intensity of darkening behaves as $1/\rho$ for small ρ) and with additional maxima corresponding to the singularities of (7.43). Sometimes experimentalists (see, e.g., [20,21]) install inside the cylindrical volume C (especially, when it is filled with a gas) a metallic mirror inclined at an angle $\pi/4$ towards the motion axis. This mirror reflects the σ_z component (7.43) of the internal energy flux in the direction perpendicular to the motion axis, thus making it possible to observe the energy flux in the radial direction outside C. The experimentalists see the pronounced maximum at the Cherenkov angle θ_c. The possible reasons for this are: i) transition radiation arising at the surface of the metallic mirror; ii) a charge deceleration inside this mirror; iii) the finite path of a charge inside C in the presence of special optical devices focusing the gamma rays emitted at the Cherenkov angle into the sole Cherenkov ring.

7.4.3. OPTICAL INTERPRETATION

A charge moving uniformly inside the dielectric cylinder C emits a light ray at the Cherenkov angle θ_1 ($\cos\theta_1 = 1/\beta n_1$) towards the charge motion axis. Let this ray intersect the cylinder surface at some point and let i be the angle of incidence (Fig. 7.3). It is easy to check that $\sin i = \cos\theta_1$. According to classical optics (see, e.g., [22,23]), the angles of incidence i, reflection i' and refraction r are inter-related as follows: $i = i'$, $\sin r = (n_1/n_2)\sin i$. It follows from Fig. 7.3 that $\sin r = \cos\theta_2$, where θ_2 is the inclination angle of the light ray moving in medium 2 towards the z axis.

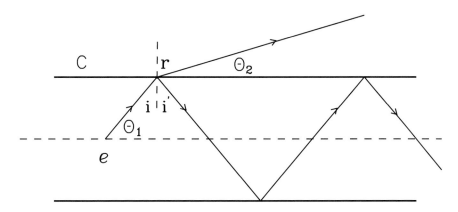

Figure 7.3.   An infinite cylindrical dielectric sample C with refractive index n_1 is surrounded by the medium with refractive index n_2. A charge moving in C emits a γ ray at the Cherenkov angle θ_1. This γ ray leaves C if $\beta n_2 > 1$. Otherwise it exhibits total internal reflection and remains within C.

Therefore $\cos\theta_2 = (n_1/n_2)\cos\theta_1 = 1/\beta n_2$. That is, if $\beta n_2 > 1$ the light ray in medium 2 propagates at the angle θ_2 towards the motion axis. Otherwise ($\beta n_2 < 1$) total internal reflection takes place. Owing to the translational symmetry of the problem the same total internal reflection takes place at all other points where a given light ray meets the cylinder surface. This means that the light ray emitted by a moving charge remains within the infinite cylindrical sample if $\beta n_2 < 1$.

The situation changes slightly if the cylindrical sample has a finite length. In order not to deal with the transition radiation (arising when the moving charge passes through the boundaries of media 1 and 2), we consider the charge motion completely confined within C (Fig. 7.4).

This situation was realized in the original Cherenkov experiments in which Compton electrons were completely absorbed in water. Usually this situation is described in terms of the so-called Tamm problem [16], where the charge moves uniformly with the velocity $\beta > n_1$ in a finite spatial interval. After a number of reflections on the surface of C, a particular light ray reaches the bottom of a cylindrical sample. It is easy to check that its angle of incidence coincides with θ_1. The angle of refraction is given by

$$\sin r' = \frac{n_1}{n_2}\sin\theta_1 = \frac{n_1}{n_2}\sqrt{1 - \frac{1}{\beta^2 n_1^2}}.$$

Obviously the light ray leaves C through its bottom if $\sin r' < 1$. This is

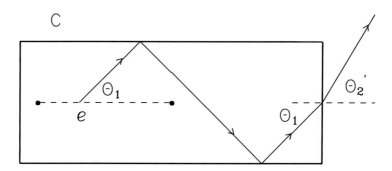

Figure 7.4. A charge moves in a finite dielectric cylindrical sample C. There is an additional possibility for the Cherenkov γ ray to leave C through its bottom (see the text).

equivalent to

$$\beta < \min(\frac{1}{n_2}, \frac{1}{\sqrt{n_1^2 - n_2^2}}).$$

It follows from this that if $n_2 > n_1/\sqrt{2}$ then the light ray passes through the bottom of C and propagates in medium 2 at the angle $\theta_2' = r'$ towards the motion axis. Let $n_2 < n_1/\sqrt{2}$. Then for $\beta < 1/\sqrt{n_1^2 - n_2^2}$ the light ray propagates in medium 2 at the same angle θ_2' towards the motion axis. On the other hand, for

$$\frac{1}{\sqrt{n_1^2 - n_2^2}} < \beta < \frac{1}{n_2}$$

total internal reflection takes place at the bottom of C as well. Therefore in this case the light ray emitted by a moving charge remains within C.

7.5. Vavilov-Cherenkov and transition radiations for a spherical sample

7.5.1. OPTICAL INTERPRETATION

Consider a dielectric sphere S of the radius R filled with a substance of refractive index n_1 and surrounded by the substance with refractive index n_2 (Fig. 7.5). Let a charge move uniformly in the spatial interval $(-z_0, z_0)$ lying completely inside S and let its velocity be such that $1/n_1 < \beta < 1/n_2$. Elementary calculations show that the Cherenkov γ ray emitted at the point

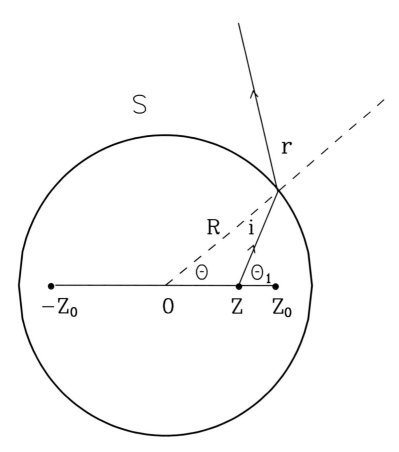

Figure 7.5. There are more chances for the Cherenkov γ ray to leave the sphere S, than the dielectric cylinder C. The reason is that the Cherenkov γ ray meets S at different angles of incidence depending on the charge position z on the motion axis. Here i and r are the angles of incidence and refraction, respectively.

z of the motion axis, propagates outside S at the angle

$$\theta_2 = \theta + \arcsin\left[\frac{n_1}{n_2}\sin(\theta_1 - \theta)\right] = \theta + \arcsin\left(\frac{n_1 z}{n_2 R}\sin\theta_1\right) \qquad (7.45)$$

towards the motion axis. Here θ_1 is defined by $\cos\theta_1 = 1/\beta_1 n$ and θ is related to the charge particular position z as follows

$$\cos\theta = \frac{z}{R}\sin^2\theta_1 + \cos\theta_1\sqrt{1 - \frac{z^2}{R^2}\sin^2\theta_1}. \qquad (7.46)$$

When a charge moves from $z = -z_0$ to $z = z_0$, $\cos\theta$ changes in the interval

$$-\frac{z_0}{R}\sin\theta_1 + \cos\theta_1\sqrt{1 - \frac{z_0^2}{R^2}\sin^2\theta_1} < \cos\theta$$

$$< \frac{z_0}{R}\sin\theta_1 + \cos\theta_1\sqrt{1 - \frac{z_0^2}{R^2}\sin^2\theta_1}$$

for $z_0 < n_2 R/(n_1 \sin\theta_1)$ and in the interval

$$-\frac{n_2}{n_1}\sin\theta_1 + \cos\theta_1\sqrt{1 - n_2^2/n_1^2} < \cos\theta$$

$$< \frac{n_2}{n_1}\sin\theta_1 + \cos\theta_1\sqrt{1 - n_2^2/n_1^2}$$

for $z_0 > n_2 R/(n_1 \sin\theta_1)$. Substituting this into (7.45) we find the angular interval in which the Cherenkov radiation differs from zero outside S. For a radius of the sphere S much larger than the motion interval ($R \gg z_0$), $\theta_2 \approx \theta_1$, that is, the Cherenkov ray propagates in medium 2 under the same angle as in medium 1. The aforesaid means that the Cherenkov radiation has more chances of leaving the sphere than the cylinder. The reason is that the Cherenkov γ ray meets the sphere surface at different angles of incidence depending on the charge position on the motion axis. However, only concrete calculations can determine the value of the radiation intensity in the medium 2.

The semi-intuitive consideration of two last section

i) shows that the observation of the Cherenkov radiation strongly depends on the boundaries surrounding the volume in which a charge moves;

ii) defines conditions under which the Cherenkov radiation can penetrate from the medium 1 with $\beta n_1 > 1$ into the medium 2 with $\beta n_2 < 1$ without exhibiting total internal reflection on their boundary.

7.5.2. EXACT SOLUTION

Green's function
Let the spatial regions inside and outside the sphere S of the radius a be filled by the substances with parameters ϵ_1, μ_1 and ϵ_2, μ_2, respectively. The Green function satisfying equations

$$(\Delta + k_1^2)G = -4\pi\delta^3(\vec{r} - \vec{r}')$$

for $r < a$ and

$$(\Delta + k_2^2)G = -4\pi\delta^3(\vec{r} - \vec{r}')$$

for $r > a$ has the same form as (2.116) but with $G_l(r, r')$ given by

$$G_l = ik_1\Theta(a - r)\Theta(a - r')j_l(k_1 r_<)h_l(k_1 r_>)$$

$$+ik_2\Theta(r - a)\Theta(r' - a)j_l(k2r_<)h_l(k_2 r_>)$$

$$+ik_1 D_l\Theta(a - r)\Theta(r' - a)j_l(k_1 r)h_l(k_2 r')$$

$$+ik_2 C_l\Theta(r - a)\Theta(a - r')j_l(k_1 r')h_l(k_2 r).$$

Here $k_1 = kn_1$ and $k_2 = kn_2$,

$$j_l(x) \quad \text{and} \quad h_l(x) = h_l^{(1)}(x)$$

are the spherical Bessel and Hankel functions; the constants C_l and D_l are defined by the boundary conditions at $r = a$. The vector potential for a charge moving along the z axis is found from the equation

$$A_z = \frac{1}{c} \int G(\vec{r}, \vec{r}')\mu(r')j_z(\omega)dV', \tag{7.47}$$

where $\mu = \mu_1$ for $r < a$ and $\mu = \mu_2$ for $r > a$. Let a charge move uniformly with a velocity v in the interval $-z_0 < z < z_0$. The Fourier component of the current density is given by

$$j_z(\omega) = \frac{e}{2\pi} \exp(iwz/v)\delta(x)\delta(y)\Theta(z + z_0)\Theta(z_0 - z)$$

in cartesian coordinates and

$$j_z(\omega) = \frac{e}{4\pi^2 r^2 \sin\theta}[\delta(\theta)\exp(\frac{ikr}{\beta}) + \delta(\theta - \pi)\exp(-\frac{ikr}{\beta})]\Theta(z_0 - r) \tag{7.48}$$

in spherical coordinates.

The Tamm problem for a charge moving inside the spherical sample
Let a charge move in a finite spatial interval $(-z_0, z_0)$ lying entirely inside the sphere S of the radius a (Fig. 7.6). The sphere is filled with substance 1 with the parameters ϵ_1 and μ_1. The observations are made in the medium with the parameters ϵ_2 and μ_2 surrounding S. Using (7.47) and (7.48) we easily find the magnetic vector potential corresponding to this problem

$$A_z = \frac{iek_2\mu_2}{2\pi c} \sum(2l + 1)P_l(\cos\theta)h_l(k_2 r)C_l$$

for $r > a$,

$$A_z = \frac{iek_1\mu_1}{2\pi c} \sum(2l + 1)P_l(\cos\theta)[j_l(k_1 r)D_l + h_l(k_1 r)J_l^{(1)}(0, z_0)],$$

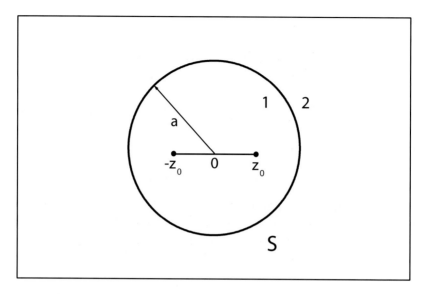

Figure 7.6. A charge moves inside a dielectric sphere S filled with the medium 1. The radiation intensity is measured outside S, in the medium 2.

for $z_0 < r < a$ and

$$A_z = \frac{iek_1\mu_1}{2\pi c} \times$$

$$\times \sum (2l+1)P_l(\cos\theta)[j_l(k_1 r)D_l + h_l(k_1 r)J_l^{(1)}(0, r) + j_l(k_1 r)H_l^{(1)}(r, z_0)]\}$$

for $r < z_0$. Here we put

$$J_l^{(1)}(x, y) = \int_x^y j_l(k_1 r')f_l(r')dr' \quad H_l^{(1)}(x, y) = \int_x^y h_l(k_1 r')f_l(r')dr',$$

$$f_l(r') = \exp\left(\frac{ikr'}{\beta}\right) + (-1)^l \exp\left(-\frac{ikr'}{\beta}\right), \quad k_1 = kn_1, \quad k_2 = kn_2.$$

The coefficients C_l and D_L are to be determined from the continuity of $\vec{E}$ and $\vec{H}$ components tangential to the sphere S. The EMF strengths contributing to the radial energy flux are equal to

$$H_\phi = -\frac{iek^2 n_2^2}{2\pi c} \sum \tilde{C}_l P_l^1 h_l(k_2 r),$$

$$E_\theta = -\frac{i}{\epsilon_2 kr} \frac{d}{dr}(rH_\phi) = -\frac{e\mu_2 n_2 k^2}{2\pi c} \sum H_l(k_2 r)P_l^1 \tilde{C}_l \qquad (7.49)$$

for $r > a$ and

$$H_\phi = -\frac{iek^2 n_1^2}{2\pi c} \sum P_l^1 [\tilde{D}_l j_l(k_1 r) + \tilde{J}_l^{(1)}(0, z_0) h_l(k_1 r)],$$

$$E_\theta = -\frac{e\mu_1 n_1 k^2}{2\pi c} \sum P_l^1 [\tilde{D}_l J_l(k_1 r) + \tilde{J}_l^{(1)}(0, z_0) H_l(k_1 r)] \qquad (7.50)$$

for $z_0 < r < a$. Here we set

$$\tilde{C}_l = C_{l-1} + C_{l+1}, \quad \tilde{D}_l = D_{l-1} + D_{l+1},$$

$$\tilde{J}_l^{(1)}(x, y) = J_{l-1}^{(1)}(x, y) + J_{l+1}^{(1)}(x, y)$$

$$= \int_x^y dr' [j_{l+1}(k_1 r') + j_{l-1}(k_1 r')] f_{l+1} dr' = (2l + 1) \int_x^y \frac{j_l(k_1 r')}{k_1 r'} f_{l+1} dr',$$

$$J_l(x) = \frac{dj_l(x)}{dx} + \frac{j_l(x)}{x} = \frac{1}{2l+1}[(l+1)j_{l-1} - lj_{l+1}],$$

$$H_l(x) = \frac{dh_l(x)}{dx} + \frac{h_l(x)}{x} = \frac{1}{2l+1}[(l+1)h_{l-1} - lh_{l+1}],$$

Imposing the continuity of H_ϕ and E_θ at $r = a$, one finds the following equations for $\tilde{C}_l$ and $\tilde{D}_l$:

$$n_2^2 \tilde{C}_l h_l(2) - n_1^2 \tilde{D}_l j_l(1) = n_1^2 h_l(1) \tilde{J}_l^{(1)}(0, z_0),$$

$$\mu_2 n_2 \tilde{C}_l H_l(2) - \mu_1 n_1 \tilde{D}_l J_l(1) = \mu_1 n_1 H_l(1) \tilde{J}_l^{(1)}(0, z_0),$$

where $1 = k_1 a$ and $2 = k_2 a$. From this one easily finds $\tilde{C}_l$:

$$\tilde{C}_l = \frac{i\mu_1}{n_2 k^2 a^2 \triangle_l} \tilde{J}_l^{(1)}(0, z_0), \qquad (7.51)$$

where

$$\triangle_l = \mu_2 n_1 j_l(1) H_l(2) - \mu_1 n_2 J_l(1) h_l(2).$$

Since EMF strengths contain only $\tilde{C}_l$ and $\tilde{D}_l$, the coefficients C_l and D_l entering the electromagnetic potentials are not needed.

At large distances one can replace the Hankel function by its asymptotic value. This gives

$$H_\phi = -\frac{ekn_2 \exp(ik_2 r)}{2\pi c} \frac{\exp(ik_2 r)}{r} S, \quad E_\theta = -\frac{ek\mu_2 \exp(ik_2 r)}{2\pi c} \frac{\exp(ik_2 r)}{r} S,$$

where

$$S = \sum i^{-l} P_l^1 \tilde{C}_l.$$

The radiation intensity per unit frequency and per unit solid angle is

$$\frac{d^2\mathcal{E}}{d\Omega d\omega} = \frac{1}{2}cr^2(E_\theta H_\phi^* + c.c.) = \frac{e^2 k^2 n_2 \mu_2}{4\pi^2 c}|S|^2. \qquad (7.52)$$

The integration over the solid angle gives the frequency distribution of radiation

$$\frac{d\mathcal{E}}{d\omega} = \frac{e^2 k^2 n_2 \mu_2}{\pi c} \sum \frac{l(l+1)}{2l+1}|\tilde{C}_l|^2. \qquad (7.53)$$

When the media inside and outside S are the same ($\epsilon_1 = \epsilon_1 = \epsilon$, $\mu_1 = \mu_2 = \mu$) one finds

$$\Delta_l = \frac{i\mu}{nk^2 a^2} \quad \text{and} \quad \tilde{C}_l = \tilde{J}_l^{(1)}(0, z_0),$$

that is, one obtains the spherical representation for the single-medium Tamm problem corresponding to the spatial interval $(-z_0, z_0)$ (see Chapter 2).

Numerical results

In Fig. 7.7 are shown angular radiation intensities (solid lines) evaluated according to (7.52) for $kz_0 = 10$, $ka = 20$, $n_1 = 2$ and $n_2 = 1$ (that is, there is a vacuum outside S) for a number of charge velocities. Side by side with them, the Tamm angular intensity (2.29) (dotted lines) corresponding to $n = n_1, L = 2z_0$ are shown. The distinction of (7.52) from the Tamm angular intensity is owed to the presence of the medium 2 outside S. The latter results in the broadening of the angular intensity distribution. This was shown qualitatively above. The corresponding frequency distributions (7.53) (solid lines) together with the Tamm frequency distributions (2.109) (dotted lines) are shown in Fig. 7.8. The latter almost coincide with the approximate intensities (2.31) except for the velocity $\beta = 0.4$ lying below the Cherenkov threshold, where the approximate Tamm frequency distribution (2.31) depends on the frequency only through $n(\omega)$. It is seen from Fig. 7.8 that the frequency distribution (7.53) oscillates around the frequency distributions (2.109) corresponding to the Tamm problem. When evaluating $d\mathcal{E}/d\omega$ we have implicitly assumed that in this frequency interval the refractive index n_1 does not depend on ω. In fact, this is a common thing in refractive media. For example, for the fresh water the refractive index is almost constant in the frequency interval $(6 \times 10^{14} < \omega < 6 \times 10^{15})$ s^{-1} encompassing the visible light region.

In Fig. 7.9 there are shown angular radiation intensities (solid lines) evaluated according to (7.52) for $kz_0 = 10$, $ka = 20$, $n_1 = 1$ and $n_2 = 2$ (that is, there is a vacuum bubble embodied into the medium 2) for a number of charge velocities. Side by side with them, the Tamm angular intensities (2.29) (dotted lines) corresponding to $n = n_1, L = 2z_0$ are shown.

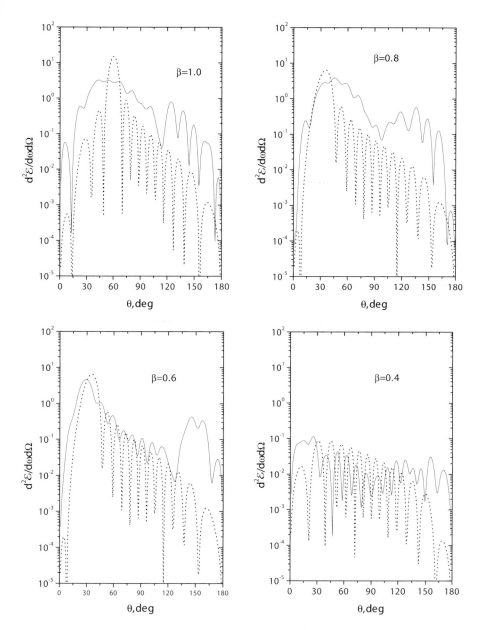

Figure 7.7.   The angular radiation intensities in e^2/c units (solid curves) for the charge motion shown in Fig. 7.6 and various charge velocities. The media parameters are $n_1 = 2, n_2 = 1$ (that is, there is vacuum outside S). Further, $kz_0 = 10, ka = 20$. The dotted curves are the Tamm angular intensities (2.29) evaluated for $kL = 2kz_0$ and $n = n_1$. The difference between these two curves is because the medium outside S is not the same as inside S. The exact angular intensities are much broader than the corresponding Tamm intensities. Probably the rise of angular intensities at large angles shown in Figs. 7.7 and 7.9 is owed to the reflection of the Vavilov-Cherenkov radiation from the internal side of S.

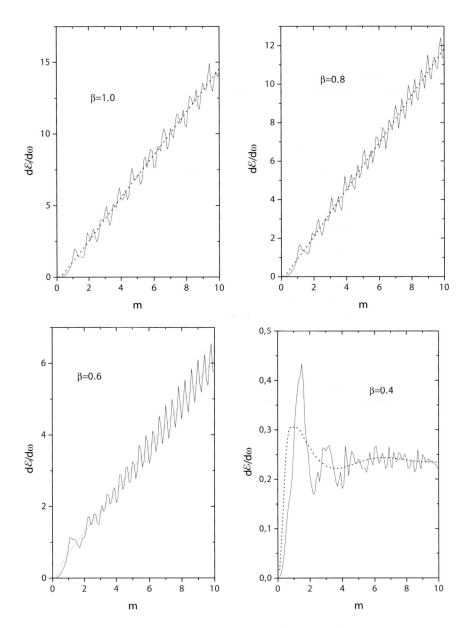

Figure 7.8. The frequency radiation intensities in e^2/c units (solid curves) for the charge motion shown in Fig. 7.6 and various charge velocities. The media parameters are the same as in Fig. 7.7. Furthermore, $kz_0 = m, ka = 2m$. The dotted curves are the Tamm frequency intensities (2.109) evaluated for $kL = 2kz_0$ and $n = n_1$. It is seen that the exact frequency intensities oscillate around the Tamm intensities.

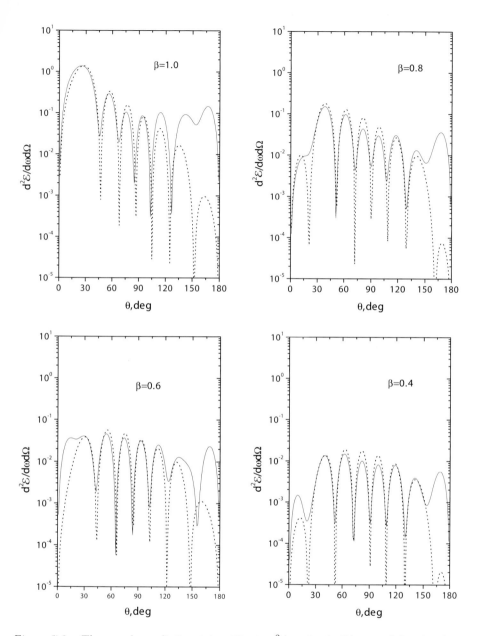

Figure 7.9. The angular radiation intensities in e^2/c units (solid curves) for the charge motion shown in Fig. 7.6 and various charge velocities. The media parameters are $n_1 = 1, n_2 = 2$ (that is, there is vacuum inside S). Furthermore, $kz_0 = 10, ka = 20$. The dotted curves are the Tamm angular intensities (2.29) evaluated for $kL = 2kz_0$ and $n = n_1$.

It is seen that the presence of a medium outside S affects not so strongly as in Fig. 7.7. The corresponding frequency distributions are shown in Fig. 7.10. Again, oscillations around the Tamm frequency distribution (2.109) are observed. Probably, they are of the same nature as oscillations of the frequency radiation intensity for the cylindrical dielectric sample (see section 7.4.2).

Probably, the rise of angular intensities at large angles shown in Figs. 7.7 and 7.9 is owed to the reflection of the Vavilov-Cherenkov radiation from the internal side of S.

The Tamm problem for a charge passing through the spherical sample
Let a charge move with a constant velocity v on the interval $(-z_0, z_0)$. There is a sphere S of radius $a < z_0$ with its center at the origin (Fig. 7.11). The space inside S is filled with the substance with the parameters ϵ_1, μ_1. Outside S there is a substance with parameters the ϵ_2, μ_2. The magnetic vector potential satisfying the equations $(\triangle + k_2^2)A_z = 0$ for $r > z_0$, $(\triangle + k_2^2)A_z = -4\pi\mu_2 j_z/c$ for $a < r < z_0$ and $(\triangle + k_1^2)A_z = -4\pi\mu_1 j_z/c$ for $r < a$ is obtained from (7.47) and (7.48). It is given by:

$$A_z = \frac{iek_2\mu_2}{2\pi c} \sum (2l+1)P_l h_l(k_2 r) \left[C_l \frac{\mu_1}{\mu_2} J_l^{(1)}(0, a) + J_l^{(2)}(a, z_0) \right],$$

for $r > z_0$,

$$A_z = \frac{iek_2\mu_2}{2\pi c} \sum (2l+1)P_l$$

$$\times \left[h_l(k_2 r) \times C_l \frac{\mu_1}{\mu_2} J_l^{(1)}(0, a) + h_l(k_2 r) J_l^{(2)}(a, r) + j_l(k_2 r) H_l^{(2)}(r, z_0) \right]$$

for $a < r < z_0$ and

$$A_z = \frac{iek_1\mu_1}{2\pi c} \sum (2l+1)P_l(\cos\theta)$$

$$\times \left[j_l(k_1 r) \frac{\mu_2}{\mu_1} D_l H_l^2(2)(a, z_0) + h_l(k_1 r) J_l^{(1)}(0, r) + j_l(k_1 r) H_l^{(1)}(r, a) \right]$$

for $r < a$. Here

$$J_l^{(2)}(x, y) = \int_x^y j_l(k_2 r') f_l(r') dr', \quad H_l^{(2)}(x, y) = \int_x^y h_l(k_2 r') f_l(r') dr'.$$

It is convenient to redefine C_l and D_l:

$$C_l' = C_l \frac{\mu_1}{\mu_2} J_l^{(1)}(0, a) + J_l^{(2)}(a, z_0), \quad D_l' = D_l \frac{\mu_2}{\mu_1} H_l^{(2)}(a, z_0).$$

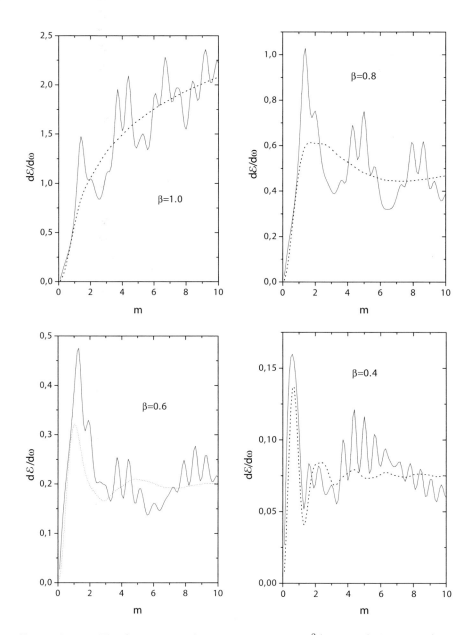

Figure 7.10.   The frequency radiation intensities in e^2/c units (solid curves) for the charge motion shown in Fig. 7.6 and various charge velocities. The media parameters are the same as in Fig. 7.9. Furthermore, $kz_0 = m, ka = 2m$. The dotted curves are the Tamm frequency intensities (2.109) evaluated for $kL = 2kz_0$ and $n = n_1$.

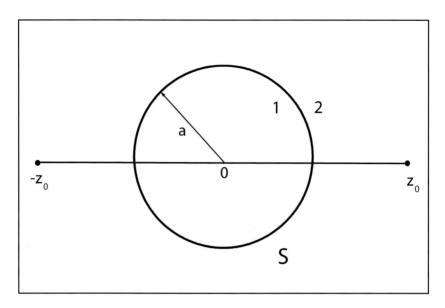

Figure 7.11. The charge motion begins and ends in medium 2. A charge passes through a sphere S filled with the medium 1. The radiation of intensity is measured outside S in medium 2.

Then,

$$A_z = \frac{iek_2\mu_2}{2\pi c} \sum (2l + 1) P_l h_l(k_2 r) C_l'$$

for $r > z_0$,

$$A_z = \frac{iek_2\mu_2}{2\pi c} \sum (2l + 1) P_l(\cos\theta)$$

$$\times [C_l' h_l(k_2 r) - h_l(k_2 r) J_l^{(2)}(r, z_0) + j_l(k_2 r) H_l^{(2)}(r, z_0)]$$

for $a < r < z_0$ and

$$A_z = \frac{iek_1\mu_1}{2\pi c} \sum (2l + 1) P_l(\cos\theta)$$

$$\times [D_l' j_l(k_1 r) + h_l(k_1 r) J_l^{(1)}(0, r) + j_l(k_1 r) H_l^{(1)}(r, a)].$$

for $r < a$.

The EMF strengths contributing to the radial energy flux are

$$H_\phi = -\frac{iek_2^2}{2\pi c} \sum \tilde{C}_l P_l^1 h_l(k_2 r), \quad E_\theta = -\frac{ek^2\mu_2 n_2}{2\pi c} \sum H_l(k_2 r) P_l^1 \tilde{C}_l \quad (7.54)$$

for $r > z_0$,

$$H_\phi = -\frac{iek_2^2}{2\pi c}\sum \tilde{P}_l^1[\tilde{C}_l h_l(k_2 r) - h_l(k_2 r)\tilde{J}_l^{(2)}(r, z_0) + j_l(k_2 r)\tilde{H}_l^{(2)}(r, z_0)],$$

$$E_\theta = -\frac{ek^2\mu_2 n_2}{2\pi c}$$

$$\times \sum \tilde{P}_l^1[\tilde{C}_l H_l(k_2 r) - H_l(k_2 r)\tilde{J}_l^{(2)}(r, z_0) + J_l(k_2 r)\tilde{H}_l^{(2)}(r, z_0)] \qquad (7.55)$$

for $a < r < z_0$ and

$$H_\phi = -\frac{iek_1^2}{2\pi c}\sum \tilde{P}_l^1[\tilde{D}_l j_l(k_1 r) + h_l(k_1 r)\tilde{J}_l^{(1)}(0, r) + j_l(k_1 r)\tilde{H}_l^{(1)}(r, a)],$$

$$E_\theta = -\frac{ek^2\mu_1 n_1}{2\pi c}$$

$$\times \sum \tilde{P}_l^1[\tilde{D}_l J_l(k_1 r) + H_l(k_1 r)\tilde{J}_l^{(1)}(0, r) + J_l(k_1 r)\tilde{H}_l^{(1)}(r, a)] \qquad (7.56)$$

for $r < a$. Here

$$\tilde{J}_l^{(2)}(x, y) = J_{l-1}^{(2)}(x, y) + J_{l+1}^{(2)}(x, y), \quad \tilde{H}_l^{(2)}(x, y) = H_{l-1}^{(2)}(x, y) + H_{l+1}^{(2)}(x, y),$$

$$\tilde{J}_l^{(1)}(x, y) = J_{l-1}^{(1)}(x, y) + J_{l+1}^{(1)}(x, y), \quad \tilde{H}_l^{(1)}(x, y) = H_{l-1}^{(1)}(x, y) + H_{l+1}^{(1)}(x, y),$$

$$\tilde{C}_l = C_{l-1}' + C_{l+1}', \quad \tilde{D}_l = D_{l-1}' + D_{l+1}'.$$

Equating E_θ and H_ϕ at $r = a$, one obtains the following equations for $\tilde{C}_l$ and $\tilde{D}_l$:

$$n_2^2 h_l(2)\tilde{C}_l - n_1^2 j_l(1)\tilde{D}_l$$

$$= n_1^2 h_l(1)\tilde{J}_l^{(1)}(0, a) + n_2^2[h_l(2)\tilde{J}_l^{(2)}(a, z_0) - j_l(2)\tilde{H}_l^{(2)}(a, z_0)],$$

$$\mu_2 n_2 H_l(2)\tilde{C}_l - \mu_1 n_1 J_l(1)\tilde{D}_l$$

$$= \mu_1 n_1 H_l(1)\tilde{J}_l^{(1)}(0, a) + n_2\mu_2[H_l(2)\tilde{J}_l^{(2)}(a, z_0) - J_l(2)\tilde{H}_l^{(2)}(a, z_0)]\}.$$

Here we put $1 = k_1 a$ and $2 = k_2 a$. For example, $j_l(1) \equiv j_l(k_1 a)$, etc.. From this one easily obtains $\tilde{C}_l$:

$$\tilde{C}_l = \frac{1}{\triangle_l}\{\frac{i\mu_1}{n_2 k^2 a^2}\tilde{J}_l^{(1)}(0, a) + \tilde{J}_l^{(2)}(a, z_0)[\mu_2 n_1 j_l(1)H_l(2) - \mu_1 n_2 J_l(1)h_l(2)]$$

$$-\tilde{H}_l^{(2)}(a, z_0)[\mu_2 n_1 j_l(1)J_l(2) - \mu_1 n_2 J_l(1)j_l(2)]\}$$

$$= \frac{i}{\triangle_l} \{ \frac{\mu_1}{n_2 k^2 a^2} \tilde{J}_l^{(1)}(0, a) + \tilde{J}_l^{(2)}(a, z_0)[\mu_2 n_1 j_l(1) N_l(2) - \mu_1 n_2 J_l(1) n_l(2)]$$

$$- \tilde{N}_l^{(2)}(a, z_0)[\mu_2 n_1 j_l(1) J_l(2) - \mu_1 n_2 J_l(1) j_l(2)] \}. \tag{7.57}$$

Here $\triangle_l = n_1 \mu_2 j_l(1) H_l(2) - \mu_1 n_2 J_l(1) h_l(2)$. Again, we do not need C_l and D_l entering the vector potential, since the EMF field strengths (and the radiation intensity) depend only on $\tilde{C}_l$ and $\tilde{D}_l$. At large distances ($kr \gg 1$), one has

$$H_\phi \approx -\frac{ekn_2}{2\pi cr} \exp(ikn_2 r) S, \quad E_\theta \approx -\frac{ek\mu_2}{2\pi cr} \exp(ikn_2 r) S,$$

where

$$S = \sum i^{-l} \tilde{C}_l P_l^1. \tag{7.58}$$

Correspondingly, the energy flux through a sphere of the radius r is

$$\frac{d^2\mathcal{E}}{d\omega d\Omega} = \frac{1}{2} cr^2 (E_\theta H_\phi^* + c.c.) = \frac{e^2 k^2 n_2 \mu_2}{4\pi^2 c} |S|^2. \tag{7.59}$$

Integration over the solid angle gives the frequency distribution of radiation

$$\frac{d^2\mathcal{E}}{d\omega} = \frac{e^2 k^2 n_2 \mu_2}{\pi c} \sum \frac{l(l+1)}{2l+1} |\tilde{C}_l|^2. \tag{7.60}$$

The single-medium Tamm problem is obtained either in the limit $ka \to 0$ or when media 1 and 2 are the same.

Numerical results
In Fig. 7.12 there are shown angular radiation intensities (solid lines) evaluated according to (7.59) for $kz_0 = 20$, $ka = 10$, $n_1 = 2$ and $n_2 = 1$ (that is, there is a vacuum outside the sphere S filled with a substance with $n_1 = 2$) for a number of the charge velocities. Side by side with them the Tamm angular intensities (2.29) (dotted lines) corresponding to $n = n_1$, $L = 2a$ are shown. In fact, it is the usual thing in the Vavilov-Cherenkov radiation theory to associate the observed radiation with that part of the charge trajectory where $\beta n > 1$. It the case treated it lies within the sphere S. We observe a rather poor agreement of exact intensity (7.59) with the Tamm intensity (2.29). An experimentalist studying, e.g., an electron passing through the dielectric sphere S, will not see the pronounced Cherenkov maximum at $\theta = \theta_c$ ($\cos \theta_c = 1/\beta n$), and on these grounds will not identify the Cherenkov radiation and the charge velocity. For $\beta = 0.4$ we have not presented the Tamm intensity. The reason is that for this velocity the Tamm intensities arising from the charge motion in the intervals

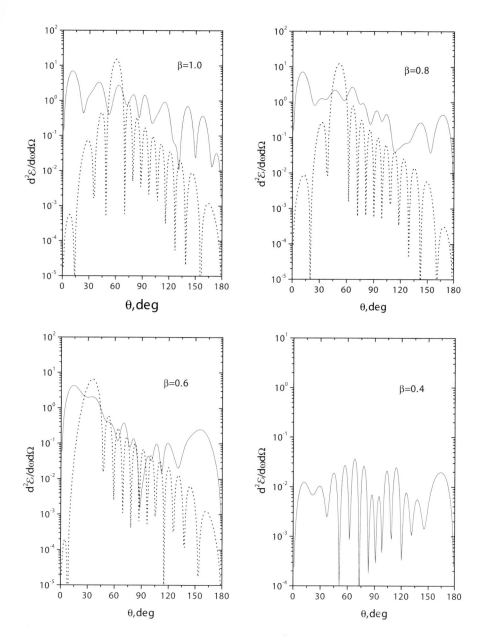

Figure 7.12. The angular radiation intensities in e^2/c units (solid curves) for the charge motion shown in Fig. 7.11 and various charge velocities. The medium inside S is dielectric. The media parameters are $n_1 = 2, n_2 = 1$ (that is, there is vacuum outside S). Furthermore, $ka = 10$, $kz_0 = 20$. The dotted curves are the Tamm angular intensities (2.29) evaluated for $kL = 2ka$ and $n = n_1$. The non-coincidence of exact angular intensities with the corresponding Tamm intensities (especially for $\beta = 1$ and $\beta = 0.8$) and, in particular, the absence of a pronounced maximum at $\cos\theta = 1/\beta n$ demonstrates that the applicability of the Tamm formula for describing the radiation arising from the charge passage through the dielectric sample is somewhat in doubt.

$0 < r < a$ (medium 1) and $a < r < z_0$ (medium 2) are of the same order. It is not clear to us how to combine the corresponding Tamm amplitudes. In any case, Eqs. (7.59) and (7.60) give the exact solution of the problem treated, whilst the Tamm intensities are needed only for the interpretation purposes.

The corresponding frequency distribution (7.60) also differs appreciably from that of the Tamm (2.109) (Fig. 7.13).

In Fig. 7.14 there are shown angular radiation intensities (solid lines) evaluated according to (7.59) for $kz_0 = 20$, $ka = 10$, $n_1 = 1$ and $n_2 = 2$ (that is, the vacuum bubble inside S surrounded by a substance with $n_2 = 2$) for a number of charge velocities. Side by side with them the Tamm angular intensities (2.29) (dotted lines) corresponding to $n = n_2$, $L = 2(z_0 - a)$ are shown. In the case treated, the part of the charge trajectory where $\beta n > 1$ lies outside the sphere S. We observe a satisfactory agreement of the exact intensity (7.59) with the Tamm intensity (2.29). An experimentalist studying, e.g., an electron passing through the dielectric sphere S will see a pronounced Cherenkov maximum at $\theta = \theta_c$ ($\cos \theta_c = 1/\beta n$). The corresponding frequency distribution (7.60) does not differ appreciably from the Tamm distribution (2.109) (Fig. 7.15).

7.5.3. METALLIC SPHERE

On the surface of an ideal conductor the tangential components of the electric field strength vanish [23]. For a metallic sphere of radius a this leads to the disappearance of E_θ. This defines $\tilde{C}_l$

$$\tilde{C}_l = \tilde{J}_l^{(2)}(a, z_0) - \frac{J_l(2)}{H_l(2)} \tilde{H}_l^{(2)}(a, z_0)$$

$$= \frac{i}{H_l(2)} [N_l(2)\tilde{J}_l^{(2)}(a, z_0) - J_l(2)\tilde{N}_l^{(2)}(a, z_0)]. \tag{7.61}$$

Then the angular and frequency distributions are given by (7.58)-(7.60), but with $\tilde{C}_l$ given by (7.61).

Numerical results
Let there be a vacuum outside S. The corresponding angular distributions (7.59) (solid lines) are compared in Fig. 7.16 with the Tamm angular intensities (2.29) (dotted lines) evaluated for $L = 2(z_0 - a)$ and $n = n_2$. Since $\beta n \leq 1$ outside S, the angular intensities are quite small. The corresponding frequency distributions (7.60) (solid lines) and those of Tamm (2.109) (dotted lines) are shown in Fig. 7.17. Their agreement is rather poor.

Let there be a medium with refractive index $n_2 = 2$ outside S. The corresponding angular and frequency distributions are shown in Figs. 7.18 and

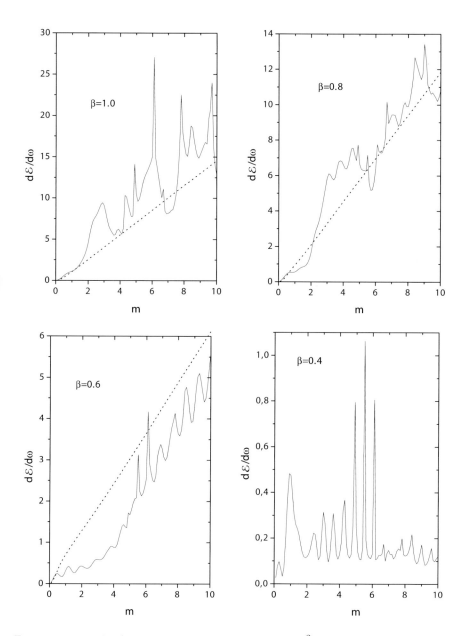

Figure 7.13.   The frequency radiation intensities in e^2/c units (solid curves) for the charge motion shown in Fig. 7.11 and various charge velocities. The media parameters are the same as in Fig. 7.12. Furthermore, $ka = m, kz_0 = 2m$. The dotted curves are the Tamm frequency intensities (2.109) evaluated for $kL = 2kz_0$ and $n = n_1$.

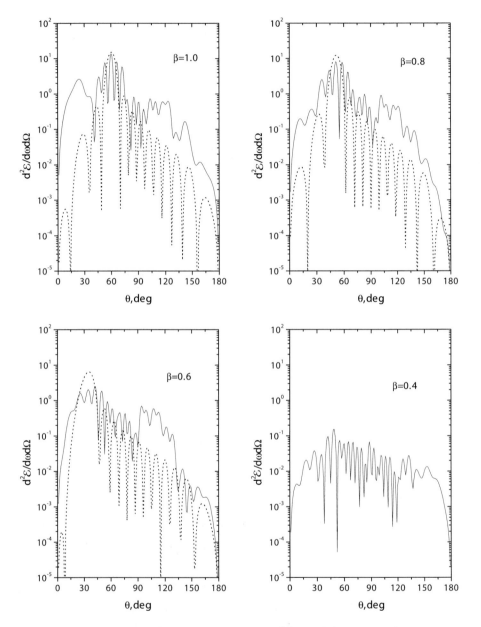

Figure 7.14.   The angular radiation intensities in e^2/c units (solid curves) for the charge motion shown in Fig. 7.11 and various charge velocities. The media parameters are $n_1 = 1, n_2 = 2$ (that is, there is a vacuum inside S). Furthermore, $ka = 10, kz_0 = 20$. The dotted curves are the Tamm angular intensities (2.29) evaluated for $kL = 2k(z_0 - a)$ and $n = n_2$.

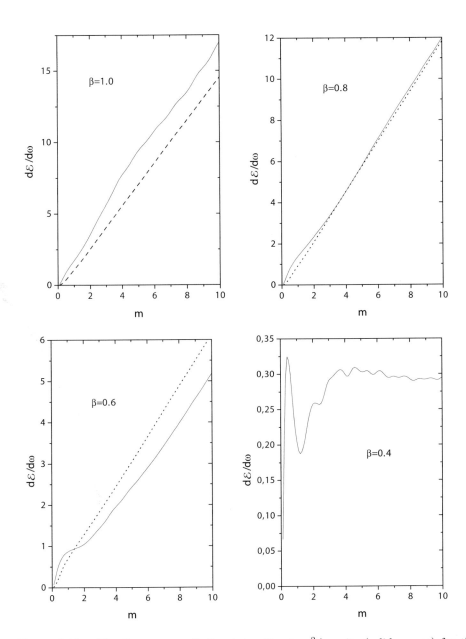

Figure 7.15.   The frequency radiation intensities in e^2/c units (solid curves) for the charge motion shown in Fig. 7.11 and various charge velocities. The media parameters are the same as in Fig. 7.14. Furthermore, $ka = m, kz_0 = 2m$. The dotted curves are the Tamm frequency intensities (2.109) evaluated for $kL = 2k(z_0 - a)$ and $n = n_2$.

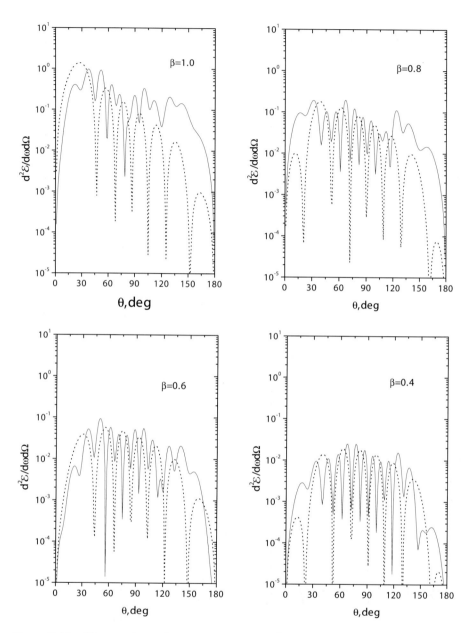

Figure 7.16. The angular radiation intensities in e^2/c units (solid curves) for the charge motion shown in Fig. 7.11 and various charge velocities. The medium inside S is an ideal metallic substance (conductor). The medium refractive index outside S is $n_2 = 1$ (vacuum). Furthermore, $ka = 10, kz_0 = 20$. The dotted curves are the Tamm angular intensities (2.29) evaluated for $kL = 2k(z_0 - a)$ and $n = n_2$.

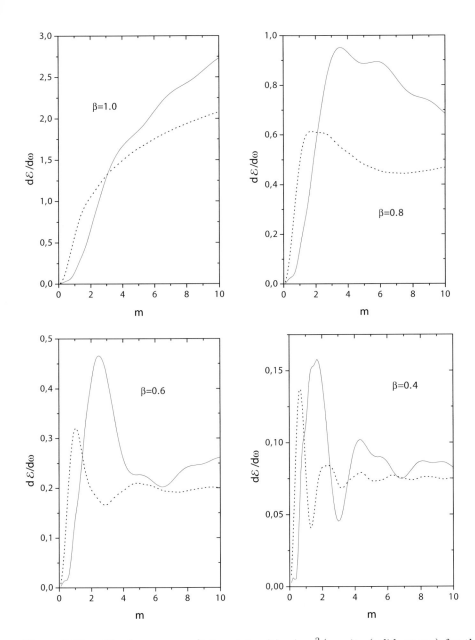

Figure 7.17.   The frequency radiation intensities in e^2/c units (solid curves) for the charge motion shown in Fig. 7.11 and various charge velocities. The medium inside S is ideal metallic substance. The medium refractive index outside S is $n_2 = 1$ (vacuum). Furthermore, $ka = m$, $kz_0 = 2m$. The dotted curves are the Tamm frequency intensities (2.109) evaluated for $kL = 2k(z_0 - a)$ and $n = n_2$.

7.19, respectively. We observe the satisfactory agreement with the Tamm intensities evaluated for $L = 2(z_0 - a)$ and $n = n_2$.

7.6. Discussion on the transition radiation

The formulae obtained in previous two sections describe the VC radiation, the radiation arising from the charge instantaneous acceleration and deceleration and the transition radiation arising from a charge passing from one medium to another. To separate the contribution of the transition radiation, one should subtract (according, e.g., to [11]) the field strengths corresponding to the inhomogeneous solution of the Maxwell equations from the total field strengths. In the treated case, the field strengths corresponding to the Tamm problem should be subtracted (they are written out in section 2.6 of the Chapter 2). This leads to the following redefinition of the $\tilde{C}_l$ coefficients:

$$\tilde{C}_l \to \tilde{C}_l - \sqrt{\frac{n_1\mu_1}{n_2\mu_2}} \, \tilde{J}_l^{(1)}(0, z_0)$$

for the motion shown in Fig. 7.6,

$$\tilde{C}_l \to \tilde{C}_l - \sqrt{\frac{n_1\mu_1}{n_2\mu_2}} \, \tilde{J}_l^{(1)}(0, a) - J_l^{(2)}(a, z_0)$$

for the charge motion through the dielectric sphere (Fig. 7.11) and

$$\tilde{C}_l \to \tilde{C}_l - J_l^{(2)}(a, z_0)$$

for the charge motion through the metallic sphere (Fig. 7.11). These newly defined $\tilde{C}_l$ being substituted into (7.52), (7.53), (7.59) and (7.60) give transition radiation intensities. Since the observable radiation intensities are the total ones presented in Figs. 7.7-7.10 and 7.12-7.19, we do not evaluate the transition radiation intensities here.

In the physical literature there are semi-intuitive interpretations of the transition radiation and the radiation in the Tamm problem in terms of instantaneous acceleration and deceleration, and in terms of semi-infinite charge motions terminating at one side of the media interface and beginning at the other one. Their insufficiencies (see below) enable us not to apply them to the consideration of the Vavilov-Cherenkov and transition radiations on the spherical sample. In any case, exact solutions and numerical calculations presented above contain all the necessary information for the analysis of experimental data.

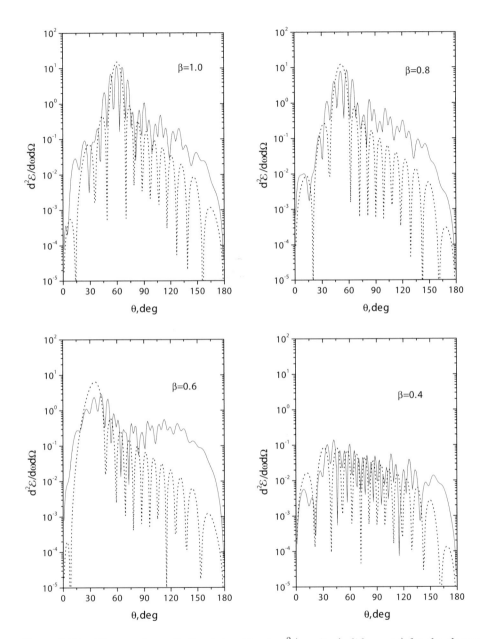

Figure 7.18. The angular radiation intensities in e^2/c units (solid curves) for the charge motion shown in Fig. 7.11 and various charge velocities. The medium inside S is an ideal metallic substance. The medium refractive index outside S is $n_2 = 2$. Furthermore, $ka = 10$, $kz_0 = 20$. The dotted curves are the Tamm angular intensities (2.29) evaluated for $kL = 2k(z_0 - a)$ and $n = n_2$.

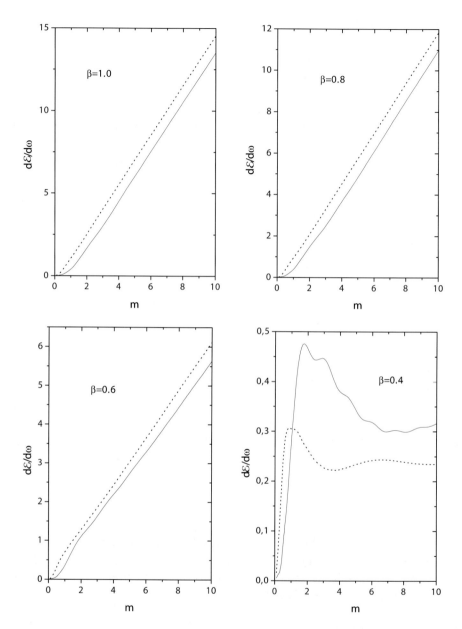

Figure 7.19. The frequency radiation intensities in e^2/c units (solid curves) for the charge motion shown in Fig. 7.11 and various charge velocities. The medium inside S is an ideal metallic substance. The medium refractive index outside S is $n_2 = 2$. Furthermore, $ka = m$, $kz_0 = 2m$. The dotted curves are the Tamm frequency intensities (2.109) evaluated for $kL = 2k(z_0 - a)$ and $n = n_2$.

7.6.1. COMMENT ON THE TRANSITION RADIATION

Interpretation of the transition radiation in terms of instantaneous velocity jumps

Sometimes the transition radiation is interpreted as a charge uniform motion with a velocity v in medium 1, its sudden stop in medium 1 at the border with medium 2, the sudden start of motion in medium 2 and the charge uniform motion in medium 2 with the velocity v [10,12-14]. It is suggested that the main contribution to the radiation intensity comes from the above-mentioned instantaneous jumps of the charge velocity. The radiation intensity arising from the charge sudden stop in medium 1 is taken in the form

$$\frac{d^2\mathcal{E}}{d\omega d\Omega} = \frac{e^2}{4\pi^2 c}\left[\frac{\vec{\beta} \times \vec{n}_r}{1 - n_1(\vec{\beta}\vec{n}_r)}\right]^2, \tag{7.62}$$

where $\vec{\beta} = \vec{v}/c$, $\vec{n}_r$ is the unit radius vector of the observational point and n_1 is the refractive index of medium 1.

On the other hand, the exact calculations were made in [24] (see also Chapter 5) for the following decelerated motion along the z axis:

$$z(t) = z_1 + v_1(t - t_1) - \frac{1}{2}a(t - t_1)^2,$$

$$v(t) = v_1 - a(t - t_1), \quad t_1 < t < t_2, \tag{7.63}$$

which begins at the instant t_1 at a spatial point z_1 with a velocity v_1 and ends at the instant t_2 at a spatial point z_2 with a velocity v_2. The time interval $t_2 - t_1$ of the motion and deceleration a are easily expressed through z_1, z_2, v_1, and v_2

$$t_2 - t_1 = 2\frac{z_2 - z_1}{v_1 + v_2}, \quad a = \frac{1}{2}\frac{v_1^2 - v_2^2}{z_2 - z_1}. \tag{7.63'}$$

It was shown in [24] that for a fixed wavelength λ, the intensity of radiation tends to zero for $k(z_2 - z_1) \to 0$ $(k = 2\pi/\lambda)$. This certainly disagrees with (7.62) which differs from zero for any interval of motion. To clarify the situation we turn to the derivation of (7.62).

The derivation of (7.62)

For simplicity, we consider first a charge motion in vacuum closely following Landau and Lifshitz treatise [25]. Its authors begin with the equations

$$\vec{H} = (\vec{n}_r \times \vec{E}), \quad \vec{E} = -\frac{1}{c}\dot{\vec{A}}$$

which are valid in the wave zone (the dot above $\vec{A}$ means the differentiation w.r.t. the laboratory time). For the Fourier transform of $\vec{H}$ one finds

$$\vec{H}_\omega = -\frac{1}{2\pi c} \int\limits_{-\infty}^{\infty} (\vec{n}_r \times \dot{\vec{A}}) \exp(i\omega t) dt. \tag{7.64}$$

Now, if $\vec{A} \neq 0$ for $t_1 < t < t_2$, then for $\omega(t_2 - t_1) \ll 1$ one can put $\exp(i\omega t) \approx 1$, thus obtaining

$$\vec{H}_\omega = -\frac{1}{2\pi c} \int \vec{n}_r \times \frac{\partial \vec{A}}{\partial t} = -\frac{1}{2\pi c} \vec{n}_r \times (\vec{A}_2 - \vec{A}_1). \tag{7.65}$$

Here $\vec{A}_2 = \vec{A}(t = t_2)$ and $\vec{A}_1 = \vec{A}(t = t_1)$. Furthermore, the authors of [25] replace $\vec{A}_1$ and $\vec{A}_2$ by the Liénard-Wiechert potentials. This gives

$$\vec{H}_\omega = \frac{e}{2\pi cr} \left[\frac{\beta_2 \times n_r}{1 - (\vec{\beta}_2 \vec{n}_r)} - \frac{\beta_1 \times n_r}{1 - (\vec{\beta}_1 \vec{n}_r)} \right]. \tag{7.66}$$

The radiation intensity per unit frequency and per unit solid angle is

$$\frac{d^2\mathcal{E}}{d\omega d\Omega} = cr^2 |\vec{H}_\omega|^2 = \frac{e^2}{4\pi^2 c} \left[\frac{\beta_2 \times n_r}{1 - (\vec{\beta}_2 \vec{n}_r)} - \frac{\beta_1 \times n_r}{1 - (\vec{\beta}_1 \vec{n}_r)} \right]^2. \tag{7.67}$$

Now if the final velocity is zero (7.67) coincides with (7.62).

Resolution of the paradox
We rewrite the integral entering (6.4) in the form

$$\int \frac{\partial \vec{A}}{\partial t} dt = \int \frac{\partial \vec{A}(t(t'))}{\partial t'} dt' = \vec{A}_2 - \vec{A}_1, \quad \vec{A}_2 = \vec{A}(t_2), \quad \vec{A}_1 = \vec{A}(t_1), \tag{7.68}$$

where t' is a charge retarded (proper) time. The laboratory times t_1 and t_2 expressed through the retarded times for the one-dimensional motion along the z axis are given by

$$t_1 = t_1' + \frac{1}{c}[\rho^2 + (z - z_1')^2]^{1/2}, \quad t_2 = t_2' + \frac{1}{c}[\rho^2 + (z - z_2')^2]^{1/2}, \tag{7.69}$$

where $z_1' = z'(t_1')$ and $z_2' = z'(t_2')$ are the charge positions at the times t_1' and t_2'.

Now let the charge proper time t' be uniquely related to its position z'. Then for $z_1' = z_2'$ one has $t_1' = t_2'$, $t_1 = t_2$, and therefore, $\vec{A}_2 = \vec{A}_1$,

$\vec{H}_\omega = 0$ and $d^2\mathcal{E}/d\omega d\Omega = 0$. We illustrate this using the motion law (7.63) as an example (note that t and z entering into (7.63) are the charge proper (retarded) time t' and its position z'). For this motion t' is uniquely related to z':

$$t' = t_1 + 2v_1 \frac{z_2 - z_1}{v_1^2 - v_2^2} \left[1 - \left(1 - \frac{z' - z_1}{z_2 - z_1} \frac{v_1^2 - v_2^2}{v_1^2} \right)^{1/2} \right]. \tag{7.70}$$

It follows from this that $t' = t_1$ for $z' = z_1$ and $t' = t_2$ for $z' = z_2$. According to (7.63'), $t_2 = t_1$ for $z_2 = z_1$. Therefore $\vec{A}_2 = \vec{A}_1$ for $t_1 = t_2$, and $\vec{H}_\omega$ given by (7. 65) vanishes in the limit $k(z_2 - z_1) \to 0$ in accordance with [24].

The main assumptions for the vanishing of $\vec{H}_\omega$ are:

i) the discontinuous charge motion with the velocity jumps can be viewed as a limiting case of a continuous motion without the velocity jumps when the length along which the velocity changes from v_1 to v_2 tends to zero;

ii) the retarded (proper) time of the charge is uniquely related to its position.

We conclude: the interpretation of the transition radiation in terms of the charge instantaneous acceleration and deceleration at the border of two media is not sufficient if the discontinuous charge motion can be treated as a limiting case of the continuous charge motion. In any case, the discontinuous charge motion cannot be realized in nature: it is a suitable idealization of the continuous charge motion.

In general, $\vec{A}(t_2)$ does not coincide with $\vec{A}(t_1)$ if the proper time of the charge is not uniquely related to its position. Consider, for example, an immovable elementary (infinitesimal) time dependent source. Then $\vec{A}(t_2) \neq \vec{A}(t_1)$ and $\vec{H}_\omega \neq 0$. Another possibility of obtaining $\vec{A}(t_2) \neq \vec{A}(t_1)$ is to take into account the internal degrees of freedom of a moving charged particle (for example, its spin flip on the path between z_1 and z_2 can give $\vec{A}(t_2) \neq \vec{A}(t_1)$).

Interpretation of the transition radiation
in terms of the charge semi-infinite motions

In [10,11], the transition radiation was associated with the charge radiation on the semi-infinite intervals $(-\infty, 0)$ and $0, \infty$ lying in media 1 and 2, respectively. We analyse this situation using the vector potential as an example.

The vector potential corresponding to the charge semi-infinite motion in medium 1 is given by

$$A_z = \frac{e\mu_1}{2\pi c} \int\limits_{-\infty}^{0} \frac{dz'}{R} \exp(i\psi),$$ (7.71)

where $\psi = kz'/\beta + k_1 R$, $k_1 = kn_1$, $R = \sqrt{\rho^2 + (z - z')^2}$. In the quasi-classical approximation one finds

$$A_z = \frac{e\mu_1}{2\pi ckr} \frac{1}{1 - \beta n_1 \cos\theta}$$ (7.72)

for $\beta < \beta_1 = 1/n_1$. For $\beta > \beta_1$

$$A_z = (7.72)$$

for $\theta < \theta_1$ and

$$A_z = (7.72) + A_T^{(1)}$$

for $\theta > \theta_1$. Here

$$A_T^{(1)} = \frac{e\mu_1}{2\pi c} \exp\left(\frac{i\pi}{4}\right) \sqrt{\frac{2\pi\beta\gamma_1}{kr\sin\theta}} \exp\left[\frac{ikr}{\beta}\left(\cos\theta + \frac{\sin\theta}{\gamma_1}\right)\right],$$

$$\gamma_1 = 1/\sqrt{|1 - \beta_1^2|}, \quad \cos\theta_1 = \frac{1}{\beta_1}.$$ (7.73)

Since A_T decreases as $1/\sqrt{kr}$, the radiation intensity is much larger in the $\theta > \theta_1$ angular region.

Similarly, the vector potential corresponding to the charge motion in medium 2 is given by

$$A_z = -\frac{e\mu_2}{2\pi ckr} \frac{1}{1 - \beta_2 \cos\theta}$$ (7.74)

for $\beta < \beta_2$ ($\beta_2 = 1/n_2$). For $\beta > \beta_2$,

$$A_z = (7.74)$$

for $\theta > \theta_2$ and

$$A_z = (7.74) + A_T^{(2)}$$

for $\theta < \theta_2$. Here

$$A_T^{(2)} = \frac{e\mu_2}{2\pi c} \exp\left(\frac{i\pi}{4}\right) \sqrt{\frac{2\pi\beta\gamma_2}{kr\sin\theta}} \exp\left[\frac{ikr}{\beta}\left(\cos\theta + \frac{\sin\theta}{\gamma_2}\right)\right],$$

$$\gamma_2 = 1/\sqrt{|1 - \beta_2^2|}, \quad \cos\theta_2 = \frac{1}{\beta_2}. \qquad (7.75)$$

Usually, the terms $A_T^{(1)}$ and $A_T^{(2)}$ are dropped in standard considerations of the transition radiation. Their interference with (7.72) and (7.74) leads to the oscillations of the radiation intensity in the $\theta > \theta_1$ angular region for the charge semi-infinite motion $(-\infty, 0)$ in medium 1 and in the $\theta < \theta_2$ angular region for the charge semi-infinite motion $(0, \infty)$ in medium 2.

A further procedure in obtaining intensities of the transition radiation is the evaluation of EMF strengths corresponding to (7.72) and (7.74) and their superposition with the corresponding Fresnel coefficients. Sometimes the secondary photon re-scatterings at the boundary of media 1 and 2 (for the dielectric plate) are taken into account.

Since we have at hand the exact solution for a charge moving inside and outside the dielectric or metallic sphere, these tricks are not needed: they are automatically taken into account in closed expressions for radiation intensities.

Physical meaning of $A_T^{(1)}$ and $A_T^{(2)}$ terms

To clarify the physical meaning of the $A_T^{(1)}$ and $A_T^{(2)}$ terms, we consider the case when media 1 and 2 are the same. The vector potential corresponding to the infinite motion $(-\infty, \infty)$ then reduces to the sum of vector potentials corresponding to semi-infinite motions in media 1 and 2:

$$A_z = 0 \qquad (7.76)$$

for $\beta < 1/n$ and

$$A_z = \frac{e\mu}{2\pi c} \exp\left(\frac{i\pi}{4}\right) \sqrt{\frac{2\pi\beta\gamma_n}{kr\sin\theta}} \exp\left[\frac{ikr}{\beta}\left(\cos\theta + \frac{\sin\theta}{\gamma_n}\right)\right] \qquad (7.77)$$

for $\beta > 1/n$. Here $\gamma_n = 1/\sqrt{|1 - \beta_n^2|}$, $\beta_n = \beta n$.

However, this is the asymptotic form $(\rho \to \infty)$ of the Cherenkov vector potential corresponding to the charge infinite medium

$$A_z = \frac{e\mu}{\pi c} K_0\left(\frac{k\rho}{\beta\gamma_n}\right)$$

for $\beta < 1/n$ and

$$A_z = \frac{ie\mu}{2c} \exp\left(\frac{ikz}{\beta}\right) H_0^{(1)}\left(\frac{k\rho}{\beta\gamma_n}\right) \qquad (7.78)$$

for $\beta > 1/n$ (see Chapter 2). This means that the terms $A_T^{(1)}$ and $A_T^{(2)}$ describe the Cherenkov radiation for the semi-infinite charge motions in media

1 and 2, respectively. This is also confirmed by the exact solution corresponding to the semi-infinite charge motion in the dispersion-free medium found in [26,27] in the time representation. Indeed, the spatial regions where the Cherenkov radiation differs from zero are just the same where the terms $A_T^{(1)}$ and $A_T^{(2)}$ differ from zero.

It is easy to check that the values of A_z given by (7.72) are defined by the boundary point $z' = 0$ in (7.71), whilst the values of the terms $A_T^{(1)}$ and $A_T^{(2)}$ are defined by stationary points z' lying in the intervals $(-\infty, 0)$ and $(0, \infty)$, respectively.

We can see that the interpretation of the transition radiation in terms of semi-infinite motions in the intervals $(-\infty, 0)$ and $(0, \infty)$ is sufficient only for $\beta < 1/n$. On the other hand, for $\beta > 1/n$, the Cherenkov terms $A_T^{(1)}$ and $A_T^{(2)}$ should be taken into account.

7.6.2. COMMENT ON THE TAMM PROBLEM

For the Tamm problem (uniform charge motion in a restricted spatial interval), the vector potential is given by

$$A_z = \frac{e\mu_1}{2\pi c} \int_{-z_0}^{z_0} \frac{dz'}{R} \exp(i\psi), \qquad (7.79)$$

It is easily evaluated in the quasi-classical approximation. For $z < \rho\gamma_n - z_0$ and $z > \rho\gamma_n + z_0$ one gets

$$A_z^{\text{out}} = -\frac{ie\mu\beta\sin\theta}{2\pi ck} \Big\{ \frac{1}{r_2 - \beta_n(z - z_0)} \exp\left[\frac{ik}{\beta}(\beta n r_2 + z_0)\right]$$

$$- \frac{1}{r_1 - \beta_n(z + z_0)} \exp\left[\frac{ik}{\beta}(\beta n r_1 - z_0)\right] \Big\}. \qquad (7.80)$$

Here $r_1 = \sqrt{\rho^2 + (z + z_0)^2}$ and $r_2 = \sqrt{\rho^2 + (z - z_0)^2}$. Inside the interval $\rho\gamma_n - z_0 < z < \rho\gamma_n + z_0$, the vector potential is equal to

$$A_z^{\text{in}} = A_z^{\text{out}} + A_z^{Ch}, \qquad (7.81)$$

where

$$A_z^{Ch} = \frac{e\mu}{2\pi c} \exp\left(\frac{ikz}{\beta}\right) \sqrt{\frac{2\pi\beta\gamma_n}{kr\sin\theta}} \exp\left(i\frac{\pi}{4}\right) \exp\left(\frac{ikr\sin\theta}{\beta\gamma_n}\right).$$

It is seen that A_z^{out} is infinite at $z = \rho\gamma_n \pm z_0$. Therefore, the radiation intensity should have maxima at $z = \rho\gamma_n \pm z_0$, with a kind of plateau for $\rho\gamma_n - z_0 < z < \rho\gamma_n + z_0$ and a sharp decrease for $z < \rho\gamma_n - z_0$ and $z > \rho\gamma_n + z_0$. At observational distances much larger than the motion length

$$r_1 - \beta_n(z + z_0) \approx r(1 - \beta_n\cos\theta), \qquad r_2 - \beta_n(z - z_0) \approx r(1 - \beta_n\cos\theta),$$

$$\beta_n r_1 - z_0 = \beta_n r - z_0(1 - \beta_n\cos\theta), \qquad \beta_n r_2 + z_0 = \beta_n r + z_0(1 - \beta_n\cos\theta).$$

Then

$$A_z^{out} = \frac{e\mu\beta}{\pi ckr}\exp(iknr)\frac{\sin[\omega t_0(1 - \beta n\cos\theta)]}{1 - \beta n\cos\theta} \tag{7.82}$$

coincides with the Tamm vector potential A_z^T entering into (2.29). Inside the interval $\rho\gamma_n - z_0 < z < \rho\gamma_n + z_0$,

$$A_z^{in} = A_z^T + A_z^{Ch}. \tag{7.83}$$

We observe that the infinities of A_z^{out} have disappeared as a result of the approximations involved. It is seen that for $kr \gg 1$ the A_z^{Ch} and A_z^T behave as $1/\sqrt{kr}$ and $1/kr$, respectively.

It follows from this that the radiation intensity in the spatial regions $z > \rho\gamma_n + z_0$ and $z < \rho\gamma_n - z_0$ is described by the Tamm formula (2.29). On the other hand, inside the spatial region $\rho\gamma_n - z_0 < z < \rho\gamma_n + z_0$, the radiation intensity differs appreciably from the Tamm intensity. In fact, the second term in A_z^{in} is much larger than the first (A_z^T) for $kr \gg 1$ (since they decrease as $1/\sqrt{kr}$ and $1/kr$ for $kr \to \infty$, respectively.) It is easy to check that on the surface of the sphere of the radius r the intervals $z < \rho\gamma_n - z_0$, $\rho\gamma_n - z_0 < z < \rho\gamma_n + z_0$ and $z > \rho\gamma_n + z_0$ correspond to the angular intervals $\theta > \theta_1$, $\theta_2 < \theta < \theta_1$ and $\theta < \theta_2$, where θ_1 and θ_2 are defined by

$$\cos\theta_1 = -\frac{\epsilon_0}{\beta_n^2\gamma_n^2} + \frac{1}{\beta_n}\left[1 - \left(\frac{\epsilon_0}{\beta_n\gamma_n}\right)^2\right]^{1/2},$$

$$\cos\theta_2 = \frac{\epsilon_0}{\beta_n^2\gamma_n^2} + \frac{1}{\beta_n}\left[1 - \left(\frac{\epsilon_0}{\beta_n\gamma_n}\right)^2\right]^{1/2}. \tag{7.84}$$

Here $\epsilon_0 = z_0/r$. For $r \gg z_0$

$$\theta_1 = \theta_c + \frac{\epsilon_0}{\beta_n\gamma_n}, \qquad \theta_2 = \theta_c - \frac{\epsilon_0}{\beta_n\gamma_n},$$

where θ_c is defined by $\cos\theta_c = 1/\beta n$. Therefore inside the angular interval $\theta_2 < \theta < \theta_1$ there should be observed a maximum of the radiation intensity with its amplitude proportional to the observational distance r. In the limit $r \to \infty$, the above θ interval decreases and for the radiation intensity one gets the delta singularity at $\cos\theta = 1/\beta n$ (in addition to A_z^T). However, the

θ integral from it is finite. Although $\Delta\theta = \theta_1 - \theta_2 = 2\epsilon_0/\beta_n\gamma_n$ is very small for $r \gg z_0$, the length of an arc on the observational sphere (on this arc the radiation intensity differs from the Tamm intensity) is finite: it is given by $2z_0/\beta_n\gamma_n$. It would be interesting to observe this deviation experimentally (see Chapter 9).

From the previous consideration it follows that A_z^{Ch} is a part of the Cherenkov shock wave enclosed between the straight lines $z = -z_0 + \rho\gamma_n$ and $z = z_0 + \rho\gamma_n$ with its normal inclined at the angle θ_c towards the motion axis. In the quasi-classical approximation the stationary point $z' = z - \rho\gamma_n$ of (7.79) lying inside the motion interval $(-z_0, z_0)$ defines the value of A_z^{Ch}. On the other hand, for the A_z^{out} the stationary point of (7.79) lies outside the interval of the charge motion and the value of (7.79) is defined by the initial and final points of the motion interval. Therefore A_z^{out} is somehow related to the beginning and end of the motion.

It was suggested in [28,29] that the origin of A_z^{out} is due to the BS shock waves arising from the charge acceleration at the beginning and its deceleration at the end of the motion. However, if one replaces the instantaneous velocity jumps by the smooth change of the velocity then tends the width of the transition region (where the velocity changes smoothly) to zero then the contribution of this region to the radiation intensity also tends to zero [24]. There are no velocity jumps for this smoothed problem and, therefore, A_z^{out} cannot be associated with instantaneous velocity jumps. However, there are acceleration jumps at the beginning and the end of motion and at the instants when the accelerated motion meets the uniform motion. Thus A_z^{out} can still be associated with acceleration jumps. To clarify the situation, the Tamm problem with absolutely continuous charge motion (for which the velocity itself and all its time derivatives are absolutely continuous functions of time) was considered in [30,31] (see also Chapter 5). It was shown there that the relatively slow decrease of A_z^{out} for $\theta > \theta_1$ and $\theta < \theta_2$ is replaced by the exponential damping. In the past, for the charge motion in vacuum, the exponential damping in the whole angular region was recognized in [32-35].

The same considerations as for the semi-infinite motion show that the instantaneous velocity jumps at the beginning and the end of motion do not contribute to the radiation intensity, provided they can be viewed as the limiting cases of the smooth charge motion in the limit when the lengths of the accelerated (decelerated) pieces of the charge trajectory tend to zero.

We conclude: the instantaneous velocity jumps at the beginning and end of the motion do not contribute to the radiation intensity provided, they can be viewed as a limiting case of the smooth charge motion in the limit when the lengths of the accelerated (decelerated) pieces of the charge trajectory tend to zero. This means that the above-mentioned attempts [28,29] to interpret the radiation intensity given by the Tamm formula

(2.29) in terms of the charge instantaneous acceleration and deceleration are insufficient.

We summarize the discussion on the transition radiation: i) the interpretation of the transition radiation and the Tamm problem in terms of instantaneous acceleration and decceleration is not sufficient;

ii) the usual interpretation of the radiation arising when the charge crosses the boundary between two media in terms of semi-infinite charge motions is valid only if $\beta < 1/n_1$ and $\beta < 1/n_2$. Otherwise, this interpretation should be supplemented by Cherenkov-like terms;

iii) there is no need for the artificial means mentioned in the previous two items in the exactly solvable case treated corresponding to the transition and Cherenkov radiation on a spherical sample.

We briefly review the content of this chapter:

1. It has been analysed how finite dimensions of a moving charge affect the frequency spectrum of the radiated energy. It has been shown that the frequency spectrum extends up to ka, where k and a are the wave number and the typical dimension of a moving charge, respectively.

2. It has been shown how a charge should move in a medium if, in the absence of an external force, all its energy losses were owed to the Vavilov-Cherenkov radiation. Analytic formulae for the charge velocity are obtained for a charge of finite dimensions moving in a dispersion-free medium, for a point-like charge moving in a dispersive medium, and for the point-like charge moving in a medium with ionization losses.

3. There have been discussed complications with the observation of the Vavilov-Cherenkov radiation when a charge moves in a medium in which the Vavilov-Cherenkov radiation condition holds, whilst the observations of the radiated energy are made in another medium in which this condition is not satisfied. It has been shown that the radiation spectrum is discrete for a charge moving inside a dielectric sample with a velocity greater than the velocity of light in medium. It is desirable to observe this discreteness experimentally.

4. It has been found the electromagnetic field arising from the charge motion confined to a dielectric sphere S which is surrounded by another dielectric medium with electrical properties different from those inside S. It has been studied how differences of media properties inside and outside S affect the angular and frequency radiation intensities for various charge velocities. In general, these differences lead to the broadening of the angular spectrum and to the appearance of oscillations in the frequency spectrum. Probably, they have the same nature as the discreteness of the radiation spectrum for the dielectric sample mentioned in a previous item. 5. It has been also considered the radiation of a charge whose motion begins and

ends in medium 2 and which passes through a dielectric sphere S filled with medium 1 or through a metallic sphere. The evaluated energy flux includes the VC and transition radiations as well as those originating from the beginning and end of motion. To our best knowledge transition radiation for the spherical interface is considered here for the first time. It is shown that when medium 2 outside S is a vacuum and medium 1 inside S has a refractive index n_1 satisfying $\beta n_1 > 1$, the angular and frequency radiation intensities cannot always be interpreted in terms of the Tamm intensities corresponding to the charge motion inside S (as is usually believed).

6. It has been proved that the interpretation of the transition radiation in terms of the instantaneous end of the charge motion in one medium and its instantaneous start in the other is not valid if the above motion with sudden velocity jumps can be considered as a limiting case of the smooth charge motion. It is shown that the interpretation of the transition radiation in terms of semi-infinite motions with instantaneous end of the charge motion in one medium and with its instantaneous start in the other one [10,11] should be supplemented by the VC radiation terms. Certainly, these remarks are related only to the interpretation of the transition radiation, not to the exact solutions obtained for the plane interface, e.g., in [11].

The content of this chapter is partly based on [36,37]

References

1. Smith G.S. (1993) Cherenkov Radiation from a Charge of Finite Size or a Bunch of Charges *Amer. J. Phys.*, **61**, pp. 147-155.
2. Villaviciencio M., Roa-Neri J.A.E. and Jimenez J.L. (1996) The Cherenkov Effect for Non-Rotating Extended Charges *Nuovo Cimento*, **B 111**, pp. 1041-1049.
3. Frank I.M. and Tamm I.E. (1937) Coherent Radiation of Fast Electron in Medium, *Dokl. Akad. Nauk*, **14**, pp. 107-113.
4. Jackson J.D (1975) *Classical Electrodynamics*, Wiley New York.
5. Fermi E. (1940) The Ionization Loss of Energy in Gases and in Condensed Materials, *Phys. Rev*, **57**, pp. 485-493.
6. Afanasiev G.N. and Kartavenko V.G. (1998) Radiation of a Point Charge Uniformly Moving in a Dielectric Medium *J. Phys. D: Applied Physics*, **31**, pp.2760-2776.
7. Afanasiev G.N., Kartavenko V.G. and Magar E.N. (1999) Vavilov-Cherenkov Radiation in Dispersive Medium *Physica*, **B 269**, pp. 95-113.
8. Afanasiev G.N., Eliseev S.M and Stepanovsky Yu.P. (1999) Semi-Analytic Treatment of the Vavilov-Cherenkov Radiation *Physica Scripta*, **60**, pp. 535-546.
9. Stevens T.E., Wahlstrand J.K., Kuhl J. and Merlin R. (2001) Cherenkov Radiation at Speeds below the Light Threshold: Photon-Assisted Phase Matching *Science*, **291**, pp. 627-630.
10. Ginzburg V.L. and Frank I.M. (1946) Radiation of a Uniformly Moving Electron due to its Transition from one Medium to Another *Zh. Eksp. Theor. Phys.*, **16**, pp. 15-28.
11. Ginzburg V.L. and Tsytovich V.N. (1984) *Transition radiation and transition scattering*, Nauka, Moscow.
12. Wartski L., Roland, Lasalle J., Bolore M. and Filippi G. (1975) Interference Phe-

nomenon in Optical Transition Radiation and its Application to Particle Beam Diagnostics and Multiple-Scattering Measurements *J. Appl. Phys.*, **46**, pp.3644-3653.

13. Ruzicka J. and Zrelov V.P. (1993) Optical Transition Radiation in a Transparent Medium and its Relation to the Vavilov-Cherenkov Radiation *Czech. J. Phys.*, **43**, pp. 551-567.

14. Hrmo A. and Ruzicka J. (2000) Properties of Optical Transition Radiation for Charged Particle Inclined Flight through a Plate of Metal *Nucl. Instr. Meth.*, **A 451**, pp. 506-519.

15. Frank I.M., (1988) *Vavilov-Cherenkov Radiation*, Nauka, Moscow, in Russian.

16. Tamm I.E. (1939) Radiation Induced by Uniformly Moving Electrons, *J. Phys. USSR*, **1, No 5-6**, pp. 439-461.

17. Buskirk F.R. and Neighbours J.R. (1983) Cherenkov Radiation from Periodic Electron Bunches *Phys. Rev.*, **A 28**, pp. 1531-1538.

18. Cherenkov P.A. (1944) Radiation of Electrons Moving in Medium with Superluminal Velocity, *Trudy FIAN*, **2, No 4**, pp. 3-62.

19. Ginzburg V.L. and Frank I.M. (1947) Radiation of Electron and Atom Moving on the Channel Axis in a dense Medium *Dokl. Akad. Nauk SSSR*, **56**, pp. 699-702.

20. Aitken D.K. et al. (1963) Transition Radiation in Cherenkov Detectors *Proc. Phys. Soc.*, **83**, pp. 710-722.

21. Ruzicka J. and Zrelov V.P., 1993, Czech. J. Phys., 43, 551.

22. Born M. and Wolf E. (1975) *Principles of Optics*, Pergamon, Oxford.

23. Landau L.D. and Lifshitz E.M, (1960) *Electrodynamics of Continuous Media*, Pergamon, Oxford.

24. Afanasiev G.N. and Shilov V.M. (2002) Cherenkov Radiation versus Bremsstrahlung in the Tamm Problem *J.Phys. D: Applied Physics*, **35**, pp. 854-866.

25. L.D. Landau and E.M. Lifshitz (1962) *The Classical Theory of Fields*, Pergamon, New York, 1962.

26. Afanasiev G.N., Beshtoev Kh. and Stepanovsky Yu.P. (1996) Vavilov-Cherenkov Radiation in a Finite Region of Space *Helv. Phys. Acta*, **69**, pp. 111-129.

27. Afanasiev G.N., Kartavenko V.G. and Stepanovsky Yu.P. (1999) On Tamm's Problem in the Vavilov-Cherenkov Radiation Theory *J.Phys. D: Applied Physics*, **32**, pp. 2029-2043.

28. Zrelov V.P. and Ruzicka J. (1989) Analysis of Tamm's Problem on Charge Radiation at its Uniform Motion over a Finite Trajectory *Czech. J. Phys.*, **B 39**, pp. 368-383.

29. Zrelov V.P. and Ruzicka J. (1992) Optical Bremsstrahlung of Relativistic Particles in a Transparent Medium and its Relation to the Vavilov-Cherenkov Radiation *Czech. J. Phys.*, **42**, pp. 45-57.

30. Afanasiev G.N., Shilov V.M., Stepanovsky Yu.P. (2002) New Analytic Results in the Vavilov-Cherenkov Radiation Theory *Nuovo Cimento*, **B 117**, pp. 815-838;

31. Afanasiev G.N., Shilov V.M., Stepanovsky Yu.P. (2003) Numerical and Analytical Treatment of the Smoothed Tamm Problem *Ann.Phys. (Leipzig)*, **12**, pp. 51-79

32. Abbasov I.I. (1982) Radiation Emitted by a Charged Particle Moving for a Finite Interval of Time under Continuous Acceleration and Deceleration *Kratkije soobchenija po fizike FIAN*, **No 1**, pp. 31-33; English translation: (1982) *Soviet Physics-Lebedev Institute Reports* **No1**, pp.25-27.

33. Abbasov I.I. (1985) Radiation of a Charged Particle Moving Uniformly in a Given Bounded Segment with Allowance for Smooth Acceleration at the Beginning of the Path, and Smooth Deceleration at the End *Kratkije soobchenija po fizike FIAN*, **No 8**, pp. 33-36. English translation: (1985) *Soviet Physics-Lebedev Institute Reports*, **No 8**, pp. 36-39.

34. Abbasov I.I., Bolotovskii B.M. and Davydov V.A. (1986) High-Frequency Asymptote of Radiation Spectrum of the Moving Charged Particles in Classical Electrodynamics *Usp. Fiz. Nauk*, **149**, pp. 709-722. English translation: Sov. Phys. Usp., 29 (1986), 788.

35. Bolotovskii B.M. and Davydov V.A. (1981) Radiation of a Charged Particle with

Acceleration at a Finite Path Length *Izv. Vuzov, Radiofizika*, **24** , pp. 231-234.

36.   Afanasiev G.N., Shilov V.M. and Stepanovsky Yu.P. (2003) Questons Concerning Observation of the Vavilov-Cherenkov Radiation *J.Phys.*, **D 36**, pp. 88-102.

37.   Afanasiev G.N., Kartavenko V.G. and Stepanovsky Yu.P. (2003) Vavilov-Cherenkov and Transition Radiations on the Dielectric and Metallic Spheres *J. Math. Phys.* **44**, pp. 4026-4056.

CHAPTER 8

SELECTED PROBLEMS OF THE SYNCHROTRON RADIATION

8.1. Introduction

Synchrotron radiation (SR) is such a well-known phenomenon that it seems to be almost impossible to add anything essential in this field.

Schott, probably, was the first who extensively studied SR. His findings were summarized in the encyclopedic treatise *Electromagnetic Radiation* [1]. He expanded the electromagnetic field (EMF) into a Fourier series and found solutions of the Maxwell equations describing the field of a charge moving in vacuum along a circular orbit. The infinite series of EMF strengths had a very poor convergence in the most interesting case $v \sim c$. Fortunately Schott succeeded in an analytical summation of these series and obtained closed expressions for the radiation intensity averaged over the azimuthal angle ([1], p.125).

Further development is owed to the Moscow State University school (see, e.g., the books [2-5] and review [6]) and to Schwinger et al. [7] who considered the polarization properties of SR and its quantum aspects.

The instantaneous (i.e., taken at the same instant of proper time) distribution of SR on the surface of the observational sphere was obtained by Bagrov et al [8,9] and Smolyakov [10]. They showed that the instantaneous distribution of SR in a vacuum possesses the so-called projector effect (that is, the SR has the form of a beam which is very thin for $v \sim c$).

Much less is known about SR in a medium. The papers by Schwinger, Erber et al [11,12] should be mentioned in this connection. Yet, they limited themselves to the EMF in a spectral representation and did not succeed in obtaining the electromagnetic field strengths and radiation flux in the time representation. It should be noted that Schott's summation procedure does not work if the charge velocity exceeds the velocity of light in medium.

The formulae obtained by Schott and Schwinger are valid at the distances r much larger than the radius a of a charge orbit.

In modern electron and proton accelerators this radius reaches few hundred meters and few kilometers, respectively. This means that large observational distances are unachievable in experiments performed on modern accelerators and the formulae describing the radiation intensity at moderate distances and near the charge orbit are needed. In the past, time-averaged radiation intensities in the near zone were studied in [13-15]. However, their consideration was based on the expansion of field strengths in powers of a/r. The convergence of this expansion is rather poor in the neighbourhood of a charge orbit.

SR has numerous applications in nuclear physics (nuclear reactions with γ quanta), solid state physics (see, e.g., [16]), astronomy [17,18], etc. There are books and special issues of journals devoted to application of SR [19-21].

The goal of this Chapter consideration is to study SR in vacuum and medium. In the latter case the charge velocity v can be less or greater than the velocity of light in medium c_n.

The plan of our exposition is as follows. Section 8.2 is devoted to the SM in vacuum. In subsection 8.2.1 we present the main mathematical formulae for synchrotron radiation. In subsection 8.2.2 we evaluated the electromagnetic energy fluxes radiated for the period of the charge motion in three mutually orthogonal directions (radial, azimuthal and polar) on the observational spheres with radii greater and smaller than the charge orbit radius. Subsection 8.2.3 is devoted to the investigation of the instantaneous radiation in vacuum. It is shown that it has a more complicated structure than was known up to now. The new formula for the intensity of radiation generalizing the Schwinger formula for arbitrary observational distances and velocities is obtained. The results of this section may be applied to astrophysical problems associated, e.g., with sunspots, the Crab nebula, Jupiter's radiation belts, etc. [17,18]. The synchrotron radiation in medium is treated in section 8.3. The necessary mathematical preliminaries are given in subsection 8.3.1. The explicit expressions for EMF strengths at arbitrary distances are given in subsection 8.3.2. In section 8.3.3 the spatial distribution of EMF singularities on the surface of the observational sphere for the case $v > c_n$ is analysed. Their relation to the singularities of the instantaneous Cherenkov cone attached to a rotating charge is discussed in subsection 8.3.4. Subsection 8.3.5 is devoted to the consideration of SR in the wave zone. It turns out that the position of EMF singularities for $v > c_n$ drastically depends on the radius of the observational sphere. At a fixed instant of laboratory time they fill a spiral-like surface. The space-time distributions of different polarization components are analysed. The spatial domains where they vanish and where they are infinite are determined for various charge velocities and radial distances. The brief account of the results obtained is given in section 8.4.

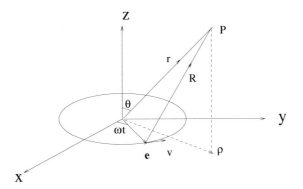

Figure 8.1.   Schematic presentation of the synchrotron motion. $P(r, \theta, \phi)$ is the observational point.

8.2. Synchrotron radiation in vacuum.

8.2.1. INTRODUCTORY REMARKS

Consider a point charge moving uniformly in vacuum along the circular orbit of radius a lying in the plane $z = 0$ (Fig. 8.1): $x = a \cos \omega_0 t$, $y = a \sin \omega_0 t$.

Charge and current densities and corresponding electromagnetic potentials are given by

$$\rho = e\delta(z)\delta(x - a\cos\omega_0 t)\delta(y - a\sin\omega_0 t),$$

$$\vec{j} = \rho v \vec{n}_\phi, \quad \vec{n}_\phi = \vec{n}_y \cos\omega_0 t - \vec{n}_x \sin\omega_0 t, \quad v = a\omega_0,$$

$$\Phi = e \int \frac{\rho(\vec{r}', t')}{R}\delta(t' - t + R/c)dV'dt',$$

$$\vec{A} = \frac{e}{c} \int \frac{\vec{j}(\vec{r}', t')}{R}\delta(t' - t + R/c)dV'dt'. \tag{8.1}$$

Here $R = |\vec{r} - \vec{r}'|$, $\Omega' = \omega_0 t' - \phi$, c is the velocity of light in vacuum. Substituting ρ and $\vec{j}$ here one obtains

$$\Phi = e \int \frac{dt'}{R}\delta(t' - t + R/c), \quad A_x = -e\beta \int \frac{dt'}{R}\sin\omega_0 t'\delta(t' - t + R/c),$$

$$A_y = e\beta \int \frac{dt'}{R}\cos\omega_0 t'\delta(t' - t + R/c),$$

$$R = [r^2 + a^2 - 2ra\sin\theta\cos\Omega']^{1/2}. \tag{8.2}$$

To clarify the applicability of Schott's formula we limit ourselves to the evaluation of electric potential Φ. Using the relation

$$\delta(\tau) = \frac{\omega_0}{2\pi}(1 + 2\sum_{m=1}^{\infty}\cos m\omega_0\tau), \tag{8.3}$$

we obtain

$$\Phi = \frac{e\omega_0}{2\pi}\int\frac{dt'}{R}[1 + 2\sum_{m=1}^{\infty}\cos m\omega_0(t' - t + R/c)]. \tag{8.4}$$

The following approximations are usually made in (8.4):
i) Outside the cosine R is replaced by r ;
ii) Inside the cosine R is replaced by $r - a\sin\theta\cos(\omega_0 t' - \phi)$.
After these approximations Φ is integrated in a closed form:

$$\Phi = \frac{e}{r} + \frac{2e}{r}\sum_m J_m(mk_0 a\sin\theta)\cos m\chi,$$

$$\chi = k_0 r + \phi - \omega_0 t - \frac{\pi}{2}, \quad k_0 = \frac{\omega_0}{c}.$$

In the same way one obtains the vector potential. Differentiating potentials, one finds field strengths and, finally, the electromagnetic energy flux through the observational sphere of radius r.

In the wave zone the EMF strengths are given by [1]

$$E_\theta = H_\phi = -\frac{2e\beta}{ar}\cot\theta\sum_{m=1}^{\infty}mJ_m(m\beta\sin\theta)\sin m\chi,$$

$$H_\theta = -E_\phi = \frac{2e\beta^2}{ar}\sum_{m=1}^{\infty}mJ'_m(m\beta\sin\theta)\cos m\chi \tag{8.5}$$

(J'_m means the derivative of J_m w.r.t. its argument).

The radial energy flux averaged over the period of rotation is

$$\frac{dW}{d\Omega} = r^2\frac{c}{4\pi}\int(E_\theta H_\phi - H_\theta E_\phi)d\phi = \sum_{m=1}^{\infty}W_m(\theta), \tag{8.6}$$

$$W_m(\theta) = \frac{e^2 c\beta^2}{2\pi a^2}m^2[\cot^2\theta J_m^2(m\beta\sin\theta) + \beta^2 J_m'^2(m\beta\sin\theta)].$$

The sum over m in (8.6) is evaluated analytically:

$$\frac{dW}{d\Omega} = \frac{e^2 c\beta^4}{32\pi a^2}(F_\sigma + F_\pi),$$

$$F_\sigma = \frac{4 + 3\beta^2 \sin^2 \theta}{(1 - \beta^2 \sin^2 \theta)^{5/2}}, \quad F_\pi = \cos^2 \theta \frac{4 + \beta^2 \sin^2 \theta}{(1 - \beta^2 \sin^2 \theta)^{7/2}},$$

where F_σ and F_π are the so-called s and π components of polarization (see sect. 8.3.5).

In the ultra-relativistic limit $(1 - \beta^2 \ll 1)$, using the asymptotic behaviour of the Bessel functions, one finds for $W_m(\theta)$ [2-6]

$$W_m(\theta) = \frac{e^2 c}{6\pi^3 a^2} m^2 [\delta^2 K_{2/3}^2(m\delta^{3/2}/3) + \delta \cot^2 \theta K_{1/3}^2(m\delta^{3/2}/3)]. \quad (8.7)$$

Here $K_\nu(x)$ is the modified Bessel function and $\delta = 1 - \beta^2 \sin^2 \theta$. We now elucidate under what physical conditions the approximations i) and ii) are satisfied.

The approximation i) means that the observational distance r is much larger than the orbit radius a. For a typical orbit radius $a \approx 1$ m, the approximation i) will work for $r \geq 5$m. However, in modern accelerators a reaches few hundred meters. In this case, approximation i) will not be satisfied at realistic distances.

Even worse is the situation with the second approximation. We write out the argument of the cosine function: $\omega(t' - t + R/c)$. Here $\omega = m\omega_0$ is the observable frequency. We develop R up to the second order of a^2:

$$R = r \left\{ 1 - \frac{a}{r} \sin \theta \cos(\omega_0 t' - \phi) + \frac{a^2}{2r^2} [1 - \sin^2 \theta \cos^2(\omega_0 t' - \phi)] \right\}. \quad (8.8)$$

Thus Schott's formulae are valid if the last term in (8.8), which is of the order $\omega a^2/cr$, is small compared to 2π (since this term is inside the cosine), i.e., one should have $\omega a^2/car \ll 2\pi$. We rewrite this equation using wavelength $\lambda = 2\pi c/\omega$:

$$\frac{a^2}{r\lambda} \ll 1 \quad (8.9)$$

For $a \approx 1$ m and $\lambda \approx 4 \times 10^{-5}$ cm (the optical region), the l.h.s. of (8.9) compares with 1 for $r \approx 100$ km. It is the strong violation of the approximation ii) that enables out to seek another approach to the problem treated.

Equations (8.6) and (8.7) were also obtained by Schwinger [7] without approximations i) and ii). However, his method of derivation includes the use of retarded and advanced EMF (the latter conflicts with causality) and the *ad hoc* omission of terms with definite symmetry properties.

Other methods of treating SR in a vacuum without using approximations i) and ii) are owed to the explicit formulae for EMF generated by a charge in arbitrary motion (see, e.g., [17,22]:

$$\vec{E} = \frac{e}{(R - \vec{\beta}\vec{R})^3} \{(1 - \beta^2)(\vec{R} - \vec{\beta}R) + \frac{1}{c}[\vec{R} \times ((\vec{R} - \vec{\beta}R) \times \dot{\vec{\beta}})]\},$$

$$\vec{H} = \frac{1}{R}(\vec{R} \times \vec{E}). \tag{8.10}$$

Here $\vec{R}$ is the vector going from the retarded position of the charge to the observational point $P(r, \theta, \phi)$; $\vec{\beta} = \vec{v}/c$ and $\dot{\vec{\beta}} = \dot{\vec{v}}/c$ are taken at the retarded time t'. We apply these equations to a charge moving along the circular orbit of radius a. The energy radiated by this charge per unit of laboratory time, into the unit solid angle of the sphere of the radius r is given by

$$\frac{d^2W}{d\Omega dt} = \frac{c}{4\pi}r^2(\vec{E} \times \vec{H})_r. \tag{8.11}$$

Expressing in this equation the retarded time t' through the laboratory one t via the relation:

$$c(t - t') = R = [r^2 + a^2 - 2ra\sin\theta\cos(\omega_0 t' - \phi)]^{1/2}, \tag{8.12}$$

we obtain the spatial distribution of radiation at the fixed instant of laboratory time t. This is essentially the idea of the present consideration. Equation (8.12) can be rewritten in another form

$$\chi = \chi' + \frac{\beta}{\epsilon_0}(1 - \tilde{R}), \tag{8.13}$$

where

$$\tilde{R} = (1 + \epsilon_0^2 - 2\epsilon_0\sin\theta\cos\chi')^{1/2}, \quad \chi = \phi - \omega(t - r/c), \quad \chi' = \phi - \omega t', \quad \epsilon_0 = \frac{a}{r}.$$

In the case treated a charge moves along the circular orbit of the radius a lying in the plane $z = 0$:

$$\xi_x = a\cos\omega t', \quad \xi_y = a\sin\omega t', \quad \xi_z = 0$$

(see Fig. 8.1).

In a manifest form, the spherical components of $\vec{E}$ look like

$$E_r = \frac{e}{ra}\tilde{E}_r, \quad E_\theta = \frac{e}{ra}\tilde{E}_\theta, \quad E_\phi = \frac{e}{ra}\tilde{E}_\phi, \tag{8.14}$$

where the dimensionless field strengths are

$$\tilde{E}_r = \frac{\epsilon_0}{Q^3}[1 - \beta\tilde{R}\sin\theta\sin\chi' - \beta^2\sin^2\theta\cos^2\chi' - \epsilon_0(1 - \beta^2)\sin\theta\cos\chi'],$$

$$\tilde{E}_\theta = \frac{\cos\theta}{Q^3}[\beta^2\cos\chi' - \epsilon_0\beta(\beta\sin\theta\cos^2\chi' + \tilde{R}\sin\chi') - \epsilon_0^2(1 - \beta^2)\cos\chi'],$$

$$\tilde{E}_\phi = \frac{1}{Q^3}[\beta^2(\beta\tilde{R}\sin\theta - \sin\chi') - \beta\epsilon_0\cos\chi'(\tilde{R} - \beta\sin\theta\sin\chi') + \epsilon_0^2(1-\beta^2)\sin\chi'].$$

Here $Q = \tilde{R} - \beta\sin\theta\sin\chi'$; r, θ, and ϕ define the position of the observational point. The spherical components of the Poynting vector

$$\vec{S} = \frac{c}{4\pi}(\vec{E}\times\vec{H}) = \frac{c}{4\pi R}[\vec{R}E^2 - \vec{E}(\vec{E}\vec{R})]$$

are given by

$$S_r = \frac{ce^2}{4\pi r^2 a^2}\tilde{S}_r, \quad S_\theta = \frac{ce^2}{4\pi r^3 a}\tilde{S}_\theta, \quad S_\phi = \frac{ce^2}{4\pi r^3 a}\tilde{S}_\phi,$$

where $\tilde{S}_r$, $\tilde{S}_\theta$ and $\tilde{S}_\phi$ are the corresponding dimensionless components:

$$\tilde{S}_r = \frac{1 - \epsilon_0\sin\theta\cos\chi'}{\tilde{R}}\tilde{E}^2 - \tilde{E}_r\epsilon_0\frac{1-\beta^2}{Q^2},$$

$$\tilde{S}_\theta = -\frac{\cos\theta\cos\chi'}{\tilde{R}}\tilde{E}^2 - \tilde{E}_\theta\frac{1-\beta^2}{Q^2},$$

$$\tilde{S}_\phi = \frac{\sin\chi'}{\tilde{R}}\tilde{E}^2 - \tilde{E}_\phi\frac{1-\beta^2}{Q^2}. \tag{8.15}$$

When obtaining (8.15) it was taken into account that

$$R_r = r - a\sin\theta\cos\chi', \quad R_\theta = -a\cos\theta\cos\chi', \quad R_\phi = a\sin\chi',$$

$$\vec{E}\vec{R} = e(1-\beta^2)r\tilde{R}/Q^2.$$

At large distances ($r \gg a$)

$$E_r \approx O(r^{-2}), \quad H_r \approx O(r^{-2}), \quad H_\phi = E_\theta = \frac{e\beta^2}{ra}\frac{\cos\theta\cos\chi'}{q^3},$$

$$E_\phi = -H_\theta = \frac{e\beta^2}{ra}\frac{\beta\sin\theta - \sin\chi'}{q^3}, \quad S_\theta = S_\phi \approx O(r^{-3}),$$

$$S_r = \frac{c}{4\pi}(E_\phi^2 + E_\theta^2) = \frac{c}{4\pi}\frac{e^2\beta^4}{r^2 a^2 q^6}[\cos^2\theta\cos^2\chi' + (\beta\sin\theta - \sin\chi')^2]. \tag{8.16}$$

Here $q = 1 - \beta\sin\theta\sin\chi'$. Obviously $S_r d\sigma_r$, $S_\theta d\sigma_\theta$ and $S_\phi d\sigma_\phi$ are energies radiated per unit of laboratory time through the surface elements $d\sigma_r = r^2\sin\theta d\theta d\phi$, $d\sigma_\theta = r\sin\theta dr d\phi$, $d\sigma_\phi = r dr d\theta$ attached to the sphere of the radius r and oriented in radial, meridional and azimuthal directions, respectively. Correspondingly,

$$\frac{d^3\mathcal{E}}{\sin\theta d\theta d\phi dt} = r^2 S_r, \quad \frac{d^3\mathcal{E}}{\sin\theta dr d\phi dt} = r S_\theta, \quad \frac{d^3\mathcal{E}}{dr d\theta dt} = r S_\phi$$

are the energies per unit of laboratory time related to the rectangles with sides $(d\theta, \sin\theta d\phi)$, $(dr, \sin\theta d\phi)$ and $(dr, d\theta)$, respectively.

8.2.2. ENERGY RADIATED FOR THE PERIOD OF MOTION

We are interested in energies flowing through the above surface elements for the period of charge motion.

$$\sigma_r = \frac{d^2\mathcal{E}}{\sin\theta d\theta d\phi} = r^2\int_0^T S_r dt, \quad \sigma_\theta = \frac{d^2\mathcal{E}}{\sin\theta dr d\phi} = r\int_0^T S_\theta dt,$$

$$\sigma_\phi = \frac{d^2\mathcal{E}}{dr d\theta} = r\int_0^T S_\phi dt, \quad T = 2\pi/\omega.$$

From (8.12) we find

$$dt = \frac{Q}{\tilde{R}}dt' = -\frac{Q}{\tilde{R}\omega}d\chi'.$$

Then,

$$\sigma_r = \frac{d^2\mathcal{E}}{\sin\theta d\theta d\phi} = \frac{e^2}{4\pi a\beta}\int_0^{2\pi} \tilde{S}_r\frac{Q}{\tilde{R}}d\chi',$$

$$\sigma_\theta = \frac{d^2\mathcal{E}}{\sin\theta dr d\phi} = \frac{e^2}{4\pi r^2\beta}\int_0^{2\pi} \tilde{S}_\theta\frac{Q}{\tilde{R}}d\chi',$$

$$\sigma_\phi = \frac{d^2\mathcal{E}}{dr d\theta} = \frac{e^2}{4\pi\beta r^2}\int_0^{2\pi} \tilde{S}_\phi\frac{Q}{\tilde{R}}d\chi'. \tag{8.17}$$

The χ' integration runs from 0 to 2π. For large distances ($\epsilon_0 \to 0$) one gets

$$\sigma_\theta \to 0, \quad \sigma_\phi \to 0, \quad \sigma_r \to \frac{e^2\beta^3}{4a}\frac{1}{(1-\beta^2\sin^2\theta)^{5/2}}$$

$$\times\left[2 + \beta^2\sin^2\theta - \sin^2\theta\frac{1-\beta^2}{1-\beta^2\sin^2\theta}\left(1 + \frac{1}{4}\beta^2\sin^2\theta\right)\right]. \tag{8.18}$$

Equation (8.18) coincides with that given in [22].

The surface integral from the radial energy flux

$$\int \sigma_r d\Omega = \frac{4\pi\beta^3\gamma^4}{3a}$$

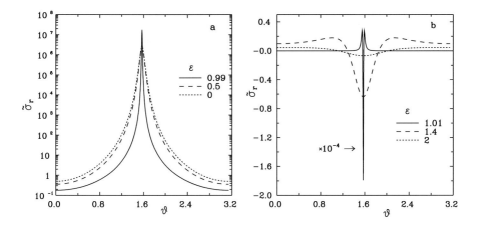

Figure 8.2. Dimensionless distributions of the radial energy flux $\tilde{\sigma}_r$ for the period of motion as a function of the polar angle θ for $\beta = 0.999$, for the radii of an observational sphere greater (a) and smaller (b) than radius a of the charge orbit. It is seen (a) that the radial distribution for $\epsilon = 0.5$ practically coincides with that for $\epsilon = 0$ (this corresponds to an infinite observational distance). For $r < a$ (b), $\tilde{\sigma}_r$ is large only in the neighbourhood of a charge orbit ($\epsilon = 1.01$). The increasing of $\tilde{\sigma}_r$ near the charge orbit is owed to the proximity of a charge and is usually called the focusing effect (see the text).

is equal to the energy radiated by a moving charge during the time $T = 2\pi/\omega$. It follows from (8.17) that σ_r, σ_θ and σ_ϕ have different dimensions, and therefore cannot be compared between themselves. To make this possible we introduce dimensionless intensities

$$\tilde{\sigma}_r = \sigma_r/(e^2/a), \quad \tilde{\sigma}_\phi = \sigma_\phi/(e^2/a^2), \quad \tilde{\sigma}_\theta = \sigma_\theta/(e^2/a^2).$$

The radial energy flux $\tilde{\sigma}_r$ emitted during the period of a charge motion is shown in Fig. 8.2 as a function of a polar angle θ. The calculations were made for the radii of an observational sphere r larger (Fig. 8.2 a) and smaller (Fig. 8.2 b) than radius a of the charge orbit. It is seen that with the increase of radius r of the observational sphere, $\tilde{\sigma}_r$ reaches its asymptotic value (8.18) for $\epsilon_0 \approx 0.5$ (Fig. 8.2 a). On the other hand, for r smaller than a, $\tilde{\sigma}_r$ falls very rapidly with decrease of r (Fig. 8.2 (b)). The increase of the radial energy flux in the neighbourhood of a charge orbit ($\epsilon_0 \to 1$) is owed to the proximity of the observational point to a moving charge. This fact was called the 'focusing' effect in [15].

The azimuthal energy flux $\tilde{\sigma}_\phi$ emitted for the period of the charge motion is shown in Fig. 8.3 as function of the polar angle θ. In accordance with Schwinger's results it is large in the immediate neighbourhood of the charge trajectory ($\epsilon_0 = 0.99$ and $\epsilon_0 = 1.01$). For large observational distances it decreases as ϵ_0^2. On the observational spheres lying inside the

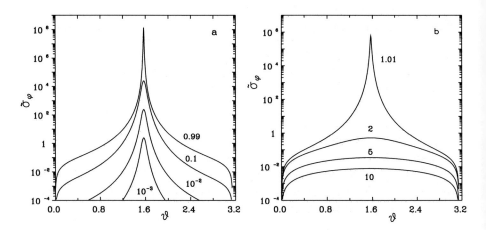

Figure 8.3.   Distributions of the dimensionless azimuthal energy flux $\tilde{\sigma}_\phi$ for the period of motion as a function of the polar angle θ for $\beta = 0.999$, for the radii of the observational sphere greater (a) or smaller (b) than a. Numbers on curves are $\epsilon = a/r$. For large observational distances $\tilde{\sigma}_\phi$ falls like $1/r^2$. From the comparison with Fig. 8.2 it follows that $\tilde{\sigma}_\phi \gg \tilde{\sigma}_r$ in the neighbourhood of a charge orbit ($\epsilon \sim 1$) and $\tilde{\sigma}_\phi \ll \tilde{\sigma}_r$ at large distances. This reconciles Schwinger's and Schott's predictions.

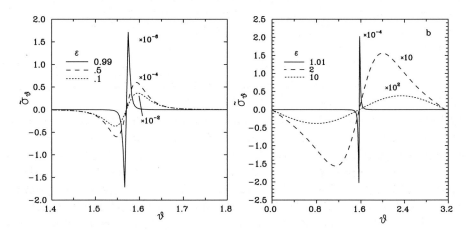

Figure 8.4.   Distributions of the dimensionless polar energy flux $\tilde{\sigma}_\theta$ for the period of motion as a function of the polar angle θ for $\beta = 0.999$, for the radii of the observational sphere greater (a) or smaller (b) than a. For large observational distances they decrease as $1/r^2$. From the comparison with Figs. 8.1 and 8.2 it follows that $\tilde{\sigma}_\theta$ is much smaller than $\tilde{\sigma}_\phi$ and $\tilde{\sigma}_r$.

charge orbit the dependence $\tilde{\sigma}_\phi$ is rather flat for $\epsilon_0 > 2$.   The polar energy flux $\tilde{\sigma}_\theta$ emitted for the period of the charge motion is shown in Fig. 8.4 as a function of the polar angle θ. Owing to the presence of the factor $\cos\theta$ in

$\tilde{S}_\theta$ (see 8.15), $\tilde{\sigma}_\theta$ exhibits a characteristic oscillation in the neighbourhood of $\theta = \pi/2$. It is easy to check that $\int \sigma_\theta d\theta = 0$. In general, polar intensities $\tilde{\sigma}_\theta d\theta$ are much smaller than $\tilde{\sigma}_\phi$ and $\tilde{\sigma}_r$. Comparison of Figs. 8.2 (a) and 8.3 demonstrates that the focusing effect is more pronounced for the energy flux in the azimuthal direction. This is essentially the Schwinger result, according to which a charge moving with a velocity $v \sim c$ radiates mainly in the direction of its motion. Figs. 8.2 (b) and 8.3 (b) demonstrate that focusing effect takes place also for $r < a$.

What is new in this section? The radial energy flux at arbitrary distances for $r > a$ was studied previously in [13-15], [18]. To the best of our knowledge, the energy fluxes in other directions and radial energy flux for $r < a$ were never studied before.

8.2.3. INSTANTANEOUS DISTRIBUTION OF SYNCHROTRON RADIATION

Up to now we have studied the spatial distribution of the energy radiated for a period of the motion. Now we intend to study its instantaneous distribution in the laboratory reference frame at a given instant of laboratory time. In the past, the instantaneous distribution of the radiated power at large distances was studied in the reference frame attached to a moving charge [23-26]. The instantaneous intensity in the radial direction was identified with S_r defined in (8.16). However, all quantities in this equation are referred to a fixed instant of proper time t' of a moving charge (since $\chi' = \phi - \omega t'$). Owing to equation (2.2) different spatial points in S_r correspond to different instants of laboratory time t. The physical meaning of this intensity is not very clear. We are interested in finding the intensity at a given instant of laboratory time. For this purpose, for a given instant of laboratory time t we find t' from the equation (8.12) at a given spatial point x, y, z. Substituting t' thus obtained into the field strengths we find EMF at the spatial point x, y, z at the given instant t of laboratory time. By varying x, y, z we obtain the spatial distribution of the EMF at the given instant of laboratory time. This is essentially the computing procedure used below.

Infinities of field strengths
First we note that the denominators Q entering (8.14) have zero only at $\beta = 1$, $\theta = \pi/2$, $\cos \chi' = \epsilon_0$. The corresponding value of χ is equal to

$$\chi = \arccos \epsilon_0 + (1 - \sqrt{1 - \epsilon_0^2})/\epsilon_0. \qquad (8.19)$$

In particular, for large observational distances, $\epsilon_0 \approx 0$, $\chi' \approx \pi/2$, $\chi \approx \pi/2$. In the neighbourhood of the charge orbit $\epsilon_0 \approx 1$, $\chi' \approx 0$, $\chi \approx 1$. For the intermediate distance $\epsilon_0 = 0.5$ one finds $\chi' = \pi/3$, $\chi = \pi/3+2(1-\sqrt{0.75}) \approx$ 1.3. Thus zeroes of Q fill the interval $1 < \chi < \pi/2$. Obviously field strengths are infinite at those spatial points where Q vanishes. Physically this may be understood in the framework of the Schwinger approach [7] according to which a charge moving along a circular trajectory with the velocity $v \sim c$ radiates in the direction of its motion. The equation of this radiation line is

$$y - a \sin \omega t' = - \cot \omega t' (x - a \cos \omega t').$$

Or, in spherical coordinates,

$$\cos(\phi - \omega t') = \frac{a}{r}, \quad \phi - \omega t' = \arccos \frac{a}{r}$$

(it was set here $\theta = \pi/2$ since a charge moves in the equatorial plane). Substituting this equation into (8.13) one finds

$$\chi = \arccos \epsilon_0 + \frac{\beta}{\epsilon_0}(1 - \sqrt{1 - \epsilon_0^2}).$$

For $\beta = 1$ this coincides with (8.19).

Extremes of the Q function
For $\beta \neq 1$ and $\theta \neq \pi/2$ the denominator Q entering the field strengths does not vanish. Yet it may take minimal and maximal values corresponding to maximal and minimal values of field strengths, respectively. In the next two sections we study the positions of Q extremes in the plane $\theta = const$ (the parallel plane) and in the plane $\phi = const$ (the meridional plane)

Extremes of field strengths in parallel planes. To find extremes of the functions Q relative to the azimuthal angle ϕ we differentiate Q by ϕ for $r, t,$ and θ fixed and take into account that

$$c\frac{dt'}{d\phi} = -\frac{ar \sin \theta \sin \chi'}{R - \beta r \sin \theta \sin \chi'} = -\frac{a \sin \theta \sin \chi'}{Q}.$$

Then equating $dQ/d\phi$ to zero one has

$$a \sin \chi' = R\beta \cos \chi', \qquad (8.20)$$

or, in dimensionless variables,

$$\epsilon_0 \sin \chi' = \tilde{R}\beta \cos \chi', \quad \tilde{R} = (1 + \epsilon_0^2 - 2\epsilon_0 \sin \theta \cos \chi')^{1/2}.$$

This leads to the following third-order equation:

$$\cos^3 \chi' - b\cos^2 \chi' + \frac{\epsilon_0}{2\beta^2 \sin \theta} = 0, \quad b = \frac{1 + \epsilon_0^2(1 + 1/\beta^2)}{2\epsilon_0 \sin \theta}. \qquad (8.21)$$

This equation has three real roots

$$\cos \chi_1' = \frac{b}{3}\left(2\cos\frac{\psi}{3}+1\right), \quad \cos\chi_2' = \frac{b}{3}\left(1-\cos\frac{\psi}{3}+\sqrt{3}\sin\frac{\psi}{3}\right),$$

$$\cos\chi_3' = \frac{b}{3}\left(1-\cos\frac{\psi}{3}-\sqrt{3}\sin\frac{\psi}{3}\right). \tag{8.22}$$

Here

$$\cos\psi = 1 - 54\frac{\epsilon_0^4\beta^4\sin^2\theta}{[\beta^2(1+\epsilon_0^2)+\epsilon_0^2]^3}, \quad 0 < \psi < \pi.$$

Since $(\cos\chi')_1 > 1$, it is unphysical. Furthermore, it follows from (8.21) that $\cos\chi_2' > 0$ and $\cos\chi_3' < 0$. Owing to (8.20), $\sin\chi'$ has the same sign as $\cos\chi'$. Therefore

$$\chi_2' = \arccos\left[\frac{b}{3}\left(1-\cos\frac{\psi}{3}+\sqrt{3}\sin\frac{\psi}{3}\right)\right],$$

$$\chi_3' = \pi + \arccos\left|\frac{b}{3}\left(1-\cos\frac{\psi}{3}-\sqrt{3}\sin\frac{\psi}{3}\right)\right|$$

lie in the first and third quadrants, respectively. From the definition of Q it follows that χ_2' and χ_3' correspond to the minimum and maximum of Q and to the maximum and minimum of field strengths, respectively. We rewrite equation (8.13) in the form

$$\chi_2 = \chi_2' + \frac{\beta}{\epsilon_0}(1-\tilde{R}_2), \quad \chi_3 = \chi_3' + \frac{\beta}{\epsilon_0}(1-\tilde{R}_3),$$

$$\tilde{R}_{2,3} = (1+\epsilon_0^2 - 2\epsilon_0\sin\theta\cos\chi_{2,3}')^{1/2}. \tag{8.23}$$

These equations define χ_2 and χ_3 (corresponding to the fixed r and θ) for which the field strengths are maximal and minimal, respectively.

The dependences $\chi_2(\theta)$ and $\chi_3(\theta)$ given by (8.23) for $\beta = 0.999$, on the observational spheres of various radii are shown in Fig. 8.5. A particular curve defines the position of the field strength maxima and minima on a sphere of a particular radius. However, the numerical value of extremum along each of these curves depends on θ. To evaluate the absolute minimum and maximum of Q, we substitute (8.20) and (8.23) into Q

$$Q_{2,3} = \tilde{R}_{2,3}\left(1 - \frac{\beta^2}{\epsilon_0}\sin\theta\cos\chi_{2,3}'\right)$$

$$= \tilde{R}_{2,3}\left\{1 - \frac{\beta^2}{6\epsilon_0^2}[1+\epsilon_0^2(1+1/\beta^2)]\left(1-\cos\frac{\psi}{3}\pm\sqrt{3}\sin\frac{\psi}{3}\right)\right\}.$$

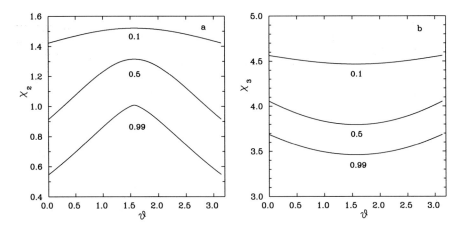

Figure 8.5. Lines on which Q is minimal (a) and maximal (b) for $\beta = 0.999$ and different radii of the observational sphere. Numbers on curves mean ϵ. Along each of these curves the absolute minimum (a) and maximum (b) of Q are reached at $\theta = \pi/2$.

Differentiating by θ we find that absolute minimum (for Q_2) and maximum (for Q_3) are reached at $\theta = \pi/2$. Then, the first of the equations

$$\chi_2 = \chi_2' + \frac{\beta}{\epsilon_0}(1 - \tilde{R}_2),$$

$$\chi_3 = \chi_3' + \frac{\beta}{\epsilon_0}(1 - \tilde{R}_3), \qquad (8.24)$$

(where $\cos\chi_2'$ and $\cos\chi_3'$ are obtained from (8.22) by setting $\theta = \pi/2$ in them), generalizes Schwinger's formula for arbitrary β and ϵ_0.

The azimuthal positions of absolute minimum and maximum of Q as a function of radius of the observational sphere are shown in Fig. 8.6 for various charge velocities. It is seen that χ_2 and χ_3 fill the intervals $(0, \pi/2)$ and $(\pi, 3\pi/2)$, respectively.

Finally, in Fig. 8.7 it is shown how the function Q^{-1} behaves in the equatorial plane $\theta = \pi/2$. Obviously the maxima of field strengths and radiation intensity coincide with those of Q^{-1}.

It should be mentioned that Eq. (8.24) defining χ_2 may be interpreted in three ways. First, for fixed t, Eq. (8.24) defines how the azimuthal position of the maximum of Q^{-1} changes with r. Clearly this dependence has a spiral-like structure. Second, for fixed r, Eq.(8.24) defines how the azimuthal position of the maximum of Q^{-1} changes with t. Obviously this dependence is linear. Third, for fixed ϕ, Eq.(8.24) defines how the radial position of the maximum of Q^{-1} changes with t. Obviously, r linearly rises with t.

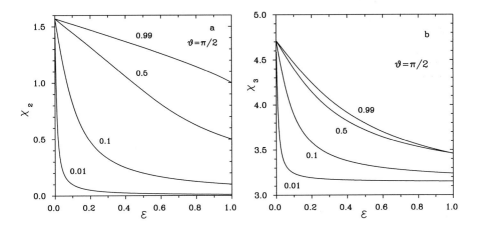

Figure 8.6. Azimuthal position of the absolute minimum (a) and maximum (b) of Q as a function of radius of the observational sphere. Numbers on curves mean charge velocities β. It is seen that the absolute minima and maxima of Q fill the regions $0 < \chi < \pi/2$ and $\pi < \chi < 3\pi/2$, respectively.

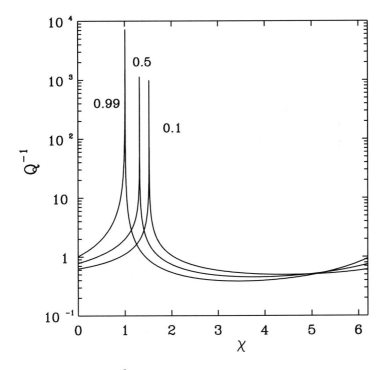

Figure 8.7. Behaviour of Q^{-1} in the plane $\theta = \pi/2$ for $\beta = 0.999$ and various radii of the observational sphere. Numbers on curves mean ϵ.

Consider particular cases.

1) $\epsilon_0 \to 0$. This corresponds to an observational point on the sphere with a radius $r \gg a$. Then

$$\chi_2 = \chi_2' \to \frac{\pi}{2}, \quad \phi_2 \to \frac{\pi}{2} + \omega\left(t - \frac{r}{c}\right),$$

$$\chi_3 = \chi_3' \to \frac{3\pi}{2}, \quad \phi_3 \to \frac{3\pi}{2} + \omega\left(t - \frac{r}{c}\right). \tag{8.25}$$

Therefore at large distances the minimum and maximum of Q are reached at the planes $\chi_2 = \pi/2$ and $\chi_3 = 3\pi/2$, respectively. The corresponding values of Q are equal to $Q_2 = 1 - \beta \sin\theta$ and $Q_3 = 1 + \beta \sin\theta$. The absolute minimum and maximum of Q are reached at the points $\chi_2 = \pi/2, \theta = \pi/2$ and $\chi_3 = 3\pi/2, \theta = \pi/2$, respectively. This is demonstrated in Fig. 8.7, from which it follows that, indeed, for $\epsilon_0 = 0.1$, Q reaches the minimal and maximal values approximately at these points.

2) $\beta \to 0$. This corresponds to a charge which is permanently at rest at the point $x = a, y = z = 0$. Eqs.(8.21) and (8.22) then give $\chi_2 = \chi_2' = \phi_2 = 0$, $\chi_3 = \chi_3' = \phi_3 = \pi$. These values correspond to the nearest and most remote meridional planes on the sphere of the radius r, respectively.

3) $\epsilon_0 \to 1$, $\theta = \pi/2$, $\beta \to 1$. This corresponds to the observational point on the charge trajectory. Then,

$$\chi_2' \to 0, \quad \chi_2 \to 1, \quad \phi_2 \to \omega t, \quad \chi_3' \to \frac{4\pi}{3},$$

$$\chi_3 \to \frac{4\pi}{3} + 1 - \sqrt{3}, \quad \phi_3 \to \omega t + \frac{4\pi}{3} - \sqrt{3}.$$

Again, this is supported by Fig. 8.7 which shows that for $\epsilon_0 = 0.99$, Q reaches the minimal and maximal values at these points.

The dimensionless instantaneous radial and azimuthal energy fluxes (8.15) taken along the curves χ_2 with minimal Q defined by Eq. (8.23) and depicted in Fig. 8.5(a) are shown in Figs. 8.8(a) and (b). It is seen that in the neighbourhood of the charge orbit, the $\tilde{S}_\phi$ component of the Poynting vector dominates. This may be shown analytically. For simplicity let $\epsilon_0 = 1$, $\beta = 1$, whilst $\theta = \pi/2 + \delta_\theta$. Then,

$$\sin\theta \approx 1 - \delta_\theta^2/2, \quad \cos\psi \approx -1 + 2\delta_\theta^2, \quad \cos\chi_2' \approx 1 - \delta_\theta/\sqrt{3},$$

$$\sin\chi_2' \approx \sqrt{2}\delta_\theta/\sqrt{3}, \quad \tilde{R} \approx \sqrt{2}(\delta_\theta/\sqrt{3})^{1/2}, \quad Q \approx \sqrt{2}(\delta_\theta/\sqrt{3})^{3/2}.$$

We observe that for $\beta = 1$ and $\epsilon_0 = 1$

$$\tilde{S}_r \sim |\delta_\theta|/\sqrt{3}, \quad S_\theta \sim \delta_\theta, \quad \tilde{S}_\phi \sim \sqrt{2}|\delta_\theta\sqrt{3}|^{1/2}$$

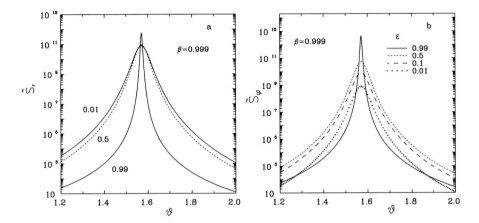

Figure 8.8.   Instantaneous radial (a) and azimuthal (b) energy fluxes along the curves with minimal Q shown in Fig. 5a. Numbers on curves are ϵ. It is seen that $\tilde{S}_\phi \gg \tilde{S}_r$ near the charge orbit ($\epsilon_0 = 0.99$).

(the same singular factor $E^2/\tilde{R}$ is omitted). Hence, it follows that in the neighbourhood of a charge orbit, $\tilde{S}_\phi$ is much larger than $\tilde{S}_r$ and S_θ (since $|\delta_\theta|^{1/2} \gg |\delta_\theta|$ for $|\delta_\theta| \ll 1$).

The dominance of $\tilde{S}_\phi$ over $\tilde{S}_r$ near the charge orbit, and $\tilde{S}_r$ over $\tilde{S}_\phi$ at large distances may be understood as follows. Following Schwinger [7] assume that for $r \sim a$ all energy is radiated along the vector $\vec{n} = \cos\omega t \vec{n}_y - \sin\omega t \vec{n}_x$ tangential to a charge orbit. An energy flux (lying on the continuation of $\vec{n}$)) then intersects the sphere S_r of the radius r at the azimuthal angle $\phi = \omega t + \arccos(a/r)$. The scalar product of the radial unit vector belonging to S_r with the unit vector lying on the continuation of $\vec{n}$ (along which the energy flux propagates) is $(\vec{n}_r \vec{n}) = \sin(\phi - \omega t)$. At large distances $\phi - \omega t = \arccos(a/r) \approx \pi/2$ and $(\vec{n}_r \vec{n}) \approx 1$. Therefore, at large distances Schwinger's flux has mainly the radial component.

We now evaluate the radial and azimuthal energy fluxes in the equatorial plane $\theta = \pi/2$. For the radii of the observational sphere not too close to a charge orbit, S_r is positive for all χ (Fig. 8.9(a)). However, in the neighbourhood of a charge orbit, $\tilde{S}_r$ may be negative in some region of χ (the energy flows into the observational sphere there). This is demonstrated in Fig. 8.9(b), where the region with $\tilde{S}_r < 0$ is shown by the dotted line. The reason for this is evident from Eqs. (8.15). It is seen that $\tilde{S}_r$ consists of two terms. The second term is compared with the first one in the neighbourhood of a charge orbit where $\epsilon_0 \approx 1$.

On the other hand, both terms in $\tilde{S}_\phi$ are of the same order. Therefore, one may expect that $\tilde{S}_\phi$ may take negative values in some region of θ for

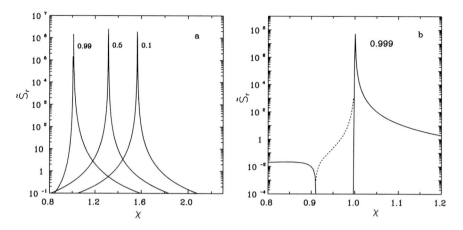

Figure 8.9.   Instantaneous radial energy fluxes in the plane $\theta = \pi/2$ for $\beta = 0.999$ and radii of the observational sphere not too close to the charge orbit (a) and in its immediate neighbournood (b). Numbers on curves mean $\epsilon = a/r$. In the latter case (b) energy flux may take negative values shown by the dotted line.

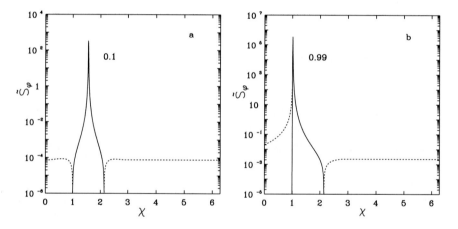

Figure 8.10.   Instantaneous azimuthal energy fluxes in the plane $\theta = \pi/2$ for $\beta = 0.999$ at a large distance from the charge orbit (a) and near it (b). In both cases azimuthal energy fluxes take negative values (shown by dotted lines) in some angular regions. Numbers on curves mean ϵ_0.

arbitrary radius of the observational sphere. It is shown in Figs. 10(a) and 10(b) that regions of χ where $\tilde{S}_\phi$ is negative, exist both for large ($\epsilon_0 = 0.1$) and small ($\epsilon_0 = 0.99$) observational distances. Again, the regions with $\tilde{S}_\phi < 0$ are shown by the dotted lines. Although the instantaneous radial and azimuthal EMF fluxes may acquire negative values in some angular regions, their time averages are positive. Figs. 8.2 (a) and 8.3 demonstrate this.

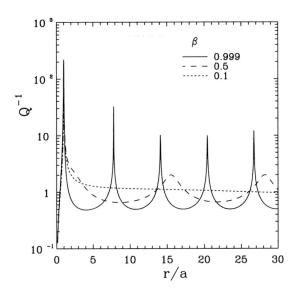

Figure 8.11. Radial distribution of Q^{-1} defining maxima of field strengths for a fixed laboratory time in the equatorial plane $\theta = \pi/2$ for a number of charge velocities. The period of oscillations is $2\pi/\beta$.

The dependence of Q^{-1} on the radius in the equatorial plane $\theta = \pi/2$ at a fixed instant of laboratory time t is shown in Fig. 8.11. The oscillations with the period $2\pi/\beta$ are observed. The dependences of Q^{-1} on the laboratory time t in the equatorial plane $\theta = \pi/2$ for the fixed radius are shown in Fig. 8.12 (a) for a large observational distance ($\epsilon_0 = 0.1$), and in Fig. 8.12 (b) in the neighbourhood of a charge orbit ($\epsilon_0 = 0.99$). Again, oscillations with the period $2\pi/\beta$ are observed. Both these cases are described by the following two formulae:

$$Q^{-1} = \frac{1}{c(t - t')/r + \beta \sin \theta \sin(\beta ct'/a - \phi)},$$

$$c(t - t') = [r^2 + a^2 - 2ra \sin \theta \cos(\beta ct'/a - \phi)]^{1/2}.$$

Extremes of field strengths in meridional planes. Now we find the minimum of Q relative to θ for χ fixed. For this purpose one should solve the equation $dQ/d\theta = 0$. Taking into account that

$$\frac{d\chi'}{d\theta} = -\frac{\beta \cos \theta \cos \chi'}{\tilde{R} - \beta \sin \theta \sin \chi'},$$

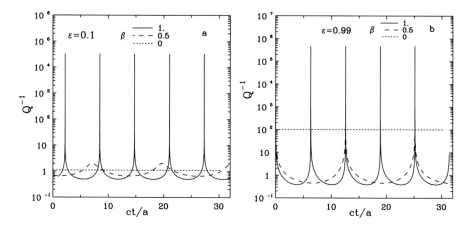

Figure 8.12.   Time dependences of Q^{-1} at a fixed radial point lying in $\theta = \pi/2$ plane at a large distance (a) and near (b) the charge orbit. The period of oscillations is $2\pi/\beta$.

we find the following relation

$$-\beta \cos \theta \sin \theta \frac{\cos \chi'}{\tilde{R} - \beta \sin \theta \sin \chi'} = \cos \theta \frac{\epsilon_0 \cos \chi' + \beta \tilde{R} \sin \chi'}{\epsilon_0 \sin \chi' - \beta \tilde{R} \cos \chi'}. \qquad (8.26)$$

This equation is satisfied trivially for $\theta = \pi/2$. In this case

$$Q = \tilde{R} - \beta \sin \chi',$$

$$\chi = \chi' + \frac{\beta}{\epsilon_0}(1 - \tilde{R}), \quad \tilde{R} = (1 + \epsilon_0^2 - 2\epsilon_0 \cos \chi')^{1/2}.$$

To see whether Q reaches the maximum or minimum at this χ', one should find $d^2Q/d\theta^2$ at $\theta = \pi/2$. It is given by

$$\frac{d^2 Q}{d\theta^2}(\theta = \pi/2) = \frac{\Delta}{\tilde{R} - \beta \sin \chi'}, \quad \Delta = \epsilon_0 \cos \chi' + \beta \tilde{R} \sin \chi' - \beta^2.$$

Obviously $Q(\theta)$ has a minimum or maximum $\theta = \pi/2$ for Δ greater or smaller than zero, respectively (since $R - \beta \sin \chi'$ is always positive). Correspondingly, Q^{-1} and field strengths have a maximum or minimum there. The value of χ is found from the equation $\chi = \chi' + \beta(1 - R)/\epsilon_0$.

Consider particular cases.

For $\epsilon_0 \to 0$ (large distances) $\Delta = \beta(\sin \chi' - \beta)$. Therefore Q, as a function of θ, has a minimum at $\theta = \pi/2$ for $\sin \chi' > \beta$ and maximum for $\sin \chi' < \beta$. The corresponding χ is given by $\chi = \chi' + \beta \cos \chi'$. Therefore for large distances and $\beta \approx 1$, Δ is negative everywhere except for the

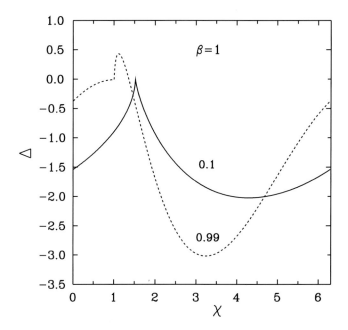

Figure 8.13.   Azimuthal angular dependence of the parameter Δ in the equatorial plane $\theta = \pi/2$ on the sphere of a particular radius. Q^{-1} may take maximal values in the region of χ where $\Delta > 0$ and minimal values in the region of χ where $\Delta < 0$. Numbers on curves mean $\epsilon_0 = a/r$.

neighbourhood of $\chi' = \pi/2$. Correspondingly, the maxima of Q^{-1} and field strengths should be near $\chi \approx \chi' = \pi/2$.

For $\epsilon_0 \approx 1$ (i.e., near the charge orbit), Δ and χ are reduced to:

$$\Delta = \cos\chi' + 2\beta \sin\chi' \sin(\chi'/2) - \beta^2, \quad \chi = \chi' + \beta(1 - 2\sin(\chi'/2)).$$

We see that $\Delta > 0$ for $1 < \chi < 1.36$ and $\Delta < 0$ in other regions of χ. Therefore Q acquires the minimum at $\theta = \pi/2$ only for $1 < \chi < 1.36$.

The dependences $\Delta(\chi)$ for large observational distances ($\epsilon_0 = 0.1$) and in the neighbourhood of the charge orbit ($\epsilon_0 = 0.99$) are shown in Fig. 8.13. They are in complete agreement with the analytical results just obtained. This is also confirmed by Figs. 8.7 and 8.9 where $Q^{-1}(\chi)$ and field strengths are shown in the equatorial plane $\theta = \pi/2$. We see that maxima of $Q^{-1}(\chi)$ and field strengths lie in the neighbourhood of $\chi = \pi/2$ for $\epsilon_0 \to 0$, whilst its minima are outside this region. The position of the $Q(\theta = \pi/2)$ extremum as a function of the observational sphere radius are shown in Figs. 8.6(a) and (b).

We conclude: for $\theta = \pi/2$ the positions and values of Q extremes coincide with those found at the beginning of this section.

For $\theta \neq \pi/2$ equation (8.26) reduces to

$$\beta \tilde{R} \sin \chi' = \beta^2 \sin \theta - \epsilon_0 \cos \chi'. \tag{8.27}$$

Or,

$$\cos^3 \chi' - b \cos^2 \chi' + c = 0,$$

where

$$b = \frac{1 + \epsilon_0^2(1 + 1/\beta^2)}{2\epsilon_0 \sin \theta}, \quad c = \frac{1 + \epsilon_0^2 - \beta^2 \sin^2 \theta}{2\epsilon_0 \sin \theta}.$$

Three roots of this equation are given by

$$(\cos \chi')_1 = \frac{b}{3}\left(2\cos\frac{\psi'}{3} + 1\right), \quad (\cos \chi')_2 = \frac{b}{3}\left(1 - \cos\frac{\psi'}{3} + \sqrt{3}\sin\frac{\psi'}{3}\right),$$

$$(\cos \chi')_3 = \frac{b}{3}\left(1 - \cos\frac{\psi'}{3} - \sqrt{3}\sin\frac{\psi'}{3}\right). \tag{8.28}$$

Here

$$\cos\psi' = 1 - 54\epsilon_0^2\sin^2\theta\frac{1 + \epsilon_0^2 - \beta^2\sin^2\theta}{(1 + \epsilon_0^2 + \epsilon_0^2/\beta^2)^3}.$$

It is easy to check that $(\cos\chi')_1 > 1$, and therefore it is unphysical. We observe that $(\cos\chi')_2 \geq 0$ and $(\cos\chi')_3 \leq 0$. It follows from (8.27) that $(\sin\chi')_3 \geq 0$. Therefore

$$\chi_3' = \pi - \arccos\left|\frac{b}{3}\left(1 - \cos\frac{\psi'}{3} - \sqrt{3}\sin\frac{\psi'}{3}\right)\right| \tag{8.29}$$

lies in the second quadrant and corresponds to the minimum of Q.

Furthermore (see again (8.27)), if

$$\beta^2\sin\theta - \frac{1}{3}\epsilon_0 b\left(1 - \cos\frac{\psi'}{3} + \sqrt{3}\sin\frac{\psi'}{3}\right) > 0,$$

then $\sin\chi_2' > 0$, $\cos\chi_2' > 0$, and

$$\chi_2' = \arccos\frac{b}{3}\left(1 - \cos\frac{\psi'}{3} + \sqrt{3}\sin\frac{\psi'}{3}\right) \tag{8.30}$$

is in the first quadrant. On the other hand, if

$$\beta^2\sin\theta - \frac{1}{3}\epsilon_0 b\left(1 - \cos\frac{\psi'}{3} + \sqrt{3}\sin\frac{\psi'}{3}\right) < 0,$$

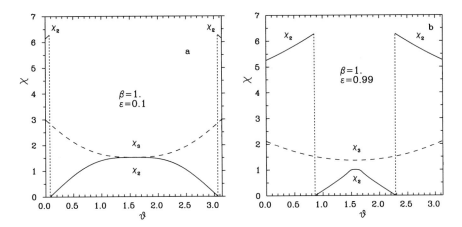

Figure 8.14. Position of the Q^{-1} extremes on a sphere of large radius (a) and near the charge orbit (b). There are two lines of extremes: χ_2 and χ_3. The former consists of two branches connected by the dotted line.

then $\sin \chi_2' < 0$, $\cos \chi_2' > 0$ and

$$\chi_2' = 2\pi - \arccos \frac{b}{3}\left(1 - \cos \frac{\psi'}{3} + \sqrt{3}\sin \frac{\psi'}{3}\right). \qquad (8.31)$$

is in the fourth quadrant.

The lines on which Q are minimal are obtained from the equations

$$\chi_2 = \chi_2' + \frac{\beta}{\epsilon_0}(1 - \tilde{R}_2) \quad \text{and} \quad \chi_3 = \chi_3' + \frac{\beta}{\epsilon_0}(1 - \tilde{R}_3), \qquad (8.32)$$

where $\tilde{R}_{2,3} = (1 + \epsilon_0^2 - 2\epsilon_0 \sin \theta \cos \chi_{2,3}')^{1/2}$, and $\cos \chi_2'$ and $\cos \chi_3'$ are defined by (8.28).

The dependences (8.32) for $\beta = 1$ and the large radius of the observational sphere ($\epsilon_0 = 0.1$) and near the charge orbit ($\epsilon_0 = 0.99$) are presented in Fig. 8.14 (a) and 8.14 (b). For the fixed χ they define the angle θ for which Q is minimal. We see on these figures the lines χ_2 and χ_3 shown by the solid and broken lines, respectively. In accordance with (8.30) and (8.31) the curve χ_2 consists of two branches connected by the dotted vertical lines. Keep in mind that these curves do not describe the extreme of Q at $\theta = \pi/2$. The behaviour of Q^{-1} along these curves is shown in Fig. 8.15 (a) (large distances) and in Fig. 8.15 (b) (near the charge orbit). How to deal with the curves presented in Fig. 8.14? Take, e.g., Fig. 8.14 (b). It shows at which θ the minimal value of Q is reached for the given χ near the charge orbit. Now we compare Fig. 8.14 (b) with Fig. 8.16 (a), where the dependences $Q^{-1}(\theta)$ in a number of meridional planes in the neighbourhood

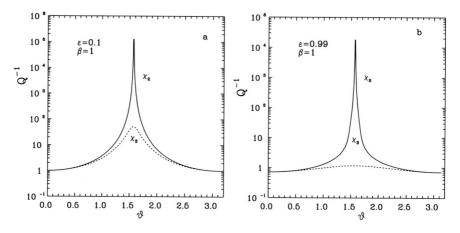

Figure 8.15. Distribution of Q^{-1} along the curves χ_2 and χ_3 shown in Fig. 8.14 and lying on a sphere of large radius (a) and near the charge orbit (b).

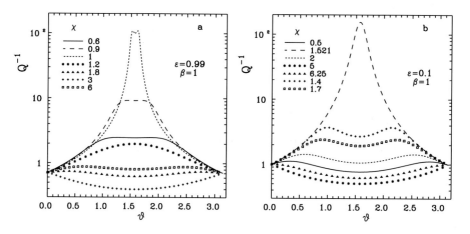

Figure 8.16. θ-dependences of Q^{-1} in different meridional planes in the neighbourhood of a charge orbit (a) and at large distances from it (b).

of a charge orbit are presented. Let χ be 0.6. Then Fig. 8.14 (b) tells us that Q has minima at $\theta \approx 1.25$ and $\theta \approx 1.85$. This is confirmed by Fig. 8.16 (a) in which one sees the maxima of $Q^{-1}(\theta)$ at the same θ. When χ increases, two maxima approach each other (Fig. 8.16 (a), $\chi = 0.9$ and $\chi = 1$). For some χ, when the horizontal line intersects the curve χ_2 only once, these maxima fuse. For larger χ the horizontal line does not intersect either the χ_2 or χ_3 curves. In Fig. 8.16 (a) one observes that for $\chi = 1.2$ there are no maxima of Q^{-1} for $\theta \neq \pi/2$ (as we have mentioned, the extremes of Q at $\theta = \pi/2$ are not described by Eq.(8.30) and Fig. 8.14)). For larger χ, the horizontal line begins to intersect the χ_3 curve. Two maxima of $Q^{-1}(\theta)$ again appear (Fig. 8.16 (a), $\chi = 1.8$). For larger χ, the intersection of the horizontal line with χ_3 disappears, only the minimum at $\theta = \pi/2$ remains (Fig. 8.16 (a), $\chi = 3$). For still larger χ the horizontal line begins to intersect the second branch of the χ_2. Two maxima of $Q^{-1}(\theta)$ again appear (Fig. 8.16 (a), $\chi = 6$). We see that the instantaneous distribution of intensities has a rather complicated and unexpected structure. For example, it is usually believed that radiation intensity is maximal in the equatorial $\theta = \pi/2$ plane. Our consideration shows that this is not always so.

For completeness, we present in Fig. 8.16 (b) the dependences $Q^{-1}(\theta)$ in a number of meridional planes at large distances ($\epsilon_0 = 0.1$).

Consider particular cases.

1) Let $\epsilon_0 \to 0$. Then,

$$\cos \psi' = 1 - 54\epsilon_0^2 \sin^2 \theta(1 - \beta^2 \sin^2 \theta), \quad \psi' = 6\sqrt{3}\epsilon_0 \sin \theta \sqrt{1 - \beta^2 \sin^2 \theta},$$

$$\cos \chi_2' = \sqrt{1 - \beta^2 \sin^2 \theta}, \quad \cos \chi_3' = -\sqrt{1 - \beta^2 \sin^2 \theta},$$

$$\sin \chi_2' = \sin \chi_3' = \beta \sin \theta, \quad \chi_2' = \arccos \sqrt{1 - \beta^2 \sin^2 \theta},$$

$$\chi_3' = \pi - \arccos \sqrt{1 - \beta^2 \sin^2 \theta}, \quad b = \frac{1}{2\epsilon_0 \sin \theta}.$$

Therefore two lines where Q is minimal appear at large distances. They are defined by equations (8.32):

$$\chi_2 = \beta \sin \theta \sqrt{1 - \beta^2 \sin^2 \theta} + \arccos \sqrt{1 - \beta^2 \sin^2 \theta}$$

and

$$\chi_3 = \pi - \beta \sin \theta \sqrt{1 - \beta^2 \sin^2 \theta} - \arccos \sqrt{1 - \beta^2 \sin^2 \theta}.$$

Approximately these curves resemble those shown in Fig. 8.14(a) corresponding to $\epsilon_0 = 0.1$ (according to (8.30) and (8.31), the second branch of χ_2 disappears in the limit $\epsilon_0 \to 0$). The corresponding values of Q along these curves are given by $Q_2 = Q_3 = 1 - \beta^2 \sin^2 \theta$. Their minima are reached at $\theta = \pi/2$. For $\beta = 1$, χ_2 and χ_3 are transformed into

$$\chi_{2,3} = \sin \theta |\cos \theta| + \theta$$

and

$$\chi_{2,3} = \sin \theta |\cos \theta| - \theta,$$

respectively.

2) Let $\beta \to 0$. Then

$$\cos\psi' = 1 - 54\frac{1+\epsilon_0^2}{\epsilon_0^4}\beta^6\sin^2\theta, \quad \psi' = 6\sqrt{3}\frac{\sqrt{1+\epsilon_0^2}}{\epsilon_0^2}\beta^3\sin\theta,$$

$$\cos\chi_2' = \frac{\sqrt{1+\epsilon_0^2}}{\epsilon_0}\beta, \quad \cos\chi_3' = -\frac{\sqrt{1+\epsilon_0^2}}{\epsilon_0}\beta,$$

$$\chi_2 = \chi_2' = \frac{3\pi}{2}, \quad \chi_3 = \chi_3' = \frac{\pi}{2}.$$

In this case $Q_2 \approx 1 + \beta\sin\theta \approx 1$ and $Q_3 \approx 1 - \beta\sin\theta \approx 1$.

We see that an instantaneous intensity of SR has a very intricate structure. However, after averaging over the period of motion these intricacies disappear (see Figs. 8.2-8.4). The main question is how to detect the instantaneous intensity which rotates along the surface of observational sphere with a velocity $v \sim c$. Fortunately, there is a notable exception. An instantaneous SR is observed in astronomical experiments [17, 18, 27, 28]. Since the radius of the orbit along which the charge moves is large (e.g., for Jupiter it is about 10^5 km), the period of its rotation is also large and, therefore, instantaneous SR is observable.

8.3. Synchrotron radiation in medium

8.3.1. MATHEMATICAL PRELIMINARIES

The essence of the present approach is to find retarded times from the equation

$$t - t' = R/c_n. \tag{8.33}$$

Let t_i be these roots. Then, for the circular motion in medium

$$\delta(t' - t + R/c_n) = \sum_i \delta(t' - t_i)|1 + \beta_n r\sin\theta\sin\Omega_i/R_i|^{-1}, \quad \beta_n = v/c_n$$

$$\Omega_i = \omega_0 t_i - \phi, \quad R_i = [r^2 + a^2 - 2ar\sin\theta\cos\Omega_i]^{1/2}.$$

The electromagnetic potentials are given by

$$\Phi = \frac{e}{\epsilon_0}\sum_i\frac{1}{|Q_i|}, \quad A_\phi = e\mu\beta\sum_i\frac{\cos\Omega_i}{|Q_i|}, \quad A_\rho = -e\mu\beta\sum_i\frac{\sin\Omega_i}{|Q_i|},$$

$$A_r = \sin\theta A_\rho, \quad A_\theta = \cos\theta A_\rho, \quad Q_i = R_i + \beta_n r\sin\theta\sin\Omega_i. \tag{8.34}$$

To evaluate field strengths, one should differentiate these expressions w.r.t. the space and time variables taking into account that retarded times t_i also depend on the observational point. From the equation

$$c_n(t - t_i) = R_i$$

one finds

$$\frac{dt_i}{dt} = \frac{R_i}{Q_i}, \quad c_n\frac{dt_i}{dr} = -\frac{r - a\sin\theta\cos\Omega_i}{Q_i},$$

$$c_n\frac{dt_i}{d\theta} = ra\frac{\cos\theta\cos\Omega_i}{Q_i}, \quad c_n\frac{dt_i}{d\phi} = ra\frac{\sin\theta\sin\Omega_i}{Q_i}.$$

8.3.2. ELECTROMAGNETIC FIELD STRENGTHS

The following expressions are valid at arbitrary distances at the fixed instant of laboratory time t

$$E_\phi = \frac{e}{\epsilon}\sum_i \frac{1}{|Q_i|^3}[\frac{R_i}{a}\beta_n^2(R_i\sin\Omega_i + \beta_n r\sin\theta)$$

$$-\beta_n R_i\cos\Omega_i - \sin\Omega_i(a - r\beta_n^2\sin\theta\cos\Omega_i)],$$

$$E_\theta = \frac{e}{\epsilon}\cos\theta\sum_i \frac{1}{|Q_i|^3}[\frac{R_i}{a}\beta_n(\beta_n R_i\cos\Omega_i + a\sin\Omega_i)$$

$$-\cos\Omega_i(a - r\beta_n^2\sin\theta\cos\Omega_i)],$$

$$E_r = \frac{e}{\epsilon}\sum_i \frac{1}{|Q_i|^3}[\beta_n R_i\sin\theta\sin\Omega_i + r(1 - \beta_n^2\sin^2\theta\cos^2\Omega_i)$$

$$+a(\beta_n^2 - 1)\sin\theta\cos\Omega_i],$$

$$H_\phi = \frac{e}{a}\beta r\cos\theta\sum_i \frac{1}{|Q_i|^3}(\beta_n R_i\cos\Omega_i + a\sin\Omega_i), \quad (8.35)$$

$$H_\theta = -e\beta\sum_i \frac{1}{|Q_i|^3}[\beta_n\frac{r}{a}(R_i\sin\Omega_i + \beta_n r\sin\theta)$$

$$+ \sin\theta(a - r\beta_n^2\sin\theta\cos\Omega_i) - r\cos\Omega_i],$$

$$H_r = e\beta\cos\theta\sum_i \frac{1}{|Q_i|^3}(a - r\beta_n^2\sin\theta\cos\Omega_i).$$

The radial energy flux is

$$S_r = \frac{c}{4\pi}(E_\theta H_\phi - H_\theta E_\phi). \quad (8.36)$$

To obtain the radial energy flux (8.36) we should at first evaluate the EMF strengths (8.35). For this, for a given space-time point $\vec{r}, t$, we should find retarded times t_i from (8.33) and substitute them into (8.35). Varying $\vec{r}, t$, we find space-time distribution of the EMF strengths. This is essentially the numerical procedure adopted in the next sections. But first we try to obtain qualitative results without numerical calculations.

8.3.3. SINGULARITIES OF ELECTROMAGNETIC FIELD

We are especially interested in finding the position of singularities of electromagnetic potentials and field strengths. They are given by

$$Q_i = R_i + \beta_n r \sin\theta \sin\Omega_i = 0. \tag{8.37}$$

We observe that t_i (or $\Omega_i = \omega_0 t_i - \phi$) satisfies two equations ((8.33) and (8.37)). We try now to exclude t_i (or Ω_i) from them, thus obtaining space-time distribution of singularities without solving the transcendental equation (8.33).

   This procedure was invented by Schott [1]. Later it was applied to the study of creation and time evolution of Cherenkov shock waves in accelerated rectilinear motion [29]. From (8.37) we find

$$\cos\Omega_{1,2} = \frac{a}{r\beta_n^2 \sin\theta} \pm (1 - \frac{\sin^2\theta_c}{\sin^2\theta})^{1/2}, \quad \sin\Omega_{1,2} = -\sqrt{1 - \cos^2\Omega_{1,2}}, \tag{8.38}$$

where

$$\sin\theta_c = \frac{1}{\beta_n}\sqrt{1 + \frac{a^2}{r^2}(1 - \frac{1}{\beta_n^2})}.$$

The careful analysis shows that the above-mentioned singularities exist only if $\beta_n > 1$. In this case the singularities are located in the angular region $\sin\theta > \sin\theta_c$, $r > a/\beta_n$. Since $\cos\Omega_1 = \cos\Omega_2 = a/(r\beta_n^2 \sin\theta_c)$ for $\sin\theta = \sin\theta_c$ (this corresponds to $\theta = \arcsin\theta_c$ and $\theta = \pi - \arcsin\theta_c$), two branches corresponding to the $\pm$ signs in (8.38) represent, in fact, one closed curve lying on the sphere surface.

   As $\cos\Omega_1$ is always greater than zero and $\sin\Omega_1$ is always less than zero, Ω_1 lies in the fourth quadrant:

$$\Omega_1 = 2\pi - \omega_1, \tag{8.39}$$

where ω_1 is in the first quadrant:

$$\omega_1 = \arccos\Omega_1, \quad \sin\omega_1 = \sqrt{1 - \cos^2\Omega_1}.$$

Since $\sin\Omega_2$ is always less than zero Ω_2 lies in the fourth quadrant:

$$\Omega_2 = 2\pi - \omega_2, \tag{8.40}$$

when $\cos\Omega_2 > 0$, and in the third quadrant

$$\Omega_2 = \pi + \omega_2, \tag{8.41}$$

when $\cos \Omega_2 < 0$. Here w_2 lies in the first quadrant:

$$\cos w_2 = |\cos \Omega_2|, \quad \sin w_2 = \sqrt{1 - \cos^2 \Omega_2}.$$

It turns out that $\cos \Omega_2 > 0$ for $a/\beta_n < r < a\gamma_n$, $(\gamma_n = |1 - \beta_n^2|^{-1/2})$ and all angles in the interval $\sin \theta_c < \sin \theta < 1$. On the other hand, for $r > a\gamma_n > a/\beta_n$ one has $\cos \Omega_2 > 0$ for $\sin \theta_c < \sin \theta < \sin \theta'_c$, and $\cos \Omega_2 < 0$ for $\sin \theta'_c < \sin \theta < 1$. Here $\sin \theta'_c = \sqrt{1 + \epsilon^2}/\beta_n$, $\epsilon = a/r$.

We rewrite Eq. (8.33) in the form

$$\Omega = \Omega_i + \beta_n R_i/a, \tag{8.42}$$

where $\Omega = w_0 t - \phi$. Substituting Ω_i from (8.39)-(8.41), we obtain Ω as a function of the angle θ and of the radius r.

In r, θ, ϕ variables Eq.(8.42) realizes the singularity surface at the instant t of laboratory time. From the independence of the r.h.s. of (8.42) of the azimuthal angle ϕ and the invariance of its l.h.s. under the simultaneous change $t \to t + \delta t$, $\phi \to \phi + w_0 \delta t$ it follows that the singularity surface at the instant $t \to t + \delta t$ is obtained from that at the instant t by rotation of the latter through the angle $w_0 \delta t$. For r and t fixed, Eq.(8.42) defines the position of the singularity on the sphere of the radius r at the instant t of laboratory time.

For θ and ϕ fixed, Eq. (8.42) defines the radius of the sphere on which the singularity with angles θ, ϕ is located at the instant t of laboratory time. Since there are two values of Ω_i satisfying (8.37) (see Eqs.(8.38)-(8.41)), there are two such spheres.

The singular contour (8.42) exists only for $r \geq a/\beta_n$. On the sphere of the radius $r = a/\beta_n$ it contracts to one point

$$\theta = \frac{\pi}{2}, \quad w_0 t - \phi = \arccos \frac{1}{\beta_n} + \frac{1}{\gamma_n}.$$

On the sphere of radius a (along the equator of which the charge moves) two branches of the singular contour (8.42) are given by

$$w_0 t - \phi = \Omega_{1,2} + \beta_n R_{12}/a, \quad \cos \Omega_{1,2} = \frac{1}{\beta_n^2 \sin \theta} \pm \left(1 - \frac{2\beta_n^2 - 1}{\beta_n^4 \sin^2 \theta}\right)^{1/2}.$$

In particular, there are two singular points on the equator itself:

$$w_0 t - \phi = 0, \quad w_0 t - \phi = \arccos \left(\frac{2}{\beta_n^2} - 1\right) + \frac{2}{\gamma_n}.$$

The first of them coincides with the position of a moving charge. For $\beta_n \to 1$ both these points coincide with the position of the charge.

Consider particular cases.

i) Let $\beta_n \gg 1$ (this case is instructive, yet unrealistic since always $\beta < 1$). Eq.(8.38) then gives

$$\sin \theta_c \approx 0, \quad 0 < \theta < \pi, \quad \Omega_{12} = 0, \quad \text{or} \quad \pi.$$

This means that the singularity contour coincides with the meridians $\phi = \omega_0 t$ and $\phi = \omega_0 t - \pi$ lying on the observational sphere.

ii) Let the charge velocity coincide with the velocity of light in medium ($\beta_n = 1$). Then,

$$\sin \theta_c = \sin \theta = 1, \quad \cos \Omega_1 = \cos \Omega_2 = a/r.$$

Since $\cos \Omega_1 = \cos \Omega_2 > 0$ and $\sin \Omega_1 = \sin \Omega_2 < 0$, $\Omega_1 (= \Omega_2)$ lies in the fourth quadrant. Correspondingly, Eq. (8.42) takes the form

$$\theta = \frac{\pi}{2}, \quad \Omega = \omega_0 t - \phi = 2\pi - \arccos \frac{a}{r} + \sqrt{\frac{r^2}{a^2} - 1}, \qquad (8.43)$$

that is, the singularities of the electromagnetic potentials and field strengths degenerate into one point lying in the equatorial plane. If, in addition, $r \to \infty$, then

$$\theta = \frac{\pi}{2}, \quad \Omega = \omega_0 t - \phi = \frac{3}{2}\pi + \frac{r}{a}. \qquad (8.44)$$

Again, this equation may be interpreted in two ways. For r, t fixed, it defines the singularity position on the sphere of radius r at a fixed instant of laboratory time t:

$$\theta = \frac{\pi}{2}, \quad \phi = \omega_0 t - \frac{3}{2}\pi - \frac{r}{a}.$$

For θ, ϕ fixed this equation gives the radius of the sphere on which the singularity lies:

$$r = c_n t - a(\phi - 3\pi/2).$$

We observe that for $\beta_n = 1$ there is only one sphere on which the singularity lies.

It is essential that Eq. (8.33) has only an odd number of roots (see [1], pp. 83-87). In what follows we limit ourselves to the velocities $\beta_n \leq 2$. In this case Eq.(8.44) has one root for $\beta_n < 1$ and three roots for $1 < \beta_n \leq 2$.

8.3.4. DIGRESSION ON THE CHERENKOV RADIATION

As we have seen, for the charge velocity greater than the velocity of light in medium, SR has singularities on the observational sphere. It would be tempting to associate them with the Cherenkov cone attached to a moving charge.

But first we remember the main facts on the Cherenkov radiation. Let a point charge moves in a medium, along the z axis, with a velocity $v > c_n$. Then retarded times t' satisfy the equation

$$c_n(t - t') = R, \quad R = [\rho^2 + (z - vt')^2]^{1/2}, \quad \rho^2 = x^2 + y^2. \tag{8.45}$$

Two roots of this equation are given by (see,e.g., [30])

$$c_n t'_{1,2} = -\frac{c_n t - \beta_n z \pm r_m}{\beta_n^2 - 1}, \quad r_m = [(z - vt)^2 - (\beta_n^2 - 1)\rho^2]^{1/2}. \tag{8.46}$$

The singularities of EMF satisfy equation

$$R = \beta_n(z - vt'). \tag{8.47}$$

Now we proceed in the same way as for SR: excluding retarded time t' from equations (8.45) and (8.47) we obtain equation for the position of the EMF singularities at the fixed instant of laboratory time t:

$$\rho = \frac{vt - z}{\sqrt{\beta_n^2 - 1}}, \quad z < vt, \tag{8.48}$$

which coincides with the instantaneous position of the Cherenkov cone.

Now we turn back to the synchrotron motion. The following question arises:

Does the intersection of the instantaneous Cherenkov cone with the observational sphere of the radius r give SR singularities studied earlier in this section?

For definiteness let the charge at the laboratory time $t = 0$ be on the x axis, at a distance a from the origin, with its velocity directed along the y axis. Then Eq. (8.48) defining the instantaneous Cherenkov cone is reduced to

$$\frac{-y}{\sqrt{\beta_n^2 - 1}} = [(x - a)^2 + z^2]^{1/2}, \quad y < 0.$$

To find the singularity contour on the observational sphere of radius r, we insert into this equation $x = r \sin\theta \cos\phi$, $y = r \sin\theta \sin\phi$, $z = r \cos\theta$. This gives

$$-\sin\theta \sin\phi = [(\sin\theta \cos\phi - \epsilon)^2 + \cos^2\theta]^{1/2}\sqrt{\beta_n^2 - 1},$$

where θ and ϕ lie on the observational sphere.

Consider particular cases.

For $\beta_n \gg 1$ the Cherenkov cone degenerates into the singularity line lying behind the moving charge and intersecting the observational sphere at the point $\theta = \pi/2$, $\phi = 3\pi/2$. In contrast, the singularity contour of synchrotron radiation for $\beta_n \gg 1$ coincides with meridians $\phi = 0$ and $\phi = \pi$ of the observational sphere.

For $\beta_n \to 1$ the Cherenkov cone is almost perpendicular to the motion axis. The singularity contour on the observational sphere is defined as $\phi \approx 2\pi - \delta$, $\delta \ll 1$ and $0 < \theta < \pi$. On the other hand, the singularity contour of synchrotron radiation for $\beta_n \approx 1$ degenerates into one point lying in the equatorial plane (see (8.43)).

We see that, contrary to intuitive expectations, the singularity contours of synchrotron radiation on the observational sphere do not coincide with singularities of the instantaneous Cherenkov cone attached to a moving charge.

What are the reasons for this? The main differences between the Cherenkov and synchrotron motions should be mentioned:

i) For the rectilinear Cherenkov motion, there are always two retarded times for any charge velocity $v > c_n$. On the other hand, for the synchrotron motion the number of retarded times is always odd and increases with the increase of β_n.

ii) For the rectilinear Cherenkov motion there is no EMF outside the Cherenkov cone. On the other hand, the EMF of SR differs from zero everywhere, taking infinite values on the singularity contour (for $\beta_n > 1$).

iii) The frequency spectrum of Cherenkov radiation is continuous. In contrast, the frequency spectrum of SR is discrete: only those frequencies ω are emitted and observed which are integer multiples of ω_0 ($\omega = m\omega_0$).

These differences result in different spatial distributions of synchrotron and Cherenkov radiations.

8.3.5. ELECTROMAGNETIC FIELD IN THE WAVE ZONE

Electromagnetic field strengths
In the wave zone where $r \gg a$ one has

$$\Phi = \frac{e}{\epsilon r} \sum_i \frac{1}{|Z_i|}, \quad A_\phi = \frac{e\mu\beta}{r} \sum_i \frac{\cos\Omega_i}{|Z_i|}, \quad Z_i = 1 + \beta_n \sin\theta \sin\Omega_i,$$

$$A_\rho = -\frac{e\mu\beta}{r} \sum_i \frac{\sin\Omega_i}{|Z_i|}, \quad E_\phi = \frac{e\beta^2\mu}{a} \frac{1}{r} \sum_i \frac{\sin\Omega_i + \beta_n \sin\theta}{|Z_i|^3},$$

$$E_\theta = \frac{e\beta^2\mu \cos\theta}{a} \sum_i \frac{\cos\Omega_i}{|Z_i|^3}, \quad E_r = \frac{e}{\epsilon r^2} \sum_i \frac{1}{|Z_i|}, \tag{8.49}$$

$$H_\phi = \frac{e\beta\beta_n}{a} \frac{\cos\theta}{r} \sum_i \frac{\cos\Omega_i}{|Z_i|^3}, \qquad H_\theta = -\frac{e\beta\beta_n}{a} \frac{1}{r} \sum_i \frac{\sin\Omega_i + \beta_n\sin\theta}{|Z_i|^3},$$

$$H_r = \frac{e\beta}{r^3} \sum_i \frac{a - r\beta_n^2 \sin\theta\cos\Omega_i}{|Z_i|^3}, \qquad Z_i = 1 + \beta_n\sin\theta\sin\Omega_i,$$

$$S_r = \frac{\mu c e^2 \beta^4 n}{4\pi a^2 r^2} \left[\cos^2\theta \left(\sum_i \frac{\cos\Omega_i}{|Z_i|^3}\right)^2 + \left(\sum_i \frac{\sin\Omega_i + \beta_n\sin\theta}{|Z_i|^3}\right)^2\right]. \quad (8.50)$$

We see that at large distances E_r and H_r fall as $1/r^2$. Therefore their contribution to the energy flux were negligible if EMF strengths were not singular.

The singularities of electromagnetic field in the wave zone
We write out equations defining retarded times Ω_i in the wave zone

$$\Omega_r \equiv \omega_0 t - \phi - \beta_n r/a = \Omega_i - \beta_n \sin\theta\cos\Omega_i \quad (8.51)$$

and the position of singularities (this is valid only for $\beta_n > 1$).

$$1 + \beta_n\sin\theta\sin\Omega_i = 0. \quad (8.52)$$

Then

$$\sin\Omega_i = -\frac{1}{\beta_n\sin\theta}, \quad \cos\Omega_i = \pm\sqrt{1 - \left(\frac{1}{\beta_n\sin\theta}\right)^2}, \quad \sin\theta > \frac{1}{\beta_n}, \quad (8.53)$$

that is, Ω_i lies in fourth quadrant if the $+$ sign is chosen in (8.52) (branch 1) and in third quadrant for the $-$ sign there (branch 2).
Therefore

$$\Omega_1 = 2\pi - \arcsin\frac{1}{\beta_n\sin\theta} \quad \text{and} \quad \Omega_2 = \pi + \arcsin\frac{1}{\beta_n\sin\theta}.$$

Equation (8.51) is then separated into two parts

$$\Omega_r = 2\pi - \arcsin\frac{1}{\beta_n\sin\theta} - \sqrt{(\beta_n\sin\theta)^2 - 1},$$

$$\Omega_r = \pi + \arcsin\frac{1}{\beta_n\sin\theta} + \sqrt{(\beta_n\sin\theta)^2 - 1} \quad (8.54)$$

corresponding to two branches of the singularity contour on the sphere of radius r at the fixed instant of time t. This contour is closed since the branches 1 and 2 are intersected when $\sin\theta = 1/\beta_n$ which corresponds to the angles $\theta = \arcsin(1/\beta_n)$ and $\theta = \pi - \arcsin(1/\beta_n)$.

On the other hand, for θ, ϕ fixed Eq. (8.54) defines two radial distances on which the singularity lies. These distances increase with time.

In the equatorial plane ($\theta = \pi/2$), (8.54) is reduced to

$$\Omega_r = \pi + \arcsin \frac{1}{\beta_n} + \sqrt{\beta_n^2 - 1},$$

$$\Omega_r = 2\pi - \arcsin \frac{1}{\beta_n} - \sqrt{\beta_n^2 - 1}. \tag{8.55}$$

These equations define two azimuthal singularity points in plane $\theta = \pi/2$ (for the given r and t).

It follows from (8.54) that the singularity contour shifts as a whole on the surface of the sphere as the observational time t rises. This is not surprising because of the motion treated is periodic. Much more surprising is that this contour shifts as a whole when the observation is made on the neighbouring spheres. In fact, if we shift r in (8.54) by the amount $a\pi/\beta_n$ (which is small compared with r in the wave zone where $a \ll r$) then the singular contour shifts to the opposite side of the sphere ($\phi \to \phi - \pi$).

For the sake of clarity we consider in some detail the case $\beta_n \to 1$. The singularity contour at the given instant of time, on the sphere of a particular radius then shrinks to one point

$$\Omega_i^{(1)} = \Omega_i^{(2)} = \frac{3}{2}\pi, \quad \omega_0 t - \phi - r/a = \frac{3}{2}\pi, \quad \theta = \frac{\pi}{2}.$$

At the given instant of time t, the singularity curve S has a spiral form as a function of radius r:

$$\phi = \omega_0 t - \frac{r}{a} - \frac{3\pi}{2}. \tag{8.56}$$

As time goes, this curve rotates as a whole without changing its form.

Let an observer be at the point $P(\theta = \pi/2, \phi = \phi_0)$ on the sphere of radius r_0. At some instant t_0, when the singularity curve (8.56) reaches the observer, he detects an instantaneous flash of light. These intersections with the singularity curve, and therefore instantaneous flashes of light, will be repeated periodically with period $T = 2\pi/\omega_0$.

The question arises of how to observe the spiral form of the radiated energy flux. One should place two γ quanta detectors (say, D_1 and D_2) placed on the same singularity curve S and tuned on the coincidence. The signals from D_1 and D_2 will then reach the analysing device if the flux of SR is along the singularity curve S.

Three reservations should be mentioned. First, the distribution of SR along the above singularity curve S is valid in a uniform medium (say, gas). However, usually, SR is observed through the window in the body of the synchrotron. This may destroy the above picture. Second, SR flux everywhere differs from zero, taking a maximal value when the singularity curve passes through the detector. Therefore the detecting device should have some threshold. Third, to our best knowledge the typical detector registers the photons with definite energy $\mathcal{E} = \hbar m w_0$, not EMF energy fluxes (8.36) and (8.50) composed from the EMF strengths taken at the fixed instant of laboratory time t and containing the sum over the whole frequency spectrum. Our experience in dealing with Cherenkov radiation shows that spatial distributions of radiation in $\vec{r}, t$ and $\vec{r}, w$ representations may be quite different [31].

The same spiral-like behaviour of the radiation intensity holds when the charge velocity is less than the velocity of light in medium. In fact, according to (8.50) the dependence of the radial energy flux on r enters through the overall factor $1/r^2$ and through the phase Ω_i. If we shift r by an amount δr small compared with r, then in the wave zone all the changes reduce to the change of the phase factor Ω_i. According to (8.51) the dependence of Ω_i on r enters through $\Omega_r = wt - \phi - \beta_n r/a$, which is invariant under the simultaneous change

$$r \to r + \delta r, \quad \phi \to \phi - \delta\phi, \quad \delta\phi = \beta_n \delta r/a.$$

This means that the angular distributions of S_r on the spheres with radii r and $r + \delta r$ taken at the same laboratory time t will be the same except for the shift on the angle $\delta\phi$. Or, in other words, the change of the sphere radius leads to the azimuthal shifting of the radial flux distribution as a whole without changing its form.

Similarly, invariance of Ω_i under the simultaneous change

$$t \to t + \delta t, \quad \phi \to \phi + w_0 \delta t$$

leads to the rotation of the radiation flux distribution as a whole without changing its form.

The conservation of the angular dependence of S_r has no place in the near zone, where $r \sim a$. The reason is that for finite distances the dependence on r enters non-trivially in the definition of field strengths (see Eq.(8.35)).

An interesting question is: how are these spiral-like surfaces formed? The rotating charge emits photons with definite frequency $w = m w_0$ which propagate along straight lines. On the other hand, the direct solution of the Maxwell equations (without using the frequency representation) gives the EMF of a spiral-like structure for an uniformly rotating charge. Therefore, the superposition of Fourier components of the EMF should give an EMF having a spiral-like spatial structure at a fixed instant of laboratory time.

For the Cherenkov radiation, for which the exact analytic formulae are available, the transformation from the spectral components of the EMF (which differ from zero everywhere) into the EMF in the time representation (having the form of a Cherenkov cone) may be checked step by step [31,32]. At this instant, we have not succeeded in doing the same procedure for SR.

An important fact proved in [33] by direct calculation is that the total flux $r^2 \int S_r d\Omega$ does not depend on the radius r of the observational sphere. Although this is almost trivial (this follows from the continuity equation for the density of energy and momentum), the direct check is useful for controlling approximations.

The polarization components
Usually the radial flux S_r is separated in two parts corresponding to the so-called π and σ polarizations:

$$S_r = S_\pi + S_\sigma, \quad S_\pi = \frac{ce^2\mu\beta^4 n}{4\pi a^2 r^2} \cos^2\theta (\sum_i \frac{\cos\Omega_i}{|Z_i|^3})^2,$$

$$S_\sigma = \frac{ce^2\mu\beta^4 n}{4\pi a^2 r^2} (\sum_i \frac{\sin\Omega_i + \beta_n \sin\theta}{|Z_i|^3})^2. \tag{8.57}$$

S_π corresponds to $E_\phi = 0$, $E_\theta \neq 0$, whilst $E_\phi \neq 0$, $E_\theta = 0$ for S_σ.

We now consider the behaviour of S_π and S_σ in the wave zone, on a sphere of the radius r. Concerning the zeroes and singularities of polarizations it was known only up to now that the polarization S_π vanishes at $\theta = \pi/2$.

Disappearance of π polarization. The component S_π vanishes if either $\theta = \pi/2$ or $\cos\Omega_i = 0$.
  1) In the first case

$$S_\sigma = \frac{ce^2\mu\beta^4 n}{4\pi a^2 r^2} \left(\sum_i \frac{\sin\Omega_i + \beta_n}{|1 + \beta_n \sin\Omega_i|^3} \right)^2, \tag{8.58}$$

where, according to (8.51), Ω_i are found from the equation

$$\Omega_r = \Omega_i - \beta_n \cos\Omega_i. \tag{8.59}$$

These equations define S_σ in the plane $\theta = \pi/2$. In this plane, according to (8.58), S_σ disappears for

$$\sin\Omega_i = -\beta_n \tag{8.60}$$

which corresponds to

$$\Omega_r = 2\pi - \arcsin \beta_n - \beta_n\sqrt{1 - \beta_n^2}, \quad \Omega_r = \pi + \arcsin \beta_n + \beta_n\sqrt{1 - \beta_n^2}. \quad (8.61)$$

This means that both S_π and S_σ vanish in two points (8.61) lying in the plane $\theta = \pi/2$. This is possible only for $\beta_n < 1$.

When $\beta_n > 1$, S_σ, according to (8.58), has no zeroes, but is infinite for $\sin \Omega_i = -1/\beta_n$, which corresponds to the points

$$\Omega_r = 2\pi - \arcsin \frac{1}{\beta_n} - \sqrt{\beta_n^2 - 1}, \quad \Omega_r = \pi + \arcsin \frac{1}{\beta_n} + \sqrt{\beta_n^2 - 1} \quad (8.62)$$

lying in the $\theta = \pi/2$ plane.

2) In the second case S_π vanishes for

$$\cos \Omega_i = 0 \quad (8.63)$$

which corresponds to

$$\Omega_i = \pi/2 \quad \text{and} \quad \Omega_i = 3\pi/2. \quad (8.64)$$

For $\beta_n < 1$, this, according to (8.51), leads to

$$\Omega_r = \pi/2 \quad \text{and} \quad \Omega_r = 3\pi/2 \quad (8.65)$$

(under the modulus 2π). The disappearance of S_π for Ω_r given by (8.65) is rigorously valid only for $\beta_n < 1$. For $\beta_n > 1$ Eq. (8.51), with $\Omega_r = \pi/2$ or $\Omega_r = 3\pi/2$, may have solutions Ω_i different from $\Omega_i = \pi/2$ and $\Omega_i = 3\pi/2$. As the summation in S_π is performed over all roots of (8.51) it may not disappear for such values of Ω_r. However, since in all real media where SR can exist (gases) $\beta_n \approx 1$, the additional roots of (8.51) will be close to $\Omega_i = \pi/2$ and $\Omega_i = 3\pi/2$, respectively. Therefore for β_n only slightly greater than 1, S_π should have deep minima in the neighbourhood of $\Omega_r = \pi/2$ and $\Omega_r = 3\pi/2$.

We evaluate S_σ at the points (8.65) where S_π disappears for $\beta_n < 1$. They are given by

$$S_\sigma(\Omega_r = \frac{\pi}{2}) = \frac{ce^2\mu\beta^4 n}{4\pi a^2 r^2} \frac{1}{(1 + \beta_n \sin\theta)^4},$$

$$S_\sigma(\Omega_r = \frac{3\pi}{2}) = \frac{ce^2\mu\beta^4 n}{4\pi a^2 r^2} \frac{1}{(1 - \beta_n \sin\theta)^4} \quad (8.66)$$

These expressions are exactly valid for $\beta_n < 1$. It is seen that S_σ nowhere vanishes or takes infinite values. For $\theta = \pi/2$ it has minimum for $\Omega_r = \pi/2$ and a maximum for $\Omega_r = 3\pi/2$.

For $\beta_n > 1$, S_σ given by (8.66) should be supplemented at the points (8.65) by the terms corresponding to additional solutions of Eq. (8.51) (different from $\Omega_i = \pi/2$ and $\Omega_i = 3\pi/2$). In any case, for $\beta_n > 1$, S_σ nowhere vanishes or takes infinite values in the meridional $\Omega_r = \pi/2$ plane, and is infinite at $\sin\theta = 1/\beta_n$ in the $\Omega_r = 3\pi/2$ plane.

Disappearance of σ polarization.   According to (8.57) the polarization S_σ vanishes if $\sin\Omega_i = -\beta_n \sin\theta$, which defines two lines on the sphere surface:

$$\Omega_r^{(1)} = 2\pi - \arcsin(\beta_n \sin\theta) - \beta_n \sin\theta\sqrt{1 - \beta_n^2 \sin^2\theta},$$

$$\Omega_r^{(2)} = \pi + \arcsin(\beta_n \sin\theta) + \beta_n \sin\theta\sqrt{1 - \beta_n^2 \sin^2\theta}, \qquad (8.67)$$

where $0 < \theta < \pi$ for $\beta_n < 1$ and

$$0 < \theta < \arcsin(1/\beta_n) \quad \text{and} \quad \pi - \arcsin(1/\beta_n) < \theta < \pi$$

for $\beta_n > 1$. On these lines

$$S_\pi = \frac{ce^2 \mu \beta^4 n}{4\pi a^2 r^2} \cos^2\theta \frac{1}{(1 - \beta_n^2 \sin^2\theta)^4}. \qquad (8.68)$$

Again, these equations are exact only for $\beta_n < 1$. For $\beta_n > 1$, one should solve (8.51) with Ω_r given by (8.67) in its l.h.s. The additional solutions Ω_i of this equation will contribute to (8.68). In any case, S_σ will be infinite for $\sin\theta = 1/\beta_n$.

8.3.6. NUMERICAL RESULTS FOR SYNCHROTRON MOTION IN A MEDIUM

Singularity contours

In this section radii r and radial energy fluxes $r^2 S_r$ will be expressed in units of a and ce^2/a^2, respectively.

In Fig. 8.17, the singularity contours (8.54) are shown for $\beta_n = 2$; 1.1 (a), $\beta_n = 1.01$; 1.001 (b) and for $\beta_n = 1.000001$ (c). The calculations were made in the wave zone where Eqs.(8.51), (8.52), and (8.54) are valid. It is seen that in the (Ω_r, θ) plane the singularity contour shrinks to the point $(\Omega_r = 3\pi/2, \theta = \pi/2)$ for $\beta_n \to 1$. This coincides with the $\beta_n \to 1$ limit of Eq. (8.54).

The form of the singularity contours (8.42) for $\beta_n = 2$ and $\beta_n = 1.1$ on spheres of various radii is shown on Fig. 8.18. The minimal value of the

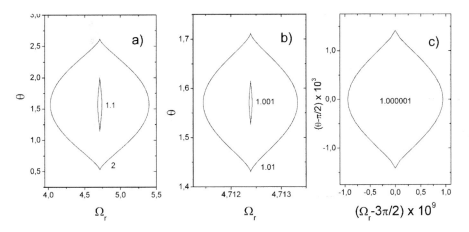

Figure 8.17. Angular positions of the singularity contour on the sphere for a number of charge velocities $v > c_n$. Singularity contours contract to the point $\theta = \pi/2$, $\Omega_r = 3\pi/2$ when $v \to c_n$. Numbers on contours are $\beta_n = v/c_n$.

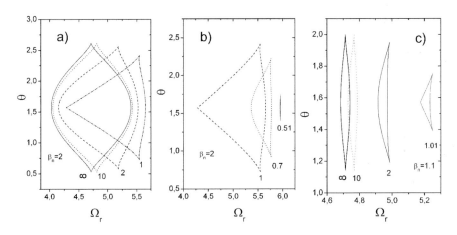

Figure 8.18. Angular positions of singularity contours for various radii of the observational sphere. The dimension of the singularity contour approaches zero when the sphere radius takes the minimal value $r = a/\beta_n$. In the wave zone the singularity contour is concentrated near $\Omega_r = 3\pi/2$ plane. Numbers on contours are r/a.

sphere radius for which the singularity contour still exists is $r = a/\beta_n$. For this value of r,

$$\theta = \theta_c = \pi/2, \quad \cos \Omega_1 = \cos \Omega_2 = 1/\beta_n,$$

$$\Omega_r^{(1)} = \Omega_r^{(2)} = 2\pi - 1 - \arccos(1/\beta_n) + \sqrt{\beta_n^2 - 1}.$$

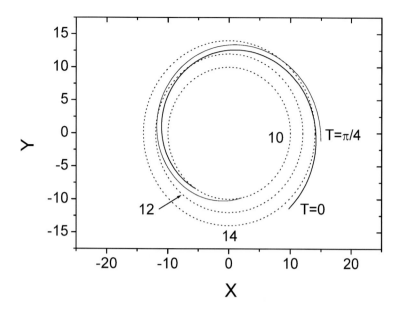

Figure 8.19. Spiral behaviour of singularity contours for the time instants $T = 0$ and $T = \pi/4$ and $\beta_n = 1.000001$. As time advances, the singularity contour rotates as a whole without changing its form. On a particular sphere (dotted curve) the singularity is at the place where it is intersected by the spiral contour. Numbers on dotted lines mean the sphere radius r/a.

For $\beta_n = 2$ this is approximately equal to 5.97. Figure 8.18 (b) confirms this.

The simultaneous (i.e., taken at the same instant of laboratory time t) spatial distributions of the singularity contour corresponding to $\theta = \pi/2$ and $\beta_n = 1.000001$ are shown in Fig. 8.19. They are of spiral structure. On the particular sphere (shown by a dotted line) the radiation intensity is infinite at the place where this sphere is intersected by a spiral surface (for the chosen β_n this surface is indistinguishable from the spiral curve, whilst the intersection region with a particular sphere reduces to a point). It is seen that the maximum of the radiation intensity occupies the different angular positions at different radii. It shifts as a whole as a function of time. The two spiral curves shown in Fig. 8.19 correspond to times $T = 0$ and $T = \pi/4$. Here $T = \omega_0 t$.

Polarization components

Consider now how the radial energy flux is distributed over the sphere surface in the wave zone. Concrete calculations were made with dimensionless

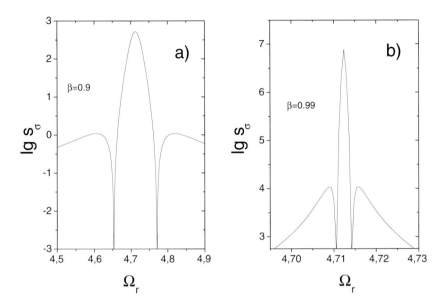

Figure 8.20. Space distribution of the polarization S_σ in the plane $\theta = \pi/2$ (where $S_\pi = 0$) for $n = 1.00001$, which corresponds to $\beta_n < 1$. In this plane S_σ vanishes at Ω_r given by (8.61). It is concentrated near the plane $\Omega_r = 3\pi/2$ for $\beta_n \to 1$.

intensities

$$S_r = S_\pi + S_\sigma, \quad S_\pi = \frac{\mu \beta^4 n}{4\pi} \cos^2 \theta \left(\sum_i \frac{\cos \Omega_i}{|Z_i|^3} \right)^2,$$

$$S_\sigma = \frac{\mu \beta^4 n}{4\pi} \left(\sum_i \frac{\sin \Omega_i + \beta_n \sin \theta}{|Z_i|^3} \right)^2 \tag{8.69}$$

which are obtained from the intensities S_π, S_σ and S_r given by (8.57) by multiplying them by the factor $r^2 a^2/ce^2$. We consider only the dielectric medium ($\mu = 1$).

In Fig. 8.20, for $\beta_n < 1$, there is shown the dependence of S_σ polarization on the angle Ω_r in the equatorial plane $\theta = \pi/2$, where $S_\pi = 0$. This figure illustrates Eq. (8.61) according to which: i) the polarization S_σ disappears for Ω_r given by (8.61) and ii) it is concentrated near the plane $\Omega_r = 3\pi/2$ as β_n approaches 1.

In Fig. 8.21 the same dependence of the polarization S_σ on the angle Ω_r is shown for $\beta_n > 1$. This figure shows that the polarization S_σ is infinite

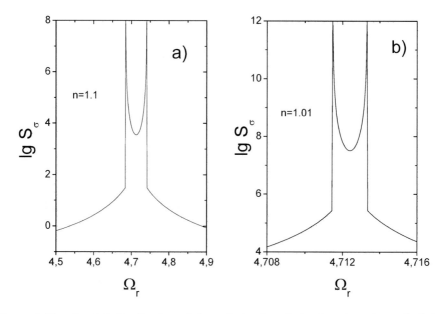

Figure 8.21.   Spatial distribution of the polarization S_σ in the plane $\theta = \pi/2$ (where $S_\pi = 0$) for $\beta = 0.999991$ which corresponds to $\beta_n > 1$. S_σ is infinite at Ω_r given by (8.62) and is concentrated near the plane $\Omega_r = 3\pi/2$ for $\beta_n \to 1$.

for Ω_r given by (8.62) and that it is concentrated near $\Omega_r = 3\pi/2$ plane as β_n approaches 1.

Fig 8.22 (a) illustrates that the polarization S_π, for the fixed $\sin\theta = 0.95$ and $\beta_n < 1$, rigorously disappears for $\Omega_r = \pi/2$ and $\Omega_r = 3\pi/2$. Part (b) of the same figure shows that for $\beta_n > 1$ the polarization S_π has deep a minimum at the same Ω_r. The singularities of S_π are at Ω_r given by (8.54) where one should put $\sin = 0.95$. This gives $\Omega_r^1 \approx 4.7$ and $\Omega_r^2 \approx 4.72$. This illustrates Fig. 8.22 (c), where the behaviour of S_π in the neighbouhood of $\Omega_r = 3\pi/2$ is presented.

The dependence of S_σ on the polar angle θ in the meridional plane $\Omega_r = 3\pi/2$ (where $S_\pi = 0$) is shown in Fig. 8.23 (a) for $\beta_n < 1$. In accordance with the second equation (8.66) S_σ has a maximum at $\theta = \pi/2$. Its behaviour in the plane $\Omega_r = \pi/2$ (where S_π also vanishes) is shown in Fig. 8.23 (b). From the first equation (8.66) it follows that S_σ has a minimum at $\theta = \pi/2$.

For $\beta_n > 1$ the dependence of S_σ on θ in the plane $\Omega_r = 3\pi/2$ is shown in Fig. 8.24 (a). The second equation (8.66) tells us that S_σ is infinite at $\sin\theta = 1/\beta_n$, which corresponds to $\theta \approx 1.14$ rad and $\theta \approx 2$ rad. The dependence of S_σ on θ in the meridional plane $\Omega_r = \pi/2$ is shown in

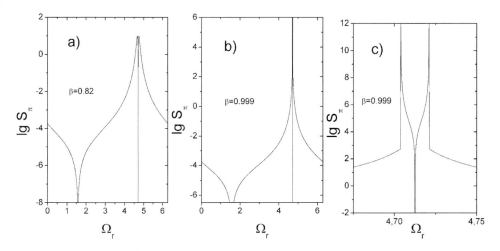

Figure 8.22. Spatial distribution of the polarization S_π as a function of the azimuthal angle Ω_r for $\sin\theta = 0.95$ and $n = 1.1$; (a): For $\beta_n < 1$, $S_\pi = 0$ at $\Omega_r = \pi/2$ and $\Omega_r = 3\pi/2$; (b): For $\beta_n > 1$, S_π has deep minima at $\Omega_r = \pi/2$ and $\Omega_r = 3\pi/2$ and infinities at Ω_r given by (8.54); (c): The behaviour of S_π polarization near $\Omega_r = 3\pi/2$ plane for $\beta_n > 1$.

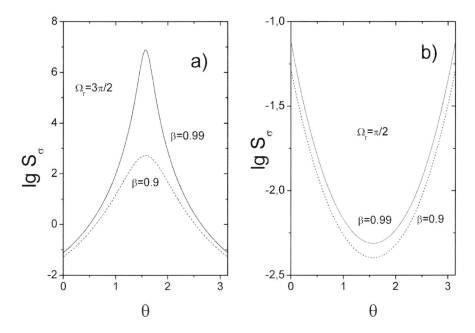

Figure 8.23. The θ-dependence of the polarization S_σ in the meridional planes $\Omega_r = \pi/2$ and $\Omega_r = 3\pi/2$ and $n = 1.00001$ for the case $v < c_n$. In both these planes $S_\pi = 0$. It is seen that S_σ is everywhere finite. For $\theta = \pi/2$ it has a maximum in the $\Omega_r = 3\pi/2$ plane and minimum in the $\Omega_r = \pi/2$ plane.

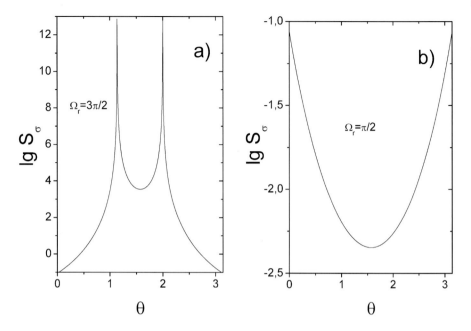

Figure 8.24. The θ-dependence of the polarization S_σ in the $\Omega_r = \pi/2$ and $\Omega_r = 3\pi/2$ meridional planes for $n = 1.1$ and $\beta = 0.999991$, which corresponds to $v > c_n$. S_σ takes infinite values only in the $\Omega_r = 3\pi/2$ plane at $\sin\theta = 1/\beta_n$. In the plane $\Omega_r = \pi/2$, S_σ has a minimum at $\theta = \pi/2$.

Fig. 8.24 (b). In agreement with the first Eq.(8.66), S_σ has a minimum at $\theta = \pi/2$. The absence of singularities of S_σ in the plane $\Omega_r = \pi/2$ means that they are located in other meridional planes. Fig. 8.18 (c) shows that in the wave zone the singularities of S_σ lie in the meridional plane $\Omega_r = 3\pi/2$. The contours on which S_σ vanishes are shown in Fig. 8.25. The solid and dotted lines correspond to $\beta_n < 1$ and $\beta_n > 1$, respectively. The singularities of the intensity of the SR for $\beta_n > 1$ are located in the region $-\arcsin(1/\beta_n) < \theta < \arcsin(1/\beta_n)$, in the neighbourhood of $\Omega_r = 3\pi/2$. For the case $\beta_n = 1.1$, the same as in Fig. 8.25, the singularity contour is presented in Fig. 8.17 (a).

Intensity of synchrotron radiation at finite distances
Up to now we have considered the radial flux distribution in the wave zone (except for Figs. 8.17 and 8.18). However, the typical radii of synchrotron orbits vary from a few to a hundred meters for electron synchrotrons and from hundred meters to 1 kilometer for the proton synchrotrons. In view of such large radii of synchrotron orbit the measurement of SR in the wave zone is very problematic.

When considering SR intensities and the position of singularities in the

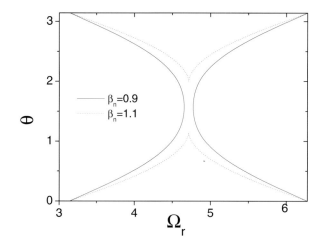

Figure 8.25.   The contours on the sphere surface where S_σ vanishes for $\beta_n < 1$ (solid curve) and $\beta_n > 1$ (dotted curve). The singularity contour lies inside the 'hole' formed by four contours of zeroes. Its position is shown in Fig. 8.17 (a).

wave zone, a suitable combination of variables was $\Omega_r = \omega_0 t - \phi - \beta_n r/a$ (see Eq. (8.51)). However, Eq. (8.42), valid at arbitrary distances, contains $\Omega = \omega_0 t - \phi$. To reconcile the choice of the variables in the wave zone and at arbitrary distances, we rewrite (8.42) in the following equivalent form

$$\Omega_r = \omega_0 t - \phi - \beta_n r/a = \Omega_i + \beta_n(R_i - r)/a.$$

This equation together with (8.39)-(8.41) defines the position of singularities in (Ω_r, θ, r) variables at finite distances.

To see how the radial flux distributions change with a decreasing radius of the observational sphere, we consider the dimensionless radial flux distributions

$$\frac{r^2 a^2}{ce^2} S_r,$$

where S_r is given by (8.36).

In Fig. 8.26, for $\beta_n < 1$ there are shown instantaneous S_σ intensities in the equatorial plane $\theta = \pi/2$ on spheres of various radii r. In particular, Fig. 8.26 (a) illustrates that intensities S_σ have almost the same height, but their positions in the equatorial plane change with r, tending to $\Omega_r = 3\pi/2$ in the wave zone. Fig. 8.26 (c) shows that the intensity S_σ for $r/a = 100$ is shifted relative to the intensity in the wave zone approximately on 0.1 rad.. The intensity S_σ in the nearest vicinity of the charge trajectory ($r/a = 1.01$) is shown in Fig. 8.26 (c). It is seen that in the near zone the radial energy flux may be negative in some angular region. The same takes place in vacuum (see Fig. 8.9 (b)).

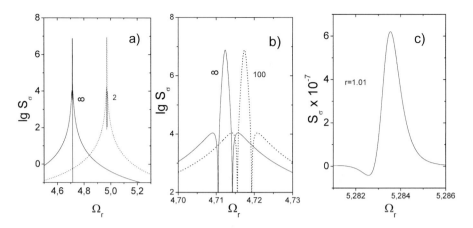

Figure 8.26. The behaviour of the polarization S_σ in the equatorial plane $\theta = \pi/2$ for $n = 1.00001$ and $\beta = 0.99$ (which corresponds to $\beta_n < 1$) on the spheres of various radii. (a): As r increases S_σ shifts as a whole, concentrating around the plane $\Omega_r = 3\pi/2$ in the wave zone; (b): S_σ polarization for $r/a = 100$ is shifted relative to that in the wave zone by 0.1 rad.; (c): In the neighbourhood of the charge orbit S_σ may be negative in some angular region. Numbers on curves are r/a.

For $\beta_n > 1$, the instantaneous intensities S_σ in the equatorial plane $\theta = \pi/2$ on spheres of various radii r are presented in Fig. 8.27. Similarly to $\beta_n < 1$, the position of the intensities S_σ tend to $\Omega_r = 3\pi/2$ for $r \to \infty$.

The instantaneous intensities S_σ in the meridional plane $\Omega_r = 3\pi/2$ (where S_π disappears) are shown for $\beta_n < 1$ in Fig. 8.28. It is seen that these intensities are concentrated near the equatorial plane $\theta = \pi/2$ as one approaches the wave zone.

The same S_σ intensities in the meridional plane $\Omega_r = 3\pi/2$ are presented for $\beta_n > 1$ in Fig. 8.29. One may observe that for large radii ($r/a = 100, \infty$) there are singularities of the intensity of SR, whilst for smaller radii ($r/a = 1.01, 2, 10$) they disappear. To see the reason for their absence turn to Fig. 8.18 (c) in which the position of singularities for $\beta_n \approx 1.1$ and radii the same as in Fig. 8.29 are presented. We see that, indeed, for $r/a = 1.01$, 2, and 10 the singularity contours do not intersect the meridional plane $\Omega_r = 3\pi/2$.

It follows from Fig. 8.18 that the 'focusing effect' the existence of which was claimed for $\beta_n < 1$ in [16] takes place also for $\beta_n > 1$: the angular region of θ occupied by SR intensities, being zero for $r = a/\beta_n$, increase with the increase of r up to some value of r_0. Further increasing of r does not change the dimension of the θ singularity interval. For the case $\beta_n = 1.1$ shown in Fig. 8.18 (c) this takes place for $r/a \approx 2$.

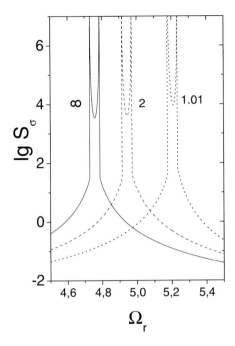

Figure 8.27.   Behaviour of the polarization S_σ in the plane $\theta = \pi/2$ for $n = 1.1$ and $\beta = 0.9999991$ (which corresponds to $\beta_n > 1$) on spheres of various radii. When r changes S_σ shifts as a whole without changing its form. Numbers on curves are r/a.

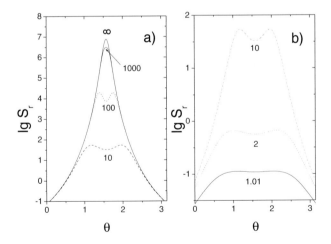

Figure 8.28.   The θ-dependence of the radial energy flux in the meridional plane $\Omega_r = 3\pi/2$ for $n = 1.00001$ and $\beta = 0.99$ (which corresponds to $\beta_n < 1$) on the spheres of various radii. Numbers on curves are r/a.

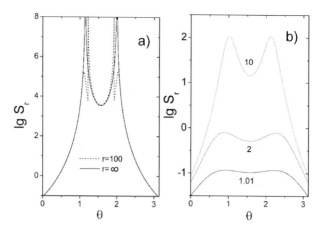

Figure 8.29. The θ-dependence of the radial energy flux in the meridional plane $\Omega_r = 3\pi/2$ for $n = 1.1$ and $\beta = 0.999991$ (which corresponds to $\beta_n > 1$) on spheres of various radii. Numbers on curves are r/a.

8.4. Conclusion

We briefly summarize the main results obtained in this Chapter.

For the synchrotron motion in vacuum:

1. We have evaluated radial, azimuthal and polar EMF fluxes averaged over the motion period. The calculations have been performed for arbitrary velocities and distances, for the observational points lying both inside and outside the charge orbit. It turns out that azimuthal energy flux is much larger than the radial flux near the charge orbit and is much smaller than the radial flux at large distances. This reconciles Schwinger's and Schott's approaches.

2. The instantaneous radial and azimuthal EMF fluxes were evaluated for various distances and charge velocities. They have a number of unexpected properties. In particular, they may acquire negative values in some angular regions. However, their time averaged values are always positive. Analytical expressions are obtained for the instantaneous positions of minima and maxima of field strengths. They generalize the famous Schwinger formula for arbitrary distances and velocities.

Another interesting observation was made in [34]. The angular distribution of the energy radiated for the period of motion in the radial direction is given by (8.18). To find its maximal value we find where the θ derivative of (8.18) vanishes. It turns out that the maximum of (8.18) is reached at $\theta = 0$ for $0 < \beta < \beta_1$, at $\theta = \pi/2$ for $\beta_2 < \beta < 1$, and for θ defined by

$$\sin^2 \theta = \frac{2}{3\beta^2(1 + 3\beta^2)}[\sqrt{15}(2 + 4\beta^2 + 9\beta^2)^{1/2} - 6 - 3\beta^2]^{1/2}$$

for $\beta_1 < \beta < \beta_2$. Here

$$\beta_1 = \frac{1}{\sqrt{7}}, \quad \beta_2 = \sqrt{\frac{2}{3}}(\sqrt{6} - 2)^{1/2}.$$

Therefore there is a smooth transition of the position of the maximum of radiation from $\theta = 0$ for $0 < \beta < \beta_1$ to $\theta = \pi/2$ for $\beta_2 < \beta < 1$.

For the charge motion in a medium:

1. The space-time distributions of the intensity of synchrotron radiation are obtained for the cases in which the charge velocity v is greater or smaller than the velocity of light in medium. It has been shown that at any fixed instant of laboratory time, the distribution of SR intensity has a spiral structure which rotates as a whole without changing its form. The experiment is proposed to test its existence.

2. For $v > c_n$ it has been found that the singularities of EMF for the synchrotron radiation differ drastically from the singularities of the instantaneous Cherenkov cone attached to a rotating charge.

3. Space-time distributions of different components of polarizations have been studied both for $v < c_n$ and $v > c_n$. Spatial regions in which they vanish and in which they are infinite are determined.

4. The intensity of synchrotron radiation has studied both in far and near zones. The dependence of the radiated energy flux distribution on the observational distance has been also studied.

References

1. Schott G.A. (1912) *Electromagnetic Radiation*, Cambridge Univ.Press, Cambridge.
2. Sokolov A.A. and Ternov I.M. (1974) *The Relativistic Electron* Nauka, Moscow (in Russian).
3. Ternov I.M., Mikhailin V.V. and Khalilov V.R. (1985) *Synchrotron radiation and its applications*, Moscow Univ. Publ., Moscow, (in Russian).
4. Ternov I.M. and Mikhailin V.V. (1986) *Synchrotron radiation. Theory and experiment.*, Energoatomizdat, Moscow, (in Russian).
5. Bordovitsyn V.A. (Ed.) (1999) *Synchrotron radiation theory and its developments. In memory of I.M. Ternov*, World Scientific, Singapore,.
6. Ternov I.M. (1995) Synchrotron Radiation *Usp. Fiz. Nauk*, **165**, pp. 429-456.
7. Schwinger J. (1949) On the Classical Radiation of Accelerated Electrons *Phys.Rev.*,**A 75**, pp. 1912-1925.
8. Bagrov V.G., (1965) Indicatrix of the Charge Radiation External Field According to Classical Theory it Optika i Spectroscopija, **28, No 4**, pp. 541-544, (In Russian).
9. Sokolov A.A., Ternov I.M. and Bagrov V.G., 1966, Classical theory of synchrotron radiation, in: Synchrotron Radiation (Eds.:Sokolov A.A. and Ternov I.M.),pp. 18-71 (Moscow, Nauka), in Russian.
10. Smolyakov N.V. (1998) Wave-Optical Properties of Synchrotron Radiation *Nucl. Instr. and Methods*,**A 405**, pp. 235-238.
11. Schwinger J., Tsai W.Y. and Erber T. (1976) Classical and Quantum Theory of Synergic Synchrotron-Cherenkov Radiation, *Ann. of Phys.*, **96**, pp.303-352.
12. Erber T., White D., Tsai W.Y. and Latal H.G. (1976) Experimental Aspects of Synchrotron-Cherenkov Radiation, *Ann. of Phys.*, **102**, pp. 405-447.

13. Villaroel D. and Fuenzalida V. (1987) A Study of Synchrotron Radiation near the Orbit, *J.Phys. A: Mathematical and General*, **20**, pp. 1387-1400.
14. Villaroel D. (1987) Focusing Effect in Synchrotron Radiation, *Phys.Rev.*, A **36**, pp. 2980-2983.
15. Villaroel D. and Milan C. (1987) Synchrotron Radiation along the Radial Direction *Phys.Rev.*, **D38**, pp. 383-390.
16. Ovchinnikov S.G. (1999) Application of Synchrotron Radiation to the Study of Magnetic Materials *Usp. Fiz. Nauk*, **169**, pp. 869-887.
17. Jackson J.D., (1975) *Classical Electrodynamics*, New York, Wiley.
18. Ryabov B.P. (1994) Jovian S emission: Model of Radiation Source *J. Geophys. Res.*, **99, No E4**, pp. 8441-8449.
19. Synchrotron Radiation (1979) (Kunz C.,edit.), Springer, Berlin.
20. (1995) *Nuclear Instr. & Methods*, **A 359**, No 1-2.
21. (1998) *Nuclear Instr. & Methods*, **A 405**, No 2-3 .
22. L.D. Landau and E.M. Lifshitz (1971) *The Classical Theory of Fields*, Reading, Massachusetts, Pergamon, Oxford and Addison-Wesley.
23. Risley J.S., Westerveld W.B. and Peace J.R. (1982) Synchrotron Radiation at Close Distances to the orbital ring *J. Opt. Soc. Am.*, **72**, pp. 943-946.
24. Bagrov V.G., Optics and Spectroscopy, 28, No 4, 541 (1965), In Russian.
25. Sokolov A.A., Ternov I.M. and Bagrov V.G. (1966) Classical theory of synchrotron radiation, in: *Synchrotron Radiation* (Eds.:Sokolov A.A. and Ternov I.M.), pp. 18-71, Nauka, Moscow, in Russian.
26. Tomboulian D.H. and Hartman P.L. (1956) Spectral and Angular Distribution of Ultraviolet Radiation from the 300-Mev Cornell Synchrotron *Phys. Rev.*, **102**, pp. 1423-1447.
27. Hillier R. (1984) *Gamma Ray Astronomy*, Clarendon Press, Oxford.
28. Stecker F.W. (1971) *Cosmic Gamma Rays*, Momo Book Corp., Baltimore.
29. Afanasiev G.N., Eliseev S.M. and Stepanovsky Yu.P. (1998) Transition of the Light Velocity in the Vavilov-Cherenkov Effect *Proc. Roy. Soc. London*, **A 454**, pp. 1049-1072;
    G.N. Afanasiev and V.G. Kartavenko (1999) *Cherenkov-like shock waves associated with surpassing the light velocity barrier Canadian J. Phys.*, **77**, pp. 561-569.
30. Afanasiev G.N., Beshtoev Kh. and Stepanovsky Yu.P. (1996) Vavilov-Cherenkov Radiation in a Finite Region of Space *Helv. Phys. Acta*, **69**, pp. 111-129.
31. Afanasiev G.N., Kartavenko V.G. and Stepanovsky Yu.P. (1999) On Tamm's Problem in the Vavilov-Cherenkov Radiation Theory *J.Phys. D: Applied Physics*, **32**, pp. 2029-2043;
32. Afanasiev G.N., Kartavenko V.G. and Magar E.N. (1999) Vavilov-Cherenkov Radiation in Dispersive Medium *Physica*, **B 269**, pp. 95-113; Afanasiev G.N., Eliseev S.M and Stepanovsky Yu.P. (1999) Semi-Analytic Treatment of the Vavilov-Cherenkov Radiation *Physica Scripta*, **60**, pp. 535-546.
33. Rivera R. and Villaroel D., (2000) Synchrotron Radiation and Symmetries *Am.J.Phys.*, **68**, pp. 41-48.
34. Bagrov et al., (2002) New results in the classical theory of synchrotron radiation in *Proc. XIV Russian Conf. on the Application of Synchrotron Radiation* pp.8-17, Novosibirsk.

CHAPTER 9

SOME EXPERIMENTAL TRENDS IN THE
VAVILOV-CHERENKOV RADIATION THEORY

9.1. Fine structure of the Vavilov-Cherenkov radiation

The classical Tamm-Frank theory [1] explaining the main properties of the Vavilov-Cherenkov (VC) effect [2,3] is based on the assertion that a charge moving uniformly in medium with a velocity v greater than the velocity of light c_n in medium radiates spherical waves from each point of its trajectory [4]. The envelope to these spherical waves propagating with the velocity c_n is the Cherenkov cone with its normal inclined at the angle θ_c towards the motion axis. Here $\cos\theta_c = 1/\beta_n$, $\beta_n = \beta n$, $\beta = v/c$, $c_n = c/n$ (c is the velocity of light in vacuum and n is the medium refractive index).

The radiation of a charge moving uniformly in a finite spatial interval inside the medium is usually studied in the framework of the so-called Tamm problem [5]. Under certain approximations Tamm obtained a remarkably simple formula which is frequently used by experimentalists for the identification of the charge velocity (see, e.g., [6]).

Ruzicka and Zrelov [7] when analyzing the angular spectrum of the radiation arising in the Tamm problem came to paradoxical result that this spectrum can be interpreted as an interference of two BS shock waves arising at the beginning and the end of the charge motion. There was no place for the Cherenkov radiation in their analysis based on the use of the Tamm approximate formula.

Tamm himself thought that his formula describes both the Cherenkov radiation and bremsstrahlung.

To resolve this controversy, the exact solution of the Tamm problem was obtained in [8] (in the time representation, for a dispersion-free medium). Its properties were investigated in some detail in [9,10]. It was shown there that side by side with BS shock waves the Cherenkov shock wave (CSW, for short) exists. The results obtained in [8-10] resolve the mentioned above inconsistency between [5] and [7] in the following way: although the Tamm problem describes both the Cherenkov radiation and bremsstrahlung, its approximate version (i.e., the Tamm formula) does not describe the CSW properly.

According to [8-10], when a charge moves in the interval $(-z_0, z_0)$ the CSW is enclosed between the moving charge and the straight line L_1 originating from the point $-z_0$ corresponding to the beginning of motion and inclined at the angle θ_c towards the motion axis (see Chapter 2). The CSW is perpendicular to L_1. When a charge stops at an instant t_0 the CSW detaches from it and propagates between the L_1 and the straight line L_2 originating from the z_0 point corresponding to the end of the motion and inclined at the same angle θ_c towards the motion axis. The positions of the shock waves BS_1, BS_2 and the CSW at a fixed instant of time are shown in Fig. 9.1(a). For an arbitrary instant of time $t > t_0$, the CSW is always tangential to both shock waves BS_1 and BS_2 and is perpendicular both to L_1 and L_2. The length of the CSW (coinciding with the distance be-

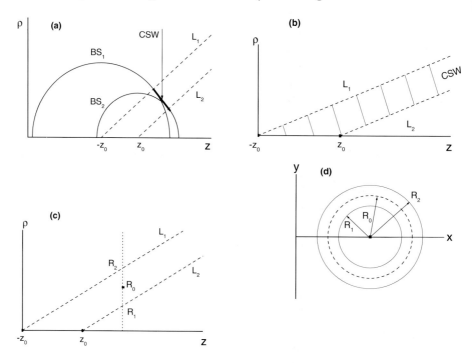

Figure 9.1. (a): The position of the Cherenkov shock wave (CSW) and the bremsstrahlung shock waves arising at the beginning (BS_1) and the end (BS_2) of the charge motion at a fixed instant of time. The CSW is enclosed between straight lines L_1 and L_2 originating from the points corresponding to the boundaries of the motion interval; (b): The propagation of the CSW between the straight lines L_1 and L_2; (c): In an arbitrary z =const plane perpendicular to the motion axis, the CSW, in the ϕ =const plane, cuts off a segment of the same length $R_2 - R_1$ for any z; (d) Because of the axial symmetry of the problem, the CSW in the z =const plane, cuts off a ring with internal and external radii R_1 and R_2, respectively. The width $R_2 - R_1$ of the Cherenkov ring and the energy released in it do not depend on the position z of the observational plane.

tween L_1 and L_2) is $L/(\beta_n\gamma_n)$, where $L = 2z_0$ is the motion interval and $\gamma_n = 1/\sqrt{|1 - \beta_n^2|}$. As time advances, the CSW propagates between L_1 and

L_2 with the velocity of light in medium (Fig. 9.1 (b)). The shock waves BS_1 and BS_2 are not shown in this figure.

In the spectral representation (since transition to it involves the time integration) one obtains spatial regions lying to the left of L_1 and to the right of L_2 to which the BS_1 and BS_2 shock waves are confined, and the spatial region between L_1 and L_2 to which BS_1, BS_2 and the CSW are confined. Let the measurements of the radiation intensity be made in the plane perpendicular to the motion axis z. The CSW then cuts out in each of the $z =$const planes the segment with its length $\delta\rho = L/\gamma_n$ independent of z and with its center at $R_0 = z/\gamma_n$ (Fig. 9.1 (c)). This picture refers to a particular $\phi =$const plane (ϕ is the angle in the $z =$const plane).

Since the problem treated is the axially symmetric problem, the intersection of the CSW with the $z =$const plane looks like a ring with minor and major radii equal to $R_1 = R_0 - L/2\gamma_n$ and $R_2 = R_0 + L/2\gamma_n$, respectively (Fig. 9.1 (d)). This qualitative consideration implies only the possible existence of a Cherenkov ring of the finite width. To find the distribution of the radiation intensity within and outside it, numerical calculations are needed.

When the ratio of the motion interval to the observed wavelength is very large (this is usual in Cherenkov-like experiments) the Tamm formula has a sharp delta function peak within the Cherenkov ring. Owing to this it cannot describe a quite uniform distribution of the radiation intensity inside the Cherenkov ring.

It should be mentioned that by the 'shock waves' used throughout this Chapter we do not mean the usual shock waves used, e.g., in acoustics or hydrodynamics where they are the solutions of essentially nonlinear equations. The Maxwell equations describing the charge motion in medium are linear, yet they can have solutions (when the charge velocity is greater than the velocity of light in medium) with properties very similar to the true shock waves. For example, there is no electromagnetic field outside the Cherenkov cone, an infinite electromagnetic field on its surface, and a quite smooth field inside the Cherenkov cone. The analog of the Cherenkov cone in acoustics is the Mach cone.

We see that due to the approximations involved, an important physics has dropped out from the consideration. It is our goal to analyze the experimental and theoretical aspects of this new physics. For this we obtain the exact (numerical) and approximate (analytical) theoretical radiation intensities describing a charge motion in finite spatial interval and compare them with existing experimental data. Theoretical intensities (exact and analytical) predict the existence of the CSW of finite extension manifesting as a plateau in the radiation intensity and of the BS shock wave manifesting as the intensity bursts at the ends of this plateau. It turns out that the theoretical (numerical and analytical) and experimental intensities are in satisfactory agreement with each other, but disagree sharply with the Tamm formula.

The observation of the above shock waves encounters certain difficulties when the focusing devices are used which collect radiation from the part of the charge trajectory lying inside the radiator into the single ring, thus projecting the VC radiation and bremsstrahlung into the same place. The typical experimental setup with a lens radiator and the corresponding Cherenkov ring are shown in Fig. 9.2. In its left part 1 means the proton

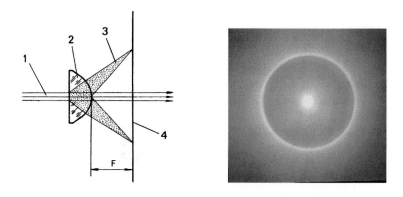

Figure 9.2. Left: The scheme of an experiment with a lens radiator; 1 is the proton beam, 2 is the lens radiator, 3 is the focused VC radiation, 4 is the plane photographic film placed perpendicular to the motion axis, F is the focal distance for paraxial rays; Right: the black and white photographic print from the photographic film shown on the left.

beam with the energy 657 MeV and diameter 0.5 cm, 2 is the lens radiator with refractive index 1.512 and the focal distance 2.27 cm (for paraxial rays), 3 is the focused VC radiation ($\theta_{Ch} = 35.17^0$), 4 is a plane photographic film (18 × 24 cm). On the right side there is a black and white photographic print of the photographic film shown on the left. It has the form of a narrow ring.

To see how the VC radiation and bremsstrahlung are distributed in space we turn to experiments in which the VC radiation was observed without using the focusing devices. These successful (although qualitative) experiments were performed by V.P. Zrelov (unpublished) in 1962 when preparing illustrations for the monograph [11] devoted to the VC radiation and its applications. We have processed these experimental data. The results are presented in the next section.

One may wonder why we apply the recently developed theoretical methods for the description of rather ancient experiments. The reason is that these experiments are the only ones in which the Cherenkov radiation was studied with rather thick dielectric samples, without using the special focusing devices.

9.1.1. SIMPLE EXPERIMENTS WITH 657 MEV PROTONS

The first 1962 experiment
The 657 MeV ($\beta = 0.80875$) proton beam of the phasotron in the JINR Laboratory of Nuclear Problems was used. The experimental setup is shown in Fig. 9.3. The collimated proton beam (1) of diameter 0.5 cm was directed

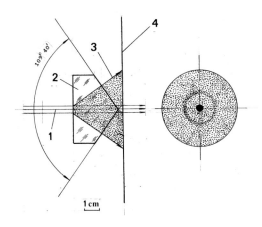

Figure 9.3. The experimental setup of the experiment discussed (Zrelov 1962). The proton beam (1) passing through the conical plexiglass radiator (2) induces the VC radiation (3, shaded region) propagating in the direction perpendicular to the cone surface. The observations are made in the plane photographic film (4) placed perpendicular to the motion axis.

to the conical polished plexiglass radiator (2) ($n = 1.505$ for $\lambda = 4 \times 10^{-5}$ cm). The apex angle of 109.7^0 of the cone enabled the VC radiation (3) to go out from the radiator in a direction perpendicular to the cone surface. The radiation was detected by the plane colour 18×24 cm photographic film placed perpendicular to the beam at a distance of 0.3 cm from the cone apex. Nearly 10^{12} protons passed through the conical radiator. The black and white photographic print and the corresponding photometric curve (from which the beam background was subtracted) are shown in the left and right parts of Fig. 9.4, respectively. The photometric curve

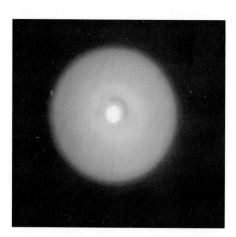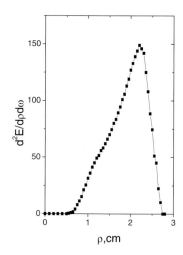

Figure 9.4. Left: The black and white photographic print from the photographic film shown in Fig.9.3; Right: The photometric curve corresponding to the left part. One observes the increment of the radiation intensity at $\rho \approx 2.25$ cm which corresponds to the Cherenkov ray emitted from the point where the proton beam enters the radiator.

describes the distribution $d\mathcal{E}(\rho)/d\rho$ of the energy released inside the ring of finite width. More accurately, $d\rho \cdot d\mathcal{E}(\rho)/d\rho$ is the energy released in an elementary ring with minor and major radii ρ and $\rho + d\rho$, respectively. It is seen from this figure that the increment of the radiation intensity takes place at a radius $\rho = 2.25$ cm corresponding to the radiation emitted at the Cherenkov angle θ_c from the boundary point where the charge enters the radiator.

The second 1962 experiment

In another experiment performed in the same year 1962 the maxima of the radiation intensity corresponding to the radiation from the boundary points of the radiator are more pronounced. The experimental setup is shown in Fig. 9.5. The radiator was chosen in the form of a crystalline quartz cube of side 1.5 cm. The proton beam (1) passed through the cube (2) along the axis connecting opposite vertices. In this case the VC radiation went out through the three cube sides inclined at an angle $\psi = 35.26^0$ towards the motion axis. As in the first experiment, the plane colour photographic film was placed perpendicular to the beam axis, at a distance of $L = 2.35$ cm from the cube vertex. This guaranteed a smaller (as compared to a previous experiment) proton beam background in the region of the VC radiation. The direction of the rays (4) of the VC radiation through one particular side G of the cube is shown. The black and white photographic print and

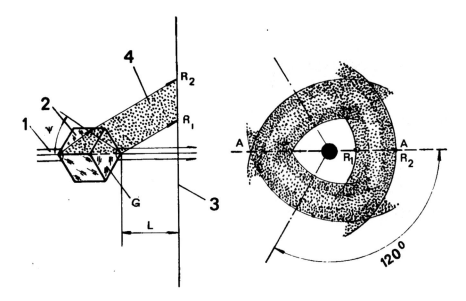

Figure 9.5.   The experimental setup of another experiment (Zrelov 1962). The proton beam (1) propagates through the quartz cube (2) along the axis connecting the opposite vertices of the cube. The observations are made in the plane photographic film (3) placed behind the quartz cube perpendicular to the motion axis; (4) is the direction of the Cherenkov rays passing through one of the cube sides.

the corresponding photometric curve measured along the direction A-A (Fig. 9.5) are shown in Fig. 9.6. To make rough estimates, we averaged the crystalline quartz refractive index over the directions of ordinary and non-ordinary wave vectors, thus obtaining $n = 1.55$ for $\lambda = 5 \times 10^{-5}$ cm. The corresponding Cherenkov angle was $\theta_c = 37.09^0$. In this case the rays of VC radiation emitted from the cube vertices should be at the radii $R_1 \approx 1.4$ cm and $R_2 \approx 2.3$ cm in the photographic film perpendicular to the motion axis. There is a rather pronounced maximum of radiation in Fig. 9.6 only at $R_2 \approx 2.3$ cm which corresponds to the γ ray emitted from the cube vertex at which the proton beam enters the radiator.

Theoretical consideration and numerical calculations presented below show that the just mentioned maxima of radiation intensity should indeed take place and they are owed to the discontinuities at the beginning and the end of the charge motion interval.

9.1.2. MAIN COMPUTATIONAL FORMULAE

In the past, the finite width of the Cherenkov rings on an observational sphere S of finite radius r was studied numerically in [12], and analytically

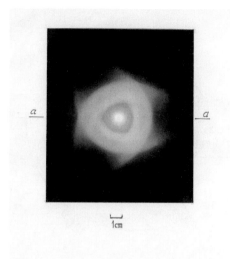
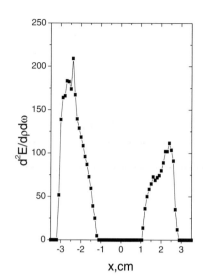

Figure 9.6.   Left: The black and white photographic print from the photographic film shown in Fig. 9.5; Right: The photometric curve corresponding to the left part along the direction $a-a$; x means the distance along $a-a$. The increments of the radiation intensity at radii $R_2 \approx 2.3$ cm and $R_1 \approx 1.4$ cm correspond to the Cherenkov rays emitted at the vertices where the beam enters and leaves the cube, respectively. The radiation intensity for negative x describes the superposition of the VC radiations passing through two sides of cube (2). The radiation maxima relating to the ends of the Cherenkov rings are more pronounced than in Fig. 9.4.

and numerically in [13] (see also Chapters 2 and 5).

It was shown there that the angular region to which the Cherenkov ring is confined is large for small r and diminishes with increasing of r. However, the width of the band corresponding to the Cherenkov ring remains finite even for infinite values of r. Since the measurements in the experiment discussed were made in the plane perpendicular to the motion axis (which we identify with the z axis), we should adjust formulae obtained in [12,13] to the case treated.

The exact formula

In the spectral representation the non-vanishing z component of the vector potential corresponding to the Tamm problem is given by

$$A_z(x, y, z) = \frac{e\mu}{2\pi c}\alpha_T, \tag{9.1}$$

where

$$\alpha_T = \int_{-z_0}^{z_0} \frac{dz'}{R} \exp(i\psi), \quad \psi = k\left(\frac{z'}{\beta} + nR\right), \quad R = [\rho^2 + (z - z')^2]^{1/2},$$

$$\rho^2 = x^2 + y^2, \quad k = \frac{\omega}{c}, \tag{9.2}$$

and μ is the magnetic permeability (in the subsequent concrete calculations we always put $\mu = 1$).

The field strengths corresponding to this vector potential are

$$H_\phi = \frac{ekn\rho}{2\pi c} \int dz' \exp(i\psi)\frac{1}{R^2}\left(-i + \frac{1}{k_n R}\right),$$

$$E_\rho = \frac{iek\mu\rho}{2\pi c} \int dz' \exp(i\psi)\frac{z - z'}{R^3}\left(1 + \frac{3i}{k_n R} - \frac{3}{k_n^2 R^2}\right), \quad k_n = kn$$

(we do not write out the z component of the electric strength since it does not contribute to the z component (along the motion axis) of the energy flux which is of interest for us).

The energy flux emitted in the frequency interval $d\omega$ and passing through the circular ring with radii ρ and $\rho + d\rho$ lying in the $z =$const plane is equal to

$$d\omega d\rho \frac{d^2\mathcal{E}}{d\rho d\omega},$$

where

$$\frac{d^2\mathcal{E}}{d\rho d\omega} = 2\pi\rho\frac{c}{2}(E_\rho H_\phi^* + c.c.) = \frac{e^2 k^2 n\mu\rho^3}{2\pi c}(I_c I_c' + I_s I_s'). \tag{9.3}$$

Here we put

$$I_c = \int dz' \frac{1}{R^2}\left(\cos\psi_1 - \frac{\sin\psi_1}{k_n R}\right),$$

$$I_c' = \int dz' \frac{z - z'}{R^3}\left[\left(1 - \frac{3}{k_n^2 R^2}\right)\cos\psi_1 - 3\frac{\sin\psi_1}{k_n R}\right],$$

$$I_s = \int dz' \frac{1}{R^2}\left(\sin\psi_1 + \frac{\cos\psi_1}{k_n R}\right),$$

$$I_s' = \int dz' \frac{z - z'}{R^3}\left[\left(1 - \frac{3}{k_n^2 R^2}\right)\sin\psi_1 + 3\frac{\cos\psi_1}{k_n R}\right],$$

$$\psi_1 = \frac{kz'}{\beta} + k_n(R - r), \quad r^2 = \rho^2 + z^2.$$

The Tamm approximate formula

Imposing the conditions: i) $R \gg z_0$ (this means that the observational distance is much larger than the motion interval); ii) $k_n R \gg 1$, $k_n = \omega/c_n$ (this means that the observations are made in the wave zone); iii) $nz_0^2/2r\lambda \ll \pi$, $\lambda = 2\pi c/\omega$ (this means that the second-order terms in the expansion of R should be small compared with π since they enter the phase ψ_1; λ is the observed wavelength), Tamm [5] obtained the following expression for the magnetic vector potential

$$A_z = \frac{e\mu}{\pi n\omega r} \exp(iknr)q,$$

$$q = \frac{1}{1/\beta_n - \cos\theta} \sin\left[\frac{kLn}{2}\left(\frac{1}{\beta_n} - \cos\theta\right)\right]. \tag{9.4}$$

Here $L = 2z_0$ is the motion interval and $\beta_n = \beta n$, $\beta = v/c$. Using this vector potential one easily evaluates the quantity similar to (9.3)

$$S_z(T) = \frac{d^2\mathcal{E}}{d\rho d\omega}(T) = \frac{2e^2\mu z\rho^3}{\pi ncr^5}q^2, \tag{9.5}$$

where $\cos\theta = z/r$ and $r = \sqrt{\rho^2 + z^2}$. The value of (9.5) at $\cos\theta = 1/\beta_n$ is given by

$$S_z(T)|_{\cos\theta=1/\beta_n} = \frac{e^2\mu k^2 L^2}{2\pi cn^4\beta^5\gamma_n^3 z}, \quad \gamma_n = \frac{1}{\sqrt{|1-\beta_n^2|}}. \tag{9.6}$$

For large kL (9.5) is reduced to

$$S_z(T)|_{kL\gg1} = \frac{e^2\mu kL}{c}\left(1 - \frac{1}{\beta_n^2}\right)\delta\left(\rho - \frac{z}{\gamma_n}\right). \tag{9.7}$$

Integration over ρ gives the energy flux through entire $z =$ const plane

$$\frac{d\mathcal{E}}{d\omega}(TF) = \frac{e^2\mu kL}{c}\left(1 - \frac{1}{\beta_n^2}\right), \quad k = \frac{\omega}{c} \tag{9.8}$$

which is independent of z and coincides with the Tamm-Frank value [1].

Tamm himself evaluated the energy flux per unit solid angle and per unit frequency through a sphere of infinite radius

$$\frac{d^2\mathcal{E}}{d\Omega d\omega}(T) = \frac{e^2\mu}{\pi^2 nc}q^2 \sin^2\theta. \tag{9.9}$$

This famous formula obtained by Tamm refers to the spectral representation and is frequently used by experimentalists for identification of the charge velocity.

The Fresnel approximation

This approximation is valid if the terms quadratic in z' in the expansion of R inside the ψ_1 are taken into account whilst the cubic terms are neglected. The condition for the validity of the Fresnel approximation (in addition to items i) and ii) of the Tamm formula) is $nz_0^3/2r^2\lambda \ll 1$. In this approximation,

$$\frac{d^2\mathcal{E}}{d\rho d\omega}(F) = \frac{e^2\mu k\rho z}{2cr^2}[(S_+ - S_-)^2 + (C_+ - C_-)^2]. \tag{9.10}$$

Here

$$C_\pm = C(z_\pm), \quad S_\pm = S(z_\pm), \quad z_\pm = \sqrt{\frac{k_n r}{2}}\sin\theta\left(\frac{1-\beta_n\cos\theta}{\beta_n\sin^2\theta} \pm \frac{z_0}{r}\right),$$

$C(x)$ and $S(x)$ are the Fresnel integrals defined as

$$S(x) = \sqrt{\frac{2}{\pi}}\int_0^x dt\sin t^2, C(x) = \sqrt{\frac{2}{\pi}}\int_0^x dt\cos t^2.$$

From the asymptotic behaviour of the Fresnel integrals

$$S(x) \sim \frac{1}{2} - \frac{1}{\sqrt{2\pi}}\frac{\cos x^2}{x}, \quad C(x) \sim \frac{1}{2} + \frac{1}{\sqrt{2\pi}}\frac{\sin x^2}{x}$$

as $x \to \infty$, and their oddness ($C(-x) = -C(x)$, $S(-x) = -S(x)$) it follows that for large kr (9.10) has a kind of plateau (if $\rho_2 - \rho_1 \ll \rho$)

$$\frac{e^2\mu k\rho z}{cr^2}, \tag{9.11}$$

for $\rho_1 < \rho < \rho_2$, where ρ_1 and ρ_2 correspond to the vanishing of the arguments of the Fresnel integrals. For $r \gg z_0$, they are reduced to

$$\rho_{1,2} = \sqrt{\beta_n^2 - 1}(z \mp z_0).$$

Outside the plateau, for a fixed z and $\rho \to \infty$, (9.10) decreases as $1/\rho^2$ coinciding with the Tamm formula (9.5). Mathematically the existence of a plateau is because for $\rho_1 < \rho < \rho_2$ the Fresnel integral arguments z_+ and z_- have different signs. At the Cherenkov threshold ($\beta = 1/n$)

$$z_\pm = \sqrt{\frac{k_n r}{2}}\sin\theta\left(\frac{1}{2\cos^2(\theta/2)} \pm \frac{z_0}{r}\right)$$

have the same sign for $r > L$ and the radiation intensity for $kr \gg 1$ and $r > L$ should be small (as compared to the plateau value (9.11)) everywhere.

These asymptotic expressions are not valid at $\rho = \rho_1$ and $\rho = \rho_2$. At these points the radiation intensities are obtained directly from (9.10)

$$\frac{d^2\mathcal{E}}{d\rho d\omega}(\rho = \rho_1) = \frac{e^2 \mu k z \rho_1}{2cr_1^2}$$

$$\times \left\{ \left[C\left(\sqrt{\frac{2kn}{r_1}} z_0 \sin \theta_1 \right) \right]^2 + \left[S\left(\sqrt{\frac{2kn}{r_1}} z_0 \sin \theta_1 \right) \right]^2 \right\},$$

$$\frac{d^2\mathcal{E}}{d\rho d\omega}(\rho = \rho_2) = \frac{e^2 \mu n k z \rho_2}{2cr_2^2}$$

$$\times \left\{ \left[C\left(\sqrt{\frac{2kn}{r_2}} z_0 \sin \theta_2 \right) \right]^2 + \left[S\left(\sqrt{\frac{2kn}{r_2}} z_0 \sin \theta_2 \right) \right]^2 \right\}, \qquad (9.12)$$

where r_1, r_2, θ_1 and θ_2 are defined as

$$r_1 = \sqrt{\rho_1^2 + z^2}, \quad r_2 = \sqrt{\rho_2^2 + z^2}, \quad \cos\theta_1 = z/r_1, \quad \cos\theta_2 = z/r_2.$$

For $kz_0^2/z \gg 1$, one gets

$$\frac{d^2\mathcal{E}}{d\rho d\omega}(\rho = \rho_1) = \frac{e^2 \mu k z \rho_1}{4cr_1^2}, \quad \frac{d^2\mathcal{E}}{d\rho d\omega}(\rho = \rho_2) = \frac{e^2 \mu n k z \rho_2}{4cr_2^2}, \qquad (9.13)$$

that is four times smaller than (9.11) taken at the same points.

For $kz_0^2/r \ll 1$ the radiation intensity (9.10) outside the Cherenkov ring coincides with that given by the Tamm formula (9.5).

Frequency distribution Integrating (9.11) over ρ from ρ_1 to ρ_2 (suggesting that outside this interval, the radiation intensity (9.10) is negligible), one gets the frequency distribution of the radiated energy

$$\frac{d\mathcal{E}}{d\omega}(F) = \frac{e^2 \mu k L}{c}\left(1 - \frac{1}{\beta_n^2}\right), \quad k = \frac{\omega}{c}, \qquad (9.14)$$

which coincides with the Tamm-Frank frequency distribution (9.8).

Energy radiated in the given frequency interval per unit radial distance Integrating (9.11) over ω from ω_1 to ω_2, one obtains the spatial distribution of the energy emitted in the frequency interval (ω_1, ω_2). It is equal to

$$\frac{d\mathcal{E}}{d\rho}(F) = \frac{e^2 \mu \rho z}{2c^2 r^2}(\omega_2^2 - \omega_1^2) \tag{9.15}$$

for $\rho_1 < \rho < \rho_2$ and zero outside this interval. When performing the ω integration we have disregarded the ω dependence of the refractive index n. This is valid for a quite narrow frequency interval.

The total energy radiated in the given frequency interval. Integration of (9.14) over ω or (9.15) over ρ gives the total energy emitted in the frequency interval (ω_1, ω_2)

$$\mathcal{E} = \frac{e^2 \mu L}{2c^2}(\omega_2^2 - \omega_1^2)\left(1 - \frac{1}{\beta_n^2}\right). \tag{9.16}$$

(Again, the medium dispersion has been neglected).

Quasi-classical (WKB) approximation
To make easier the interpretation of the numerical calculations presented in the next section, we apply the quasi-classical approximation (the stationary phase method) for the evaluation of the vector potential (9.1). For $\rho < (z - z_0)/\gamma_n$ and $\rho > (z + z_0)/\gamma_n$ (that is, below L_2 or above L_1) one has

$$A_z(BS) = A_1(BS) - A_2(BS), \tag{9.17}$$

where

$$A_1(BS) = \frac{ie\mu\beta}{2\pi ck}\frac{1}{R_1}\exp(i\psi_1), \quad A_2(BS) = \frac{ie\mu\beta}{2\pi ck}\frac{1}{R_2}\exp(i\psi_2),$$

$$R_1 = \frac{1}{r_1 - \beta_n(z + z_0)}, \quad R_2 = \frac{1}{r_2 - \beta_n(z - z_0)},$$

$$\psi_1 = k\left(nr_1 - \frac{z_0}{\beta}\right), \quad \psi_2 = k\left(nr_2 + \frac{z_0}{\beta}\right),$$

$$r_1 = \sqrt{\rho^2 + (z + z_0)^2}, \quad r_2 = \sqrt{\rho^2 + (z - z_0)^2}.$$

It is seen that for $\beta > 1/n$, A_z^{out} is infinite at $\rho = (z - z_0)/\gamma_n$ and $\rho = (z + z_0)/\gamma_n$, that is, at the border with the CSW. There are no singularities in A_z^{out} for $\beta < 1/n$. Expanding r_1 and r_2 entering ψ_1 and ψ_2 up to the first order in z_0 ($r_1 = r + z_0 \cos\theta$, $r_2 = r - z_0 \cos\theta$) and setting $r_1 = r$ and $r_2 = r$ in R_1 and R_2 one finds

$$A_z^T = \frac{e\mu q}{\pi cknr}\exp(iknr) \tag{9.18}$$

which coincides with the Tamm vector potential (9.4). Owing to the approximations involved the singularities of $A_1(BS)$ and $A_2(BS)$ compensate each other and the vector potential (9.18) becomes finite at all angles. Thus, $A_z(BS)$ is the quasi-classical analogue of the Tamm vector potential.

On the other hand, in the spatial region $(z - z_0)/\gamma_n < \rho < (z + z_0)/\gamma_n$ (that is, between L_2 and L_1) one has

$$A_z = A_z(BS) + A_z(Ch),   \tag{9.19}$$

where $A_z(BS)$ is the same as in (9.17) while

$$A_z(Ch) = \frac{e\mu}{2\pi c} \exp(i\psi_{Ch}) \sqrt{\frac{2\pi\beta\gamma_n}{k\rho}}$$

$$\times \Theta[\rho - (z - z_0)/\gamma_n]\Theta[(z + z_0)/\gamma_n - \rho],   \tag{9.20}$$

where $\Theta(x)$ is the step function and

$$\psi_{Ch} = \frac{kz}{\beta} + \frac{\pi}{4} + \frac{k\rho}{\beta\gamma_n}.$$

It should be noted that $A_z(Ch)$ exists only if $\beta > 1/n$. Otherwise ($\beta < 1/n$), the vector potential is given by (9.17) in the whole angular region.

One can ask on what grounds we have separated the vector potential into the Cherenkov ($A_z(Ch)$) and bremsstrahlung ($A_z(BS)$) parts? First, $A_1(BS)$ and $A_2(BS)$ exist below and above the Cherenkov threshold while $A_z(Ch)$ exists only above it. This is what is intuitively expected for the VC radiation and bremsstrahlung. Second, $A_z(Ch)$ originates from the stationary point of the integral α_T (see Eq. (9.1)) lying inside the motion interval $(-z_0, z_0)$. For $A_1(BS)$ and $A_2(BS)$ the stationary points lie outside this interval, and their values are determined by the boundary points ($\pm z_0$) of the motion interval. Again, this is intuitively expected since the VC radiation is owed to the charge radiation in the interval $(-z_0, z_0)$ whilst the bremsstrahlung is determined by the points ($\mp z_0$) corresponding to the beginning and the end of motion, respectively. Third, to clarify the physical meaning of $A_z(Ch)$, we write out the vector potential corresponding to the unbounded charge motion. It is equal to

$$A_z = \frac{e\mu}{\pi c} \exp(\frac{ikz}{\beta}) K_0(\frac{k\rho}{\beta\gamma_n})$$

for $\beta < 1/n$ and

$$A_z = \frac{ie\mu}{2c} \exp(\frac{ikz}{\beta}) H_0^{(1)}(\frac{k\rho}{\beta\gamma_n})   \tag{9.21}$$

for $\beta > 1/n$. Since this vector potential tends to (9.20) as $\rho \to \infty$, $A_z(Ch)$ in (9.19) is a piece of the unbounded vector potential (9.21) confined to the region $(z - z_0)/\gamma_n < \rho < (z + z_0)/\gamma_n$.

It is seen that for $kr \to \infty$, $A_z(BS)$ and $A_z(Ch)$ decrease as $1/kr$ and $1/\sqrt{kr}$, respectively. This means that at large distances, $A_z(Ch)$ dominates in the region $(z - z_0)/\gamma_n < \rho < (z + z_0)/\gamma_n$. Thus A_z has a kind of plateau inside this interval with infinite maxima at its ends (quasi-classics does not work at these points) and sharply decreases outside it. The corresponding quasi-classical field strengths are given by

$$E = E(BS) + E(Ch), \quad H = H(BS) + H(Ch), \tag{9.22}$$

$$H(BS) = H_1(BS) - H_2(BS), \quad E(BS) = E_1(BS) - E_2(BS),$$

$$H_1(BS) = \frac{e\beta\rho}{2\pi ckr_1 R_1^2}(k_n R_1 + i)\exp(i\psi_1),$$

$$H_2(BS) = \frac{e\beta\rho}{2\pi ckr_2 R_2^2}(k_n R_2 + i)\exp(i\psi_2),$$

$$E_1(BS) = -\frac{e\beta\rho}{2\pi c\epsilon k^2 r_1^2 R_1^2}\exp(i\psi_1)$$
$$\times \left[(1 - iknr_1)(1 - iknR_1)\frac{z + z_0}{r_1} + \frac{r_1}{R_1}(2 - iknR_1)\left(\frac{z + z_0}{r_1} - \beta_n\right)\right],$$

$$E_2(BS) = -\frac{e\beta\rho}{2\pi c\epsilon k^2 r_2^2 R_2^2}\exp(i\psi_2)$$
$$\times \left[(1 - iknr_2)(1 - iknR_2)\frac{z - z_0}{r_2} + \frac{r_2}{R_2}(2 - iknR_2)\left(\frac{z - z_0}{r_2} - \beta_n\right)\right]$$

$$H(Ch) = -\frac{e}{2\pi c}\sqrt{\frac{2\pi\beta\gamma_n}{k\rho}}\frac{1}{2\rho}\left(\frac{2ik\rho}{\beta\gamma_n} - 1\right)\exp(i\psi_{Ch}), \quad E(Ch) = \frac{1}{\epsilon\beta}H(Ch).$$

Here ϵ is the electrical permittivity ($n^2 = \epsilon\mu$). It should be noted that when evaluating field strengths we have not differentiated step functions entering (9.20). If this were done the δ functions at the ends of the Cherenkov ring would appear. Owing to the breaking of the WKB approximation at these points, the vector potentials and field strengths are singular there and the inclusion of the δ functions just mentioned does not change anything.

The energy flux along the motion axis is

$$S_z = \frac{d^2\mathcal{E}}{d\rho d\omega}(WKB) = \pi\rho c(EH^* + HE^*) \tag{9.23}$$

In (9.22) and (9.23), $E \equiv E_\rho$ and $H \equiv H_\phi$ (in order not to overload formulae, we have dropped the indices of E_ρ and H_ϕ).

We estimate the height of the plateau to which mainly $H(Ch)$ and $E(Ch)$ contribute. It is given by

$$S_z(plateau) = \pi\rho c[E(Ch)H^*(Ch) + H(Ch)E^*(Ch)] \approx \frac{e^2\mu k}{c\beta_n^2\gamma_n} \qquad (9.24)$$

Since S_z is negligible outside this plateau and since infinities at the ends of the Cherenkov ring are unphysical (they are owed to the failure of the WKB method at these points) the frequency distribution is obtained by multiplying (9.24) by the width of the Cherenkov ring

$$\frac{d\mathcal{E}}{d\omega}(WKB) = \frac{e^2k\mu}{c\beta_n^2\gamma_n}\frac{L}{\gamma_n} = \frac{e^2\mu kL}{c}(1 - \frac{1}{\beta_n^2}). \qquad (9.25)$$

This coincides with the Tamm-Frank formula (9.8). It is rather surprising that quite different angular distributions corresponding to the Tamm intensity (9.5), to the Fresnel intensity (9.10) and the quasi-classical intensity (9.23) give the same frequency distribution (9.8).

9.1.3. NUMERICAL RESULTS

In Fig. 9.7 the radiation intensities are presented for various distances δz of the observational plane (δz is the distance from the point $z = z_0$ corresponding to the end of motion). We observe the qualitative agreement of the exact radiation intensity (9.3) with the Fresnel intensity (9.10). Both of them disagree sharply with the Tamm intensity (9.5) which does not contain the CSW responsible for the appearance of plateau in (9.3) and (9.10). Fig. 9.7 (d) demonstrates that at large observational distances ($\delta z = 100$ cm) the Tamm radiation intensity approaches the exact intensity outside the Cherenkov ring.

In Fig. 9.8 the magnified versions of exact radiation intensities corresponding to $\delta z = 0.3$ cm and $\delta z = 1$ cm are presented. In accordance with quasi-classical predictions, one sees the maxima at the ends of the interval $(z - z_0)/\gamma_n < \rho < (z + z_0)/\gamma_n$.

In Section 9.1 it was mentioned about the special optical devices focusing the rays directed at the Cherenkov angle into one ring. In the case treated it is the plateau shown in Figs. 9.7 and 9.8 and the BS peaks at its ends that are focused into this ring. The remaining part of BS will form the tails of the focused total radiation intensity. For such a compressed radiation distribution the Tamm formula probably has a greater range of applicability.

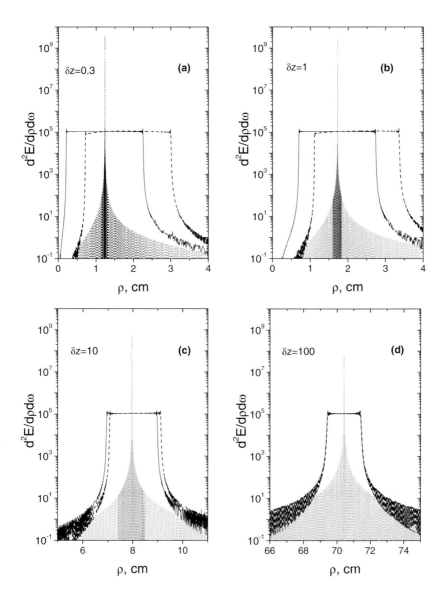

Figure 9.7.   Theoretical radiation intensities in a number of planes perpendicular to the motion axis for the experimental setup shown in Fig. 9.3; δz means the distance (in cm) from the cone vertex to the observational plane. The solid, dashed, and dotted curves refer to the exact, Fresnel, and Tamm intensities, respectively. In this figure and the following figures the radiation theoretical intensities are in e^2/cz_0 units.

9.1.4. DISCUSSION

Vavilov-Cherenkov radiation and bremsstrahlung on the sphere
In the original and in nearly all subsequent publications about the Tamm problem, the radiation intensity was considered on the surface of a sphere

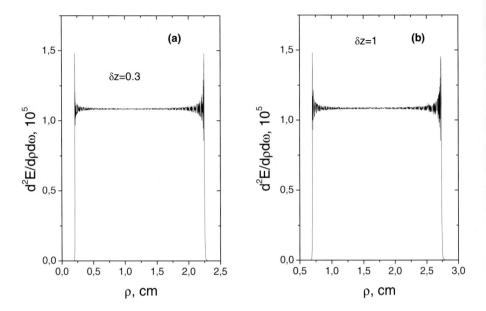

Figure 9.8.   Exact theoretical radiation intensities in the $\delta z = 0.3$ cm and $\delta z = 1$ cm planes.

of radius r much larger than the motion interval $L = 2z_0$. It is easy to check that on the surface of the sphere of finite radius r, the intervals

$$\rho > (z + z_0)/\gamma_n, \quad (z - z_0)/\gamma_n < \rho < (z + z_0)/\gamma_n, \quad \text{and} \quad \rho < (z - z_0)/\gamma_n$$

correspond to the angular intervals

$$\theta > \theta_1, \quad \theta_2 < \theta < \theta_1, \quad \text{and} \quad \theta < \theta_2,$$

where θ_1 and θ_2 are defined by

$$\cos\theta_1 = -\frac{\epsilon_0}{\beta_n^2 \gamma_n^2} + \frac{1}{\beta_n}[1 - (\frac{\epsilon_0}{\beta_n \gamma_n})^2]^{1/2}$$

and

$$\cos\theta_2 = \frac{\epsilon_0}{\beta_n^2 \gamma_n^2} + \frac{1}{\beta_n}[1 - (\frac{\epsilon_0}{\beta_n \gamma_n})^2]^{1/2}. \qquad (9.26)$$

Here $\epsilon_0 = z_0/r$. For $r \gg z_0$

$$\theta_1 = \theta_c + \frac{\epsilon_0}{\beta_n \gamma_n}, \quad \theta_2 = \theta_c - \frac{\epsilon_0}{\beta_n \gamma_n},$$

where θ_c is defined by $\cos\theta_c = 1/\beta n$. In this case, the Tamm formula (9.9) is valid for $\theta < \theta_2$ and $\theta > \theta_1$, that is, in nearly the whole angular region. It should be added that the existence of the Cherenkov shock wave on the sphere is masked by the smallness of the angular region to which it is confined. It seems at first that on an observational sphere of infinite radius there is no room for CSW. This is not so. Although $\Delta\theta = \theta_1 - \theta_2 = 2\epsilon_0/\beta_n\gamma_n$ is very small for $r \gg z_0$, the length of an arc corresponding to $\Delta\theta$ in a particular $\phi =$const plane of the sphere S is finite: it is given by $\mathcal{L} = 2z_0/\beta_n\gamma_n$ and does not depend on the sphere radius r for $r \gg z_0$. Owing to the axial symmetry of the problem, on the observational sphere S the region to which the VC radiation is confined looks like a band of finite width $\mathcal{L}$. Thus the observation of the Cherenkov ring on the sphere is possible if the detector dimension is smaller than $\mathcal{L}$.

Vavilov-Cherenkov radiation and bremsstrahlung in the plane perpendicular to the motion axis

The separation of the VC radiation and the BS looks more pronounced in the plane perpendicular to the motion axis. We illustrate this using the quasi-classical intensities as an example.

In Fig. 9.9 (a) we present the quasi-classical intensity (9.23) for $\delta z = 0.3$ cm. We observe perfect agreement between it and the exact intensity shown in Fig. 9.8 (a) everywhere except for the boundaries of the region to which the VC radiation is confined. The quasi-classical approximation is unique in the sense that contributions of the VC radiation and the BS are clearly separated in the vector potential (9.19) and field strengths (9.22). To see the contribution of the BS, we omit $A_z(Ch), E(Ch)$, and $H(Ch)$ in these relations by setting them equal to zero. The resulting intensity describing BS is shown in Fig. 9.9 (b). It disagrees sharply with the Tamm intensity (9.5). From the smallness of the BS intensity everywhere except for the boundaries of the Cherenkov ring it follows that oscillations of the total radiation intensity inside the Cherenkov ring are owed to the interference of the VC radiation and the BS.

On the nature of the bremsstrahlung shock waves in the Tamm problem

Some words should be added on the nature of BS shock waves discussed above. In [7] they were associated with velocity jumps at the beginning and end of motion.

On the other hand, the smoothed Tamm problem was considered in [14] in the time representation. In it the charge velocity v changes smoothly from zero up to some value $v_0 > c_n$ with which it moves in some time interval. Later v decreases smoothly from v_0 to zero. It was shown there that at the instant when v coincides with the velocity c_n of light in medium, a complex

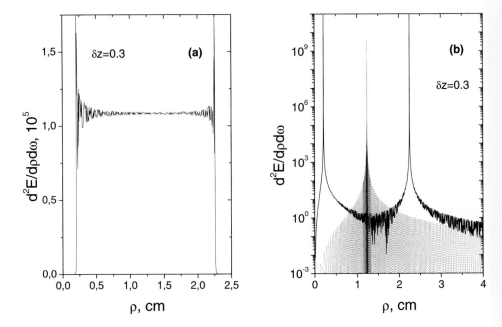

Figure 9.9. (a): Quasi-classical radiation intensity in the plane $\delta z = 0.3$ cm. It coincides with the exact intensity shown in Fig. 9.8 (a) everywhere except for the boundary points of the Cherenkov ring where the quasi-classical intensities are infinite owing to the breaking of the WKB approximation; (b): The quasi-classical bremsstrahlung intensity (solid curve) and the Tamm intensity (dotted curve) in the plane $\delta z = 0.3$ cm. The sharp disagreement between them is observed.

arises consisting of the CSW with its apex attached to a moving charge, and the shock wave SW_1 closing the Cherenkov cone (and not coinciding with the BS_1 shock wave originating at the beginning of motion). The inclination angle of the normal to SW_1 towards the motion axis (defining the direction in which SW_1 propagates) varies smoothly from 0 at the motion axis up to the Cherenkov angle θ_c at the point where SW_1 intersects the Cherenkov cone. Therefore, the radiation produced by the SW_1 fills the angular region $0 < \theta < \theta_c$. As time advances, the dimensions of the above complex grow since its apex moves with the velocity $v > c_n$, whilst the shock wave SW_1 propagates with the velocity c_n. In the past, the existence of radiation arising at the Cherenkov threshold and directed along the motion axis was suggested in [15].

Since in the original Tamm problem the charge velocity changes instantly from 0 to v_0, the CSW and SW_1 are not separated in subsequent instants of time too. They are marked as CSW in Fig. 9.1 (a,b).

The smoothed Tamm problem was also considered in [10] in the spectral representation. It was shown there that when a length of motion along which a charge moves non-uniformly tends to zero, its contribution to the total radiation intensity also tends to zero. There are no velocity jumps for the smoothed problem, and therefore the BS cannot be associated with instantaneous velocity jumps. However, there are acceleration jumps at the beginning and end of motion and at the instants when the accelerated motion meets the uniform motion. Thus BS can still be associated with acceleration jumps. To clarify the situation the Tamm problem with absolutely continuous charge motion (for which the velocity itself and all its time derivatives are absolutely continuous functions of time) was considered in [16]. It was shown there that a rather slow decrease in the radiation intensity outside the above plateau is replaced by the exponential damping (in the past, for the charge motion in vacuum, the exponential damping for all angles was recognized in [17-20]). It follows from this that the authors of [7] were not entirely wrong if by the BS shock waves used by them, one understands the mixture of the shock waves mentioned above and originating from the discontinuities of velocity, acceleration, other higher velocity time derivatives, and from the transition through the medium light barrier.

This is also confirmed by the consideration of radiation intensities for various charge velocities. Figure 9.10 (a) demonstrates that the position of the maximum of radiation intensity approaches the motion axis, whilst its width diminishes as the charge velocity approaches the Cherenkov threshold ($\beta = 1/n \approx 0.665$). The radiation intensities presented in Fig. 9.10 (b) show their behaviour just above ($\beta = 0.67$) and below ($\beta = 0.66$) the Cherenkov threshold. It is seen that the maxima of the under-threshold and the over-threshold intensities differ by 10^5 times. Far from the maximum position they approach each other. The radiation intensity at the Cherenkov threshold shown in Fig. 9.10 (c) is three orders smaller than that corresponding to $\beta = 0.67$. The calculations in Figs. 9.10 (a-c) were performed using the Fresnel approximate intensity (9.10) which is in good agreement with the exact intensity (9.3) for the treated position ($\delta z = 10$ cm) of the observational plane (as Fig. 9.7 demonstrates).

To see manifestly how the bremsstrahlung changes when one passes through the Cherenkov threshold we present in Fig. 9.10 (d) the quasi-classical radiation BS intensities evaluated for $\beta = 0.67$ (in this case the VC radiation was removed by hand from (9.22) similarly as was done for Fig. 9.9 (b)) and $\beta = 0.66$. The position of the observational plane is ($\delta z = 0.3$ cm). Again, we observe the sharp decrease in the BS intensities in the neighbourhood of their maxima when one passes the Cherenkov barrier. This confirms that the BS shock waves used in [7] are the mixture of the shock waves mentioned above for the charge velocity above the Cherenkov threshold. For the charge velocity below the Cherenkov threshold only the

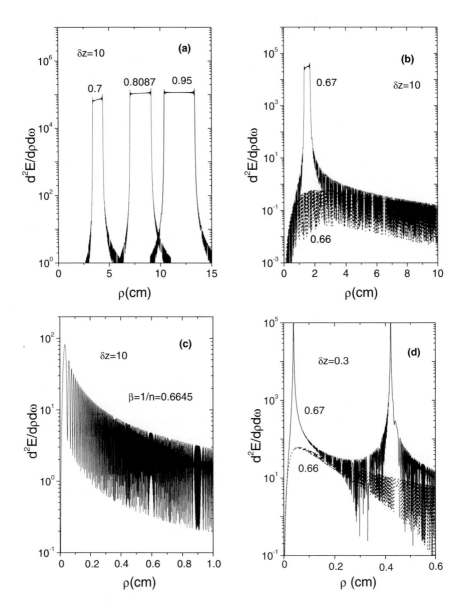

Figure 9.10. (a): Radiation intensities for a number of charge velocities above the Cherenkov threshold in the observational plane $\delta z = 10$ cm. As the charge velocity approaches the velocity of light in medium, the position of the Cherenkov ring approaches the motion axis whilst its width diminishes; (b): Radiation intensities for the charge velocity slightly above (0.67) and below (0.66) the Cherenkov threshold ($1/n \approx 0.6645$) in the plane $\delta z = 10$ cm; (c): Radiation intensity at the Cherenkov threshold in the plane $\delta z = 10$ cm. In accordance with theoretical predictions it is much smaller than above the threshold; (d): Quasi-classical BS intensities for the charge velocity slightly above and below the Cherenkov threshold in the plane $\delta z = 0.3$ cm.

BS shock waves originating from the discontinuities of velocity, acceleration and other higher velocity time derivatives survive. They are much smaller than the singular shock wave originating when the charge velocity coincides with the velocity of light in medium.

Comparison with experiment
Strictly speaking, the formulae obtained above and describing the fine structure of the Cherenkov rings are valid if the observations are made in the same medium where a charge moves. Because of this, the plateau of the radiation intensity and its bursts at the ends of this plateau cannot be associated with the transition radiation which appears when a charge intersects the boundary between two media.

Turning to the comparison with experiment, we observe that it corresponds to the charge moving subsequently in air, in the medium, and, finally, again in air. According to [21], the contribution of the transition radiation which arises at the boundary of the medium with air is approximately 100 times smaller than the contribution of VC radiation. Since the uniformly moving charge does not radiate in air, where $\beta n < 1$, and radiates in medium, where $\beta n > 1$, the observer inside the medium associates the radiation with the instantaneous appearance and disappearance of a charge at the medium boundaries and with its uniform motion inside the medium. We quote, e.g., Jelly ([22], p.59):

> A situation alternative to that of a particle of constant velocity traversing a finite slab may arise in the following way; suppose instead that we have an infinite medium and that a charged particle, initially at rest at a point A, is rapidly accelerated up to a constant velocity (above the Cerenkov threshold) which it maintains until, at a point B, it is brought abruptly to rest. If, as in the first case, the distance $AB = d$, the output of Cerenkov radiation will be the same as before. In this case, there will be radiation at the two points A and B; this will be now identified as a form of acceleration radiation. This and transition radiation are essentially the same; the intensities work out the same in both cases and it is only convention which decides which term shall be used.

This justifies the applicability of the Tamm problem for the description of the discussed experiments.

Comparing theoretical and experimental intensities we see that:

i) theoretical intensities have a plateau (Figs. 9.7-9.10), whilst the experimental intensities have a triangle form (Figs. 9.4, 9.6);

ii) the observed radiation peaks at the boundaries of the Cherenkov rings are not so pronounced as the predicted ones.

The triangle form of the observed radiation intensities can be associated with the non-existence of the instantaneous velocity jumps in realistic cases.

To prove this, we evaluated the radiation intensities for the smooth Tamm problem (Fig. 3.11) for a number of intervals of the charge uniform motion (Fig. 9.11). We observe the existence of the radiation intensity plateau,

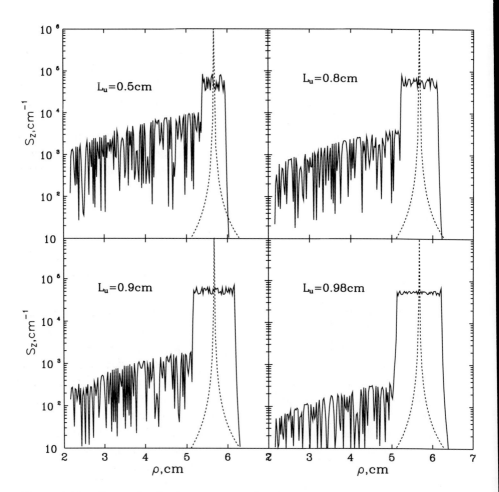

Figure 9.11. Spectral radiation intensities for the smooth Tamm problem for various lengths L_u of the charge uniform motion. The total motion interval is 1 cm, the distance of the observation plane from the end point of the motion interval is 4.5 cm, the observed wavelength $\lambda = 5.893 \times 10^{-5}$ cm, the medium refractive index $n = 1.512$, the uniform charge velocity $\beta_0 = v_0/c = 1$. The plateau in the radiation intensity corresponds to the CSW in Fig. 3.13 (d). The sudden drop of the radiation intensity to the right of this plateau is due to the absence of the singular shock waves above L_1. The oscillations of the radiation intensities to the left of plateau are due to the interference of the singular shock waves SW_1 and SW_2 in the region below L_2. The dotted curves are the Tamm approximate radiation intensities corresponding to the charge uniform motion on the interval L_u. They are fastly oscillating functions. To make them visible, we draw them through their maxima.

its sudden drop to the right of the plateau and its moderate decreasing to the left of plateau. The sudden drop of the radiation intensity to the right of the plateau takes place in the space region where only the non-singular BS shock waves associated with the beginning and the end of a charge motion exist. In Fig. 3.13 (d), this space region lies above L_1. The plateau corresponds to the space region lying between L_2 and L_1 where the singular wave CSW (Cherenkov shock wave) and SW_1 (singular shock wave associated with the transition of the light velocity barrier at the accelerated part of the charge trajectory) exist. The moderate decrease in the radiation intensity to the left of the plateau is due to the existence of SW_1 and SW_2 shock waves (the latter arises at the decelerated part of the charge trajectory). The oscillations of the radiation intensity in this region are due to the interference of SW_1 and SW_2. The smallness of oscillations inside the plateau indicates that the contribution of the CSW to the radiation intensity is much larger than that of SW_1.

Turning to Fig. 3.12 (b) describing the position of the shock waves in the limiting case of the smooth Tamm problem, we see that three shock waves (BS_1, SW_1 and CSW) are intersected on the straight line L_1. Therefore, the radiation intensity should be large there. Above L_1 (this corresponds to the region lying to the right of plateau in Fig. 9.11) only the non-singular shock waves BS_1 and BS_2 contribute to the radiation intensity. Therefore, the radiation intensity should be very small there. The experimental curve shown in Fig. 9.4 partly supports these claims. However, the experimental intensity decreases smoothly in the space region where the theory predicts the existence of plateau. The picture similar to Fig. 9.4 might be possible if the focusing devices (their use is wide-spread in the Cherenkov-like experiments) projecting the γ rays emitted at the Cherenkov angle into the narrow Cherenkov ring (and transforming the plateau of the radiation intensity into this ring) were used. However, no focusing lens was used in the experiments discussed in which the Cherenkov light left the radiator along the direction perpendicular to the radiator surface.

A possible explanation of the deviation of the theoretical data from the experimental ones is due to the medium dispersion. A charge moving uniformly in medium emits all frequencies satisfying the Tamm-Frank radiation condition $\beta n(\omega) > 1$ (see, e.g.,[4]). In the experiment treated the dependence on the frequency enters through the refractive index n of the sample (where a charge moves) and through the spectral sensitivity of the photographic film placed in the observational plane perpendicular to the motion axis. Let the intervals where a charge is accelerated and decelerated be arbitrary small but finite. For large frequencies the radiation intensity in the space region to the left the plateau shown in Fig. 9.11 begins to rise This was clearly shown in [23,24] and in Chapters 5 and 7. The resulting radiation intensities are obtained from those presented in Fig. 9.11 by convoluting them with the spectral sensitivity of the photographic film and integrating over all ω. If the integrand differs from zero for large frequencies and the Tamm-Frank radiation condition is still fulfilled for it, then the arising radiation intensity will resemble the experimental curve of Fig. 9.4. However, to perform concrete calculations the knowledge of the frequency dependence of the sample refractive index and the spectral sensitivity of the photographic film is needed.

The item ii) can be understood if one takes into account that the experiments mentioned in section 2 were performed with a relatively broad proton beam (0.5 cm in diameter). This leads to the smoothing of the boundary peaks after averaging over the diameter of proton beam.

9.1.5. CONCLUDING REMARKS TO THIS SECTION

According to quantum theory [25], a charge moving uniformly in a medium with a velocity greater than the velocity of light in medium radiates γ quanta at an angle θ_c towards the motion axis ($\cos\theta_c = 1/\beta n$). It should be noted that for the uniform charge motion in an unbounded medium, a photographic plate placed perpendicular to the motion axis will be darkened with an intensity proportional to $1/\rho$ (ρ is the distance from the motion axis) without any maximum at the Cherenkov angle. Despite its increase for small ρ, the energy emitted in a particular ring of width $d\rho$ is independent of ρ. The surface of the cylinder coaxial with the motion axis will be uniformly darkened.

The Cherenkov ring can be observed only for the finite motion interval. In the $z =$const plane the width of the ring is proportional to the charge motion interval L: $\Delta R = L/\gamma_n$ ($\gamma_n = 1/\sqrt{|1 - \beta_n^2|}$, $\beta_n = \beta n$). It does not depend on the position z of the observational plane. The frequency dependence enters only through the refractive index n. The radiation emitted into a particular ring does not depend on z. For a fixed observational plane the radiation intensity oscillates within the Cherenkov ring. These oscillations are owed to the interference of bremsstrahlung and the Vavilov-Cherenkov radiation in (9.23). The large characteristic peaks at the ends of the Cherenkov ring are owed to the bremsstrahlung shock waves, which include shock waves originating from the jumps of velocity, acceleration, other higher time derivatives of a velocity, and from the transition through the medium velocity light barrier.

The finite width of the Cherenkov ring in the plane $z =$const is owed to the Cherenkov shock wave. Inside the Cherenkov ring ($R_1 < \rho < R_2$) the Tamm formula does not describes the radiation intensity at any position of the observational plane (see Fig. 9.7). Outside the Cherenkov ring ($\rho <$

R_1 and $\rho > R_2$) the exact and the Tamm radiation intensities are rather small. In this spatial region they approach each other at large distances satisfying $kz_0^2/r \ll 1$. For the experiments treated in the text, the l.h.s. of this equation is equal to unity at the distance $r \approx 1$ km. On the other hand, the exact formula (9.3) describes the radiation intensity in all spatial regions.

We conclude: the experiments performed with a relatively broad 657 MeV proton beam passing through various radiators point to the existence of diffused radiation peaks at the boundary of the broad Cherenkov rings. This supports theoretical predictions [7, 15, 26, 27] (see Chapters 2,3 and 5) on the existence of the shock waves arising when the charge motion begins and ends, and when the charge velocity coincides with the velocity of light in medium.

It is desirable to repeat experiments similar to those described in Section 2 with a charged particle beam of a smaller diameter (≈ 0.1 cm), with a rather thick dielectric sample, without using the focusing devices and for various observational distances. This should result in the appearance of more pronounced, just mentioned, radiation peaks.

9.2. Observation of anomalous Cherenkov rings

The Cherenkov radiation induced by the relativistic lead ions moving through the rarefied air was studied in [28]. In addition to the main Cherenkov ring with a radius corresponding to the lead ion velocity, additional rings were observed with radii corresponding to a velocity greater than the velocity of light in vacuum. A careful analysis of the experimental conditions was made to exclude the errors possible. The authors of [28] associated the anomalous Cherenkov rings with the existence of tachyons, hypothetical particles moving with a velocity greater than the velocity of light in vacuum. This highly intriguing question needs special consideration.

9.3. Two-quantum Cherenkov effect

The possibility of this effect was predicted by Frank and Tamm in [29]:

> We note in passing that for $v < c$ the conservation laws prohibit the emission of one particular photon as well as the simultaneous emission of a group of photons. However, for the superluminal velocity such higher order processes are possible although for them the radiation condition (4) is not necessary.

(Under this condition Tamm and Frank meant the one-photon radiation condition $\cos\theta = c/vn$). In this case, the conservation of energy and momenta does not prohibit the process in which a moving charge emits simultaneously two photons. There is no experimental confirmation for this effect up to now. We briefly review the main features of the kinematics of the two photon Cherenkov effect.

The calculations of the two-photon radiation intensity are known [30-34], but they were performed without paying enough consideration to the exact kinematical relations. The goal of this treatment is to point out that the two-photon Cherenkov effect will be enhanced for special orientations of photons and the recoil charge. This makes easier the experimental search for the 2-photon Cherenkov effect.

9.3.1. PEDAGOGICAL EXAMPLE: THE KINEMATICS OF THE ONE-PHOTON CHERENKOV EFFECT

This effect was considered quite schematically in Chapter 2. We consider here its kinematics in some detail since it clarifies the situation with the two-photon Cherenkov effect.

Let a point-like charge e having the rest mass m_0 move in medium of the refractive index n. It emits the photon with the frequency ω. The conservation of the energy and momentum gives

$$m_0 c^2 \gamma_0 = m_0 c^2 \gamma + \hbar\omega, \quad m_0 \vec{v}_0 \gamma_0 = m_0 \vec{v}\gamma + \frac{\hbar\omega n}{c}\vec{e}_\gamma. \tag{9.27}$$

Here $\hbar$ is the Plank constant, $\vec{v}_0$ and $\vec{v}$ are the charge velocities before and after emitting of the γ quanta; $\gamma = 1/\sqrt{1-\beta^2}$, $\gamma_0 = 1/\sqrt{1-\beta_0^2}$; $\vec{e}_\gamma$ and ω are the unit vector in the direction of emitted γ quanta and its frequency, n is the medium refractive index. We rewrite (9.27) in the dimensionless form

$$\gamma_0 = \gamma + \epsilon, \quad \vec{\beta}_0\gamma_0 = \vec{\beta}\gamma + \epsilon n\vec{e}_\gamma. \tag{9.28}$$

Here $\vec{\beta} = \vec{v}/c$, $\vec{\beta}_0 = \vec{v}_0/c$, $\epsilon = \hbar\omega/m_0 c^2$.

Let $\vec{v}_0$ be directed along the z axis. We project all vectors on this axis and two others perpendicular to it:

$$\vec{\beta}_0 = \beta_0\vec{e}_z, \quad \vec{\beta} = \beta[\vec{e}_z\cos\theta + \sin\theta(\vec{e}_x\cos\phi + \vec{e}_y\sin\phi)],$$

$$\vec{e}_\gamma = \vec{e}_z\cos\theta_\gamma + \sin\theta_\gamma(\vec{e}_x\cos\phi_\gamma + \vec{e}_y\sin\phi_\gamma)]. \tag{9.29}$$

Substituting (9.29) into (9.28), one gets

$$\gamma_0 = \gamma + \epsilon, \quad \beta_0\gamma_0 = \beta\gamma\cos\theta + n\epsilon\cos\theta_\gamma,$$

$$\beta\gamma\sin\theta\cos\phi + n\epsilon\sin\theta_\gamma\cos\phi_\gamma = 0,$$

$$\beta\gamma\sin\theta\sin\phi + n\epsilon\sin\theta_\gamma\sin\phi_\gamma = 0. \tag{9.30}$$

From two last equations one finds

$$\sin\theta \sin(\phi - \phi_\gamma) = 0, \quad \sin\theta_\gamma \sin(\phi - \phi_\gamma) = 0. \tag{9.31}$$

For $\sin(\phi - \phi_\gamma) \neq 0$, one finds that $\sin\theta = \sin\theta_\gamma = 0$. There are three different physical ways to satisfy this equality.

Let $\theta = \theta_\gamma = 0$. Then Eqs. (9.30) are reduced to

$$\gamma_0 = \gamma + \epsilon, \quad \beta_0\gamma_0 = \beta\gamma + n\epsilon. \tag{9.32}$$

From this one easily obtains

$$\beta = \frac{2n - \beta_0(n^2 + 1)}{n^2 + 1 - 2n\beta_0}, \quad \epsilon = \frac{2\gamma_0(\beta_0 n - 1)}{n^2 - 1}. \tag{9.33}$$

The conditions $0 < \epsilon < \gamma_0$ and $0 < \beta < \beta_0$ give

$$\frac{1}{n} < \beta_0 < \frac{2n}{1 + n^2} \tag{9.34}$$

for $n > 1$. There are no solutions for $n < 1$. In the past, the possibility of the one-photon radiation in the forward direction by a charge moving in medium was suggested by Tyapkin on the purely intuitive grounds [15]. Equations (9.32)-(9.34) tell us that this assumption is not in conflict with kinematics.

There are no solutions of (9.30) for $\theta = 0$, $\theta_\gamma = \pi$.

Finally, for $\theta = \pi$, $\theta_\gamma = 0$ one finds

$$\beta = \frac{\beta_0(n^2 + 1) - 2n}{n^2 + 1 - 2n\beta_0}, \quad \epsilon = \frac{2\gamma_0(\beta_0 n - 1)}{n^2 - 1}.$$

This solution exists only for $n > 1$, $\beta_0 > 2n/(1 + n^2)$.

Let now $\sin(\phi - \phi_\gamma) = 0$. There are no physical solutions of (9.30) if $\phi = \phi_\gamma$. It remains only $\phi = \phi_\gamma + \pi$. Then,

$$\gamma_0 = \gamma + \epsilon, \quad \beta_0\gamma_0 = \beta\gamma \cos\theta + n\epsilon \cos\theta_\gamma,$$

$$\beta\gamma \sin\theta = n\epsilon \sin\theta_\gamma. \tag{9.35}$$

These equations have the well-known solution given by Ginzburg [25]

$$\cos\theta_\gamma = \frac{1}{\beta_0 n}\left[1 + \frac{\epsilon(n^2 - 1)}{2\gamma_0}\right],$$

$$\cos\theta = \frac{\beta^2\gamma^2 + \beta_0^2\gamma_0^2 - n^2(\gamma_0 - \gamma)^2}{2\beta\gamma\beta_0\gamma_0}. \tag{9.36}$$

The conditions that the r.h.s. of these equations should be smaller than 1 and greater than -1, lead to the following conditions

$$\frac{|2n - \beta_0(n^2 + 1)|}{n^2 + 1 - 2n\beta_0} < \beta < \beta_0, \quad \epsilon < \frac{2\gamma_0(\beta_0 n - 1)}{n^2 - 1}. \tag{9.37}$$

Eqs. (9.35)-(9.37) can be realized only for $n > 1$, $\beta_0 > 1/n$.

9.3.2. THE KINEMATICS OF THE TWO-PHOTON CHERENKOV EFFECT

General formulae
The energy-momentum conservation gives

$$\gamma_0 = \gamma + \epsilon_1 + \epsilon_2, \quad \gamma_0\vec{\beta}_0 = \gamma\vec{\beta} + \epsilon_1 n_1 \vec{e}_1 + \epsilon_2 n_2 \vec{e}_2. \tag{9.38}$$

Here

$$\epsilon_1 = \frac{\hbar\omega_1}{m_0 c^2}, \quad \epsilon_2 = \frac{\hbar\omega_2}{m_0 c^2}, \quad n_1 = n(\omega_1), \quad n_2 = n(\omega_2),$$

ω_1 and ω_2 are the frequencies of the γ quanta 1 and 2, n_1 and n_2 are the corresponding refractive indices, $\vec{e}_1$ and $\vec{e}_2$ are the unit vectors along the directions of the emitted photons.

Projecting (9.38) on the same axes as above one gets

$$\gamma_0 = \gamma + \epsilon_1 + \epsilon_2, \quad \gamma_0\beta_0 = \gamma\beta\cos\theta + \epsilon_1 n_1 \cos\theta_1 + \epsilon_2 n_2 \cos\theta_2,$$

$$\beta\gamma\sin\theta\cos\phi + \epsilon_1 n_1 \sin\theta_1 \cos\phi_1 + \epsilon_2 n_2 \sin\theta_2 \cos\phi_2 = 0,$$

$$\beta\gamma\sin\theta\sin\phi + \epsilon_1 n_1 \sin\theta_1 \sin\phi_1 + \epsilon_2 n_2 \sin\theta_2 \sin\phi_2 = 0. \tag{9.39}$$

From the last two equations one gets

$$\cos(\phi_1 - \phi) = \frac{\epsilon_2^2 n_2^2 \sin^2\theta_2 - \epsilon_1^2 n_1^2 \sin^2\theta_1 - \beta^2\gamma^2 \sin^2\theta}{2\beta\gamma\epsilon_1 n_1 \sin\theta \sin\theta_1},$$

$$\cos(\phi_2 - \phi) = \frac{\epsilon_1^2 n_1^2 \sin^2\theta_1 - \epsilon_2^2 n_2^2 \sin^2\theta_2 - \beta^2\gamma^2 \sin^2\theta}{2\beta\gamma\epsilon_2 n_2 \sin\theta \sin\theta_2}. \tag{9.40}$$

For the given β_0 (initial charge velocity), β, θ, ϕ (the final charge velocity and its direction), ϵ_1, θ_1 (the frequency and the inclination angle towards the motion axis for the first photon) the first and second of Eqs. (9.39) define the frequency and the inclination angle towards the motion axis for the second photon) while Eqs. (9.40) define azimuthal angles for the 1 and 2 photons. These angles are not independent:

$$\cos(\phi_2 - \phi_1) = \frac{\beta^2\gamma^2 \sin^2\theta - \epsilon_1^2 n_1^2 \sin^2\theta_1 - \epsilon_2^2 n_2^2 \sin^2\theta_2}{2\epsilon_1 n_1 \epsilon_2 n_2 \sin\theta_1 \sin\theta_2}. \tag{9.41}$$

The conditions

$$-1 < \cos(\phi_1 - \phi) < 1, \quad -1 < \cos(\phi_2 - \phi) < 1, \quad -1 < \cos(\phi_2 - \phi_1) < 1$$

lead to the following restrictions on θ, θ_1 and θ_2:

$$\frac{|n_1\epsilon_1 \sin\theta_1 - n_2\epsilon_2 \sin\theta_2|}{\beta\gamma} \le \sin\theta \le \frac{n_1\epsilon_1 \sin\theta_1 + n_2\epsilon_2 \sin\theta_2}{\beta\gamma},$$

$$\frac{|\beta\gamma \sin\theta - n_2\epsilon_2 \sin\theta_2|}{n_1\epsilon_1} \le \sin\theta_1 \le \frac{\beta\gamma \sin\theta + n_2\epsilon_2 \sin\theta_2}{n_1\epsilon_1},$$

$$\frac{|\beta\gamma \sin\theta - n_1\epsilon_1 \sin\theta_1|}{n_2\epsilon_2} \le \sin\theta_2 \le \frac{\beta\gamma \sin\theta + n_1\epsilon_1 \sin\theta_1}{n_2\epsilon_2}. \tag{9.42}$$

The energy of the recoil charge enters only through the $\beta\gamma \sin\theta$ term. It can be excluded using the relations

$$\beta\gamma = \sqrt{(\gamma_0 - \epsilon_1 - \epsilon_2)^2 - 1},$$

$$\beta\gamma \sin\theta = [\beta^2\gamma^2 - (\gamma_0\beta_0 - \epsilon_1 n_1 \cos\theta_1 - \epsilon_2 n_2 \cos\theta_2)^2]^{1/2}. \tag{9.43}$$

For the extremely relativistic charges ($\gamma_0 \gg \epsilon_1$, $\gamma_0 \gg \epsilon_2$)

$$\sin\theta = \sqrt{\frac{2}{\gamma_0}}[\epsilon_1(n_1 \cos\theta_1 - 1) + \epsilon_2(n_2 \cos\theta_2 - 1)]^{1/2},$$

that is, $\theta \to 0$ when $\beta_0 \to 1$. It follows from this that

$$\epsilon_1(n_1 \cos\theta_1 - 1) + \epsilon_2(n_2 \cos\theta_2 - 1) \ge 0.$$

This inequality cannot be satisfied if both n_1 and n_2 are smaller than 1. In the same relativistic limit

$$\beta\gamma \sin\theta = \sqrt{2\gamma_0}[\epsilon_1(n_1 \cos\theta_1 - 1) + \epsilon_2(n_2 \cos\theta_2 - 1)]^{1/2}$$

is finite despite the large $\sqrt{\gamma_0}$ factor. This becomes evident if we rewrite the first of equations (9.42) in the form

$$|n_1\epsilon_1 \sin\theta_1 - n_2\epsilon_2 \sin\theta_2| \le \beta\gamma \sin\theta \le n_1\epsilon_1 \sin\theta_1 + n_2\epsilon_2 \sin\theta_2$$

and note that θ enters into two last inequalities (9.42) through the same combination $\beta\gamma \sin\theta$.

If the energy of one of photons is zero, Eqs. (9.40)-(9.42) are reduced to (9.35) and, consequently, to (9.36).

Particular cases

Inequalities (9.40)-(9.42) reduce to equalities when either the recoil charge moves in the same direction as the initial one ($\theta = 0$) or when one of photons moves along the direction of the initial charge ($\theta_1 = 0$ or $\theta_2 = 0$). We consider these cases separately.

A charge does not change the motion direction   Let $\theta = 0$, that is a charge does not change the motion direction. Then, from (9.42) it follows that

$$n_1 \epsilon_1 \sin \theta_1 = n_2 \epsilon_2 \sin \theta_2, \qquad (9.44)$$

whilst (9.41) gives

$$\cos(\phi_2 - \phi_1) = -1, \quad \phi_2 = \phi_1 + \pi, \qquad (9.45)$$

that is, photons fly in opposite azimuthal directions. As a result, Eqs. (9.39) reduce to

$$\gamma_0 = \gamma + \epsilon_1 + \epsilon_2, \quad \gamma_0 \beta_0 - \gamma \beta = \epsilon_1 n_1 \cos \theta_1 + \epsilon_2 n_2 \cos \theta_2,$$

$$n_1 \epsilon_1 \sin \theta_1 = n_2 \epsilon_2 \sin \theta_2. \qquad (9.46)$$

From this one easily obtains $\cos \theta_1$ and $\cos \theta_2$

$$\cos \theta_1 = \frac{(\beta_0 \gamma_0 - \beta \gamma)^2 + \epsilon_1^2 n_1^2 - \epsilon_2^2 n_2^2}{2(\beta_0 \gamma_0 - \beta \gamma) \epsilon_1 n_1},$$

$$\cos \theta_2 = \frac{(\beta_0 \gamma_0 - \beta \gamma)^2 - \epsilon_1^2 n_1^2 + \epsilon_2^2 n_2^2}{2(\beta_0 \gamma_0 - \beta \gamma) \epsilon_2 n_2}. \qquad (9.47)$$

The conditions $-1 < \cos \theta_1 < 1$ and $-1 < \cos \theta_2 < 1$ lead to the inequality which can be presented in the following two equivalent forms:

$$|\epsilon_1 n_1 - \epsilon_2 n_2| \leq \beta_0 \gamma_0 - \beta \gamma \leq \epsilon_1 n_1 + \epsilon_2 n_2,$$

$$|\beta_0 \gamma_0 - \beta \gamma - \epsilon_1 n_1| \leq n_2 (\gamma_0 - \gamma - \epsilon_1) \leq \beta_0 \gamma_0 - \beta \gamma + \epsilon_1 n_1. \qquad (9.48)$$

These inequalities can be easily resolved. For definiteness, we suggest that $n_2 > n_1$. There are the following possibilities depending on n_1, n_2, β_0 and β (see [39] for details):

$$1) \quad n_2 > 1 > n_1.$$

In this case the inequality (9.48) has solution

$$\beta_2 < \beta < \beta_0 \quad \text{for} \quad \frac{1}{n_2} < \beta_0 < \frac{2n_2}{1 + n_2^2} \qquad (9.49)$$

and

$$0 < \beta < \beta_0 \quad \text{for} \quad \beta_0 > \frac{2n_2}{1 + n_2^2}. \tag{9.50}$$

Here

$$\beta_1 = \frac{2n_1 - \beta_0(1 + n_1^2)}{1 + n_1^2 - 2n_1\beta_0}, \quad \beta_2 = \frac{2n_2 - \beta_0(1 + n_2^2)}{1 + n_2^2 - 2n_2\beta_0}.$$

When the conditions (9.49) and (9.50) are satisfied, the dimensionless energy of the first photon belongs to the interval

$$\frac{n_2(\gamma_0 - \gamma) - (\beta_0\gamma_0 - \beta\gamma)}{n_1 + n_2} \leq \epsilon_1 \leq \frac{n_2(\gamma_0 - \gamma) - (\beta_0\gamma_0 - \beta\gamma)}{n_2 - n_1}. \tag{9.51}$$

The energy of the second photon is positive if $\epsilon_2 = \gamma_0 - \gamma - \epsilon_1 > 0$. Since the inequality

$$\epsilon_1 < \frac{n_2(\gamma_0 - \gamma) - (\beta_0\gamma_0 - \beta\gamma)}{n_2 - n_1} < \gamma_0 - \gamma \tag{9.52}$$

holds when the inequalities (9.49) and (9.50) are satisfied, the positivity of ϵ_2 is guaranteed.

$$1) \quad n_2 > n_1 > 1.$$

For $n_1 < (1 + n_2^2)/2n_2$ (this corresponds to the following chain of inequalities $1/n_2 < 2n_2/(1 + n_2^2) < 1/n_1 < 2n_1/(1 + n_1^2)$) one obtains:

$$\beta_2 < \beta < \beta_0 \quad \text{for} \quad \frac{1}{n_2} < \beta_0 < \frac{2n_2}{1 + n_2^2},$$

$$0 < \beta < \beta_0 \quad \text{for} \quad \frac{2n_2}{1 + n_2^2} < \beta_0 < \frac{1}{n_1},$$

$$0 < \beta < \beta_1 \quad \text{for} \quad \frac{1}{n_1} < \beta_0 < \frac{2n_1}{1 + n_1^2}. \tag{9.53}$$

For $n_1 > (1 + n_2^2)/2n_2$ (this corresponds to the chain of inequalities $1/n_2 < 1/n_1 < 2n_2/(1 + n_2^2) < 2n_1/(1 + n_1^2)$) one finds:

$$\beta_2 < \beta < \beta_0 \quad \text{for} \quad \frac{1}{n_2} < \beta_0 < \frac{1}{n_1},$$

$$\beta_2 < \beta < \beta_1 \quad \text{for} \quad \frac{1}{n_1} < \beta_0 < \frac{2n_2}{1 + n_2^2},$$

$$0 < \beta < \beta_1 \quad \text{for} \quad \frac{2n_2}{1 + n_2^2} < \beta_0 < \frac{2n_1}{1 + n_1^2}. \tag{9.54}$$

When β and β_0 lie inside the intervals defined by (9.53) and (9.54), ϵ_1 satisfies the same inequality (9.51).

On the other hand, the inequality

$$\frac{n_2(\gamma_0 - \gamma) - (\beta_0\gamma_0 - \beta\gamma)}{n_1 + n_2} \le \epsilon_1 \le \frac{n_2(\gamma_0 - \gamma) + (\beta_0\gamma_0 - \beta\gamma)}{n_2 + n_1} \qquad (9.55)$$

holds when

$$\beta_1 < \beta < \beta_0 \quad \text{for} \quad \frac{1}{n_1} < \beta_0 < \frac{2n_1}{1 + n_1^2}$$

and

$$0 < \beta < \beta_0 \quad \text{for} \quad \beta_0 > \frac{2n_1}{1 + n_1^2}. \qquad (9.56)$$

There are no solutions of (9.48) if both n_1 and n_2 are smaller than 1. A further analysis of (9.47) and (9.48) requires the knowledge of the dispersion law $n(\omega)$.

For the nondispersive medium, these equations are greatly simplified. It turns out that ϵ_1 satisfies the inequality

$$\frac{n(\gamma_0 - \gamma) - (\beta_0\gamma_0 - \beta\gamma)}{2n} \le \epsilon_1 \le \frac{n(\gamma_0 - \gamma) + (\beta_0\gamma_0 - \beta\gamma)}{2n} \qquad (9.57)$$

which is valid under the condition

$$n(\gamma_0 - \gamma) > \beta_0\gamma_0 - \beta\gamma. \qquad (9.58)$$

In a manifest form, this equation for $n > 1$ looks like

$$\frac{2n - \beta_0(n^2 + 1)}{1 + n^2 - 2n\beta_0} \le \beta \le \beta_0, \quad \text{for} \quad \frac{1}{n} < \beta_0 < \frac{2n}{1 + n^2}$$

and

$$0 < \beta < \beta_0 \quad \text{for} \quad \beta_0 > \frac{2n}{1 + n^2}. \qquad (9.59)$$

There are no solutions of (9.57) for $n < 1$.

As a result, we obtain the following prescription for the measurement of the two-photon Cherenkov radiation. Set the charged particle detector on the motion axis. It should be tuned in such a way as to detect a particular charge velocity in the intervals (9.49), (9.50),(9.53), (9.54) or (9.56). Correspondingly, the energy of one of the photons should be chosen in the intervals (9.51) or (9.55). The energy of other photon is found from the first of Eqs. (9.38). Set the photon detectors under the polar angles given by (9.47) and, in accordance with (9.45), under opposite azimuthal angles. Since θ_1 and θ_2 are uniquely determined by β_0, β and ϵ_1, the corresponding radiation intensities should have sharp maxima at these angles. Equations (9.47)-(9.56) are useful if one is able to measure the charge velocity after emitting the gamma quanta. When only the measurements of gamma quanta energies are possible we rewrite (9.48) in the form

$$|\beta_0\gamma_0 - \beta\gamma - \epsilon_1 n_1| \leq n_2\epsilon_2 \leq \beta_0\gamma_0 - \beta\gamma + \epsilon_1 n_1 \qquad (9.60)$$

and substitute $\beta\gamma$ given by (9.43) into (9.47) and (9.60). Then, (9.47) define θ_1 and θ_2 for the given ϵ_1 and ϵ_2, while (9.60) defines the available values of ϵ_1 and ϵ_2.

This is especially clear for the extremely relativistic case when the velocities of the initial and recoil charges are very close to c ($\beta_0 \approx 1$, $\beta \approx 1$). Instead of (9.47) one gets

$$\cos\theta_1 = \frac{(\epsilon_1 + \epsilon_2)^2 + \epsilon_1^2 n_1^2 - \epsilon_2^2 n_2^2}{2(\epsilon_1 + \epsilon_2)\epsilon_1 n_1},$$

$$\cos\theta_2 = \frac{(\epsilon_1 + \epsilon_2)^2 - \epsilon_1^2 n_1^2 + \epsilon_2^2 n_2^2}{2(\epsilon_1 + \epsilon_2)\epsilon_2 n_2}. \qquad (9.61)$$

The inequality (9.48) reduces to

$$\frac{1 - n_1}{n_2 - 1}\epsilon_1 < \epsilon_2 < \frac{n_1 + 1}{n_2 - 1}\epsilon_1$$

for $n_2 > 1 > n_1$ and to

$$\frac{n_1 - 1}{n_2 + 1}\epsilon_1 < \epsilon_2 < \frac{n_1 + 1}{n_2 - 1}\epsilon_1 \qquad (9.62)$$

for $n_2 > n_1 > 1$.

For the extremely relativistic charges moving in the hypothetical nondispersive medium ($n > 1$) these equations are simplified

$$\cos\theta_1 = \frac{(\epsilon_1 + \epsilon_2)^2 + n^2(\epsilon_1^2 - \epsilon_2^2)}{2(\epsilon_1 + \epsilon_2)\epsilon_1 n}, \qquad \cos\theta_2 = \frac{(\epsilon_1 + \epsilon_2)^2 - n^2(\epsilon_1^2 - \epsilon_2^2)}{2(\epsilon_1 + \epsilon_2)\epsilon_2 n},$$

$$\frac{n - 1}{n + 1}\epsilon_1 < \epsilon_2 < \frac{n + 1}{n - 1}\epsilon_1.$$

It should be mentioned on the case $\theta = \pi$ corresponding to a recoil charge moving in the backward direction. The photon emission angles and the available photon frequencies are obtained from (9.47) and (9.48) by replacing ($\beta\gamma \rightarrow -\beta\gamma$) in them. It is seen that at least one of photons should have high energy.

One of the photons moves along the direction of the initial charge. For definiteness, let this photon be the second one ($\theta_2 = 0$). Then, it follows from (9.42) that $\beta\gamma\sin\theta = n_1\epsilon_1\sin\theta_1$. Substituting this into (9.40) one finds $\cos(\phi_1 - \phi) = -1$, $\phi_1 = \phi - \pi$, that is, the recoil charge and photon fly in opposite azimuthal directions. As a result, one gets the following equations

$$\gamma_0 = \gamma + \epsilon_1 + \epsilon_2, \quad \beta\gamma\sin\theta = n_1\epsilon_1\sin\theta_1,$$

$$\beta_0\gamma_0 = \epsilon_2 n_2 + \epsilon_1 n_1\cos\theta_1 + \beta\gamma\cos\theta.$$

From this one finds easily θ_1 and θ:

$$\cos\theta = \frac{(\gamma_0\beta_0 - \epsilon_2 n_2)^2 + \gamma^2\beta^2 - \epsilon_1^2 n_1^2}{2\gamma\beta(\gamma_0\beta_0 - \epsilon_2 n_2)},$$

$$\cos\theta_1 = \frac{(\gamma_0\beta_0 - \epsilon_2 n_2)^2 - \gamma^2\beta^2 + \epsilon_1^2 n_1^2}{2\epsilon_1 n_1(\gamma_0\beta_0 - \epsilon_2 n_2)}. \tag{9.63}$$

The conditions that r.h.s. of these equations be smaller than 1 and greater than -1, give the following inequality

$$|\gamma_0\beta_0 - \epsilon_1 n_1 - \epsilon_2 n_2| < \beta\gamma < |\gamma_0\beta_0 + \epsilon_1 n_1 - \epsilon_2 n_2|. \tag{9.64}$$

These equations are useful when one is able to measure only the photons energies. In fact, substituting $\gamma\beta$ from (9.43) into (9.63) one gets the polar angles of recoil charge and that of the first photon. Making the same substitution in (9.64), one finds the set of available ϵ_1 and ϵ_2:

$$|\gamma_0\beta_0 - \epsilon_1 n_1 - \epsilon_2 n_2| < [(\gamma_0 - \epsilon_1 - \epsilon_2)^2 - 1]^{1/2} < |\gamma_0\beta_0 + \epsilon_1 n_1 - \epsilon_2 n_2|. \tag{9.65}$$

We do not further elaborate Eq.(9.65) by presenting it in a manifest form similar as it was done for (9.48).

The measurement procedure reduces to the following one. Choose the photon energies ϵ_1 and ϵ_2. Check, whether they satisfy (9.65). Set the photon counters at the initial charge motion direction and at the angle θ_1 defined in (9.63). Since the kinematical conditions define uniquely θ_1 (similarly to the one-photon radiation), the corresponding radiation intensity will have a sharp maximum at this angle for the photons with energy ϵ_1. The counters tuned into the coincidence, will certainly detect photons arising from the two-photon Cherenkov effect.

It should be mentioned on the case $\theta_2 = \pi$ when one of photons (say, 2) moves in the backward direction. The emission angles of the recoil charge and another photon, and the available β, ϵ_1 and ϵ_2 are obtained from (9.63)-(9.5) by replacing $(n_2\epsilon_2 \rightarrow -n_2\epsilon_2)$ in them.

9.3.3. BACK TO THE GENERAL TWO-PHOTON CHERENKOV EFFECT

The situation is more complicated for the general two-photon Cherenkov radiation described by Eqs. (9.39)-(9.42). It is easy to check that only one of inequalities (9.42) is independent. It is convenient to choose the first of them rewriting it in the form

$$(n_1\epsilon_1 \sin\theta_1 - n_2\epsilon_2 \sin\theta_2)^2 \leq \beta^2\gamma^2 \sin^2\theta \leq (n_1\epsilon_1 \sin\theta_1 + n_2\epsilon_2 \sin\theta_2)^2. \quad (9.66)$$

This inequality is satisfied trivially for particular cases $\theta = 0$ and $\theta_1 = 0$. considered above. However, there are other solutions of (9.66).

Another particular case
To find this case we substitute $\beta\gamma \sin\theta$ from (9.43) to (9.66) thus obtaining the following inequality

$$\cos\theta_2^{(1)} < \cos\theta_2 < \cos\theta_2^{(2)}, \quad (9.67)$$

where

$$\cos\theta_2^{(1)} = A - R, \quad \cos\theta_2^{(2)} = A + R, \quad (9.68)$$

$$A = \frac{c_1}{2n_2\epsilon_2} \frac{\beta_0\gamma_0 - \epsilon_1 n_1 \cos\theta_1}{Z^2},$$

$$R = \frac{\beta_0\gamma_0\epsilon_1^2 n_1^2 \sin\theta_1}{\epsilon_2 n_2 Z^2} [(\cos\theta_1 - \cos\theta_1^{(1)})(\cos\theta_1^{(2)} - \cos\theta_1)]^{1/2},$$

$$\cos\theta_1^{(1)} = \frac{\epsilon_1^2 n_1^2 + \beta_0^2\gamma_0^2 - (\epsilon_2 n_2 + \beta\gamma)^2}{2\beta_0\gamma_0\epsilon_1 n_1},$$

$$\cos\theta_1^{(2)} = \frac{\epsilon_1^2 n_1^2 + \beta_0^2\gamma_0^2 - (\epsilon_2 n_2 - \beta\gamma)^2}{2\beta_0\gamma_0\epsilon_1 n_1}, \quad (9.69)$$

$$c_1 = \epsilon_1^2 n_1^2 + \epsilon_2^2 n_2^2 + (\epsilon_1 + \epsilon_2)(2\gamma_0 - \epsilon_1 - \epsilon_2) - 2\beta_0\gamma_0\epsilon_1 n_1 \cos\theta_1,$$

$$Z^2 = \epsilon_1^2 n_1^2 + \beta_0^2\gamma_0^2 - 2\beta_0\gamma_0\epsilon_1 n_1 \cos\theta_1.$$

We see that available values of θ_1 are in the interval $\cos\theta_1^{(1)} < \cos\theta_1 < \cos\theta_1^{(2)}$. The inequality (9.67) becomes an equality when $R = 0$. Aside from the trivial case $\sin\theta_1 = 0$ considered above, R vanishes for $\cos\theta_1 = \cos\theta_1^{(1)}$ or $\cos\theta_1 = \cos\theta_1^{(2)}$. The $\cos\theta_2^{(i)}$ corresponding to $\cos\theta_1^{(i)}$ are given by

$$\cos\theta_2^{(1)} = \frac{\beta_0^2\gamma_0^2 - \epsilon_1^2 n_1^2 + (\epsilon_2 n_2 + \beta\gamma)^2}{2\beta_0\gamma_0(\epsilon_2 n_2 + \beta\gamma)},$$

$$\cos\theta_2^{(2)} = \frac{\beta_0^2\gamma_0^2 - \epsilon_1^2 n_1^2 + (\epsilon_2 n_2 - \beta\gamma)^2}{2\beta_0\gamma_0(\epsilon_2 n_2 - \beta\gamma)}. \quad (9.70)$$

Obviously, the r.h.s. of (9.69) and (9.70) should be smaller than 1 and greater than -1. This defines the interval of ϵ_1 and ϵ_2 for which the solution discussed exists.

The polar angle of the recoil charge is found from the relation

$$\beta\gamma\cos\theta_i = \beta_0\gamma_0 - \epsilon_1 n_1 \cos\theta_1^{(i)} - \epsilon_2 n_2 \cos\theta_2^{(i)}, \tag{9.71}$$

where $\cos\theta_1^{(i)}$ and $\cos\theta_2^{(i)}$ are the same as in (9.69) and (9.70).

In the relativistic limit ($\epsilon_1 \ll \gamma_0$, $\epsilon_2 \ll \gamma_0$) (9.69) and (9.70) go into

$$\cos\theta_1^{(1)} = \frac{\epsilon_1 + \epsilon_2 - \epsilon_2 n_2}{\epsilon_1 n_1}, \quad \cos\theta_2^{(1)} = 1,$$

$$\cos\theta_1^{(2)} = \frac{\epsilon_1 + \epsilon_2 + \epsilon_2 n_2}{\epsilon_1 n_1}, \quad \cos\theta_2^{(2)} = -1.$$

The first and second lines of these equations coincide with the relativistic limits of $\theta_2 = 0$ and $\theta_2 = \pi$ cases considered at the end of section (9.3.2).

Relativistic limit
In the relativistic limit, (9.66) reduces to inequalities

$$\epsilon_1 + \epsilon_2 \leq n_1\epsilon_1\cos\theta_1 + n_2\epsilon_2\cos\theta_2$$

$$\epsilon_1 + \epsilon_2 \geq n_1\epsilon_1\cos\theta_1 + n_2\epsilon_2\cos\theta_2,$$

which are compatible only if

$$\epsilon_1(n_1\cos\theta_1 - 1) + \epsilon_2(n_2\cos\theta_2 - 1) = 0. \tag{9.72}$$

This equation has no solutions if both n_1 and n_2 are smaller than 1. We extract $\cos\theta_2$:

$$\cos\theta_2 = \frac{1}{n_2} + \frac{\epsilon_1(1 - n_1\cos\theta_1)}{n_2\epsilon_2}. \tag{9.73}$$

For definiteness we choose $n_2 > n_1$ and $n_2 > 1$. The right hand side of this equation should be smaller than 1 and greater than -1. This leads to the following inequality for $\cos\theta_1$:

$$\frac{1}{n_1} - \frac{\epsilon_2(n_2 - 1)}{\epsilon_1 n_1} \leq \cos\theta_1 \leq \frac{1}{n_1} + \frac{\epsilon_2(n_2 + 1)}{\epsilon_1 n_1}.$$

It is convenient to rewrite this equation in a manifest form.

Let $n_2 > n_1 > 1$.

Then, available θ_1 lie in the following intervals

$$-1 < \cos\theta_1 < 1 \quad \text{for} \quad \epsilon_2 > \epsilon_1 \frac{1+n_1}{n_2-1},$$

$$\frac{1}{n_1} - \frac{\epsilon_2(n_2-1)}{\epsilon_1 n_1} < \cos\theta_1 < 1 \quad \text{for} \quad \epsilon_1 \frac{n_1-1}{n_2+1} < \epsilon_2 < \epsilon_1 \frac{n_1+1}{n_2-1}$$

and

$$\frac{1}{n_1} - \frac{\epsilon_2(n_2-1)}{\epsilon_1 n_1} \leq \cos\theta_1 \leq \frac{1}{n_1} + \frac{\epsilon_2(n_2+1)}{\epsilon_1 n_1} \quad \text{for} \quad 0 < \epsilon_2 < \epsilon_1 \frac{n_1-1}{n_2+1}.$$

Let $n_2 > 1$, $n_1 < 1$.

Then, available values of θ_1 belong to the intervals

$$-1 < \cos\theta_1 < 1 \quad \text{for} \quad \epsilon_2 > \epsilon_1 \frac{1+n_1}{n_2-1},$$

and

$$\frac{1}{n_1} - \frac{\epsilon_2(n_2-1)}{\epsilon_1 n_1} < \cos\theta_1 < 1 \quad \text{for} \quad \epsilon_1 \frac{1-n_1}{n_2-1} < \epsilon_2 < \epsilon_1 \frac{n_1+1}{n_2-1}.$$

It follows from these equations that there is a continuum of pairs θ_1, θ_2 connected by (9.72). This means that in a general relativistic case, rather broad distributions of radiation intensities should be observed. The kinematical consideration is not sufficient now and concrete calculations are needed.

In the specific case $\theta = 0$, $\cos\theta_1$ and $\cos\theta_2$ also satisfy (9.72) but their values are fixed by (9.47).

9.3.4. RELATION TO THE CLASSICAL CHERENKOV EFFECT

We discuss now how the classical electromagnetic field strengths (which are the solutions of the Maxwell equations with classical currents in their r.h.s.) to the quantum field strengths operators. In quantum electrodynamics [35, 36] they are defined as eigenvalues of the quantum field strengths operators when they act on the so-called coherent states. The latter can be presented as an infinite sum over states with a fixed photon numbers. The coefficients at these states are related to the Fourier components of the classical currents. Therefore, classical solutions of the Maxwell equations involve contributions from states with arbitrary photon numbers. Aforesaid is valid only for the current flowing in vacuum. If one suggests that the same reasoning can be applied to the charge motion in medium, the classical formulae describing Cherenkov radiation contain contributions from the states with arbitrary photon numbers.

9.4. Discussion and Conclusion on the Two-Photon Cherenkov Effect

Using the analogy with the Doppler effect for the scattering of light by a charge moving in medium, Frank [4, 37] obtained the following condition for the emission of two photons:

$$\epsilon_1(\beta n_1 \cos \theta_1 - 1) + \epsilon_2(\beta n_2 \cos \theta_2 - 1) = 0, \tag{9.74}$$

where β is the initial charge velocity. In the relativistic limit ($\beta \approx 1$), (9.74) coincides with equation (9.72) following from the relativistic kinematics. However, for arbitrary β, (9.74) is not compatible with exact kinematical inequalities (9.66) and (9.67) for the two-photon emission and, therefore, the above analogy with the Doppler effect is not at least complete.

It turns out that highly relativistic charges are not convenient for the observation of the two-photon Cherenkov effect. As we have seen, in this case the recoil charge flies in the almost forward direction and it will be rather difficult to discriminate it from the recoil charge moving exactly in the forward direction (only for this particular kinematics the photon emission angles θ_1 and θ_2 are fixed (see (9.47)). It is desirable to choose the energy of the initial charge only slightly above the summary energy of two photons. Certainly, kinematics itself cannot tell us how frequently the recoil charge or one of the photons moves exactly in the forward direction. For this, concrete calculations are needed.

In general, to each angle θ_1 there corresponds the interval of θ_2 defined by (9.66). Only for special cases:

1) when the recoil charge moves in the same (or in opposite) direction as the initial charge; 2) when one of the photons moves along (or against) the direction of the initial charge; 3) for the orientations of the photons and recoil charge defined by (9.69)- (9.71) the directions of the recoil charge and photons are uniquely defined similarly to the single-photon Cherenkov effect. The corresponding radiation intensities should have a sharp maximum for such orientations.

This makes easier the experimental search for the 2-photon Cherenkov effect.

The content of this paper is partly grounded on Refs. [38] and [39].

References

1. Frank I.M. and Tamm I.E. (1937) Coherent Radiation of Fast Electron in Medium, *Dokl. Akad. Nauk*, **14**, pp. 107-113.
2. Cherenkov P.A. (1934) Visible luminescence of the pure fluids induced by γ rays *Dokl. Acad. Nauk SSSR*, **2**, pp. 451-454.
3. Vavilov S.I. (1934) On Possible Reasons for the Blue γ Radiation in Fluids, *Dokl. Akad, Nauk*, **2**, **8**, pp. 457-459.

4. Frank I.M. (1988) *Vavilov-Cherenkov Radiation*, Nauka, Moscow.
5. Tamm I.E. (1939) Radiation Induced by Uniformly Moving Electrons, *J. Phys. USSR*, **1, No 5-6**, pp. 439-461.
6. Lawson J.D. (1954) On the Relation between Cherenkov Radiation and Bremsstrahlung *Phil. Mag.*, **45**, pp.748-750;
   Lawson J.D. (1965) Cherenkov Radiation, "Physical" and "Unphysical", and its Relation to Radiation from an Accelerated Electron *Amer. J. Phys.*, **33**, pp. 1002-1005;
   Aitken D.K. et al. (1963) Transition Radiation in Cherenkov Detectors *Proc. Phys. Soc.*, **83**, pp. 710-722.
7. Zrelov V.P. and Ruzicka J. (1989) Analysis of Tamm's Problem on Charge Radiation at its Uniform Motion over a Finite Trajectory *Czech. J. Phys.*, **B 39**, pp. 368-383;
   Zrelov V.P. and Ruzicka J. (1992) Optical Bremsstrahlung of Relativistic Particles in a Transparent Medium and its Relation to the Vavilov-Cherenkov Radiation *Czech. J. Phys.*, **42**, pp. 45-57.
8. Afanasiev G.N., Beshtoev Kh. and Stepanovsky Yu.P. (1996) Vavilov-Cherenkov Radiation in a Finite Region of Space *Helv. Phys. Acta*, **69**, pp. 111-129.
9. Afanasiev G.N., Kartavenko V.G. and Stepanovsky Yu.P. (1999) On Tamm's Problem in the Vavilov-Cherenkov Radiation Theory *J.Phys. D: Applied Physics*, **32**, pp. 2029-2043.
10. Afanasiev G.N. and Shilov V.M. (2002) Cherenkov Radiation versus Bremsstrahlung in the Tamm Problem *J.Phys. D: Applied Physics*, **35**, pp. 854-866.
11. Zrelov V.P. (1970) *Vavilov-Cherenkov Radiation in High-Energy Physics*, vols. 1 and 2, Israel Program for Scientific Translations.
12. Afanasiev G.N., Kartavenko V.G. and Ruzicka J. (2000) Tamm's Problem in the Schwinger and Exact Approaches *J. Phys. A: Mathematical and General*, **33**, pp. 7585-7606.
13. Afanasiev G.N. and Shilov V.M. (2000) New Formulae for the Radiation Intensity in the Tamm Problem *J. Phys.D: Applied Physics*, **33**, pp. 2931-2940.
14. Afanasiev G.N. and Shilov V.M. (2000) On the Smoothed Tamm Problem *Physica Scripta*, **62**, pp. 326-330.
15. Tyapkin A.A. (1993) On the Induced Radiation Caused by a Charged Relativistic Particle Below Cherenkov Threshold in a Gas *JINR Rapid Communications*, **No 3**, pp. 26-31; Zrelov V.P., Ruzicka J. and Tyapkin A.A. (1998) Pre-Cherenkov Radiation as a Phenomenon of 'Light Barrier" *JINR Rapid Communications*, **1[87]-98**, pp. 23-25.
16. Afanasiev G.N., Shilov V.M., Stepanovsky Yu.P. (2002) New Analytic Results in the Vavilov-Cherenkov Radiation Theory *Nuovo Cimento*, **B 117**, pp. 815-838;
17. Abbasov I.I. (1982) Radiation Emitted by a Charged Particle Moving for a Finite Interval of Time under Continuous Acceleration and Deceleration *Kratkije soobchenija po fizike FIAN*, **No 1**, pp. 31-33; English translation: (1982) *Soviet Physics-Lebedev Institute Reports* **No1**, pp.25-27.
18. Abbasov I.I. (1985) Radiation of a Charged Particle Moving Uniformly in a Given Bounded Segment with Allowance for Smooth Acceleration at the Beginning of the Path, and Smooth Deceleration at the End *Kratkije soobchenija po fizike FIAN*, **No 8**, pp. 33-36. English translation: (1985) *Soviet Physics-Lebedev Institute Reports*, **No 8**, pp. 36-39.
19. Abbasov I.I., Bolotovskii B.M. and Davydov V.A. (1986) High-Frequency Asymptote of Radiation Spectrum of the Moving Charged Particles in Classical Electrodynamics *Usp. Fiz. Nauk*, **149**, pp. 709-722. English translation: Sov. Phys. Usp., 29 (1986), 788.
20. Bolotovskii B.M. and Davydov V.A. (1981) Radiation of a Charged Particle with Acceleration at a Finite Path Length *Izv. Vuzov, Radiofizika*, **24** , pp. 231-234.
21. Ruzicka J. and Zrelov V.P. (1979) Interference Effects in Transition Radiation Near the Threshold of Vavilov-Cherenkov Radiation, it Nucl. Instr. Methods **165** pp. 307-

316;

Hrmo A. and Ruzicka J. (2000) Properties of Optical Transition Radiation and Vavilov-Cherenkov Radiation for Charged Particle Inclined Flight through a Plate of Metal, *Nucl. Instr. Methods* **A451**, pp. 506-519.

22. Jelley J.V., (1958) *Cherenkov Radiation and its Applications*, Pergamon, London, New York, Paris.

23. Afanasiev G.N., Shilov V.M., Stepanovsky Yu.P. (2003) Numerical and Analytical Treatment of the Smoothed Tamm Problem *Ann.Phys. (Leipzig)*, **12**, pp. 51-79

24. Afanasiev G.N., Kartavenko V.G. and Stepanovsky Yu.P. (2003) Vavilov-Cherenkov and Transition Radiations on the Dielectric and Metallic Spheres, *Journal of Mathematical Physics*, **44**, pp. 4026-4056.

25. Ginzburg V.L. (1940) Quantum Theory of Light Radiation of Electron Moving Uniformly in Medium, *Zh. Eksp. Teor. Fiz.*, **10** pp. 589-600.

26. Afanasiev G.N., Eliseev S.M. and Stepanovsky Yu.P. (1998) Transition of the Light Velocity in the Vavilov-Cherenkov Effect *Proc. Roy. Soc. London*, **A 454**, pp. 1049-1072.

27. Afanasiev G.N. and Kartavenko V.G (1999) Cherenkov-like shock waves associated with surpassing the light velocity barrier *Canadian J. Phys.*, **77**, pp. 561-569.

28. Vodopianov A.S., Zrelov V.P. and Tyapkin A.A. (2000) Analysis of the Anomalous Cherenkov Radiation Obtained in the Relativistic Lead Ion Beam at CERN SPS *Particles and Nuclear Letters*, **No 2[99]-2000**, pp. 35-41.

29. Tamm I.E. and Frank I.M. (1944) Radiation of Electron Moving Uniformly in Refractive Medium *Trudy FIAN*, **2, No 4**. pp. 63-68.

30. Frank I.M., Tsytovich V.N. (1980) Two-Quantum Radiation of Particle Travelling Uniformly in Refracting Medium *Yad. Phys.*, **31**, pp. 974-985.

31. Tidman D.M. (1956,57) A Quantum Theory of Refractive Index, Cherenkov Radiation and the Energy Loss of a Fast Charged Particle *Nucl. Phys.*, **2**, pp. 289-346.

32. Batyghin V.V. (1965) Bremsstrahlung on Medium Electrons and Hard Vavilov-Cherenkov Radiation *Zh. Eksp. Theor. Phys.*, **49**, pp. 1637-1649; Batyghin V.V. (1965) On the Possibility of Emission of Hard Vavilov-Cherenkov Radiation *Zh. Eksp. Theor. Phys.*, **49**, pp. 272-274;

Batyghin V.V. (1968) Hard Vavilov-Cherenkov Radiation at Moderate Energies *Zh. Eksp. Theor. Phys.*, **54**, pp. 1132-1136.

33. Batyghin V.V. On the Possibility of Experimental Observation of Hard Vavilov-Cherenkov Radiation *Phys. Lett.*, **A 28**, pp. 64-65.

34. Batyghin V.V. (1968) On Influence of Medium Structure on the Bremsstrahlung Spectrum *Phys. Lett.*, **A 28**, pp.65-66.

35. Glauber R.J. (1965), Optical Coherence and Photon Statistics, pp. 65-185, In: *Quantum Optics and Electronics* (Lectures delivered at Les Houches, 1964), Eds. C De Witt, A. Blandin, and C. Cohen -Tannoudi Gordon and Breach, New York.

36. Akhiezer A.I. and Berestetzky V.B. (1981) *Quantum Electrodynamics*, Nauka, Moscow.

37. Frank I.M. (1968) Light Scattering by an Electron Moving in Refracting Medium *Yadernaya Fizika*, **7**, pp. 1100-1105.

38. Afanasiev G.N., Kartavenko V.G. and Zrelov V.P. (2003) Fine Structure of the Vavilov-Cherenkov Radiation, *Phys. Rev.*,**E68**, pp. 066501(1-12).

39. Afanasiev G.N. and Stepanovsky Yu.P. (2003) On the Kinematics of the Two-Photon Cherenkov Effect, *Nuovo Cimento*,**118 B**, pp. 699-712.

INDEX

489

Fundamental Theories of Physics

Series Editor: Alwyn van der Merwe, University of Denver, USA

Fundamental Theories of Physics

Fundamental Theories of Physics

46. P.P.J.M. Schram: *Kinetic Theory of Gases and Plasmas.* 1991   ISBN 0-7923-1392-5
47. A. Micali, R. Boudet and J. Helmstetter (eds.): *Clifford Algebras and their Applications in Mathematical Physics.* 1992   ISBN 0-7923-1623-1
48. E. Prugovečki: *Quantum Geometry.* A Framework for Quantum General Relativity. 1992   ISBN 0-7923-1640-1
49. M.H. Mac Gregor: *The Enigmatic Electron.* 1992   ISBN 0-7923-1982-6
50. C.R. Smith, G.J. Erickson and P.O. Neudorfer (eds.): *Maximum Entropy and Bayesian Methods.* Proceedings of the 11th International Workshop (Seattle, 1991). 1993   ISBN 0-7923-2031-X
51. D.J. Hoekzema: *The Quantum Labyrinth.* 1993   ISBN 0-7923-2066-2
52. Z. Oziewicz, B. Jancewicz and A. Borowiec (eds.): *Spinors, Twistors, Clifford Algebras and Quantum Deformations.* Proceedings of the Second Max Born Symposium (Wrocław, Poland, 1992). 1993   ISBN 0-7923-2251-7
53. A. Mohammad-Djafari and G. Demoment (eds.): *Maximum Entropy and Bayesian Methods.* Proceedings of the 12th International Workshop (Paris, France, 1992). 1993
          ISBN 0-7923-2280-0
54. M. Riesz: *Clifford Numbers and Spinors* with Riesz' Private Lectures to E. Folke Bolinder and a Historical Review by Pertti Lounesto. E.F. Bolinder and P. Lounesto (eds.). 1993
          ISBN 0-7923-2299-1
55. F. Brackx, R. Delanghe and H. Serras (eds.): *Clifford Algebras and their Applications in Mathematical Physics.* Proceedings of the Third Conference (Deinze, 1993) 1993
          ISBN 0-7923-2347-5
56. J.R. Fanchi: *Parametrized Relativistic Quantum Theory.* 1993   ISBN 0-7923-2376-9
57. A. Peres: *Quantum Theory: Concepts and Methods.* 1993   ISBN 0-7923-2549-4
58. P.L. Antonelli, R.S. Ingarden and M. Matsumoto: *The Theory of Sprays and Finsler Spaces with Applications in Physics and Biology.* 1993   ISBN 0-7923-2577-X
59. R. Miron and M. Anastasiei: *The Geometry of Lagrange Spaces: Theory and Applications.* 1994   ISBN 0-7923-2591-5
60. G. Adomian: *Solving Frontier Problems of Physics: The Decomposition Method.* 1994
          ISBN 0-7923-2644-X
61. B.S. Kerner and V.V. Osipov: *Autosolitons.* A New Approach to Problems of Self-Organization and Turbulence. 1994   ISBN 0-7923-2816-7
62. G.R. Heidbreder (ed.): *Maximum Entropy and Bayesian Methods.* Proceedings of the 13th International Workshop (Santa Barbara, USA, 1993) 1996   ISBN 0-7923-2851-5
63. J. Peřina, Z. Hradil and B. Jurčo: *Quantum Optics and Fundamentals of Physics.* 1994
          ISBN 0-7923-3000-5
64. M. Evans and J.-P. Vigier: *The Enigmatic Photon.* Volume 1: The Field $B^{(3)}$. 1994
          ISBN 0-7923-3049-8
65. C.K. Raju: *Time: Towards a Constistent Theory.* 1994   ISBN 0-7923-3103-6
66. A.K.T. Assis: *Weber's Electrodynamics.* 1994   ISBN 0-7923-3137-0
67. Yu. L. Klimontovich: *Statistical Theory of Open Systems.* Volume 1: A Unified Approach to Kinetic Description of Processes in Active Systems. 1995   ISBN 0-7923-3199-0;
          Pb: ISBN 0-7923-3242-3
68. M. Evans and J.-P. Vigier: *The Enigmatic Photon.* Volume 2: Non-Abelian Electrodynamics. 1995   ISBN 0-7923-3288-1
69. G. Esposito: *Complex General Relativity.* 1995   ISBN 0-7923-3340-3

Fundamental Theories of Physics

70. J. Skilling and S. Sibisi (eds.): *Maximum Entropy and Bayesian Methods.* Proceedings of the Fourteenth International Workshop on Maximum Entropy and Bayesian Methods. 1996
ISBN 0-7923-3452-3

71. C. Garola and A. Rossi (eds.): *The Foundations of Quantum Mechanics Historical Analysis and Open Questions.* 1995
ISBN 0-7923-3480-9

72. A. Peres: *Quantum Theory: Concepts and Methods.* 1995 (see for hardback edition, Vol. 57)
ISBN Pb 0-7923-3632-1

73. M. Ferrero and A. van der Merwe (eds.): *Fundamental Problems in Quantum Physics.* 1995
ISBN 0-7923-3670-4

74. F.E. Schroeck, Jr.: *Quantum Mechanics on Phase Space.* 1996
ISBN 0-7923-3794-8

75. L. de la Peña and A.M. Cetto: *The Quantum Dice.* An Introduction to Stochastic Electro-dynamics. 1996
ISBN 0-7923-3818-9

76. P.L. Antonelli and R. Miron (eds.): *Lagrange and Finsler Geometry.* Applications to Physics and Biology. 1996
ISBN 0-7923-3873-1

77. M.W. Evans, J.-P. Vigier, S. Roy and S. Jeffers: *The Enigmatic Photon.* Volume 3: Theory and Practice of the $B^{(3)}$ Field. 1996
ISBN 0-7923-4044-2

78. W.G.V. Rosser: *Interpretation of Classical Electromagnetism.* 1996
ISBN 0-7923-4187-2

79. K.M. Hanson and R.N. Silver (eds.): *Maximum Entropy and Bayesian Methods.* 1996
ISBN 0-7923-4311-5

80. S. Jeffers, S. Roy, J.-P. Vigier and G. Hunter (eds.): *The Present Status of the Quantum Theory of Light.* Proceedings of a Symposium in Honour of Jean-Pierre Vigier. 1997
ISBN 0-7923-4337-9

81. M. Ferrero and A. van der Merwe (eds.): *New Developments on Fundamental Problems in Quantum Physics.* 1997
ISBN 0-7923-4374-3

82. R. Miron: *The Geometry of Higher-Order Lagrange Spaces.* Applications to Mechanics and Physics. 1997
ISBN 0-7923-4393-X

83. T. Hakioğlu and A.S. Shumovsky (eds.): *Quantum Optics and the Spectroscopy of Solids.* Concepts and Advances. 1997
ISBN 0-7923-4414-6

84. A. Sitenko and V. Tartakovskii: *Theory of Nucleus.* Nuclear Structure and Nuclear Interaction. 1997
ISBN 0-7923-4423-5

85. G. Esposito, A.Yu. Kamenshchik and G. Pollifrone: *Euclidean Quantum Gravity on Manifolds with Boundary.* 1997
ISBN 0-7923-4472-3

86. R.S. Ingarden, A. Kossakowski and M. Ohya: *Information Dynamics and Open Systems.* Classical and Quantum Approach. 1997
ISBN 0-7923-4473-1

87. K. Nakamura: *Quantum versus Chaos.* Questions Emerging from Mesoscopic Cosmos. 1997
ISBN 0-7923-4557-6

88. B.R. Iyer and C.V. Vishveshwara (eds.): *Geometry, Fields and Cosmology.* Techniques and Applications. 1997
ISBN 0-7923-4725-0

89. G.A. Martynov: *Classical Statistical Mechanics.* 1997
ISBN 0-7923-4774-9

90. M.W. Evans, J.-P. Vigier, S. Roy and G. Hunter (eds.): *The Enigmatic Photon.* Volume 4: New Directions. 1998
ISBN 0-7923-4826-5

91. M. Rédei: *Quantum Logic in Algebraic Approach.* 1998
ISBN 0-7923-4903-2

92. S. Roy: *Statistical Geometry and Applications to Microphysics and Cosmology.* 1998
ISBN 0-7923-4907-5

93. B.C. Eu: *Nonequilibrium Statistical Mechanics.* Ensembled Method. 1998
ISBN 0-7923-4980-6

Fundamental Theories of Physics

Fundamental Theories of Physics

Fundamental Theories of Physics

142. G.N. Afanasiev: *Vavilov-Cherenkov and Synchrotron Radiation.* Foundations and Applications. 2004
                                    ISBN 1-4020-2410-X
143. L. Munteanu and S. Donescu: *Introduction to Soliton Theory: Applications to Mechanics.* 2004
                                    ISBN 1-4020-2576-9

KLUWER ACADEMIC PUBLISHERS – DORDRECHT / BOSTON / LONDON